Guide to the NATE, ESCO, and EPA-608 Certification Exams

Guide to the NATE, ESCO, and EPA-608 Certification Exams

FOURTH EDITION

Robert Featherstone
Oakland Community College

Jesse Riojas
Oakland Technical Campus Northwest Oakland Community College

Pearson

Product Management: Derril Trakalo
Content Production: Shubha Banerjee
Product Marketing: Mark Marsden
Rights and Permissions: Jenell Forschler

Please contact www.AskPearsonSupport.com with any queries on this content.

Pearson puts care into the creation of our products. Our learning materials undergo content reviews to minimize bias and ensure the content is inclusive, accurate, and relevant. Please use this link to report instances of bias and inaccuracies: https://www.pearson.com/report-bias.html

You can learn more about Pearson's commitment to accessibility at https://www.pearson.com/us/accessibility.html

Cover Image by Parilov/Shutterstock and Moo photograph/Shutterstock

Library of Congress Cataloging-in-Publication Data

Names: Featherstone, Robert, author. | Riojas, Jesse R., author.
Title: Guide to the NATE / ESCO / EPA-608 certification exams / Robert Featherstone, Oakland Community College, Jesse Riojas, Oakland Technical Campus Northwest Oakland Community College.
Other titles: NATE / ESCO / EPA-608 certification exams
Description: Fourth edition. | Harlow, England ; New York : Pearson Education, [2025] | Revised edition of: Guide to the NATE certification exams / Robert Featherstone, Jesse Riojas. 3rd ed. Upper Saddle River, N.J. : Pearson/Prentice Hall, c2010.
Identifiers: LCCN 2024007451 | ISBN 9780138255725 (paperback) | ISBN 0138255725 (paperback) | ISBN 9780138255794 (ebook)
Subjects: LCSH: Heating--Problems, exercises, etc. | Air conditioning--Problems, exercises, etc. | Ventilation--Problems, exercises, etc. | Mechanics (Persons)--Certification--United States.
Classification: LCC TH7015 .F43 2025 | DDC 697.0076--dc23/eng/20240314
LC record available at https://lccn.loc.gov/2024007451

1 2024

Pearson

ISBN-10: 0-13-825572-5
ISBN-13: 978-0-13-825572-5

Preface

The HVAC/R industry has progressively become more technical through the years. With new technology, industries are requiring employee skills to be updated to match the current changes. HVAC/R certification and competency tests are directed toward students and technicians at all skill levels. National Certification Competency Exams are being offered by many organizations, including the Air Conditioning and Refrigeration Institute (ARI), North American Technician Excellence (NATE), the Refrigeration Service Engineer Society (RSES), and the ESCO Institute (HVAC Excellence).

Guide to the NATE, ESCO, and EPA-608 Certification Exams is designed to help students and technicians study for and pass many of the exams that are offered today. The book is divided into three parts. Part I, Study Guide, includes bolded key words to help the reader remember specifics about a certain topic. Part II includes multiple-choice study questions similar to those found on certification exams, and Part III is the answer key to the multiple-choice questions found in Part II. Each part of the book includes sections. These sections are divided into specific topics of study. The reader can easily go to any topic of interest for further study. This book covers study sections that include Refrigeration and Air Conditioning Systems and Tools, Gas and Oil-Forced Air, Hot Water Heating Systems, Heat Pumps, Electrical Theory and Application, Airflow Components, and Duct Fabrication.

The student/technician will benefit more by first reading the Study Guide. Then, the reader should turn to Part II and try to answer the related questions. For example, the first statement in the Study Guide reads, "A **torx driver** has a star-shaped head and is used on torx-type screws." The reader should then turn to Part II and answer the first question.

In Section 4 of the Part I Study Guide, an EPA Section 608 Certification topic has been included and updated since the guidelines have changed in 2018. This will help the student/technician to successfully pass Types I, II, and III exams and the Universal Certification.

The Organization ESCO Institute-HVAC Excellence offers many certificates that are employment-ready exams. Upon passing exams, students and technicians are more recognized by the HVAC/R industry. Earning certificates assures employers that the candidates have gained and retained the knowledge they need to be employed.

This text includes over 2,500 questions that cover many HVAC/R topics. Included are many electrical and mechanical diagrams that will also help the reader to better understand the concepts. We hope that all of the information provided in this book will assist the student/technician in successfully passing certification tests offered by national HVAC/R organizations.

Alternative Formats Also Available

In addition to this print text, the 4th edition is available in various digital formats:

eBooks: This text is also available in multiple eBook formats. These are a great choice for busy readers that are looking to save money and purchase an electronic version of the same content. Pearson eTextbook is the only eBook format that features an interactive experience where users can answer the questions and receive immediate feedback on their selection. For more information on Pearson eTextbook, visit www.pearsonplus.com.

MyLab for Guide to the NATE, ESCO, and EPA-608 Certification Exams Without eText:

For instructors who are looking to provide their students with practice tests in a digital environment, MyLab for Guide to the NATE, ESCO, and EPA-608 Certification Exams is the way to go. Take advantage of all the tools MyLab has to offer and select from the entire bank of more than 2,500 questions to create your own practice tests as you guide students to success on their certification exams. To learn more about MyLab for Guide to the NATE, ESCO, and EPA-608 Certification Exams, visit mlm.pearson.com/northamerica/myhvaclab.

Acknowledgments

The authors extend their appreciation to their family members, friends, and above all our Lord and Savior Jesus Christ, whom we could not have done this book without.

"I can do all things through Christ who strengthens me." *Phil. 4:13*

Contents

NATE, ESCO and EPA-608 CERTIFICATION

Technicians can prepare to take many available certification tests. These certification tests are offered by various associations. North American Technicians Excellence (NATE) and the ESCO Institute HVAC Excellence provide certification tests to technicians and to all who may be eligible to take the exams. Technicians can take the exams in areas of their expertise. Knowledge Areas of Technician Expertise (KATES) encompass the knowledge of skills and tasks that technicians should possess to help pass certification tests. The following are certification tests and their acronyms offered by NATE:

HVAC EFFICIENCY ANALYST SENIOR—HVAC EA-SN
HYDRONICS—GAS INSTALLATION—HG-IN
HYDRONICS—GAS SERVICE—HG-SV
HYDRONICS—OIL INSTALLATION—HO-IN
HYDRONICS—OIL SERVICE—HO-SV
AIR CONDITIONING INSTALLATION—AC-IN
AIR CONDITIONING SERVICE—AC-SV
AIR DISTRIBUTION INSTALLATION—AD-IN
AIR DISTRIBUTION SERVICE—AD-SV
CORE INSTALLATION—CR-IN
CORE SERVICE—CR-SV
GAS FURNACE INSTALLATION—GS-IN
GAS FURNACE SERVICE—GS-SV
HEAT PUMP INSTALLATION—HP-IN
HEAT PUMP SERVICE—HP-SV
COMMERCIAL REFRIGERATION—RC-IN
COMMERCIAL REFRIGERATION—RC-SV
LIGHT COMMERCIAL REFRIGERATION—RL-IN
LIGHT COMMERCIAL REFRIGERATION—RL-SV
OIL INSTALLATION—OL-IN
OIL SERVICE—OL-SV

ESCO INSTITUTE-HVAC EXCELLENCE CERTIFICATIONS

EPA SECTION 608
BRAZING AND SOLDERING EMPLOYMENT READY
RECOVERY AND EVACUATION EMPLOYMENT READY
BASIC REFRIGERATION AND CHARGING PROCEDURES EMPLOYMENT READY
CARBON MONOXIDE SAFETY EMPLOYMENT READY
GAS HEAT
AIR CONDITIONING
HEAT PUMP
COMMERCIAL REFRIGERATION

How to Use the Acronyms

The acronyms used in the text identify the types of certification tests. Part II of this book refers to the sections of multiple-choice questions.

The question portion has six sections. Before each section, you will find the acronyms for the tests these questions pertain to. These acronyms are used to inform the technician that all questions under this topic may be found in each specific test noted. The following is an example:

PART II QUESTIONS

SECTION I—Refrigeration and Air Conditioning Tools and Components

TOPIC: Compressors (AC-SV, AC-IN, HP-SV, HP-IN, RC-SV, RC-IN, RL-SV, RL-IN.

PART I
Study Guide

SECTION 1

Refrigeration and Air Conditioning Tools and Components

This section contains information that will help technicians pass certification exams. It provides emphasis on refrigeration and air conditioning theory as well as a description of the systems' mechanical components. It also covers the mechanical components of heat pumps. With a thorough understanding of this section, the technician will be better prepared and more skilled in mechanical troubleshooting. The EPA Section 608 Study Guide has been updated and the following ESCO Institute-HVAC Excellence certifications have been added:

Brazing and Soldering Employment Ready, System Recovery and Evacuation Employment Ready, Basic Refrigeration and Charging Procedures Employment Ready, and Carbon Monoxide Safety Employment Ready.

PERSONAL SAFETY

Personal safety should always be on a technician's mind. Technicians need to be aware of all unsafe conditions that they may be asked to work under. The technician must always be conscious of the unknown. The following information on personal safety will help the technician work in a safe manner.

Wear **hard hats** to protect against falling objects.

Wear **safety glasses** to protect against flying objects, soldering materials, and chemical splashes.

Wear **gloves** to protect against scrapes, cuts, heat, or burns.

Wear **steel toe safety boots** to protect feet against falling objects and punctures.

Wear **ear plugs** to protect ears while in a high-noise-level area.

Use a **respirator** to protect against breathing hazards and suffocation.

Use a **safety harness** to protect against falls when working in an area more than 6′ above the ground.

Regularly **inspect safety equipment** and care for it according to the manufacturer's specifications.

Never alter safety equipment in any way.

Never wear loose clothing, loose gloves, jewelry, or expose long hair around machinery.

Always keep **hand tools** and equipment clean and in good working condition.

To prevent an injury, never use a **broken or defective tool.**

When **using a wrench,** it is safer to pull instead of push. This will prevent an injury.

When **lifting objects,** keep your back straight and use your leg muscles.

When **lifting an object,** always move close to it, squat down, lift slowly, and keep the object centered.

When **lifting an object,** always use the full hand; do not use fingertips.

Wear a **back support belt** to aid in lifting objects.

Always ask for assistance when lifting **heavy objects.**

Keep **floors and work area** free of oil, grease, water, and crowded items that may cause a personal injury.

To **prevent a laceration,** wear hand protection when working with sheet metal.

To **prevent serious burns,** never use gasoline to clean tools or mechanical components.

Eye protection must be used to prevent an injury when working with hammers and chisels.

To prevent a personal injury, never use a **worn chisel.**

To prevent a personal injury, **eye protection** is required when using power tools such as drills and saws.

Always keep hands and feet at a three-point connection when climbing **ladders.**

HAND TOOLS

Hand tools that technicians use on a daily basis can range from the most basic to precision-made tools. The technician must be sure that the tools used are correct for the job. The technician must also keep the tools in good working condition. The following are the types of tools that may be used in the HVAC/R field.

A **torx driver** has a star-shaped head and is used on torx-type screws.

A **nut driver** has a hexagon-shaped head and is used to remove a screw with a six-point hex-head shape.

An **adjustable wrench** has a variable open end that is used to remove a nut or bolt.

An **Allen wrench** is used to loosen or tighten a set screw.

A **box-end wrench** is a closed-end wrench that is used to tighten or loosen nuts and bolts.

The **flare-nut wrench** is used to tighten or loosen a flare-type mechanical fitting on a refrigeration or air conditioning system.

An **open-end wrench** is used to tighten or loosen a nut or bolt from the side.

The **pipe wrench** is used to assemble or disassemble threaded pipe.

A **service valve wrench** is used to manipulate service valves on a refrigeration or air conditioning system. The most common sizes are $\frac{3}{16}''$, $\frac{1}{4}''$, $\frac{5}{16}''$, and $\frac{3}{8}''$.

A **socket wrench** is used to fit on top of a nut or bolt. Common socket drives are $\frac{1}{4}''$, $\frac{3}{8}''$, and $\frac{3}{4}''$.

A **torque wrench** is used to apply a specific amount of torque (force × length arm) to a nut or a bolt. Using a torque wrench will prevent over- or undertightening.

Tongue-and-groove pliers are used to hold objects of many shapes and sizes. They are also used to loosen and tighten threaded pipe.

A **level** is used to test for uniformity of work. When held in a horizontal position, it will show if the work is level. If it is held in a vertical position, it will show if the work is plumb.

A **carpenter's square** is an L-shaped tool that is used to measure materials and check for uniformity.

A **crosscut saw** is used to cut wood against the grain. It has 6 to 12 teeth per inch that bend away from the saw blade in an alternating pattern.

A **ripsaw** is used to cut wood with the grain. It has 4 to $5\frac{1}{2}$ teeth per inch that bend away from the saw blade in an alternating pattern.

A **hacksaw** is primarily used to cut metal. It has 18 to 32 teeth per inch. Note that cutting copper tends to accumulate copper filings in the saw teeth.

A **keyhole saw** is used for cutting drywall or wood. It has 8 to 10 teeth per inch.

A **single-cut file** has one row of teeth.

A **double-cut file** has two rows of teeth that crisscross each other.

There are four types of **files:** the rasp, the rough-cut file, the smooth-cut file, and the finish file. The **rasp** is used to start a smoothing process on extremely rough-cut wood. The **rough-cut file** is used on rough cuts of wood, plastic, and metal. The **smooth-cut file** is used to continue the smoothing of a material after the use of a rough-cut file. The **finish file** will complete the final smoothing process.

A **straight-cut snip** is designed for straight and wide-radius cuts in sheet metal. The maximum recommended thickness for the straight-cut snip is 18-gauge metal.

A **right-cut snip** is designed for straight and tight-right radius cuts in sheet metal. The maximum recommended thickness for the right-cut snip is 18-gauge metal.

A **left-cut snip** is designed for straight and tight-left radius cuts in sheet metal. The maximum recommended thickness for the left-cut snip is 18-gauge metal.

Larger **scissor-type snips,** such as the straight pattern and the duckbill type, are designed for cutting straight and larger radius curves in sheet metal. The maximum recommended thickness for the straight and duckbill snip is 23-gauge metal.

A **drill** is used to bore holes in materials such as sheet metal, wood, plastic, drywall, and concrete. Drills are available in three different types of drives: the $\frac{1}{4}''$ drive, the $\frac{3}{8}''$ drive, and the $\frac{1}{2}''$ drive.

Hex-head adapters are used with a drill to install or remove hex-head sheet metal screws.

A **countersink** is a type of bit that is used to recess holes in wood and other material, so that screws can be driven flush or slightly below the surface.

Drills and bits are used for drilling into wood, metal, and masonry materials.

A **spade** is a flat bit with two sharp edges and a point. It is used to drill holes into wood.

The **auger bit** has a corkscrew shape and is used to bore holes into wood. Common sizes are $\frac{3}{8}''$, $\frac{5}{8}''$, $\frac{1}{2}''$, $\frac{3}{4}''$, and 1″ in diameter; common lengths are $7\frac{1}{2}''$, 12″, and 17″.

A **masonry drill** is used to bore holes in brick, block, concrete, stone, or mortar.

A **punch** is used to create an indention, usually in a piece of sheet metal. This indention makes an easy path for a drill or a sheet metal screw to follow.

A **tap** is used for cutting inside threads on metal material.

Dies are used to cut threads on the outside of galvanized and black iron pipe. Before using a die, always ream the inside of the pipe to remove all of the burrs.

A **claw hammer** has a slightly rounded face on the front and a claw on the back that is used to remove nails. A claw hammer ranges from 13 to 28 oz in weight.

A **ball-peen hammer** is designed for metal work. It has a flat back and a round front, which is used to form bends and creases in sheet metal.

A **masonry hammer** has a larger head that is designed for pounding block or concrete. The masonry hammer has a sharp pick on the back for splitting masonry material.

All **hand tools** need to be cleaned regularly with some type of machine oil to prevent corrosion.

BASIC CONSTRUCTION

On many occasions in the construction industry, tradespeople will be working alongside each other to complete a project. Each tradesperson must be familiar with basic construction terms and knowledge. The following will help the technician understand some of the terms used in basic construction.

A residential **roof construction** can be truss built, rafter built, or a flat roof design.

The two most common **residential roof designs** are truss built and rafter built.

A common design of a **truss-built roof** is one that is premanufactured with braces between the top and the base of the truss. Trusses are commonly constructed of 2″ × 4″, 2″ × 6″, 2″ × 8″, or 2″ × 10″ lumber or steel.

A **rafter-built roof** uses 2″ × 6″, 2″ × 8″, 2″ × 10″, or 2″ × 12″ boards. They connect the ridge of the roof (the peak) to the top of an exterior wall or knee wall (a smaller exterior wall).

A **truss-and-rafter-built roof** uses 4′ × 8′ × $\frac{1}{2}$″ plywood or oriented strand board (OSB) sheathing as a base for the shingles. This also strengthens the roof by adjoining the rafters or trusses together as one.

There are **various truss and roof designs** such as the W-web truss, the King Post truss, the Scissor truss, Gambrel roof, and Hip roof.

Organic asphalt, fiberglass asphalt, tile, and clay **shingles** are used to seal the roof and protect it from the weather.

A common residential **ceiling** is made of plaster, drywall, fiberglass inserts, or celotex tile that covers 2″ × 6″ to 2″ × 12″ wood or steel ceiling joists.

A **ceiling** is generally an 8′-high flat type or a cathedral type that rises diagonally to a peak.

A **drywall ceiling** uses 4′ × 8′ × $\frac{1}{2}$″ or 4′ × 8′ × $\frac{5}{8}$″ sheets of drywall that are connected to the ceiling joists with drywall screws or nails.

An external ceiling is **insulated** with fiberglass or cellulose fiber to reduce the heat transfer of the house. Common fiberglass insulation thicknesses are as follows: R-19 is 6″, R-25 and R-30 are 9″, and R-38 is 12″. Increasing the R-value of insulation will lower the heat transfer of the home.

A common residential **wall** uses 2″ × 4″ or 2″ × 6″ wood or steel vertical studs placed 16″ apart. At the top and bottom of the vertical studs are horizontal studs. They are made of the same material as the vertical studs. Horizontal studs are known as top plates and bottom plates.

An **exterior wall** uses 4′ × 8′ × $\frac{1}{2}$″ plywood, $\frac{1}{2}$″ OSB sheathing, $\frac{1}{2}$″ celotex, or $\frac{1}{2}$″ to 1″ dense foam sheathing to adjoin the vertical studs. It is required that the corner walls be constructed of plywood or OSB sheathing.

An **exterior wall** has the weight of the roof bearing down upon it. This is why they are considered load-bearing walls.

A **house wrap** is used on exterior walls to decrease air infiltration into the home.

The common **exterior finishes** for homes are vinyl siding, wood siding, aluminum siding, and brick.

An external wall is **insulated** with fiberglass or cellulose fiber to reduce the heat transfer of the house. The common fiberglass insulation thicknesses are as follows: R-11 and R-13 are 3$\frac{1}{2}$″ and R-19 is 6″.

A **vapor barrier** is placed on the inside of an exterior wall, behind the drywall, to reduce the quantity of moisture entering or exiting the house.

A wall that uses **drywall** as its interior finish is constructed of 4′ × 8′ × $\frac{1}{2}$″ or 4′ × 8′ × $\frac{5}{8}$″ sheets of drywall. These are connected to the wall studs with drywall screws or nails.

Drywall that is used in damp areas, such as bathrooms, is moisture resistant. This drywall is also known as green board.

An interior wall that connects the living area of a house to a garage must use a $\frac{5}{8}$″ **fire-code-type drywall.**

A residential **floor** can be constructed of 2″ × 8″ to 2″ × 12″ wood or steel floor joists, engineered truss joists, or engineered "I" joists.

Common floor joist **bridging** is made of 1″ × 2″ wood slats or metal brackets. They are placed between the floor joists, at the center or at every one-third of the floor span, to stiffen a floor.

A **subfloor** is the floor that is directly above the floor joists. It is made of $\frac{3}{4}$″ tongue and groove plywood and OSB sheathing.

An **underlayment floor** is the floor that is directly above the subfloor. It is made of $\frac{1}{4}$″ finished plywood.

A **finish floor covering** is placed above the underlayment and the subfloor. It can be made of sheet vinyl, vinyl tile, ceramic tile, parquet wood, $\frac{3}{4}$″ wood strip, or carpeting.

REFRIGERANT SAFETY

Working with refrigerants that are under pressure in a sealed system could invite injury to the technician who is not careful. Refrigerants can instantly freeze eyes, hands, and fingers if one is not careful. Also, under certain conditions, refrigerant may become harmful to breathe. The following will help the technician to stay knowledgeable and to work safely.

To prevent an injury, always **wear gloves and safety glasses** when working with refrigerants.

Refrigerants can cause **frostbite burns and suffocation** if not used properly.

To **prevent a refrigerant burn,** never try to stop a refrigerant leak with your hand.

To **prevent suffocation,** never work with refrigerants in a confined area.

Because air is lighter than refrigerant vapor, **air will be displaced** by the refrigerant.

Equipment rooms using large quantities of A1 refrigerants (refrigerants with low toxicity and low flammability) must be equipped with **refrigerant** or **oxygen deprivation alarms and sensors.**

When A1 refrigerants are in contact with an open flame, **phosgene gas** is produced.

To prevent suffocation, asphyxiation, and possible brain damage, use a **breathing apparatus** when working around toxic fumes.

A **self-contained breathing apparatus** must be available in equipment rooms where systems use a large amount of refrigerants.

A pressure regulator and an in-line pressure-relief valve should be used when **pressurizing a system** with nitrogen.

To prevent an explosion, **never connect a nitrogen cylinder and refrigerant tank** to a refrigeration system at the same time.

Never exceed the **low-side pressure rating** when pressurizing a system.

When pressurizing a system, **150 psig** of dry nitrogen is usually satisfactory for leak detection on high-pressure systems.

Never use a **halide torch** to detect leaks if the system is pressurized with nitrogen.

A mixture of **oxygen and oil** is extremely explosive.

To prevent an explosion, **never use oxygen** when pressure testing an air conditioning system for refrigerant leaks.

Technicians should always follow the **manufacturer's recommendations** for charging refrigerant into a system.

To **prevent an explosion,** never front seat (close) the discharge service valve on a compressor while it is in operation.

Care must be taken near the system's discharge line to **prevent a burn** while working on an operating system.

Always wear hand protection when handling air conditioning coils to **prevent scrapes and cuts.**

Always wear eye and hand protection when working with acidic refrigerant oil to **prevent burns.**

Never use an open flame during the leak detection of natural gas or flammable refrigerants.

To prevent the danger of excessive pressures, never **overcharge** refrigeration systems.

To **prevent refrigerant burns**, always remove the refrigerant from a system before opening it to the atmosphere.

To prevent an injury caused by a liquid entering a compressor (liquid slug), never add **liquid refrigerant** into the low side of a compressor while it is in operation.

To prevent a possible explosion, **never refill** a disposable refrigerant cylinder.

To prevent a possible explosion, **never refill** a rechargeable refrigerant cylinder over 80% of its capacity.

Always take into account the cylinder weight, known as the "**tare pounds**," when estimating the net weight of the refrigerant in a rechargeable cylinder.

To prevent an explosion, **never heat refrigerant cylinders** with a torch.

Refrigerant **containers** must always be properly labeled and color coded.

Recovery cylinders are **color coded** with a gray bottom and yellow top.

When heating a refrigerant cylinder with warm water, the temperature of the cylinder water must never exceed **125°F** or the rupture disc could open.

Refrigerant cylinders must always be approved by the Department of Transportation **(DOT)**.

Relief valves are used to protect against excessively high system pressures.

The common **types of pressure-relief devices** are fusible plug, spring-loaded relief valve, and rupture disc.

To prevent a possible explosion, never place a cap on a safety relief device.

Relief devices must never be in series with one another.

Relief devices can be installed on compressors, on components, and on refrigerant piping.

When pressurizing a low-pressure chiller, **never exceed 10 psig** on the low side of the system. This will cause the ruptured disc to open and vent the refrigerant charge.

When charging a **low-pressure chiller**, refrigerant is commonly added to the low side of the system in liquid form.

Fill limit devices are part of a recovery system. They can be sensor or float-type devices used to shut down the recovery machine when a container reaches 80% full.

Empty **rechargeable refrigerant cylinders** must be evacuated before reusing.

Never use untested **refrigerant cylinders**.

To prevent personal injury, never use **dented** or **rusted** refrigerant cylinders.

Material safety data sheets (MSDS) describe the flammability, toxicity, reactance, and health problems of a particular refrigerant.

Any **refrigerant cylinder** should be stored in a vertical position.

Always secure the cylinder's **valve cap and hood cap** when a cylinder is being stored.

Always properly **secure a cylinder** to prevent it from falling or rolling around when being transported.

REFRIGERATION THEORY

Technicians must have a good understanding of science, chemistry, heat transfer, and refrigeration theory. They should be capable of applying this knowledge to their service work. The following is information that will aid the service technician.

Refrigeration is transferring heat from one place to another.

Temperature is the intensity of molecular motion.

There are four common **temperature scales:** Fahrenheit, Rankine, Celsius, and Kelvin.

Fahrenheit is the U.S. conventional unit, and its absolute zero scale is Rankine.

On the **Fahrenheit scale,** water freezes at 32°F and boils at 212°F.

To convert **Fahrenheit** to **Rankine,** add 460°F.

Example: 60°F + 460°F = 520°Rankine.

Celsius is the SI unit, and its absolute zero scale is Kelvin.

SI stands for International System of Units, which is based on the metric system.

On the **Celsius scale,** water freezes at 0°C and boils at 100°C.

To convert **Celsius** to **Kelvin,** add 273°C.

Example: 20°C + 273°C = 293°K

To convert **Celsius** to **Fahrenheit,** use the following calculation:

(1.8 × °C) + 32 = °F

Example: (1.8 × 40°C) + 32 = 104°F

To convert **Fahrenheit** to **Celsius,** use the following calculation:

.56 × (°F − 32) = °C

Example: .56 × (104°F − 32) = 40°C

Specific heat is the amount of heat required to raise 1 lb of a substance 1°F.

Sensible heat is the heat that can be measured or sensed by a thermometer.

Latent heat is hidden heat that changes the state of a substance. As the substance changes state, its temperature remains constant even though a large amount of heat energy is added or removed.

Superheat is raising the temperature of a gas above its boiling point.

Superheating added to a refrigerant before it enters a compressor ensures that there is no liquid refrigerant return to the compressor.

Subcool is lowering the temperature of a liquid below its condensing point.

Subcooling a refrigerant in the liquid line ensures 100% liquid will enter the metering device.

Heat is a form of energy, and it is measured in Btu (British thermal unit).

A **Btu** is the amount of heat required to raise 1 lb of water 1°F.

To measure the amount of **Btu** of a substance without changing the substance's state, use the following formula:

Specific heat × temperature difference × weight in pounds = **Btu**

Example: If 2 lb of water is raised from 40 to 60°F, how many Btu are added to the water?

Btu = spec. heat × temp. diff. × wt

Specific heat of water = 1

Temperature difference = 20°F

Weight = 2 lb

1 × 20 × 2 = 40 Btu

The amount of heat that it takes to change 1 lb of 212°F water to 1 lb of 212°F steam is **970 Btu.**

The amount of heat required to change 1 lb of 32°F ice to 1 lb of 32°F water is **144 Btu.**

The three **states of matter** are solid, liquid, and gas. A **solid** exerts pressure downward. A **liquid** exerts pressure downward and outward in a container. A **gas** exerts pressure in all directions in a container.

Boyle's law states that a volume of a gas varies inversely with the absolute pressure, if the temperature remains constant. That is, as pressure increases, volume decreases, and as pressure decreases, volume increases.

Charles' law states that at a constant pressure the volume of a gas varies directly with an absolute temperature. As a gas heats, it expands. As a gas cools, it contracts.

Dalton's law states that the total pressure of a confined mixture of gases is the sum of the pressures of each of the gases in the mixture.

When the pressure on a liquid is increased, the boiling point will also **increase.**

Work = Force × Distance

Example: 500 lb × 50 ft = 25,000 ft-lb of work

Power is the rate of doing work. One horsepower is 33,000 ft-lb of work in 1 min.

1 horsepower = 746 watts

A **watt** is the measurement of power in an electrical circuit. 1 watt = 3.41 Btu.

Heat is transferred by conduction, convection, or radiation. **Conduction** is the transfer of heat from one molecule to another. **Convection** is the transfer of heat from one place to another by a fluid, commonly air or water. **Radiation** is the transfer of heat from a heat source to a solid object through space.

One ton of refrigeration effect is a specific quantity of heat that is transferred. One ton of refrigeration is equivalent to 12,000 Btu per hour or 200 Btu per minute.

Latent heat is the heat that is absorbed or extracted when a substance changes its state.

When a substance changes from a liquid to a gas, it is called the **latent heat of vaporization.** When a substance changes from a gas to a liquid, it is called the **latent heat of condensation.** When a substance changes from a solid to a gas, it is called **sublimation.**

A refrigeration system absorbs heat by absorbing the latent heat of vaporization in the **evaporator.**

A refrigeration system extracts heat by removing the latent heat of condensation from the **condenser.**

REFRIGERANTS

There are many types of refrigerants in the air conditioning and refrigeration industry. Refrigerants are classified into types based on their toxicity, flammability, and total safety. The technician must be informed of refrigerant properties and methods of usage.

The most common **HCFC refrigerant** in residential and commercial air conditioning systems is R-22.

The chemical name for **R-22** is monochlorodifluoromethane. The molecular formula for **R-22** is $CHClF_2$. **R-22** uses alkylbenzene oil and boils at −40°F at atmospheric pressure.

Refrigerant tanks must never be stored in an area above 125°F.

A **single-component refrigerant** has one refrigerant, such as R-22 or R-12, that can be charged into a system as a gas.

The following are **single-component refrigerants** are R-12, R-22, R-134a, R-123, and R-236fa.

The single-component refrigerant that is used as a replacement for R-12 is hydrofluorocarbon (HFC) **R-134a.**

An **azeotropic refrigerant** is a mixture of two or more refrigerants that perform the same as a single-component refrigerant. Azeotropic refrigerants will not separate when charged as a gas.

The **azeotropic refrigerants** are R-500, R-502, R-503, R-507, and R-410A.

The azeotropic (blended) refrigerant that is used as a replacement for R-22 is **R-410A.**

A **near-azeotropic refrigerant** is a mixture of two or more refrigerants that have different boiling points.

A near-azeotropic refrigerant must be treated as a zeotropic blend. It must never be charged into a system as a gas, or the refrigerant composition will change.

The near-azeotropic refrigerants are R-401A, R-401B, R-402A, R-402B, R-403B, R-404A, R-408A, and R-409A.

The near-azeotropic (blended) refrigerants that are used as replacements for R-12 are **R-414B, R-416A, and R-409A.**

The near-azeotropic (blended) refrigerant that is used as a replacement for R-502 is **R-408A.**

A **zeotropic refrigerant** is a mixture of two or more refrigerants that have different boiling points. A zeotropic blend must never be charged into a system as a gas or the refrigerant composition will change.

The zeotropic refrigerants are R-407A, R-407B, and R-407C.

A **ternary refrigerant** is a blend of three refrigerants that act as a zeotropic refrigerant.

The ternary refrigerants are R-401A, R-401B, R-402A, R-402B, R-403B, R-404A, R-407A, R-407B, R-407C, R-408A, R-409A, and R-410A.

Zeotropic and **near-azeotropic refrigerants** must be charged into a system as a liquid or the refrigerant will change its composition.

When working with blended refrigerants, pressure temperature charts use **dewpoint line** and **bubble line values.** The **dewpoint line** value is used to measure superheat. The **bubble line** value is used to measure subcooling.

Temperature glide is the temperature difference between the vapor and liquid state of a near-azeotropic, a zeotropic, and a ternary blend refrigerant. This change occurs at a constant pressure during evaporation and condensation.

An azeotropic refrigerant has a **temperature glide** of zero.

A near-azeotropic refrigerant has a **temperature glide** above zero up to 10°F.

A zeotropic refrigerant has a **temperature glide** of above 10°F.

Fractionation occurs when a zeotropic or a ternary refrigerant is added to a system as a gas. The refrigerants separate unevenly in the system, and this causes the refrigerant to change its composition.

The chemical name for **R-410A** is difluoromethane/pentafluoroethane. The molecular formula for **R-410A** is CH_2F_2/CF_2CF_3.

R-410A uses polyol ester oil and boils at −62.9°F at atmospheric pressure.

Pressure-temperature charts are used to determine certain pressures and temperatures of different refrigerants (see Table 1–1).

Examples:

R-22 will be at 40°F when the pressure is 68 psig (pounds per square inch gauge).

R-12 will be at 37 psig when its temperature is 40°F.

TABLE 1–1 Pressure-Temperature Chart

Pressure in psig	Temperature in °F		
	R-12	R-22	R-502
25	25	1	−7
26	27	2	−6
27	28	4	−5
28	29	5	−3
29	31	6	−2
30	32	7	−1
31	33	8	0
32	34	9	1
33	35	10	2
34	37	11	3
35	38	12	4
36	39	13	5
37	40	14	6
38	41	15	7
39	42	16	8
40	43	17	9
42	45	19	11
44	47	21	13
46	49	23	15
48	51	24	16
50	53	26	18
52	55	28	20
54	57	29	21
56	58	31	23
58	60	32	24
60	62	34	26
62	64	35	27
64	65	37	29
66	67	38	30
68	68	40	32
70	70	41	33
72	72	42	34
74	73	44	36

LUBRICANTS

Technicians must understand lubricants and their properties in order to use them correctly with proper refrigerants. The following are common terms and definitions of lubricants and their properties.

Mineral oils are often referred to as petroleum oils. These oils originate from the earth.

Synthetic oils are chemicals produced to function as lubricants.

The three most common types of **synthetic oils** are glycol, alkylbenzene, and polyol ester.

R-134a is not miscible with **mineral oil.**

Mineral oils are most commonly used with chlorofluorocarbon (CFC) refrigerants.

Mineral oils are not used with HFC alternative refrigerants.

Mineral oils are not miscible with HFC refrigerants.

Polyol ester oil is the only lubricant suitable to use with R-134a.

Alkylbenzene oil is a synthetic oil made by mixing benzene and propylene.

Alkylbenzene oil is commonly used with HCFC refrigerants and can be used with CFC refrigerants.

Polyalkylene glycol (PAG) oil refers to lubricants used in automotive air conditioning systems.

A **hygroscopic oil** has a high ability to absorb moisture.

If **moisture levels** are more than 100 ppm (parts per million) in an air conditioning or refrigeration system, corrosion will result.

Good characteristics of oil are proper viscosity, low pour point, noncorrosive, low carbon-forming tendencies, stability, low floc point, and low hygroscopic ability.

Floc point is the temperature at which oil starts to produce a waxy substance that may interfere with the refrigeration system.

Pour point is the ability of oil to flow with refrigerants at very low temperatures. A lower pour point is better.

Miscibility is the ability of a fluid to readily mix with another fluid.

Lubricants should remain miscible at all temperatures and pressures.

A **synthetic lubricant** absorbs more moisture than a mineral oil.

A **PAG oil** absorbs more moisture than a polyol ester oil.

Polyol ester oil tolerates 3 to 5% mineral oil.

It is harder to **dehydrate** a system using polyol ester oil than a system using mineral oil.

Polyol ester lubricants should be stored in their own container because moisture can permeate through certain materials.

The **viscosity** of a lubricant oil is measured by the Saybolt Universal Viscosity Test.

The **higher the viscosity,** the thicker the oil.

Flash point refers to the temperature at which vapors from the oil surface flashes will ignite with a flame, but not persistently burn.

The main **function of a lubricant** is to decrease friction and minimize wear of internal components.

An oil lubricant helps to form a **liquid seal** from the high and low side of the system.

When retrofitting a system with alternative refrigerants, it is the **technician's responsibility** to recover the oil/refrigerant mixture before charging the system with the new refrigerant and oil.

According to the Environmental Protection Agency (EPA), any **oil removed** from a system is the property of the service contractor.

COMPRESSORS

A compressor is the most vital component of an air conditioning and refrigeration system. There are many types of compressors. Each is designed to be used for a specific application. Compressor manufacturers are always working on methods to improve the efficiency of their compressors. The following are related terms for air conditioning and refrigeration compressors.

A **compressor** takes a low-pressure/low-temperature refrigerant gas from the evaporator and raises it to a high-pressure/high-temperature gas. Then it pumps the refrigerant gas into the condenser.

The most common compressors used for residential and light commercial air conditioning are the **reciprocating** and the **scroll** designs.

The **positive displacement compressor** designs are reciprocating, rotary, screw, and scroll designs.

Screw compressors use two special helical rotors that induce suction vapors and then compress them. They are used on large tonnage systems.

Scroll compressors are 10% higher in efficiency than reciprocating types. They use two spiral portions that are meshed together to compress refrigerant gas. Scroll compressors handle liquid floodback better than reciprocating compressors.

A **nonpositive displacement compressor** design is the centrifugal compressor.

Reciprocating compressors use pistons, cylinders, and valves to compress the refrigerant gas.

Compressor Ratio = Discharge psia/Suction psia (pounds per square inch absolute). An increase in compression ratio will overheat a compressor.

The **three classifications** of compressors are hermetic, semihermetic, and open.

The **hermetic compressor** is a welded, nonservicing type that is driven by an internal motor.

The **semihermetic compressor** is bolted together and is serviceable. It is driven by an internal motor.

The **open compressor** is bolted together and is serviceable. It is driven by an external motor.

The open compressor can use **two types of drives:** the belt drive and the direct drive.

The two types of **compressor lubrication** systems are splash lubrication and forced lubrication.

Compressors are cooled by the low-pressure/low-temperature refrigerant gas from the evaporator and by the condenser fan moving air over its outer shell.

Multistage compressors commonly use a two-stage thermostat to cycle a two-speed compressor. They also use two compressors connected together to operate one system.

Crankcase heaters are used to heat the oil of the compressor during the off cycle. They aid in keeping the liquid refrigerant from mixing with the oil, which could cause a flooded start.

AIR-COOLED CONDENSERS

Air-cooled condensers are commonly used in split and package-type air conditioning systems. The technician must have a good understanding of condensers and their related terms.

A **condenser** removes heat from an air conditioning system by changing the refrigerant from a high-pressure/high-temperature gas to a high-pressure/high-temperature liquid. It also de-superheats and subcools the refrigerant.

Additional superheat is found at the end of the evaporator and the heat from the motor windings. This heat, along with heat of compression, must be removed from the refrigerant before it can be condensed.

The tubes of a **condenser** are made out of copper or aluminum.

The **fins** of a condenser are made out of aluminum.

Condensers must have properly designed clearances on all sides of the unit to prevent recirculation of air over the coil.

Split-system air-cooled **condensing units** contain the compressor, the condenser coil, the condenser fan, service valves, and various electrical controls.

Packaged units contain the compressor, the condenser coil, the evaporator coil, the condenser and evaporator fans, service valves, the heating portion of the unit, and various electrical controls.

The **temperature difference** between the ambient air and the refrigerant in the condenser should be between 20 and 30°F.

The **airflow** of a condenser coil is either a side-discharge or a top-discharge type.

There are two types of **side-discharge** condensers. One type pulls air across the coil, through the fan, and out to the ambient air. The other type pulls air across the fan and pushes it through the condenser coil. If the **air** is pushed over the condenser coil, the coil must be cleaned directly in front of the condenser fan blades. If the air is pulled over the condenser coil, the coil must be cleaned at the exposed section of the coil. If **air** is trapped inside a condenser coil, the high-side pressure will increase. This causes a decrease in cooling capacity and an increase in amperage, and the excessive heat could damage the system.

Top-discharge condensers pull air from the bottom or sides of the condenser coil and push it out the top of the condenser.

WATER-COOLED CONDENSERS

Many light commercial air conditioning and refrigeration systems use water-cooled condensers. It is important for the technician to have a complete understanding of their operation.

The efficiency of a **water-cooled condenser** is affected by the temperature difference between the refrigerant and the water in the condenser and the quantity of water flowing through the system.

The **capacity of the condenser increases** as the temperature difference between the refrigerant and the water increases.

In a water-cooled condenser, an **increase of water quantity** decreases the power consumption of the compressor.

In a water-cooled condenser, a **decrease of water** saves water but increases the power consumption of the compressor.

The **fouling factor** is the measurement of heat-transfer characteristics of water. The higher the fouling factor, the poorer the heat transfer.

The **higher the fouling factor** of the water, the higher the flow rate of water needs to be.

Whenever possible, a **condenser's water** should be discharged to storm drains and storm sewers if local codes permit.

A **water-regulating valve** controls the flow of water through a water-cooled condenser.

A **water-regulating valve** can be located at the inlet or outlet of a water-cooled condenser.

A **water-regulating valve** operates on spring pressure and acts as the closing force in water-cooled condensers. In refrigeration systems, high-side pressure acts as the opening force.

The condensing temperature is commonly **10 degrees higher** than the temperature of the water leaving the condenser.

Mineral deposits and **scale** build up on the inside of a water-cooled condenser. This buildup helps insulate the water from the refrigerant tube. It also decreases the efficiency. The decrease in efficiency causes an increase in high-side pressure and water consumption.

The most common water-cooled condensers are the **tube-within-a-tube, shell-and-coil,** and **shell-and-tube.**

The **tube-within-a-tube** water-cooled condenser uses the inner tube for refrigerant and the outer tube for water.

The tube-within-a-tube condenser uses the **counterflow principle** where the cooler water is in contact with the cooler refrigerant liquid and the warmer water is in contact with the hot gas.

The **tube-within-a-tube** condenser is commonly constructed of steel and copper.

There are **two common types** of tube-within-a-tube condensers. One type is the **sealed coil,** which must be chemically cleaned. The other type has **horizontal tubes,** one on top of another. These tubes are flanged on each end. The flanged ends make it possible to manually clean the condenser.

The **shell-and-coil** water-cooled condenser has the refrigerant in the shell. The water flows through the coil and tube located inside the shell.

The shell of a **shell-and-coil** condenser acts as a liquid receiver. It holds the refrigerant charge until it is needed in the evaporator.

The **shell-and-coil** condenser cannot be manually cleaned. It must be chemically cleaned.

The **water box** is the end cap of a shell-and-tube condenser.

A **shell-and-tube** condenser can be cleaned manually by removing the water box.

INDOOR EVAPORATOR COILS

Various types and styles of evaporator coils are used on air conditioning and refrigeration equipment. It is important for the technician to have thorough knowledge of evaporator coil types and to understand their function.

An **evaporator** absorbs heat into the refrigerant by the latent heat of vaporization.

On an upflow furnace, the indoor evaporator coil must be **installed at least 2″** to **3″** above the heat exchanger to prevent the furnace from overheating.

Direct expansion evaporators have refrigerant inside the tubes of the evaporator, and the controlled medium is directly in contact with the coil.

The **tubes** of an **evaporator coil** are made of copper.

The **fins** of an **evaporator coil** are made of aluminum.

A **distributor** is a device that feeds refrigerant evenly through the circuits of an evaporator coil.

Split-system evaporator coils are located in the furnace or in the air-handling unit.

When an **uncased coil** is installed into an existing furnace, metal plates are installed around the outside of the coil to force air through it. If there is any air bypass, this will decrease the coil efficiency.

When an **evaporator coil** is installed, it is important to make sure the coil is level for proper condensate drainage.

An **"A" coil** is two coil sections connected together in the shape of an "A." This is the most common coil used in upflow residential systems today.

An **"N" coil** is three coil sections connected together in the shape of an "N." This is used when a narrow width of a coil is needed in an upflow application.

A **slant coil** is used on upflow or downflow systems where a low coil height is critical.

Upflow system evaporator coils are commonly located on the positive-pressure side of the blower after the heat exchanger.

Downflow system evaporator coils are commonly located on the positive-pressure side of the blower after the heat exchanger.

An **"H" coil** is used on horizontal flow furnaces and packaged AC units. The coil is designed so air flows over it from one side of the system to the other.

In a residential split system using a **horizontal flow system,** air is pushed over the coil. In a commercial package unit using a horizontal flow system, air is pulled over the coil.

When an evaporator coil is installed on the negative side of the blower, air is pulled over the coil. A **"P" trap** is mandatory on the drain line to prevent condensation from entering the duct system.

The minimum **drain line** size on a residential evaporator coil is $\frac{3}{4}''$.

The minimum slope for a **drain line** is $\frac{1}{8}''$ per foot of horizontal run.

The **drain line** for a cooling unit larger than 6 tons must discharge to a sanitary sewer drain, storm drain, or French drain.

The **drain line** for a cooling unit 6 tons or smaller must discharge to a gutter, roof drain, or on the roof.

Lead or aluminum piping must never be used for a **drain line.**

An **auxiliary drain pan** is required when cooling coils are located in areas where water damage can occur if the primary drain was to fail. A secondary drain pan must be $1\frac{1}{2}''$ deep and not less than 3″ larger than the air-handling unit on each side.

When using galvanized steel as an **auxiliary drain pan,** the minimum allowable thickness of the steel is 26 gauge.

A **secondary drain line** is required for evaporator coils located in attic spaces where water damage may occur.

A **secondary drain line** must be routed in a conspicuous place to serve as an alarm when the primary drain becomes plugged.

An approved alternative to a **secondary drain line** would be the use of a normally closed float switch located in the secondary drain pan that is in series with the contactor coil of the condensing unit. This will shut down the condensing unit if the primary drain line becomes plugged.

METERING DEVICES

One of the main components of a refrigeration and air conditioning system is the metering device. Although there are several types of metering devices, each used for a specific application, they all work to feed refrigerant to the evaporator. The following are types of metering devices.

A **metering device** takes a high-pressure/high-temperature subcooled liquid from the condenser and changes it to a low-pressure/low-temperature saturated liquid. It then directs the liquid into the evaporator.

Flash gas is the instantaneous evaporation of liquid refrigerant as it enters the evaporator.

A **heat exchanger** is commonly used to decrease flash gas.

The three common types of **metering devices** used in residential and light commercial air conditioning are the **capillary tube,** the **TXV,** and the **piston** or **fixed bore.**

The pressure drop of a **capillary-tube-type** metering device is determined by the tubing length and inner diameter.

A **triangular metal file** is most commonly used to cut a section of capillary tubing.

TXV stands for **thermostatic expansion valve.**

A **TXV** has three operating pressures: the sensing-bulb pressure (opening force), the evaporator pressure (closing force), and the spring pressure (closing force).

Turning the **TXV valve stem** clockwise increases superheat. Turning it counterclockwise decreases superheat.

Superheat(entering the compressor) = suction line temperature − evaporator's boiling point temperature.

Example:

Suction line temperature = 50°F

Evaporator boiling point temperature = 35°F

Superheat = 50°F − 35°F

= 15°F

An **external equalizer line** is used on a TXV when there is an excessive pressure drop through an evaporator. The equalizer line compensates for the drop in pressure through the evaporator. The line is located in the suction line 6″ to 8″ downstream from the TXV sensing bulb.

The **TXV sensing bulb** needs to be placed at the 4 o'clock and 8 o'clock positions on the suction line if it is larger than $\frac{7}{8}''$ in diameter. If the suction line is smaller than $\frac{7}{8}''$ in diameter, the sensing bulb must be located at the 12 o'clock position.

The pressure drop of a **piston** or **fixed-bore-type** metering device is determined by its orifice size and the high-side pressure.

THERMOSTATIC EXPANSION VALVES

The most popular metering device in commercial refrigeration is the thermostatic expansion valve (TXV). The TXV is designed to provide accurate control of superheat under various load conditions. The service technician must have a good understanding of how various types of thermostatic expansion valves operate and are selected for different applications. The following statements will provide technicians with a better understanding of terms, operation, and application of various thermostatic expansion valves.

Superheat can be defined as a vapor that is **above its saturation temperature** for a given pressure or the difference between the refrigerant vapor temperature and its saturation temperature.

The term **hunting** refers to unstable conditions of a TXV; it will tend to overfeed or underfeed in an attempt to find a balance point and become stable.

The main functions of the thermostatic expansion valve are to **regulate** refrigerant flow and to maintain a constant **superheat.**

When **superheat increases** at the evaporator outlet, the TXV will open more and allow an increased flow of refrigerant to pass through the valve.

The thermostatic expansion valve will **decrease refrigerant flow** when the superheat at the evaporator outlet lessens.

When a thermostatic expansion valve is used, the system's refrigerant charge is generally not as **critical** as it would be with other expansion metering devices.

The **sensing bulb** is part of the TXV and is connected to the valve by a **capillary tube** to the top of the valve's **diaphragm.**

The capillary tubing allows **bulb pressure** to flow to the top of the valve's diaphragm.

The thermostatic expansion valve uses a **pin,** a **pin carrier,** and **pushrods** that work to move the pin on and off the port.

Most thermostatic expansion valves use a **superheat spring** that can be adjusted to set the correct superheat setting desired.

The three main operating pressures of a thermostatic expansion valve are **bulb pressure, evaporator pressure,** and **spring pressure.**

The **bulb pressure** within the bulb is applied to the top of the diaphragm and causes the valve to **open,** allowing refrigerant to flow through the valve.

The **evaporator pressure and the spring pressure** together work to **close** the valve.

When the system is in **normal operation**, the sensing-bulb pressure should **equal** the spring and evaporator pressure.

In general, the function of the **sensing bulb** is to sense the temperature of the refrigerant vapor as it **leaves** the evaporator coil.

When the **bulb temperature rises**, the **bulb pressure** also will rise, which will allow the valve pin to **lift off** the valve port.

When the **bulb temperature is reduced**, the **bulb pressure** also will be **reduced.** This will allow the valve pin to move to **close off** the valve port.

Turning the **superheat adjustment stem clockwise** will allow less refrigerant flow into the evaporator, causing an **increase** of superheat.

Turning the **superheat adjustment stem counterclockwise** will allow more refrigerant flow into the evaporator, causing a **decrease** of superheat.

When there is an **increase** in heat load on the evaporator, the refrigerant will tend to evaporate at a **faster** rate.

The term **internally equalized** refers to a method of bleeding off or allowing evaporator pressure to pass through a passageway within the valve body in an attempt to cause minimal **pressure drops** across the coil.

The term **externally equalized** refers to a method for allowing evaporator pressure to bleed off or flow externally from the valve via an **equalizer line** that is connected to the suction line.

Thermostatic expansion valves that are **internally equalized** should generally be used with **single-circuit** evaporators with no more than the equivalent of **2°F** saturated temperature change.

The **equalizer line** should never be **capped off.**

The **externally** equalized thermostatic expansion valve senses **evaporator pressure** at the evaporator **outlet;** it is not influenced by a pressure drop through the evaporator.

The term **thermostatic charges** refers to the substance in the sensing bulb that responds to **suction line temperature.**

Thermostatic charges are classified into the following: **liquid charge, gas charge, liquid cross charge, gas cross charge,** and **adsorption charge.**

The **liquid charge** and **gas charge** sensing bulb is commonly charged with the **same** type of refrigerant that is in the system.

The **liquid cross charge** consists of a refrigerant mixture. The liquid cross charge TXV will always have some liquid within the bulb, capillary, and diaphragm.

The **gas cross charge** element contains a mixture of refrigerant that will **condense** into a small quantity of liquid when the TXV is in its normal operating range.

The **adsorption charge** has a noncondensable gas along with an **adsorbent material** within the sensing bulb. The adsorption charge TXV does not have an **MOP.**

The term **MOP** refers to **maximum operating pressure.**

Thermostatic expansion valves are installed by **flared** or **sweat** connections.

Whenever a refrigerant distributor is used on an evaporator coil, a TXV with an **external equalizer** should be used.

The **external equalizer line** should be connected **downstream** from the sensing bulb.

The term **de-superheating TXV** refers to a thermostatic expansion valve that has been **added** to the system. The de-superheating TXV is used in conjunction with a **hot gas or discharge bypass valve.**

The **de-superheating TXV** allows refrigerant to **cool** the hot discharge gas that enters the suction line.

Specially designed thermostatic expansion valves must be used when the system's refrigerant is **ammonia.** The construction must be of steel or steel alloys.

Refrigeration systems will generally operate with a **lower superheat** than air conditioning systems.

Thermostatic expansion valves can be **installed** in any position, although they should be as **close** to the evaporator as possible.

When a refrigerant distributor cannot be directly connected to the TXV outlet, the **maximum distance** between the valve outlet and the distributor should not exceed **24″**.

The **refrigerant distributor** is the **only** device that should be installed between the TXV outlet and the evaporator.

When the system's application calls for the TXV and the evaporator to be installed **above** the receiver, **static pressure** will always be present in the **liquid line.**

The **contact** of the TXV **sensing bulb** on the suction line should be direct on the piping and connected using **two bulb straps** on a straight portion of the piping.

The TXV sensing bulb needs to be placed at the **4 o'clock** or **8 o'clock** position on the suction line when the line is $\frac{7}{8}''$ **OD or larger.**

The TXV sensing bulb should be installed on top of the suction line when the suction line is smaller than $\frac{7}{8}''$ **OD.**

When an application calls for **brine tanks and water coolers** to be used, the sensing bulb should be **below** the **liquid surface.**

Dirt, sludge, and moisture in a system will cause severe problems of TXV operation. Some thermostatic expansion valves are of the **nonadjustable** type.

When **adjusting superheat,** as much as **30 min** might be needed in order to record a new balance point.

In general, the correct superheat will be determined by the amount of **temperature difference** between the **refrigerant temperature** and the **temperature** of the **air** or **medium** being cooled.

Superheat measurements are the most accurate means of determining **TXV performance.**

SERVICE VALVES

Service valves are used to open and close off the system's flow of refrigerant. They are also used to provide access for charging evacuation and recovering refrigerant. The following are various types of service valves.

Service valves are usually located on the outdoor condensing unit.

A **discharge service valve** is located on the outlet of the compressor. NEVER FRONT SEAT A DISCHARGE SERVICE VALVE.

A **liquid line service valve** is located in the liquid line.

Front seating the liquid line service valve blocks the flow of refrigerant to the liquid line. If the compressor is started, the system pumps down, removing the refrigerant from the liquid line and the evaporator coil. The refrigerant is then stored in the condensing unit.

A **suction service valve** is located on the inlet of the compressor.

Front seating the suction service valve blocks the flow of refrigerant to the compressor and draws a vacuum on the low-pressure gauge to test for operation of compressor valves and rings.

A **king valve** is a three-position service valve that is located on the outlet of a receiver.

A **queen valve** is a two- or three-position service valve located on the inlet of the receiver.

A **two-position service valve,** located on split-system condensing units, uses the back-seated position to open the refrigeration lines to the rest of the system for refrigerant flow and to obtain gauge readings.

A **two-position liquid line service valve** uses the front-seated position to pump a system down or test compressor valves.

A **Schrader valve** is used to check system pressures. It is permanently connected in an air conditioning condensing unit. The valve core seals the refrigerant in the system until a manifold gauge hose is connected. When a manifold gauge hose is connected, the hose will depress the valve core, and pressure will be supplied to the gauge.

Three-position service valves are used on older split systems and larger packaged air conditioning systems. Three-position service valves are located on the suction line and the liquid line.

Back seating a three-position service valve blocks the gauge port and allows the system to operate.

Midseating a three-position service valve (gauge position) allows the system to operate and direct refrigerant pressure to the gauge port.

Front seating a three-position liquid line service valve blocks the flow of refrigerant to the liquid line.

If the compressor is started, the system pumps down. This removes the refrigerant from the liquid line and the evaporator coil. The refrigerant is then stored in the condensing unit.

LEAK DETECTION

There are many different types of leak detection methods and equipment. The following are examples of leak detection methods.

An **oil trace** around copper tube fittings may be an indication of a refrigerant leak.

When testing for a **leak,** the system must be pressurized and electrically de-energized.

Air has a tendency to stay in the condenser and increase the operating high-side pressure.

If there is no pressure in the system, pressure can be increased for **leak detection** by adding nitrogen and a trace amount (usually an ounce or two) of R-22.

A pressure regulator and an in-line pressure-relief valve should be used when **pressurizing a system** with nitrogen.

Oxygen must NEVER be used to purge or pressurize a refrigeration or air conditioning system. Doing this could cause an explosion.

Pressures should never exceed the low-side **test pressure rating** on the data plate.

A **soap bubble** solution can be used to test for a suspected leak. If a leak is present, bubbles will start to form.

If manufacturer's specifications allow, **refrigerant dye** can be added to a refrigeration system and circulated for 24 h to detect a leak.

A **halide leak detector** is used with systems containing halocarbon refrigerants. It must never be used near flammable material.

When using a **halide leak detector,** the flame turns green because of the results of chlorine in the refrigerant coming in contact with a hot copper element located in the halide detector. The halide leak detector will not detect a hydrofluorocarbon (HFC) refrigerant.

An **electronic leak detector** is designed to test for extremely small refrigerant leaks. Because refrigerant is heavier than air, the detector should always be held under the tubing.

When a leak is located using an **electronic leak detector,** the detector's sound becomes high pitched and its flashing light becomes steady.

An **ultrasonic leak detector** senses a leak when the high-pitched sound waves created by a leak enter the detector.

An **ultrasonic detector** can sense a leak in a system that is under positive or negative pressure.

An **ultraviolet light,** used with a fluorescent system additive, can detect extremely small refrigerant leaks.

RECOVERY EQUIPMENT

When it becomes necessary to open a sealed refrigeration system, the refrigerants need to be recovered. The following are terms and information related to recovery equipment. Refer to Table 1–2 for examples.

TABLE 1–2 Inches of Mercury (Hg) Vacuum in Appliances

	Inches of Hg Vacuum Equipment Manufactured	
High-pressure appliance:	Before 11-15-93	After 11-15-93
HCFC-22 charge (under 200 lb)	0	0
HCFC-22 charge (200 lb and over)	4	10
Other high-pressure appliances using CFC-12, 500, 502, 114 (under 200 lb)	4	0
Other high-pressure appliances using CFC-12, 500, 502, 114 (200 lb and over)	4	15
Very high-pressure appliances (using CFC-13 and 503)	0	0

Recovery means to remove refrigerant from an appliance and store it in an external container.

The **three methods of recovery** are charge migration, use of the appliance compressor, and use of a refrigerant recovery unit.

The **three types of recovery machines** are vapor recovery units, liquid recovery units, and vapor-liquid recovery units.

There are **three classifications of equipment:** very high pressure, high pressure, and low pressure.

Very-high-pressure equipment uses refrigerant that boils below −58°F at atmospheric pressure.

High-pressure equipment uses refrigerant that boils between −58 and 50°F at atmospheric pressure.

Low-pressure equipment uses refrigerant that boils above 50°F at atmospheric pressure.

The requirements of recovery equipment are divided into size and usage.

Small appliance equipment must reduce the pressure in a system to 4″ Hg.

Recovery cylinders must be approved by the U.S. Department of Transportation (DOT). A **recovery cylinder** must not be filled to more than 80% of its capacity.

To **recycle** means to remove and clean the refrigerant from an appliance without meeting the requirements for reclamation.

To **reclaim** is to process refrigerant to a level equal to new refrigerant specifications by chemical analysis.

The reclaimed refrigerant must meet the standards of ARI 700 before it can be resold.

Recycling removes moisture, air, and acids from a refrigerant.

The filters on **recycling machines** must be changed as required. The oil separators also need the oil removed regularly.

SOLDERING AND BRAZING SAFETY

Safety precautions should always be adhered to when using torches for soldering or brazing. Always be aware of the surroundings you are working in. The following information will help the technician to work in a safe manner.

Always have a **fire extinguisher** present when using a torch.

Always wear proper **safety glasses** when soldering or brazing.

Use a **noncombustible heat shield** to protect combustible material when brazing or soldering.

When using a torch, always **wear gloves** to prevent burns.

To **prevent a burn,** never stand directly under material that is being brazed or soldered.

Always use **cadmium-free** brazing materials.

To prevent asphyxiation, soldering and brazing must be done in a **well-ventilated area.**

Oxyacetylene torches must be inspected frequently to make sure they are in proper working order.

Do not use short pieces of **brazing rods.** Burnt fingers or hands may result.

To prevent an injury, acetylene pressure must **never exceed 15 psi.** Above this pressure, acetylene becomes unstable.

To prevent burns, **never use a lighter** when igniting a torch. Only use recommended strikers.

To prevent a refrigerant burn or possible explosion, never **solder or braze** a system that is under pressure.

Always **open the system** to atmospheric pressure when soldering or brazing.

SOLDERING AND BRAZING

A large part of installations and some repairs will require a technician to perform soldering and brazing techniques. The following terms and information are related to soldering and brazing.

Soldering is accomplished when a temperature below 800°F is used to bond metals.

Brazing is accomplished when a temperature of 800°F or above is used to bond metals.

Never attempt to **solder** or **braze** a joint if a system is under pressure. This could cause the rapid escape of refrigerant that could result in an injury.

Stay Brite #8 solder is 94% tin and 6% silver. It melts at 535°F.

95-5 solder is 95% tin and 5% antimony. It melts at 450°F.

Flux is used to keep material clean while soldering.

Braze and **silver solder** melting point temperatures range from 1100 to 1480°F.

Brazing material creates a stronger joint than soft solder or silver solder.

Leaks in coils made from aluminum are commonly repaired with epoxy or aluminum solder.

Never overheat a **filter dryer** when brazing it into the system. This will damage the filter.

When brazing a joint, **capillary action** causes the brazing material to flow throughout the entire joint area.

Regulators have two gauges. One gauge measures pressure to the torch tip in psig. The other measures tank pressure in psig.

The operating pressures of an **acetylene regulator** range from 0 to 15 psig.

The operating pressures of an **oxygen regulator** range from 0 to 100 psig.

An **oxygen and acetylene regulator** must be turned clockwise to increase pressure and counterclockwise to decrease pressure.

An **oxyacetylene torch** uses a set of hoses to connect the torch handle to the oxygen and acetylene pressure regulators.

The **oxygen and acetylene** pressure gauges display the remaining tank pressure and the regulated pressure to the torch tip.

A tank of dry nitrogen and a pressure regulator are used to **purge** the refrigeration tubing. This helps prevent oxidation of the tubing during the brazing process.

Oxygen must NEVER be used to purge or pressurize a refrigeration or air conditioning system. Doing this could cause an explosion.

A **tube cutter** is used to cut hard or soft tubing. A **reaming tool** generally accompanies the tube cutter. It is used to remove small burrs from inside the tubing after it has been cut.

Small tube cutters, sometimes called "imps," are used to cut small-diameter tubing. The **reaming tool** is used to remove the burrs of the copper tubing after it has been cut.

Swaging of soft copper tubing is often done to connect the same size tubing without the use of a coupling. When swaging, the inside diameter of one tube is expanded to fit over the outside diameter of the other tube. The depth of the swage is determined by the type of swaging tool that is used.

When **bending** soft copper tubing, the minimum bending radius on smaller tubing is no less than five times the diameter of the tubing.

Example: $\frac{1}{2}''$ diameter tubing $\times\ 5 = 2\frac{1}{2}''$ minimum radius bend.

A **tube bender** is used to prevent kinks when bending soft copper tubing. When bending copper tubing, the tubing diameter must never be decreased.

When assisting in a bend, a **spring-type tube bender** can fit the inside or the outside of the tubing.

A **lever bender** can make a precision bend up to 180 degrees on soft copper tubing.

A **flare nut** and a **flare fitting** are used in the refrigeration and air conditioning field to create a leak-free mechanical joint.

A **flaring block** holds copper tubing in place when using the flaring tool.

A **flaring tool** is the tool used to create a flare by beveling the cop per tubing with the flare tool. To prevent splits in the flare, tighten and loosen the tool as the flare is being made.

ESCO INSTITUTE-HVAC EXCELLENCE BRAZING AND SOLDERING CERTIFICATION EMPLOYMENT READY (ER)

Type "L" copper tubing is used in ACR applications.

Type "K" copper tubing is the thickest wall.

Type "M" copper tubing has the thinnest wall.

ACR copper tubing is measured by outside diameter.

Copper tubing used in plumbing is measured by **nominal diameter.**

Once cut, copper tubing should be **reamed.**

Expanding the inside diameter of copper tubing is referred to as a **swage.**

Opening the end of copper tubing that will appear as a funnel is called a **flare.**

Common **torch types** are oxyacetylene and air acetylene.

Nitrogen is used to dry and purge the inside of copper tubing.

Brazing temperatures are over 840°F.

Soldering temperatures are under 840°F.

A **soldering flux** is used to help the solder flow and prevent oxidizing.

Sand cloth or wire brushes are used to **clean copper tubing.**

Dry nitrogen can be used for leak detection after a repair is made.

Soap bubbles and electronic meters are **leak detection** methods.

The types of copper used in ACR are **soft and hard drawn.**

Soft copper tubing is bent with hand **lever benders.**

Methods to join copper tubing are to flare, crimp, solder, and braze.

Hand tube cutters are used to cut soft copper tubing.

A **yoke and block** are used to make flares.

The punch, compression, and hydraulic are methods used to **swage.**

The **"MC" tank kit** is a component of a portable oxyacetylene kit.

The types of **torch flames** are neutral, carburizing, and oxidizing.

When using torches, always wear the appropriate **PPE for eye and hand protection.**

Always have a **fire extinguisher** nearby when using torches.

The **two gauges on an oxyacetylene torch kit** measure outlet and tank pressures.

The **"MC" tank (10 cubic feet)** is used with acetylene.

The **"R" tank (20 cubic feet)** is used for oxygen.

The air acetylene tank is a **"B" tank (40 cubic feet).**

Propane, Mapp gas, and air acetylene torches can be used for soldering.

A **thermal cloth or a dissipating gel** should be used to prevent fires during torch use.

Never exceed the **lowest component test pressure** when pressurizing a system.

A trace of refrigerant can be added to nitrogen to detect leaks.

After a repair has been made, it should be followed with an **evacuation.**

Pressure regulators on tanks when brazing copper to copper are set at a 1:1 ratio.

Air acetylene regulators on tanks should not be adjusted higher than 14 to 15 psig.

A **carburizing flame** that contains less acetylene is called a reducing flame.

Purging with nitrogen prevents oxidation and soot from forming during brazing.

Another term for soldering is **"sweating."**

Flux should be applied to the male section of the tubing to be soldered.

A general rule when **applying solder** is not to use more than the diameter of the tubing.

Nitrogen is considered a dry gas.

ACR refers to air conditioning and refrigeration.

MANIFOLD GAUGES

A manifold gauge set is one of the most important instruments a service technician can possess. These gauges provide the technician with information about the operation and status of the system.

A **manifold gauge set** is used to check operating pressures on a refrigeration and air conditioning system in psig (pounds per square inch gauge). The temperature is taken on the inside color-coded scale.

A **manifold gauge set** is adjusted by opening the gauges to the atmosphere and adjusting the pressure to 0 psig.

Manifold gauge sets are available in analog and digital readout.

The **center yellow hose** of a manifold gauge set is used to charge, evacuate, add oil, and recover refrigerant from a refrigeration and air conditioning system.

A **manifold gauge set** contains two gauges. One is a high-side gauge and the other is a low-side gauge. Each gauge reads positive pressures in psig. The low-side gauge can also read negative pressures in inches of mercury (Hg).

In an air conditioning system, the **high-side gauge** is commonly connected to the smaller line and the **low-side gauge** is commonly connected to the larger line.

The **blue gauge** on a manifold gauge set is used for testing the low-side pressure. It is called a **compound gauge** because it measures positive pressures up to 350 psig and negative pressures down to 30″ Hg.

The **red gauge** on a manifold gauge set is used for testing high-side pressure. It measures positive pressures from 0 to 500 psig.

Both high- and low-**pressure gauges** measure the pressure and temperature of different refrigerants. On a gauge, the pressure measurements are the outer black numbers measured in psig. The temperature measurements are the inner colored numbers measured in degrees Fahrenheit.

Four-way manifold gauges use front and side valves to evacuate and charge a system without having to switch hoses.

Low-loss fittings are used on manifold gauges to decrease the amount of refrigerant that is vented to the atmosphere.

Excessively **high discharge pressure** can be caused by air (noncondensables) in the system, an overcharge of refrigerant, a dirty condenser, recirculating ambient air, or insufficient airflow over the condenser.

A **decrease in suction pressure** can be caused by a refrigerant leak, a restricted or defective metering device, a dirty evaporator coil, or insufficient airflow over the evaporator coil.

To convert **psig** to **psia** (pounds per square inch absolute), add 14.7 to the psig.

To convert **psia** to **psig,** subtract 14.7 from the psia.

SYSTEM EVACUATION

The evacuation process removes moisture and noncondensables (air) from the sealed system. Both moisture and air will hinder satisfactory operation of the system. The following are terms related to evacuation.

Any **system opened** to the atmosphere must be completely evacuated.

A minimum evacuation level for an air conditioning system is **500 microns.**

Dehydration is the removal of moisture in a refrigeration or air conditioning system.

A **larger diameter center hose** decreases the evacuation time of a system.

Higher ambient temperatures decrease the evacuation time of a system.

If a system is **evacuated** and the pressure increases slightly and then stops, moisture is present in the system.

If a system is **evacuated** and pressure continually increases, a leak is present in the system.

A **triple evacuation** is accomplished by evacuating a system to 28″ Hg and charging the system to 0 psig three times.

A **single** or **deep evacuation** is accomplished by evacuating a system from 50 to 100 microns. The system should hold the vacuum after the pump is turned off.

A **vacuum pump** is connected to the center hose of the gauge port and removes air and moisture from a refrigeration or air conditioning system.

A **vacuum pump's oil** should be changed after every evacuation.

Vacuum pump oil tends to become full of moisture and acid after an evacuation. This affects the vacuum pump's ability to pull a deep vacuum.

Always allow air to enter a **vacuum pump** before storing, or the pump could become oil locked.

A **micron gauge** is an instrument that accurately measures the depth of a vacuum.

Both **manifold gauge valves** should be open during an evacuation.

A **thermistor vacuum gauge** is an instrument that accurately measures the depth of a vacuum in microns.

A **valve core tool** is used to remove the valve core on a Schrader valve. This decreases the evacuation time of a system.

When a system incorporates a three-way **service valve,** it must be in the midposition during the testing of the system's operating pressure, recovery, evacuation, and recharging.

ESCO INSTITUTE-HVAC EXCELLENCE SYSTEM RECOVERY AND EVACUATION CERTIFICATION EMPLOYMENT READY (ER)

Evacuate new and empty recovery tanks down to **500 microns.**

Evacuation refers to lowering the system's pressure below atmospheric pressure.

Cut out the **liquid line filter dryer** and replace it when it becomes restricted.

Hygroscopic oil can absorb moisture freely.

The pressure in a system must be reduced to at least 0 psig before an **evacuation** can be performed.

Recovery is removing refrigerant from a sealed system and transferring it into a recovery tank.

Dehydration and degassing are terms used in place of evacuation.

Install a filter dryer in the suction line after a **compressor burnout.**

Desiccant is the inside section of the filter dryer that removes moisture from the refrigeration system.

A **self-contained recovery unit** uses a compressor to remove refrigerant from the system into a recovery tank.

Never fill a recovery tank more than **80%** full by weight. The formula is WC × 0.8 × SG + TW (water capacity × 80% × specific gravity + tare weight). TW = weight of the empty cylinder.

Evaporation or dehydration uses a vacuum pump to remove moisture from a refrigeration system.

When a recovery cylinder becomes 80% full, it should be transferred to a collection center for **reclamation.**

Recovered refrigerants that have been mixed in the same recovery tank cannot be recycled or reclaimed and may be destroyed at the **owner's expense.**

The EPA Section 608 determines what the requirement **levels of evacuation** should be during a recovery process.

Locate the **micron gauge** at the manifold gauge during an evacuation to help prevent oil from entering into the micron gauge.

The EPA **Clean Air Act** Section 608 regulates CFC, HCFC, HFC, and HFO refrigerants.

The U.S. DOT (Department of Transportation) hydrostatically tests recovery cylinders **every 5 years.**

Heat pump systems use **bi-flow** liquid line filter driers.

The removal of **valve core pins** helps speed up the evacuation process.

Reclamation or reclaiming is the process of cleaning the refrigerant to meet new purity levels specified by AHRI Standard 700.

Always **purge the air** from the refrigerant manifold hoses to prevent air from entering the system.

Acid forms in a refrigeration system when air and moisture enter into a system.

Air or noncondensable gases (nitrogen-oxygen) in the system cause **carbon and sludge** to form due to the breakdown of compressor oil during excessively high condensing temperatures.

The **digital micron gauge** measures deep vacuums.

The **filter drier** captures moisture and particles in the system.

The purpose of **recovering** the system's refrigerant is to allow the system to be opened, repaired, and **recharged** with fresh clean refrigerant.

A **filter dryer** is used on a recovery machine to prevent foreign particulates from entering the recovery machine.

Add **nitrogen** to a system between multiple evacuations during a triple evacuation procedure.

To achieve a deeper vacuum, close the **ballast cap** on the vacuum pump.

To accomplish a **deeper vacuum** on a refrigeration system, use larger diameter hoses and have a larger cfm vacuum pump that has a larger vacuum pump exhaust port.

When the pressure becomes very low in a recovery tank, place the tank in an **ice water** bath to reduce the pressure and increase the speed of the recovery process.

760.000 microns is equivalent to 0 psig or 14.7 psia.

The required EPA recovery level for R-22 comfort cooling/air conditioning with **less than 200 pounds** of refrigerant is 0 psig.

Recycling refers to the process of cleaning the refrigerant by using a recycling machine that has filter driers to collect contaminants.

Two noncondensable gases that are found in refrigeration systems when gauges are not properly purged are **78% nitrogen and 21% oxygen,** which is the makeup of air.

The refrigerant that has been recovered from a refrigeration system can be used on other equipment as long as it is owned by the **same owner.**

According to the EPA, when using 3/8″ hoses for the recovery process, the length of the hoses should be **6 feet or less** to reduce *de minimis* release.

The dehydration process can be reduced by using shorter and larger hoses, and the **vacuum pump oil** must be clean.

Recovering in the **liquid state** will minimize the loss of the system's oil.

A **noncondensable** gas is a gas that will not change state to a liquid and can be trapped at the top of the condenser.

The **Clean Air Act Section 608** regulates refrigerants such as CFC, HCFC, HFC, and HFO.

System-dependent and self-contained are two methods of **refrigerant transfer known as recovery.**

During an evacuation performing a **decay test** (closing off the vacuum pump and giving time for an increase in pressure), once the 500-micron level has been reached and the vacuum pump isolated, the system's pressures should not rise more than 1500 microns.

A **passive recovery** is recovering into a nonpressurized cylinder with the help of the appliance's components or heating the appliance and lowering the temperature of the cylinder.

An **active recovery** is recovering into a recovery cylinder using a self-contained recovery pump system.

The **gas ballast** on a vacuum pump prevents moisture and contaminants from condensing in the vacuum pump's oil.

When recovering the refrigerant, a **liquid recovery** is much faster than a vapor recovery.

REFRIGERANT CHARGING

There are various methods used to add refrigerant to a system. The method may depend on the type of application and the refrigerant being used. The following are terms and related information on charging.

When the **weight method** of charging is used, an electronic charging scale weighs the proper refrigerant charge into a system.

A **charging scale** is commonly preset for the proper charge of a system in pounds and ounces, or the scale is set to zero and it measures the amount that is removed.

When the **volume method** of charging is used, a charging cylinder measures the proper refrigerant quantity into a system.

A **sight glass** is a device located in the liquid line that allows a technician to see the refrigerant. It is used to aid in properly charging a system. When a system is fully charged, the sight glass will appear clear or free of bubbles. A **sight glass** in the liquid line assists in a system's refrigerant charge.

When **liquid charging** a system, the system must be de-energized and have refrigerant added into the high side of the system through the high-side manifold gauge.

When **gas charging** a system, the system must be energized and have refrigerant added to the low side of the system through the low-side manifold gauge.

A **capillary tube system** must have an accurate refrigerant charge. Too much refrigerant will flood the coil, and too little will starve the coil.

To charge a system using the **superheat method,** the technician must first measure the indoor wet-bulb temperature, the outdoor dry-bulb temperature, and possess a superheat cross-reference chart.

The **superheat charging method** is used on a system that uses a capillary tube or piston-type metering device.

To **measure superheat,** subtract the boiling point temperature of the refrigerant (found on the low-side gauge) from the suction line temperature (within 6″ from the compressor).

To charge a system using the **subcool method,** the technician must know the high-side temperature, the liquid line temperature, and the manufacturer's subcool recommendations.

To **measure subcooling,** subtract the liquid line temperature from the high-side condensing temperature.

The **subcool charging method** is used on a system that uses a TXV-type metering device.

If an air conditioning system has an **excessive leak,** oil must be added.

ESCO INSTITUTE-HVAC EXCELLENCE BASIC REFRIGERATION AND CHARGING PROCEDURES CERTIFICATION EMPLOYMENT READY (ER)

A latent heat that changes vapor to liquid occurs in the **condenser**.

A **vapor** that has absorbed heat and increases in temperature is superheat.

A **receiver** is located after the condenser and stores refrigerant until needed in the evaporator.

The two components that always contain **liquid and vapor** are the evaporator and condenser.

The **accumulator** prevents liquid refrigerant from entering the compressor and is located in the suction line after the evaporator.

Refrigerant flowing through the liquid line filter drier is in the **liquid state** under high-side pressure.

The superheat reading entering into the compressor is **20 to 30°F.**

An **overcharged system** will have higher than normal high and low-side pressures.

Latent heat causes refrigerant to change from a liquid to a vapor in the evaporator.

The **basic refrigerant components** of the system are the compressor, evaporator, condenser, receiver, discharge line, suction line, and metering device.

When charging a capillary system, always **weigh** in the charge after a system evacuation.

Latent and sensible heat transfer from the refrigerated space to the refrigerant occurs in the **evaporator.**

A **plugged filter liquid line filter** drier will have a temperature difference of 2°F or more across the inlet and outlet.

A **restricted capillary tube** metering device will result in a starved evaporator.

The method for pinpointing a refrigerant leak in a high-pressure refrigeration system is by the use of a **soap bubble** solution.

The airflow over an air conditioning evaporator coil is **400 to 450 cfm** (cubic feet per minute).

Liquid line **solenoid valves** on an automatic pump down system are located before the metering device in the liquid line.

Refrigerants in HVAC/R systems are in a **saturated vapor** or **saturated liquid** state.

Subcooling prevents liquid line flash gas.

When a refrigeration system operates with a **low charge,** bubbles will appear in the sight glass.

The **sight glass** on a refrigeration system is located on the liquid line **before** the metering device.

The two main refrigeration components that separate the high and low sides of the system are the **compressor** and the **metering device.**

The evaporator temperature for an air conditioning coil is **40 to 50°F.**

An **automatic expansion valve** (AXV) is on systems with a constant load; when the pressure in the evaporator rises, the valve closes and the adjustable spring pressure is the opening force.

When the vapor in the condenser changes to a liquid, this is **saturated condensing pressure.**

Lowering the temperature of a liquid refrigerant below its condensing point is **subcooling.**

The **thermal bulb** of a TXV typically contains the same refrigerant type as what is in the system.

When a refrigeration system includes a TXV, the liquid receiver and sight glass always charge the system under a **high load condition.**

Refer to the manufacturer's recommendations when charging a system that uses a fixed metering device or capillary tube metering device.

The **operating pressures** of a TXV are the thermal sensing bulb, which opens the valve, and the evaporator and spring pressure, which closes the valve.

When installing a TXV metering device, the **thermal sensing bulb** should be located **after** the evaporator on the suction line and should be well insulated.

The pressure reading between the outlet of the metering device and the inlet to the compressor is **evaporator, suction, or low-side pressure.**

When referring to a pressure temperature chart, the measurements are only accurate when the refrigerant is in a **saturated state (liquid and vapor mix).**

A TXV's external equalizer tube is on large evaporators that compensate for **high-pressure drops** through the evaporator.

Capillary tube systems are critically charged systems.

One method to charge a capillary tube system is to charge by the **superheat method.**

To prevent liquid refrigerant from reaching the compressor, an **accumulator** is located on the suction line.

To increase liquid line, subcooling uses a liquid-to-suction **heat exchanger.**

Subcooling, superheated, and saturated refer to **refrigerant conditions.**

Lowering the sensible temperature of a liquid refrigerant below its saturated liquid state is **subcooled** refrigerant.

To charge a refrigeration system after a system evacuation, charge a small amount of vapor refrigerant into the high side of a system followed by **weighed liquid** charge.

Performing a charging procedure using a **charging chart** on a fixed piston orifice metering device requires the indoor wet bulb temperature, outdoor dry bulb temperature, and a superheat reading.

Higher than normal suction and discharge pressures could be the result of a loose **TXV sensing bulb** from the suction line.

Refrigerant initially passing into the condenser inlet would be in a vapor state and must be **de-superheated.**

Introducing liquid refrigerant into the suction line is a method to **suppress superheat** on low-temperature systems.

The advantage of using a TXV over a capillary tube metering device is that it can quickly adjust to changing evaporator **load conditions.**

One method that helps to increase subcooling and reduce flash gas is to employ **liquid pressure amplification.**

The **balanced port** on a TXV cancels out the effects of a high liquid pressure.

EPA SECTION 608 CERTIFICATION STUDY GUIDE

The Environmental Protection Agency (EPA) has mandated in Section 608 that all air conditioning and refrigeration technicians who will service and dispose of refrigeration and air conditioning equipment be certified. The certification also allows the technician to purchase refrigerants. The technician must be knowledgeable in general core-type information, such as ozone depletion potential, general terms and methods of recovery, and recycling and reclaiming procedures. The technician must also pass a certification exam. There are three types of certification offered. The technician will test in the certification type that will be used in their work area. There is also a Universal Certificate offered for the technician who is able to pass Core, Type I, Type II, and Type III exams. The following information will assist technicians in passing EPA Recover, Recycle, and Reclaim Examinations. As of 2018, the EPA has made several changes to the EPA Section 608 certification exam. The changes have been added to the following sections: Ozone Depletion Potential (ODP) was important in the past and still is; however, the EPA now focuses on refrigerants with a low Global Warming Potential (GWP). The previous refrigerants, CFCs, HCFCs, and HFCs, were the focus of the past exams. The new and upcoming refrigerants, HFOs and HCs, are now added to the new exam since they have "ZERO" ODP and low GWP.

CORE SECTION

The **Montreal Protocol** is a treaty regulating the production of CFCs.

CFCs neither dissolve in water nor break down into compounds that dissolve in water, so they do not rain out of the atmosphere.

It takes **one chlorine atom** to destroy 100,000 ozone molecules.

ODP stands for **Ozone Depletion Potential.** It is the measurement of the ability of CFCs and HCFCs to destroy ozone. CFCs have the highest ODP.

As of **July 1, 1992,** the Clean Air Act prohibits the venting of CFCs and HCFCs.

As of **November 14, 1994,** the Clean Air Act requires a service technician to be EPA certified.

As of **November 15, 1995,** the Clean Air Act prohibits the venting of HFCs.

When **disposing of a CFC, HCFC, HFC, or an HFO refrigerant cylinder,** recover the refrigerant and render the cylinder useless.

An appliance will need its **refrigerant recovered** before it can be discarded.

According to the EPA, it is legal to allow a small amount of refrigerant into the atmosphere. That may be caused by hose purging and equipment servicing. This is called ***de minimis*** releases.

The **temperature glide** is the different refrigerant boiling and condensing points of a blended refrigerant.

Service technicians that violate the Clean Air Act may be **fined up to $44,539** per day per violation, may lose certification, and may be required to appear in federal court.

A **reward of $10,000** is given to a person who supplies information about the venting of refrigerants.

The pressure range of the **high-side gauge** is 0 to 800 psig.

The pressure range of the compound gauge or **low-side gauge** is 30″ Hg to 350 psig.

All manifold gauges must have **low-loss fittings or shutoff valves** connected to the hoses.

It is legal to add a **trace amount of R-22 and nitrogen** when leak testing a system.

CFCs were phased out of production on **December 31, 1995.**

Fractionation is the separation of blended refrigerants caused by different refrigerant vapor pressures.

A **hygroscopic oil** has a high ability to absorb moisture.

The state of the refrigerant **leaving the compressor** is high-pressure superheated gas.

The state of the refrigerant **leaving the condenser** is high-pressure subcooled liquid.

The state of the refrigerant **leaving the metering device** is low-pressure saturated liquid.

The state of the refrigerant **leaving the evaporator** is low-pressure superheated gas.

The color code of **refrigerant gauges** is high-side red and low-side blue. The high-side red gauge reads pressure above atmospheric. The low-side blue gauge reads pressures above and below atmospheric (14.7 psi).

Recovery is the removal of refrigerant in any condition from a system to store it in an external container.

Recycling is the cleaning of a refrigerant for the purpose of reusing it. After recycling, the refrigerant is free of air, moisture, and acids.

Reclaiming is the process of cleaning refrigerant equal to new or virgin refrigerant. A chemical analysis must be done at a reclaiming station, and it must meet ARI 700 standards.

Mixing more than one refrigerant makes it **impossible to reclaim** refrigerant.

Refrigerant cylinders should be free of rust, undented, and undamaged.

The **electronic or ultrasonic leak detector** will help locate the general area of a system's leak. Soap bubbles are also used to pinpoint a system's leak.

Dehydration of a refrigeration system is done with a vacuum pump. It is done to remove air and moisture from a system.

If moisture is left in a refrigeration system, **hydrochloric and hydrofluoric acids** may form.

When **evacuating a system,** if the vacuum pump is turned off and the pressure rises to 0 psig, there is a leak present in the system.

When **evacuating a system,** if the vacuum pump is turned off when the pressure is at 29.92″ Hg and the pressure rises slightly and stops, there might be moisture present in the system. There might also be refrigerant mixed with the compressor oil.

The following will affect the **length of time** that it takes to evacuate a system:

The size of the equipment

The ambient temperature

The amount of moisture in the system

The size of the vacuum pump

The size and length of the refrigerant gauge lines

To prevent an explosion, never fill a refrigerant cylinder to more than **80% full.**

Refrigerant can cause **suffocation** because it is heavier than air and it displaces oxygen.

R-11, R-12, and R-500 blends are chlorofluorocarbon **(CFC)** types of refrigerants that were commonly used with mineral oil but have been phased out since 1990's.

R-123 and R-22 are **hydrochlorofluorocarbon (HCFC)** types of refrigerants that are commonly used with alkylbenzene oil but are also being phased out.

HFCs have no ozone depletion potential (ODP) but do have a high **global warming potential (GWP).**

HFC refrigerants such as R-134a, R-410A, R-404A, R-407C, and R-422B use **polyol ester (POE) or ester oil.**

A **ternary blend** type of refrigerant is a three-part blend and uses alkylbenzene oil.

Carbon dioxide (CO_2)-R744 is used as the **baseline measurement** for GWP and has a GWP value of 1.

R-744 (CO_2) is now compared to all of the new refrigerants approved by EPA's SNAP (Significant New Alternatives Policy).

Most HFC refrigerants have a GWP thousands of times greater than carbon dioxide. This is how SNAP decides refrigerants today.

Hydrofluoroolefin (HFO) refrigerants contain hydrogen, fluorine, and carbon atoms.

HFOs have no ozone depletion potential (ODP) but do have a very low global warming potential (GWP).

HFOs are less flammable than hydrocarbon (HC) refrigerants category A3, but most are mildly flammable. They are classified under a new category, nontoxic slightly flammable (A2L).

HFO refrigerants contain fluorine, making them less flammable than hydrocarbon refrigerants.

HFO refrigerants are miscible in POE lubricants.

R-1234zd is an HFO refrigerant used in low-pressure chillers, and R-1234ze is used in high-pressure chillers.

Hydrocarbon (HC) refrigerants are made up of hydrogen and carbon.

HCs have no ozone depletion potential (ODP) and a **very low global warming potential (GWP).**

The following are two **HC refrigerants**—isobutane (R-600a) and propane (R-290).

Propane cylinders for grilling should never be used as a refrigerant since they contain impurities that can damage refrigeration equipment.

When recovering HFO refrigerants, the refrigeration system and recovery unit are grounded.

Never apply an open flame to a flammable or nonflammable refrigerant cylinder.

A concentration of **HFO and HC** refrigerants in a confined space and an ignition source are likely to lead to an explosion.

A red color marking is required on all process tubes of both HFO and HC appliances, indicating that a flammable refrigerant is present.

An **azeotrope blend** is a combination of two or more refrigerants that, when blended, create one refrigerant that has its own characteristics so it will not fractionize, such as zeotropes or near azeotropic mixtures (NARMs).

A **system-dependent** (passive) recovery uses the appliance compressor to remove refrigerant from the system.

A **self-contained** (active) recovery uses an EPA-approved recovery machine.

The EPA recommends that hoses be equipped with **low-loss fittings or valves that manually close,** or that close automatically, to minimize refrigerant loss when hoses are disconnected.

To **safely recover refrigerants,** mechanical float devices, electronic shutoff devices, and weight methods are used.

When **pressure testing** a system, never exceed the system's low-side test pressures.

The **Department of Transportation (DOT)** must approve refrigerant cylinders before they are transported.

To **prevent an explosion,** never pressure test a system using oxygen.

When disposing of appliances with **5 to 50 lb of refrigerant,** the EPA requires the following records to be kept: type of refrigerant recovered, quantity of refrigerant recovered from disposed appliances in each calendar month, and quantity of refrigerant, by type, sent for reclamation.

Significant New Alternatives Policy (SNAP) is the EPA's new program that identifies and approves new and substitute refrigerants.

An HVACR servicing company must keep records for each disposed appliance with a charge between 5 and 50 lb for 3 years.

TYPE I CERTIFICATION

Type I Certification covers appliances that are charged at the factory with 5 lb or less of refrigerant. These include refrigerators, freezers, room air conditioners, PTAC units, dehumidifiers, ice makers, vending machines, and drinking water coolers.

A **small appliance,** according to the EPA, is a system containing 5 lb of refrigerant or less, including PTACs (package terminal air conditioners).

Under conditions of ARI 740-1993, **recovery systems manufactured *before*** November 15, 1993, must be able to remove 80% of a refrigerant charge or to achieve 4″ Hg vacuum.

Under conditions of ARI 740-1993, **recovery systems manufactured *after*** November 15, 1993, must be able to remove 90% of a refrigerant charge or to achieve 4″ Hg vacuum if the system's compressor is in operation.

Under conditions of ARI 740-1993, **recovery systems manufactured *after*** November 15, 1993, must be able to remove 80% of a refrigerant charge or to achieve 4″ Hg vacuum if the system's compressor is not operating.

Under conditions of ARI 740-1993, all **recovery systems** must be able to remove 80% of the refrigerant charge or to achieve 4″ Hg vacuum if the system's compressor is not operating.

Technicians must **certify their recovery equipment** with the EPA after August 12, 1993.

It is necessary to install a **piercing valve** on both the high and low side of an appliance that has an inoperative compressor while recovering the refrigerant. This will decrease the time of the refrigerant recovery.

The EPA does not require a **leak in small appliances** to be repaired.

The **pressure of an R-12** cylinder at 70°F ambient temperature will be 70 psig.

When the pressure of an R-22 cylinder is 121 psig, its temperature is 70°F.

When the **temperature of an R-502** cylinder is 80°F, its pressure is 161 psig.

The following refrigerants **must not be recovered** with the current recovery system; they may be vented: ammonia, methyl formate, methyl chloride, sulfur dioxide, hydrogen, water (H_2O), hydrocarbons (HC), and carbon dioxide (CO_2).

R-134a refrigerant is a replacement for R-12 refrigerant. However, at this time, there is no drop in replacement for R-12 refrigerant where oil will not have to be changed or added, and expansion valves do not have to be adjusted.

When pressurizing a system with nitrogen, a **pressure regulator and a relief valve** in line with the nitrogen cylinder is required.

Always leak test **piercing valves** after installing them onto a system.

If **refrigerants are exposed to open flames** and hot surface phosgene gas, hydrochloric and hydrofluoric acids are produced.

A burnt compressor may cause a **pungent odor** in a CFC/HCFC system.

A **self-contained breathing apparatus** is required if a large amount of R-12 has been released.

Relief valves must never be installed in series with each other.

The **color code** for refrigerant recovery tanks is yellow tops with gray bottoms.

The maximum normal charge of CFC, HCFC, HFC, or HFO refrigerants in an appliance that is allowed to use system-dependent recovery equipment is 15 lb.

TYPE II CERTIFICATION

Type II Certification covers all air conditioning equipment that uses more than 5 lb of charge of any refrigerant, except R-11, R-123, and R-113, which require Type III certification. The most common refrigerants are HCFC-22 and HFC-410A.

Type II technicians are certified to service high-pressure and very-high-pressure equipment, excluding small appliances and motor vehicles.

After the installation of a refrigeration system, the technician should first **pressure test the system using nitrogen or nitrogen with a trace amount of R-22.**

The maintenance task that must always be done to recovery/recycle equipment is to change the system's **oil and filters on a daily basis.**

If an open compressor has sat for several months, it is most likely to **leak from the shaft seal.**

Two indications that a high-pressure system has a leak are **oil traces** and high **superheat.**

As of January 1, 2019:

The EPA requires that all *comfort cooling* (air conditioning) systems with **more than 50 lb** of refrigerant be repaired when the annual leak rate exceeds **10%.**

The EPA requires that all *commercial refrigeration systems* with **more than 50 lb** of refrigerant be repaired when the annual leak rate exceeds **20%.**

The EPA requires that all *industrial process refrigeration systems* with **more than 50 lb** of refrigerant be repaired when the annual leak rate exceeds **30%.**

If an appliance with 50 or more lb of a regulated refrigerant has exceeded the threshold leak rate, the equipment owner or operator has **30 days** to repair the appliance.

If an appliance with a charge of 200 lb or more of a regulated refrigerant passes the initial verification leak test, a follow-up **verification test must be conducted within 10 days.**

If after repairing a leak in an HCFC or HFC rack refrigeration system the system fails the initial leak verification test, the technician has 30 days to complete the repair or retrofit.

If a cold storage warehouse-operating appliance charged with HCFC or HFC refrigerants leaks above the threshold rate, it will be allowed to operate without repair for 12 months before it must be retrofitted or retired.

If the cold storage warehouse appliance system **leak rate exceeds 125%** of its full charge within a year, a report must be submitted to the EPA describing the efforts made to repair the leak and action must be taken by completely repairing, retiring, mothballing (isolating the leak), or retrofitting the appliance.

The repair deadline can be extended to 18 months by temporarily shutting down the appliance, recovering the refrigerant from an isolated section or the entire system to at least atmospheric pressure, and the owner or operator using a refrigerant exempt from the venting prohibition.

The component that is directly after the condenser on a system that uses a TXV-type metering device is the receiver.

A **recovery and recycle machine** is likely to overheat when drawing a deep vacuum because the refrigerant is needed to cool the compressor.

R-134 refrigerant requires a separate set of gauges, a vacuum pump, and a recovery machine when servicing.

The following methods help **reduce the recovery time** of a system:

Packing ice around the recovery tank.

Heating the appliance.

Recovering liquid first.

During recovery, the refrigerant should be removed from **the condenser outlet** when the condenser is below the receiver.

When recovering refrigerant from a system that has a condenser on the roof and the evaporator on the first floor, the recovery should begin with the **liquid line entering the evaporator.**

After removing refrigerant from a system, the **refrigerant may be reused** in the same system or on another system owned by the same person.

Before transferring refrigerant into an empty refrigerant cylinder, the **cylinder must be evacuated.**

The use of **low-loss fittings and hoses containing hand valves** commonly aids in minimizing the release of refrigerant to the atmosphere.

When using recovery and recycling equipment manufactured ***after*** **November 15, 1993,** technicians must evacuate an appliance containing **more than 200 lb** of HCFC-22 to 10″ Hg.

When using recovery and recycling equipment manufactured ***before*** **November 15, 1993,** technicians must evacuate an appliance containing **less than 200 lb** of HCFC-22 to 0 psig.

If a leak in an appliance makes the prescribed evacuation level unattainable, the appliance can be evacuated to **0 psig.**

The repair or replacement of the following components is considered **major repair:**

Compressor

Condenser

Evaporator

Auxiliary heat exchanger

When servicing an open refrigeration system, always **replace the filter dryer.**

In a refrigeration system, **high-pressure liquid** exits in the condenser and receiver.

When evacuating a system, it is required that the vacuum pump pull a minimum of **500 microns.**

A technician must **never operate a hermetic compressor under a deep vacuum** or the compressor could overheat.

Noncondensables (air) cause **high head pressure** when they are trapped in the condenser.

A technician must **never front seat the discharge service valve** while the compressor is in operation.

ASHRAE was formed as the American Society of Heating, Refrigerating and Air-Conditioning Engineers.

Defined by ASHRAE Standard 15, a sensor and alarm are required for A1 refrigerants (R-12, R-22, and R-11) to **sense oxygen depletion.**

A refrigeration system must always be protected by a **pressure-relief device.**

To prevent freezing when charging R-12 into an evacuated system, it must be in vapor form to raise the **pressure up to 33 psig.**

The sale of CFC and HCFC refrigerants has been restricted to certified technicians since **November 14, 1994.**

Refrigerators built before 1950 often contain **sulfur dioxide** (SO_2).

TYPE III CERTIFICATION

Type III Certification covers low-pressure appliances such as centrifugal compressors. This includes chiller-type systems.

If EPA **regulations change** after a technician has been certified, it is the technician's responsibility to comply with future changes.

It is easier to have **air enter gaskets and fittings** on a low-pressure system than on a high-pressure system, because a low-pressure system operates in a vacuum.

A rupture disc opens at **15 psig** on a low-pressure system.

The easiest way to leak test a low-pressure system is to **add warm water over the coil** to raise the pressure to 10 psig.

It is important **not to exceed 10 psig** when leak testing a low-pressure system because the rupture disc, located on the evaporator, might fail.

When an electronic leak detector is used to test a water box for leaks, **test at the drain valve** after the water has been removed.

A **hydrostatic test kit** is used when leak testing a water tube on a low-pressure chiller.

As of January 1, 2019:

The EPA requires that all *comfort cooling* (air conditioning) systems with **more than 50 lb** of refrigerant be repaired when the annual leak rate exceeds **10%.**

The EPA requires that all *commercial refrigeration systems* with **more than 50 lb** of refrigerant be repaired when the annual leak rate exceeds **20%.**

The EPA requires that all *industrial process refrigeration systems* with **more than 50 lb** of refrigerant be repaired when the annual leak rate exceeds **30%.**

If an appliance with 50 or more lb of a regulated refrigerant has exceeded the threshold leak rate, the equipment owner or operator has 30 days to repair the appliance.

If an appliance with a charge of 200 lb or more of a regulated refrigerant passes the initial verification leak test, a follow-up **verification test must be conducted within 10 days.**

If after repairing a leak in an HCFC or HFC rack refrigeration system, the system fails the initial leak verification test, the technician has 30 days to complete the repair or retrofit.

If a cold storage warehouse-operating appliance charged with HCFC or HFC refrigerants leaks above the threshold rate, it will be allowed to operate without repair for 12 months before it must be retrofitted or retired.

If the cold storage warehouse appliance system **leak rate exceeds 125%** of its full charge within a year, a report must be submitted to the EPA describing the efforts made to repair the leak and action must be taken by completely repairing, retiring, mothballing (isolating the leak), or retrofitting the appliance.

The repair deadline can be extended to 18 months by temporarily shutting down the appliance, recovering the refrigerant from an isolated section or the entire system to at least atmospheric pressure, and using a refrigerant exempt from the venting prohibition.

A high-pressure cut-out is set to 10 psig on a recovery unit that is used on low-pressure systems.

When recovering R-11 or R-123 from a low-pressure system, **always recover liquid first** to decrease the recovery time.

To prevent a low-pressure system from freezing during evacuation, **water is commonly run through the chiller tubes.**

ASHRAE guideline 3-1990 states that if a **1 mmHg vacuum level rises to 2.5 mmHg** during an evacuation, the system should be leak tested.

Refrigerant is added to a low-pressure chiller system after an evacuation by the use of the **evaporator charging valve.**

After a chiller system has had a deep system evacuation, pressurizing the system with vapor up to the **36°F saturation temperature** is commonly done to decrease the possibility of system freeze-up prior to a liquid charge.

R-123 is **36°F** saturation when above **20″ Hg.**

When using recovery or recycling equipment manufactured or imported before November 15, 1993, the required level of evacuation for low-pressure equipment is **25″ Hg.**

When using recovery or recycling equipment manufactured or imported on or after November 15, 1993, the required level of evacuation for low-pressure equipment is **25 mmHg absolute.**

Low-pressure appliance:		
CFC-11, HCFC-123	25 Hg	25 mm

A **purge unit** is used on low-pressure chillers to remove noncondensables from the system.

The **purge unit** removes trapped air from a low-pressure system from the top of the condenser.

R-123 is considered a group **B-1 refrigerant.** The equipment room where the low-pressure chiller is located requires a **refrigerant sensor.**

Since there are approximately **100 lb of vapor in a common 350-ton R-11 chiller,** it is important to recover the vapor refrigerant in a chiller after the liquid has been removed.

When the **threshold limit value–time weighted average (TLV–TWA)** is exceeded, ASHRAE Standard 15-2013 requires that each machinery room activate an alarm and a mechanical ventilation system.

Universal Type Certification is given to the technician who passes the Core questions as well as Types I, II, and III Certification Exams.

REFRIGERANT PIPING

The objective of refrigerant piping is to move lubrication oil and refrigerant adequately. The main priority and concern for the system's designers and installers is to design and install refrigerant piping that will return the oil to the compressor at the same rate that it is being carried through the piping. The correct size of pipes, pitch of runs, and use of oil traps where needed will help to maintain good oil movement throughout the system and eliminate serious system problems.

The line connected between the compressor and the condenser is known as the **discharge line.**

The line connected between the evaporator and the compressor is known as the **suction line.**

The line connected between the receiver and the metering device is known as the **liquid line.**

The line connected between the metering device and the evaporator is known as the **inlet line.**

The line connected between the condenser and the receiver is known as the **condensate line.**

The line that is the **hottest** during operation of the system is known as the **discharge line.**

The line that is the **coldest** during operation of the system is known as the **suction line.**

A **horizontal** discharge line should never have a velocity of below **500 fpm.**

A **vertical riser** discharge line must be sized to have a minimum velocity of **1,000 fpm.**

Discharge line **velocities** must never exceed **4,000 fpm.**

To ensure proper oil return, a discharge line must be pitched $\frac{1}{2}''$ per **10′** of horizontal run away from the compressor toward the condenser.

A discharge line **riser** is required whenever the condenser is located **above** the compressor.

A discharge line **double riser** is needed on refrigeration systems that use **unloaded** compressors.

A double riser used in the discharge line for **full load** conditions can be used for velocities of **1,000 fpm or more.**

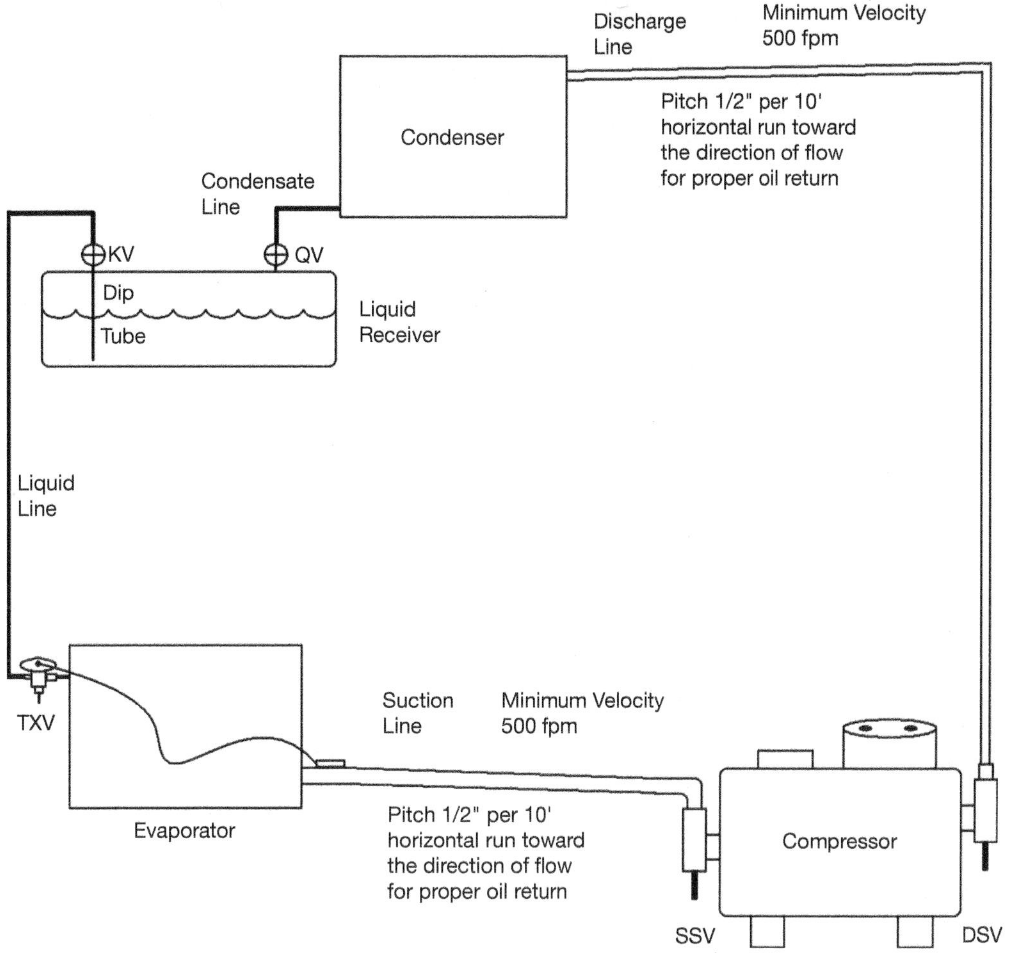

FIGURE 1–1
Refrigeration Piping Diagram

An **oil trap** is required at the base of a discharge line if the vertical riser is **8′ or more.**

A **secondary** oil trap is placed in a discharge line vertical riser if it exceeds **25′**.

A secondary oil trap must be placed in a discharge line riser approximately every **25′** interval of net lift.

The line that carries both refrigerant and oil mixture and is the least critical line is the **liquid line.**

Flash gas will form in the **liquid line** if it is undersized.

Liquid line **friction loss** is caused by the resistance to flow from liquid line to components.

Liquid line **static head losses** cause a resistance to flow due to the **weight of the refrigerant** that is flowing up a vertical riser.

Flash gas in the liquid line would be indicated by having a reduced evaporator capacity, a starved evaporator, and a noisy metering device.

A liquid-to-suction **heat exchanger** is used to reduce flash gas.

A liquid-to-suction **heat exchanger** must be located as close to the receiver as possible.

When sizing refrigerant piping lines, the most critical would be the **suction line.**

To ensure proper oil flow, a **horizontal suction line** must never have a velocity below **500 fpm.**

Vertical suction lines must be sized to have a minimum refrigerant velocity of **1,000 fpm.**

A suction line must be pitched $\frac{1}{2}''$ per **10′** of **horizontal run** away from the evaporator down toward the compressor to ensure proper oil return.

A suction line **double riser** is needed for refrigeration systems that operate with **unloaded** compressors.

When using a **double riser** in the suction line, the smaller line is used for **unloaded** conditions when the refrigerant flow is **less than 1,000 fpm.**

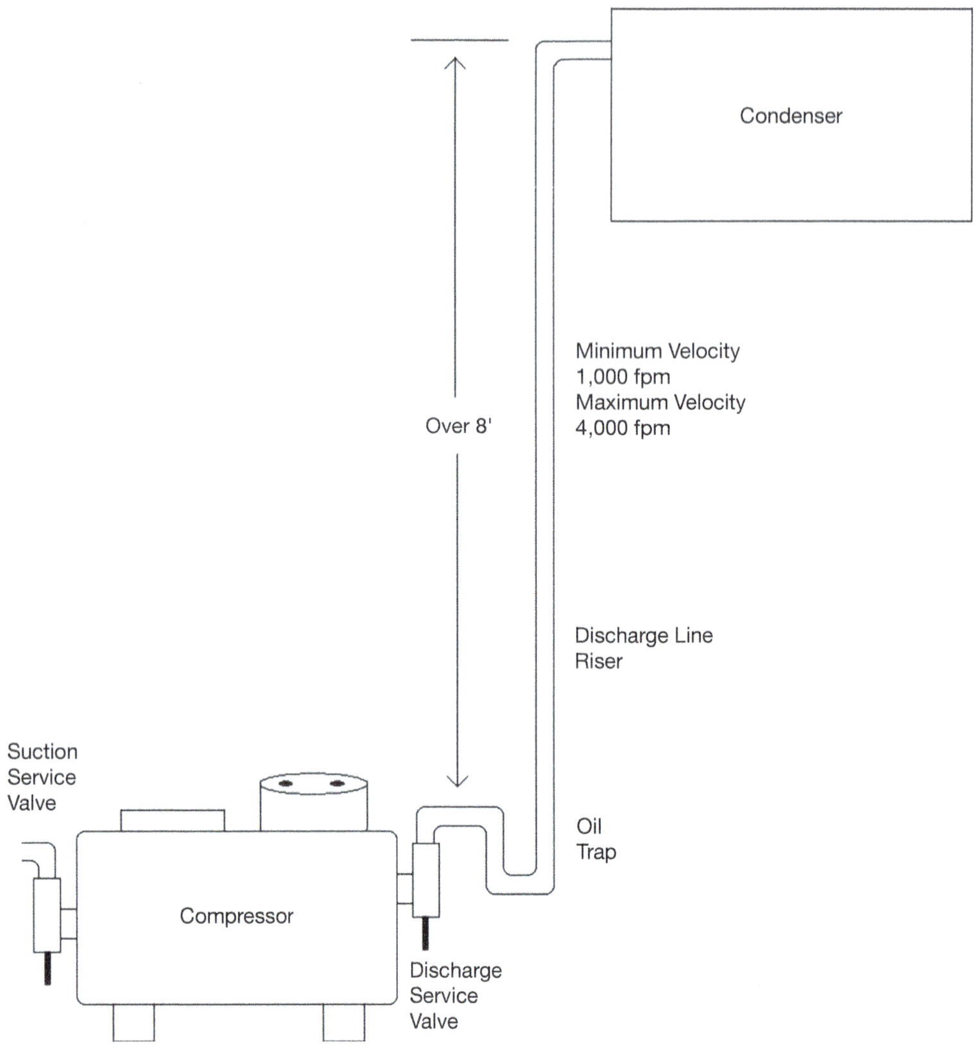

FIGURE 1–2
Discharge Line Riser with Oil Trap

A **double riser** in the suction line uses both the larger line and the smaller line for **full load** conditions.

An **oil trap** may be used at the base of a suction line if the vertical rise is **more than 8′**.

An **oversized** suction line will cause a **decrease** of oil to the compressor due to the lack of refrigerant **velocity.**

REFRIGERATION ACCESSORIES

Refrigeration accessories are devices that are installed on refrigeration and air conditioning systems to enable technicians to perform accurate system diagnosis. Some of these devices function as safety components as well as assist the service technician.

A **hot gas muffler** when needed is installed in the discharge line as close to the compressor as possible. The hot gas muffler will **minimize loud hot gas pulsations.** The muffler should be installed on the horizontal part of the piping.

The **liquid line filter drier** is used to remove water vapor and foreign matter. It is installed in the liquid line before the metering device. It is made of a molded porous core that has a **high affinity** for water vapor.

A **sight glass** is a device that is installed in the liquid line before the expansion valve. This component allows the technician to **view the flow** of refrigerant. **Bubbles** or **flash gas** can indicate a shortage of refrigerant or a restriction in the liquid line.

The **suction line filter drier** is installed close to the compressor. It is usually installed to protect a new compressor after a **burn out** has occurred. The function of this drier is to collect foreign matter and water vapor. When a system has been flushed out and a new compressor installed, it is good practice after a period of run time to replace the suction line drier. This will ensure that all contaminants have been collected.

Pressure gauges are at times installed on a system as a convenience for service. These gauges should be **compound gauges** and

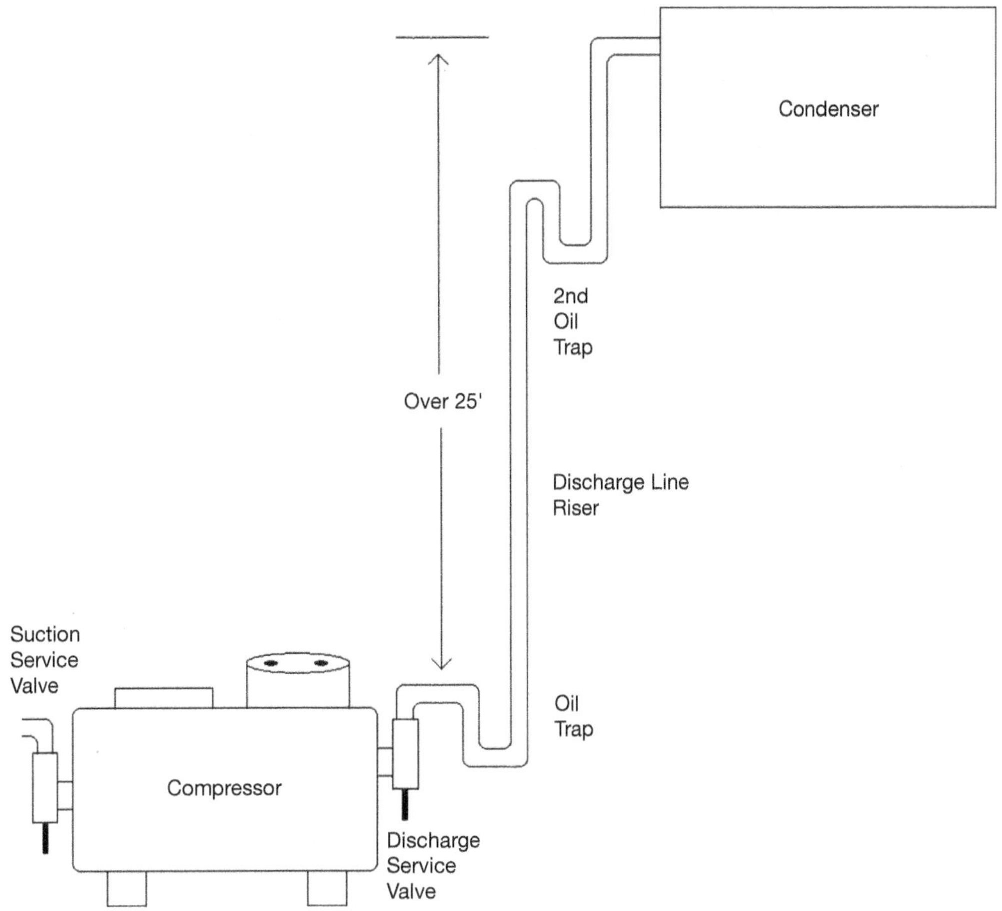

FIGURE 1–3
Discharge Line Riser with Double Oil Trap

calibrated from **0 psi** to no less than **50 to 100 psi** above the normal operating suction and discharge pressures. These gauges should be connected on the system with **throttling valves.** The throttling valves help to prevent fluctuation of the pressure gauges due to pulsating action of refrigerant gases.

Manual shutoff valves are used to shut off or divert the flow of refrigerant during service procedures. Common manual shutoff valves are designed to back seat and front seat opening ports. Many of these valves use a **seal cap** that prevents leakage. Some manual shutoff valves use a **packing gland** that helps to prevent refrigerant leakage.

Charging valves are installed in the liquid line close to the condenser outlet. When the system uses a receiver, the charging valve should be installed at the receiver's outlet side on the liquid line. Charging valves also use a threaded cap to protect against refrigerant leakage.

Receiver valves are designed with two valves. The valve at the inlet side of the receiver is referred to as the **queen valve,** and the valve at the receiver's outlet is called the **king valve.** These valves can be used during service to isolate the refrigerant in the receiver while a repair is being made or to valve off the system during a pump-down procedure.

The **liquid receiver** is designed as a storage chamber that will hold the system's refrigerant. The liquid receiver should be capable of storing mostly liquid refrigerant along with approximately **15 to 20%** room for expansion of vapors.

Relief valves are designed to relieve a system's excessive and possibly unsafe refrigerant pressures. They are considered **safety devices.** Relief valves can be located on the condenser, the liquid receiver, and sometimes on the low side of the system. Relief valves are usually of the **spring-loaded** type. When safe pressures are restored, the spring resets and reseats the outlet port opening of the valve.

Fusible plugs are also considered to be a safety device. The fusible plug uses a predetermined **melting point** that will melt when excessive and unsafe temperature/pressure occurs within the system. When the fusible plug opens at the melting point, the **entire system's** refrigerant will be **lost.**

The **check valve** is designed to allow the flow of refrigerant in **one** direction only. Check valves are mostly found on refrigeration

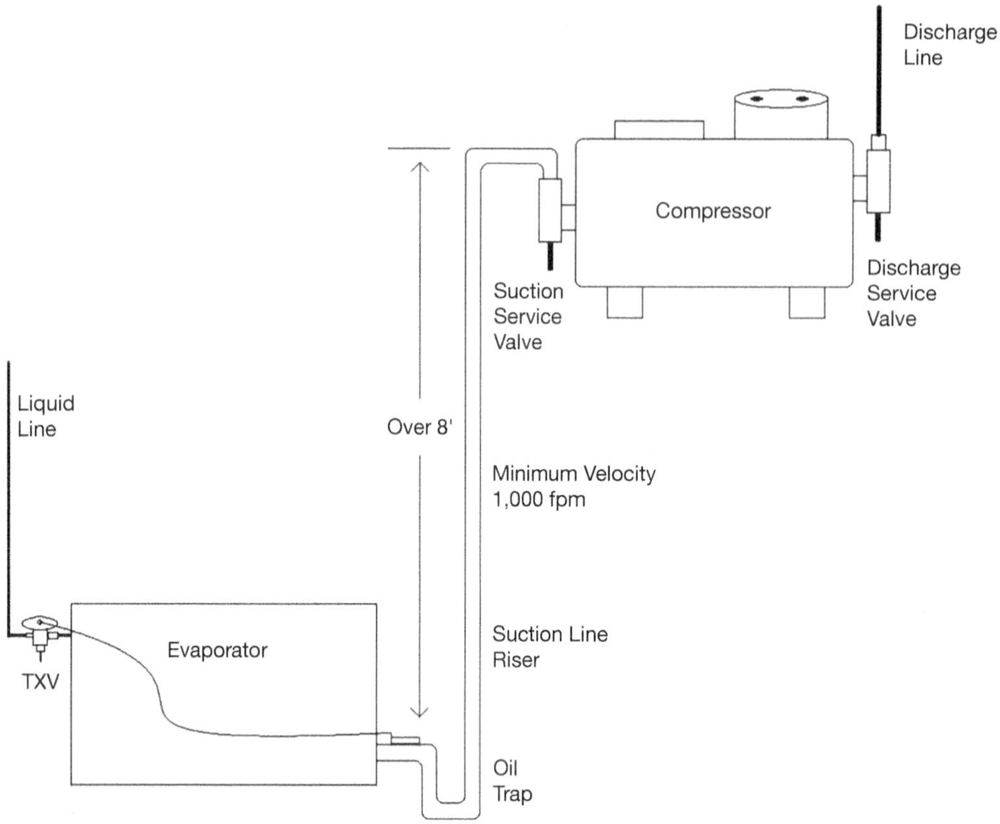

FIGURE 1–4
Suction Line Rise with Oil Trap

systems and occasionally on air conditioning applications. Check valves are usually located on **multiple evaporator** systems. The check valve would be installed on the suction line of the **coldest** evaporator.

A **solenoid valve** is a device that is used to allow refrigerant to flow to the evaporator. It is usually controlled by a **thermostat** (temperature control). The installation of a solenoid valve can be found in the liquid line before the metering device.

The **solenoid valve** consists of electrical wire that is wound in a spiral formation. When electrical current is sent through the windings, an **electromagnetic field** is created that causes a steel plunger to be pulled toward the center; as a result, it lifts a needle valve off an orifice, allowing refrigerant to flow.

A **heat exchanger** is a device that subcools the liquid refrigerant in the liquid line. A section of the liquid line is in direct contact with the suction line, subcooling the warm liquid refrigerant prior to flowing through the metering device. The main result is that it helps in preventing a high percentage of **flash gas** at the metering device. Flash gas can result in a reduction of TXV performance, lowering its capacity.

The **suction accumulator** is a component located on the suction line between the evaporator and the compressor, close to the compressor. The main function of the accumulator is to **prevent liquid floodback** to the compressor, which could result in severe internal damage.

The **refrigerant distributor** is usually connected to the outlet of a TXV. The distributor is also used with other metering devices such as with the piston type. The distributor feeds the evaporator more evenly and provides for **proper refrigerant velocity.** Without a distributor on large evaporator coils, the refrigerant flow would split into separate layers, resulting in the starving of some evaporator circuits.

The **oil separator** is located on the discharge line, close to the compressor with most refrigerants. The function of the oil separator is to separate the oil from the hot discharge gases and return it to the compressor crankcase for lubrication purposes.

The **crankcase heater** is a device that is connected around the exterior of a hermetic compressor. It is located around the lower portion of the crankcase. The function of this device is to prevent **migration** of liquid refrigerant and oil. The crankcase heater is usually operated during the compressor **off cycle.**

The **evaporator pressure regulator** (EPR) is located on the suction line close to the evaporator. **EPRs** are designed to provide modulation or a gradual throttling action. The regulator is responsive

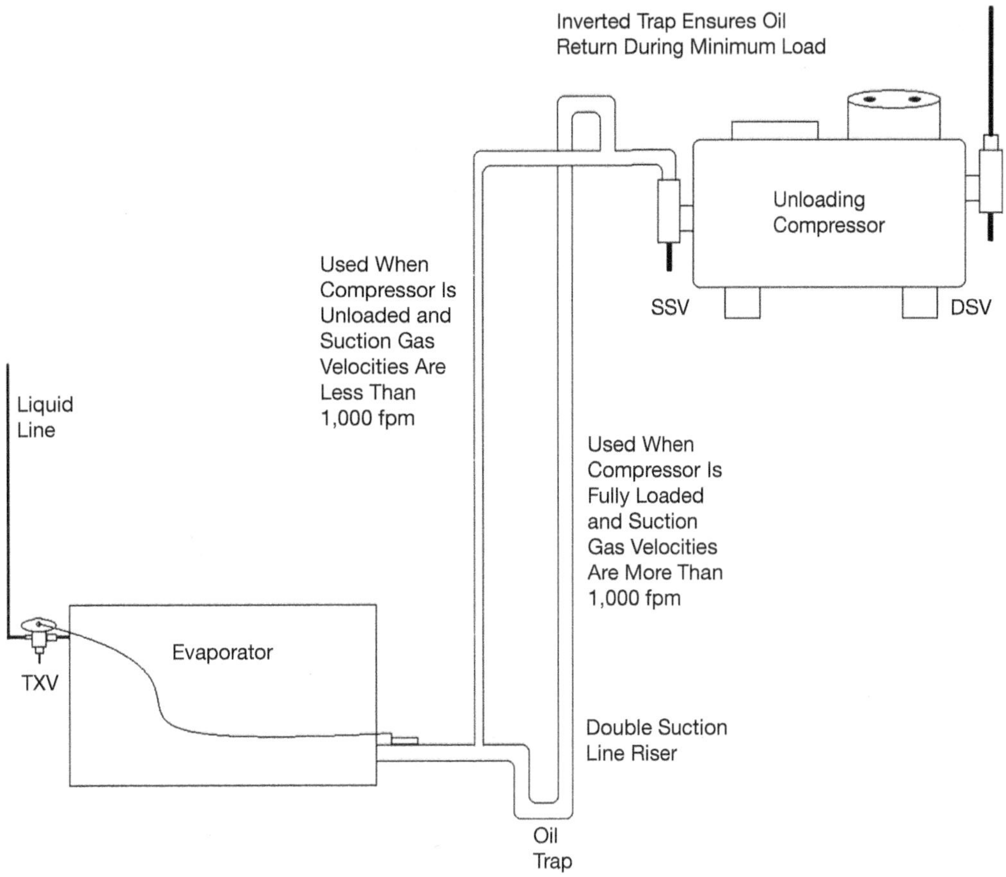

FIGURE 1–5

Double Suction Riser with Unloading Compressor

to evaporator pressure (inlet pressure). The evaporator pressure regulator limits suction pressure within the evaporator. EPRs are also known as **back pressure valves.** Most evaporator pressure regulators are installed on **multiple evaporator systems.**

The **condenser pressure regulator** (CPR) is a component located after the condenser. The function of the CPR is to maintain adequate condensing pressure during **low ambient conditions.** This device is a type of **head pressure control.** This component limits the amount of liquid refrigerant that would normally pass through the condenser. The result is a **rise** in condenser pressure.

The **oil pressure failure control** is a safety switch that is wired into the control circuit of the system. It contains two bellows: one is connected to the oil pump discharge and the other to the compressor crankcase. The function of this control is to cycle the **compressor off** when useful oil pressure falls to a predetermined minimum or if the oil pressure fails to build up to a safe operating pressure.

The **high-pressure control** is a safety switch wired within the control circuit of the system. The high-pressure control has a bellows built in as part of the control. This bellows is responsive to excessive head pressure. If head pressures were to rise to unsafe levels, the control would open the normally closed set of contacts, de-energizing the compressor.

The **low-pressure control** is used as an operating control, or it may be used as a safety device if suction pressures were to decrease to unsafe product conditions. Too low suction pressures would also result in compressor failures due to improper oil return. The low-pressure control contains a bellows built within the control. When suction pressures that are too low occur, the bellows respond by opening a normally closed set of contacts, resulting in the opening of the compressor control circuit.

The **discharge bypass valve** is a **modulating** valve that allows hot discharge gas to flow into the **low side** of the system. The valve opens on a **decrease** of suction pressure.

The **discharge bypass valve** is used on refrigeration systems to **prevent** low-side pressures from **falling below** the minimum value suggested by compressor manufacturers.

The **discharge bypass valve** should be installed on a piping line that is off the main discharge line as **close** to the compressor as possible.

The **split condenser valve** is a device that will allow discharge gas to enter one or two **condensers.**

Low ambient control valves are used to control **head pressure** during low ambient temperatures. The valves **regulate** the refrigerant flow from the condenser to the receiver and the hot discharge gas around the condenser to the receiver.

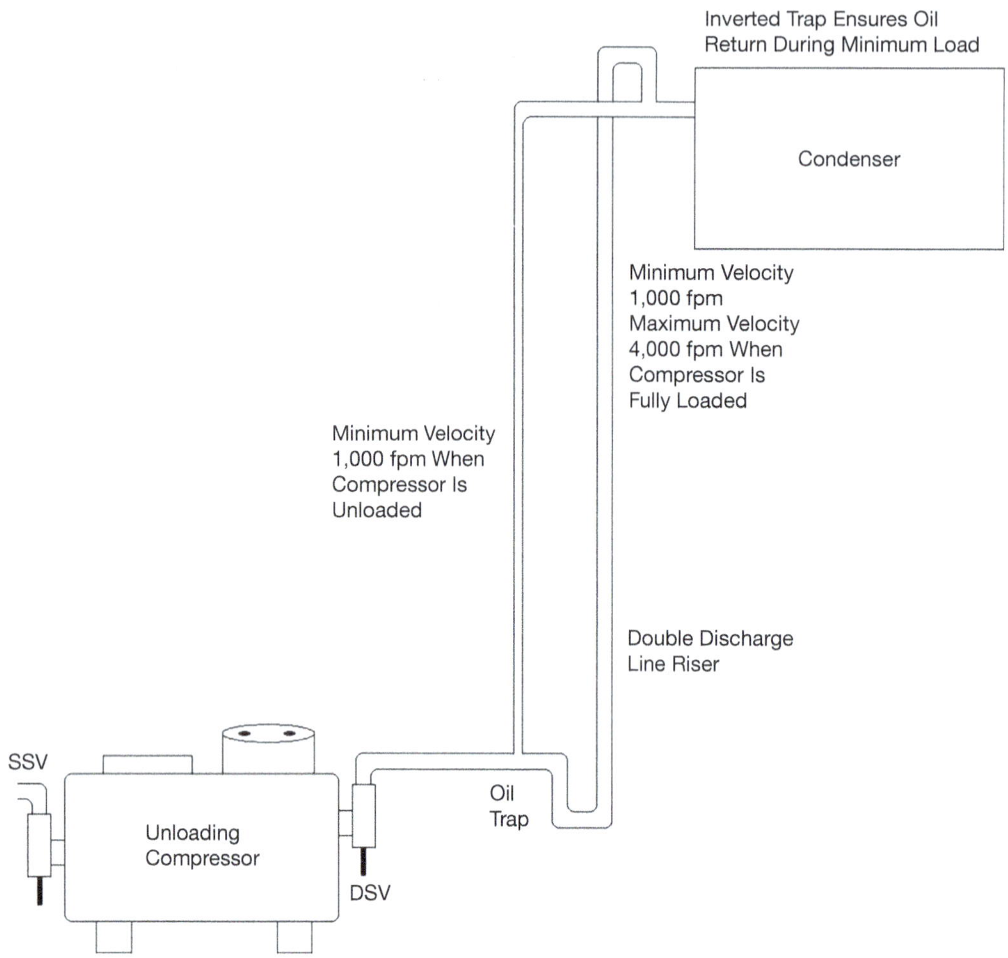

FIGURE 1–6

Double Discharge Line Riser with Unloading Compressor

MEDIUM- AND LOW-TEMPERATURE REFRIGERATION

Refrigeration technicians are called to work on various types of systems, when storage temperatures need to be maintained around −20 to +10°F. This type of refrigeration application is referred to as "low-temperature refrigeration." These systems may include upright and chest type freezers, walk-in freezers, reach-in boxes, closed display cases, open display cases, ice machines, and ultra-low-temperature ranges that include cryogenics equipment. Ultra-low-temperature refrigeration commonly operates with storage temperatures around −50°F and below. Medium-temperature refrigeration refers to storage temperatures that commonly operate around 25 to 40°F. Medium-temperature refrigeration may also include reach-in boxes, open display cases, walk-in coolers, drinking fountains, refrigerators, beverage coolers, and ice machines. Commercial refrigeration depends on design, size of equipment, and the specific application for which it will be used.

Commercial cabinets are usually constructed of porcelain and stainless steel.

The type of **insulation** for commercial cabinets is commonly polystyrene or urethane.

To prevent condensation around doors, **mullion heaters** are installed around the doors. Magnetic **door gaskets** are used to ensure against air leakage and infiltration.

The **display case** can be of the open or closed type. Either type allows the consumer to have a full view of the merchandise.

Open display cases use a cold blanket of air that is refrigerated and circulated over the merchandise. The **evaporator** is commonly installed at the bottom and back section of the unit.

Open display cases should not be installed near locations such as air ducts, exhaust fans, outdoor air sources, or entryways where outdoor air movement prevails.

Many **closed display cases** are used to store beverages, food, and dairy products. The closed display case in physical size may be ceiling high or wall to wall and have many doors.

The **evaporators** of the closed display case are usually installed on the back wall or at the top of the unit. The evaporator is the forced-air type.

Frozen food display cases usually operate near 0°F or lower. The condensing units are located either with the case or away from cases installed in a remote area. The **evaporators** are usually the forced-air type.

Restaurants, supermarkets, and meat markets commonly use **walk-in units** for refrigerating merchandise. These units use **forced-air**

evaporators, and the condensing sections can be located remotely from the cooler/freezer unit or on the top of the cabinet.

Doors of **walk-in units** must have a **safety latch** on the inside of the cabinet that will allow anyone who might be trapped to reopen the door and get out.

The **evaporators** of walk-in units should be located on the corners or on sidewalls away from the direct opening of the door.

Some units use an **evaporator fan control** that will shut off the evaporator fans once the door is opened.

The evaporator section of a **walk-in unit** has a **condensate drain pan** that needs to be piped to the outside of the cabinet. If temperatures are below freezing, the drain line must be heated, usually by resistance heat, to prevent ice plugging at the drain line.

The **piping** to commercial coolers and freezers can be of the straight **hard-pipe** type connected with copper fittings to be brazed or it can be of **precharged line sets.**

Ice-making machines are usually the **cube- or flake-**making type. Ice temperatures are in between the medium- and low-temperature range of around 10°F while making ice.

Ice-making machines use a timer clock that places the **harvest cycle** in operation. This is the time when the ice is heated off the evaporator cube section and allowed to fall into the bin section (cube-type ice machines).

Some **ice machines** use an **auger** and a **gear motor drive** that actually shaves off thin layers of ice from the evaporator.

Cube ice is commonly made by spraying water into inverted cups that are shaped in the design of a cube. The evaporators can be the horizontal or vertical type.

The **defrost (harvest)** of ice machines is commonly done by the use of **hot gas.**

Cube-making machines occasionally need cleaning and flushing out. This is done without any ice in the storage bin and with the use of **safe chemicals** and clean **circulated water.**

The **ice machine** is designed to operate where the location at ambient **temperatures** is generally 40 to 115°F.

Vending machines can be used for the refrigeration of such products as beverages, sandwiches, and even frozen products like ice cream. They are self-contained units and usually have **fractional** horsepower compressors.

Vending machines are made up of different systems consisting of the refrigeration, dispensing, and money-changing systems.

Vending machines are self-contained and equipped with an electrical cord that commonly plugs into an electrical outlet of 20 amps.

Beverage coolers are medium-temperature systems designed to keep products cool **above freezing.**

Beverage coolers need to have a **defrost cycle** because the evaporators operate below freezing temperatures.

Beverage coolers use **capillary tube** metering devices.

Beverage coolers use air-cooled packaged condensers and forced-air evaporators. Normally the cold air is circulated over the top of the products and drawn back to the evaporator at the bottom section.

Some **beverage coolers** refrigerate liquids and dispense ice into a cup. The entering water is **prechilled** and ice is made separately; **two evaporators** are used for these operations.

Beverage coolers that refrigerate liquid that will be dispensed into a cup use a **CO_2 cylinder** that injects carbonation into the drink. The cylinder is located inside the vending machine.

Water coolers (water fountains) are used to cool water for the purpose of drinking. They are used in many public buildings.

Water coolers are of two types: the **pressure type** and the type that uses a **large bottle** on the top of the cooler.

The **pressure type** requires a water-pressure regulator. The **regulator** can be adjusted for the correct arc at the bubbler.

Water coolers are self-contained units that use **fractional** horsepower compressors.

The **evaporator** of a water cooler is generally constructed of stainless steel, copper, or brass tanks. The refrigerant piping is coiled around the tank.

Ultra-low refrigeration generally starts at −10°F and below. Low temperatures are used to **fast freeze** products for commercial use, normally around −50°F.

When temperatures lower than −50°F are desired, then generally **two-stage compression** is used.

Cascade systems are used when temperatures need to be maintained around −160°F. Cascade systems operate using two or more **stages** of refrigeration; this depends on how low a temperature is required.

Some **transport refrigeration** may use a **eutectic solution (brine).** This solution is within the refrigerated plates. Other methods of refrigeration are the use of ice blocks, dry ice, and liquid nitrogen.

DEFROST SYSTEMS

There are various types of defrost systems used for specific refrigeration applications. The following are defrost types and related terms.

On a **manual shutdown defrost system,** the evaporator coil is designed to be exposed to the surrounding cabinet air.

A plate-type evaporator is most commonly used on a **manual shutdown defrost system.**

On a **manual shutdown defrost system,** the evaporator coil usually operates below 32°F.

On a **pressure-operated defrost system,** a low-pressure control is commonly used to shut the system down until the coil is free of frost.

On a **pressure-operated defrost system,** the low-pressure control is set to cut in when the evaporator coil is free of frost and has a minimum of 33°F saturation temperature.

On a **pressure-operated defrost system,** the storage temperatures are above 35°F.

On a **pressure-operated defrost system,** the evaporator coil is defrosted by de-energizing the condensing unit and keeping the evaporator fans in operation to move the 35°F air over the coil to remove frost buildup.

On a **pressure-operated defrost system,** the defrost period depends on the cut-in and cut-out settings of the low-pressure control.

A **time shutdown** defrost timer uses trip pins to set the defrost timer into defrost mode and determine its length of defrost. The trip pins are located on the defrost timer.

A **time shutdown** defrost timer uses a set of "normally closed" contacts to control the condensing unit.

On a **time shutdown defrost system,** the defrost period is initiated by a cam-operated switch that is controlled by a time clock.

On a **time shutdown defrost system,** the timer is adjusted to stop and start the compressor at selected times of the day or night.

On a **time shutdown defrost system,** the cut-in and cut-out times must be determined by the manufacturer's recommendation.

When **supplementary heat,** such as electric heaters, hot gas solenoids, and warm water, is used to aid frost removal during a defrost cycle, products are commonly kept below 29°F.

A **24-h defrost timer** is used on a system incorporating supplementary heat control with "normally open" and "normally closed" sets of contacts. The "normally open" contacts control the supplementary heat circuit and the "normally closed" contacts control the condensing unit and the evaporator fan motor.

In a **hot gas defrost cycle,** the defrost system must keep the compressor energized during defrost.

On a **24-h defrost timer** used with supplementary heat, the trip pins are used to start the defrost and an adjustable setting is used to end the defrost.

The maximum amount of **trip pins** that can be installed at one time in a 24-h defrost timer is 12, one every 2 h.

The **length of the defrost cycle** on a 24-h timer ranges from 2 to 110 min.

On an **electric heat defrost system**, a supplemental electric heater is used to remove the frost buildup on the evaporator coil.

In an **automatic water defrost system,** water may be used directly as a defrosting medium. The system's water is electrically heated. Then the water is sprayed by headers directly over the evaporator's surface, removing frost from the coil.

In an **automatic water defrost system,** excess water and melted frost flow to a drain line. During water defrost, the compressor and evaporator fans are stopped.

In a **hot gas defrost system,** hot gas exits the discharge line of the compressor and enters the inlet of the evaporator located after the **metering device.** A "normally closed" hot gas solenoid controls the refrigerant flow during the defrost cycle.

Hot gas is fed into the evaporator, raising the temperature and pressure of the coil. Because it is likely that the refrigerant may condense, a potential for **liquid slugging** of the compressor is possible and must be taken into consideration during installation. An **accumulator** is commonly added to the suction line to prevent the compressor from developing a liquid slug.

In a **hot gas defrost system,** the hot gas is more effective when initially entering the evaporator.

In a **time-initiated-temperature-terminated defrost system,** the defrost period is terminated by a termination thermostat, which may be adjustable or preset.

The **termination thermostat** is used to terminate the defrost cycle on a time-initiated-temperature-terminated defrost system and is commonly controlled by the evaporator coil temperature.

In a **time-initiated-temperature-terminated defrost system,** the termination thermostat can be a single pole, single throw, or a single pole, double throw set of contacts.

A time-initiated-temperature-terminated defrost timer commonly uses a **timer release solenoid** to change the timer switch contacts back to the cooling mode after the coil has been defrosted.

In a single pole, double throw **termination thermostat,** the switch contacts that control the timer release solenoid are commonly a "close on rise" type. The contacts that "open on rise" usually control the evaporator fan motor.

In a single pole, single throw **termination thermostat,** the switch contacts control a timer release solenoid. The timer release solenoid is energized when the coil is defrosted and the position of the contacts is closed.

The **timer release solenoid,** in conjunction with the termination thermostat, will stop the defrost cycle prior to defrost timer settings if the evaporator is clear of frost.

> ***Example:*** If the **defrost timer** is set for a 35-min defrost and the evaporator coil is defrosted after 20 min, the termination switch energizes the timer release solenoid, causing the system to switch back into the cooling mode.

In a **time-initiated-temperature-terminated defrost system,** the action of the timer release solenoid causes the defrost timer switches to change back to the cooling mode. This starts the condensing unit. At the same time, it de-energizes the defrost circuit and the termination thermostat circuit.

COMMERCIAL REFRIGERATION ELECTRICAL DIAGRAMS—STUDY SECTION

Sequence of Operation Figure 1–7A–B Diagrams

With the temperature-controlled medium-temperature cooler, Figure 1–7A and 1–7B, the cooling thermostat is typically set to 35°F. As the thermostat closes, calling for cooling, the single-phase compressor is energized by a current relay and a start capacitor. The condenser fan, being wired in parallel with the compressor, also energizes, thus moving air over the air-cooled condenser coil.

The evaporator fan is energized 24 hours a day, circulating cooler box air over the evaporator coil. Typically, mild frost may build up on the evaporator coil but will be removed during the off cycle of the compressor. As the cooling thermostat opens, the evaporator fan stays energized, moving 35 to 38°F air over the evaporator coil. Since this air is above freezing, accumulated frost on the evaporator will melt and the coil will be kept clear. This is considered an off-cycle defrost system.

Sequence of Operation Figure 1–8A–B Diagrams

With the pressure-controlled medium-temperature cooler, the low-pressure switch is typically set to a cut-out pressure that correlates to about 15 to 20°F below the cooler box temperature of 35°F. As the pressure switch closes, due to an increase of low-side pressure, the single-phase compressor is energized by a current relay and a start capacitor. The condenser fan, being wired in parallel with the compressor, also energizes, thus moving air over the air-cooled condenser coil.

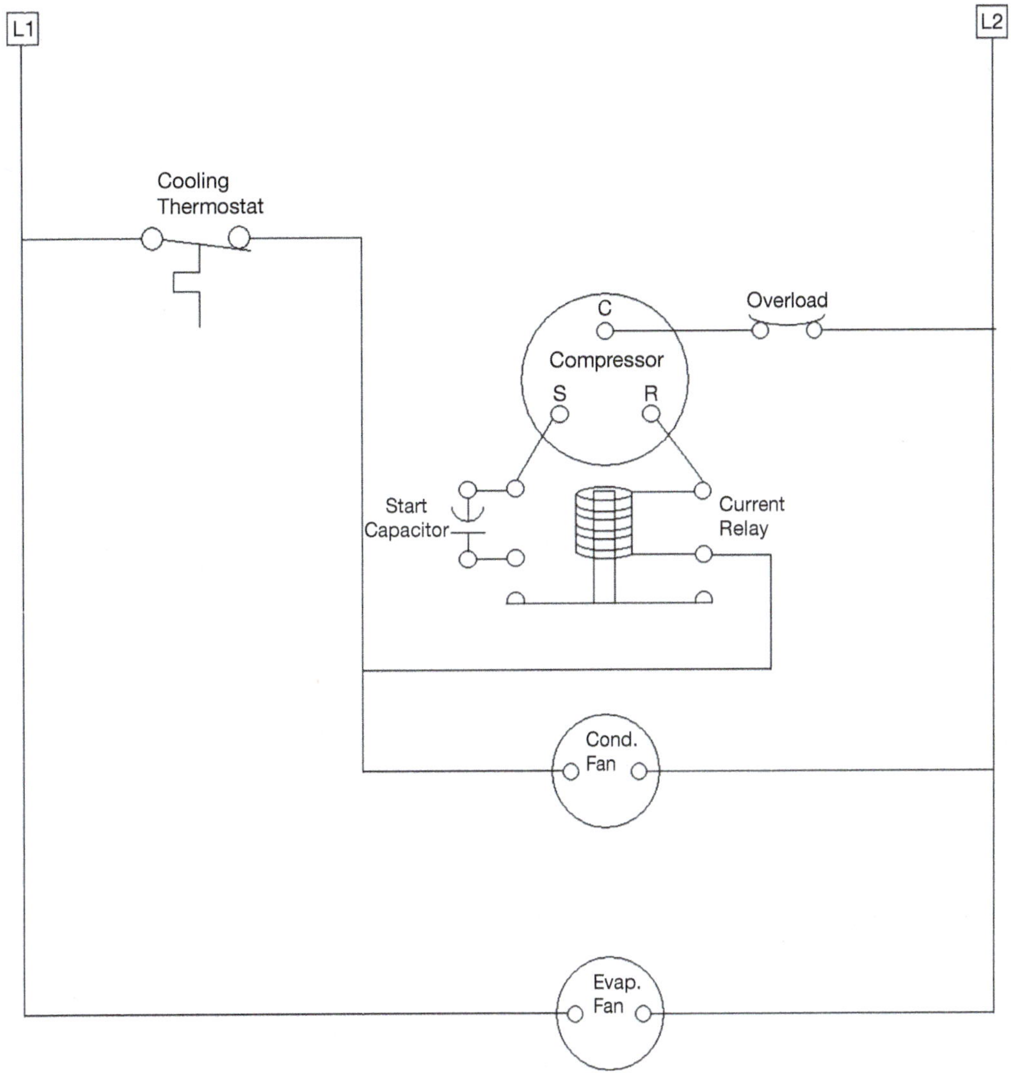

FIGURE 1–7A

Temperature-Controlled Medium-Temperature Cooler with a Capacitor Start Compressor Using a Current Relay

The evaporator fan is energized 24 hours a day, circulating cooler box air over the evaporator coil. Typically, mild frost may build up on the evaporator coil but will be removed during the off cycle of the compressor. As the low-pressure switch opens, the evaporator fan stays energized, moving 35 to 38°F air over the evaporator coil. Since this air is above freezing, accumulated frost on the evaporator will melt and the coil will be kept clear. This is considered an off-cycle defrost system.

Sequence of Operation Figure 1–9A–B Diagrams

With the pressure-controlled medium-temperature cooler, the low-pressure switch is typically set to a cut-out pressure that correlates to about 15 to 20°F below the cooler box temperature of 35°F. As the pressure switch closes, due to an increase of low-side pressure, the single-phase compressor is energized by a potential relay and a start capacitor. The condenser fan, being wired in parallel with the compressor, also energizes, thus moving air over the air-cooled condenser coil.

The evaporator fan is energized 24 hours a day, circulating cooler box air over the evaporator coil. Typically, mild frost may build up on the evaporator coil but will be removed during the off cycle of the compressor. As the low-pressure switch opens, the evaporator fan stays energized, moving 35 to 38°F air over the evaporator coil. Since this air is above freezing, accumulated frost on the evaporator will melt and the coil will be kept clear. This is considered an off-cycle defrost system.

Sequence of Operation Figure 1–10A–B Diagrams

With the medium-temperature automatic pump-down cooler, an automatic pump-down cycle is added to the electrical control system. This is typically found when condensing units are located outdoors and outdoor ambient temperatures could drop below cooler box

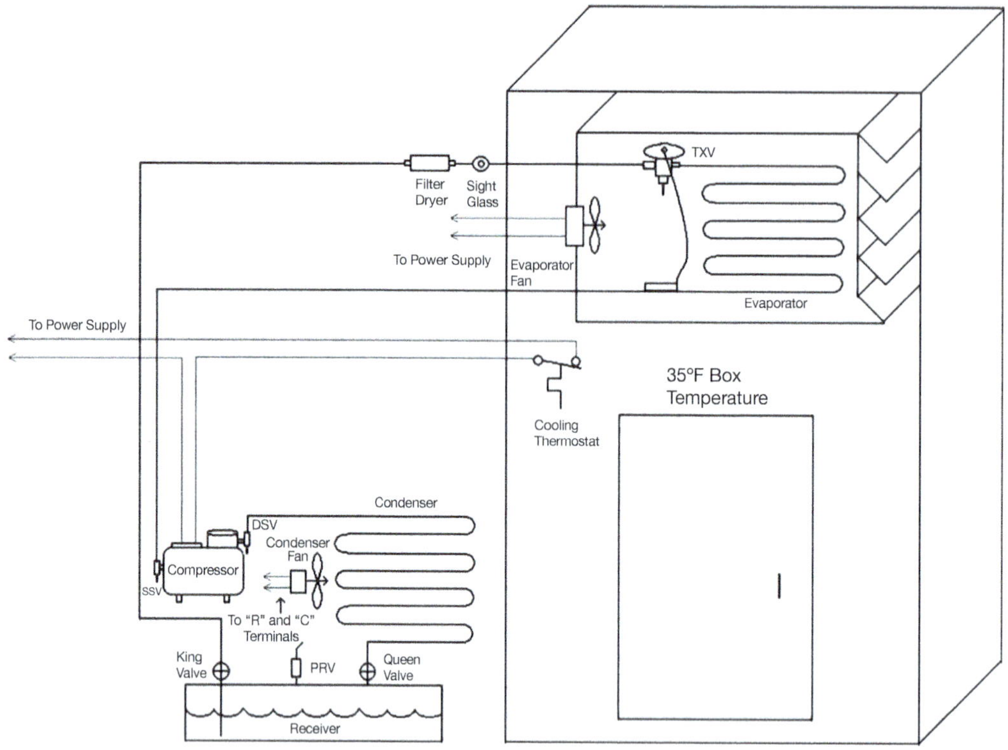

FIGURE 1–7B

Temperature-Controlled Medium-Temperature Cooler

temperatures. This type of control system prevents liquid refrigerant migration back to the compressor during the system's off cycle by pumping all of the system's refrigerant into the condenser and receiver during shutdown.

In operation, the cooling thermostat controls a liquid line solenoid and the low-pressure switch controls the compressor. The cooling thermostat may be set to a 35°F cut-out point with a 38°F cut-in point. The low-pressure switch may have a cut-out point of 2 psig with a cut-in point of 20 psig. During operation, the cooling thermostat is closed, keeping the liquid line solenoid energized, which allows refrigerant to flow through the liquid line and to the expansion valve and from the expansion valve to the evaporator. This keeps low-side/evaporator pressure above the cut-out point of the low-pressure switch. The low-pressure switch stays closed and keeps the compressor energized. During shutdown, the box temperature reaches 35°F, which is the cooling thermostat's cut-out temperature; this causes the cooling thermostat to open the electrical circuit to the liquid line solenoid. At this time, the liquid line solenoid closes and stops the flow of liquid to the expansion valve and evaporator coil. All of the refrigerant is pumped out of the evaporator, suction line, and part of the liquid line and stored in the condenser and receiver. The compressor remains in operation until the low-side/evaporator pressure reaches the 2-psig cut-out set point on the low-pressure control. At this time, the compressor shuts down and the evaporator fan continues to operate, circulating cooler box air.

During the off cycle, the cooler box temperature will eventually increase to the cut-in temperature of the thermostat, which is 38°F. At this time, the thermostat closes and energizes the liquid line solenoid. This will send liquid refrigerant to the expansion valve and the evaporator coil, increasing low-side pressure. The low-pressure switch will close when the low-side pressure reaches the cut-in set point of 20 psig. At this time, the compressor will start and the cooling cycle will begin.

As the pressure switch closes, due to an increase of low-side pressure, the single-phase compressor is energized by a potential relay and a start capacitor. The condenser fan, being wired in parallel with the compressor, also energizes, thus moving air over the air-cooled condenser coil.

Typically, mild frost may build up on the evaporator coil but will be removed during the off cycle of the compressor. As the low-pressure switch opens, the evaporator fan stays energized, moving 35 to 38°F air over the evaporator coil. Since this air is above freezing, accumulated frost on the evaporator will melt and the coil will be kept clear. This is considered an off-cycle defrost system with an automatic pump-down cycle.

Sequence of Operation Figure 1–11A–B Diagrams

With the time shutdown defrost medium-temperature cooler, the cooling thermostat is typically set to 35°F. As the thermostat closes, calling for cooling, the single-phase compressor is energized by a current relay and a start capacitor. The condenser fan, being wired in parallel with the compressor, also energizes, thus moving air over the air-cooled condenser coil. The evaporator fan is energized 24 hours per day, circulating cooler box air over the evaporator coil.

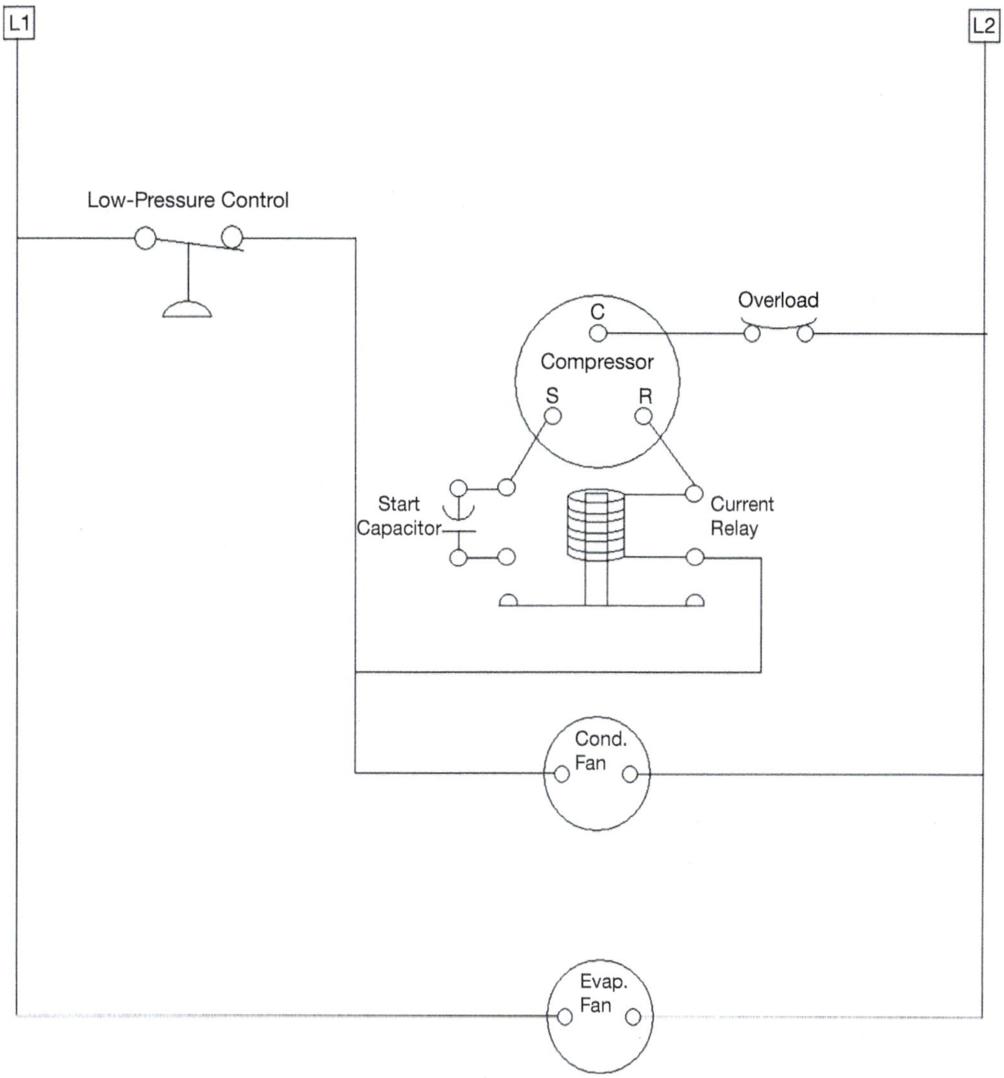

FIGURE 1–8A

Pressure-Controlled Medium-Temperature Cooler with a Capacitor Start Compressor Using a Current Relay

A defrost timer is added to the system to ensure that no frost will build up on the evaporator coil. Frost buildup will prevent heat transfer by insulating the evaporator coil during the cooling cycle; this will decrease the efficiency of the refrigeration system. The defrost timer motor operates 24 hours per day and controls a normally closed set of contacts that is wired in series with the cooling thermostat. These contacts stay closed during the cooling cycle but open and force a timed shutdown of the compressor every 12 to 24 hours. This compressor shutdown time ensures that no frost will build up on the evaporator coil, which keeps the system operating efficiently.

Typically, mild frost may build up on the evaporator coil during the cooling cycle but will be removed during the off cycle of the compressor. As the cooling thermostat or the N.C. defrost timer contacts open, the compressor shuts down while the evaporator fan stays energized, moving 35 to 38°F air over the evaporator coil. Since this air is above freezing, accumulated frost on the evaporator will melt and the coil will be kept clear. This is considered a time shutdown defrost system.

Sequence of Operation Figure 1–12A–B Diagrams

With the time shutdown medium-temperature automatic pump-down cooler, an automatic pump-down cycle is added to the electrical control system. This is typically found when condensing units are located outdoors and outdoor ambient temperatures could drop below cooler box temperatures. This type of control system prevents liquid refrigerant migration back to the compressor during the system's off cycle by pumping all of the system's refrigerant into the condenser and receiver during shutdown.

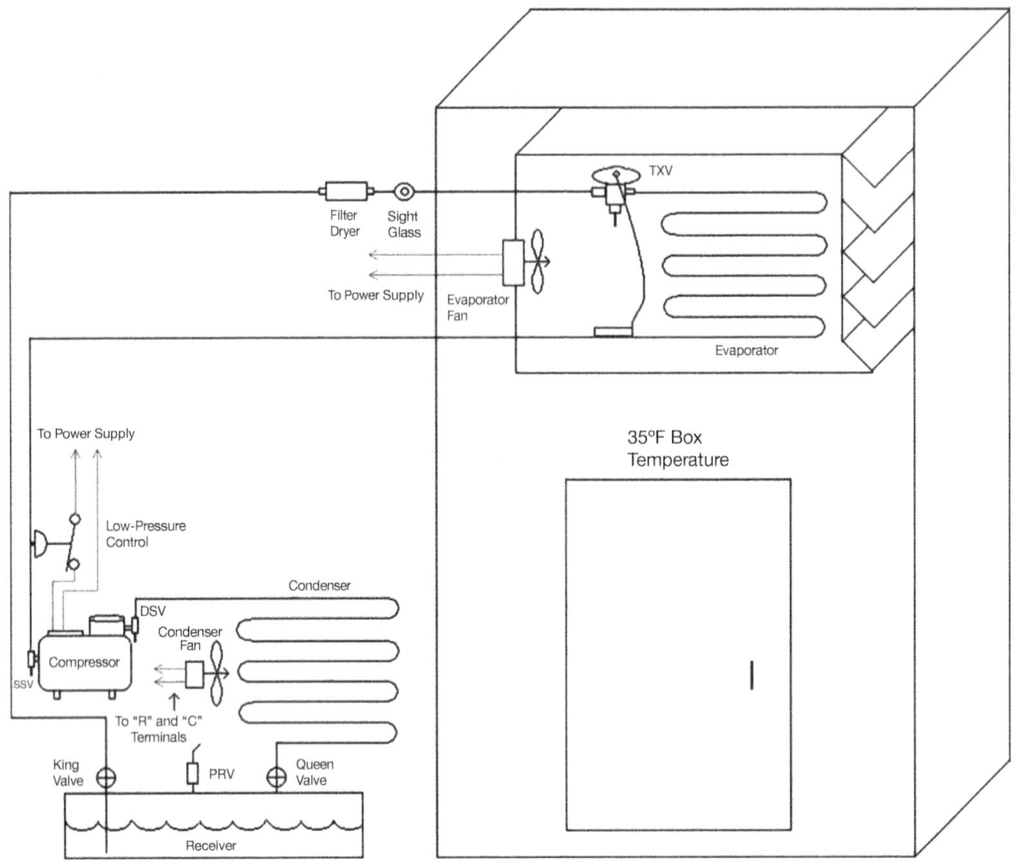

FIGURE 1–8B

Pressure-Controlled Medium-Temperature Cooler

In operation, the cooling thermostat controls a liquid line solenoid and the low-pressure switch controls the compressor. The cooling thermostat may be set to a 35°F cut-out point with a 38°F cut-in point. The low-pressure switch may have a cut-out point of 2 psig with a cut-in point of 20 psig. During operation, the cooling thermostat is closed, keeping the liquid line solenoid energized, which allows refrigerant to flow through the liquid line and to the expansion valve and from the expansion valve to the evaporator. This keeps low-side/evaporator pressure above the cut-out point of the low-pressure switch. The low-pressure switch stays closed and keeps the compressor energized. During shutdown, the box temperature reaches 35°F, which is the cooling thermostat's cut-out temperature; this causes the cooling thermostat to open the electrical circuit to the liquid line solenoid. At this time, the liquid line solenoid closes and stops the flow of liquid to the expansion valve and evaporator coil. All of the refrigerant is pumped out of the evaporator, suction line, and part of the liquid line and stored in the condenser and receiver. The compressor remains in operation until the low-side/evaporator pressure reaches the 2-psig cut-out set point on the low-pressure control. At this time, the compressor shuts down and the evaporator fan continues to operate, circulating cooler box air.

During the off cycle, the cooler box temperature will eventually increase to the cut-in temperature of the thermostat, which is 38°F. At this time, the thermostat closes and energizes the liquid line solenoid. This will send liquid refrigerant to the expansion valve and the evaporator coil, increasing low-side pressure. The low-pressure switch will close when the low-side pressure reaches the cut-in set point of 20 psig. At this time, the compressor will start and the cooling cycle will begin.

As the pressure switch closes, due to an increase of low-side pressure, the single-phase compressor is energized by a potential relay and a start capacitor. The condenser fan, being wired in parallel with the compressor, also energizes, thus moving air over the air-cooled condenser coil.

A defrost timer is added to the system to ensure that no frost will build up on the evaporator coil. Frost buildup will prevent heat transfer by insulating the evaporator coil during the cooling cycle; this will decrease the efficiency of the refrigeration system. The defrost timer motor operates 24 hours per day and controls a normally closed set of contacts that is wired in series with the cooling thermostat. These contacts stay closed during the cooling cycle but open and force a timed shutdown of the compressor every 12 to 24 hours. When these N.C. contacts open, the circuit is broken to the cooling thermostat and liquid line solenoid and the system goes into an automatic pumpdown. This happens during every defrost cycle. This compressor shutdown time ensures that no frost will build up on the evaporator coil, which keeps the system operating efficiently.

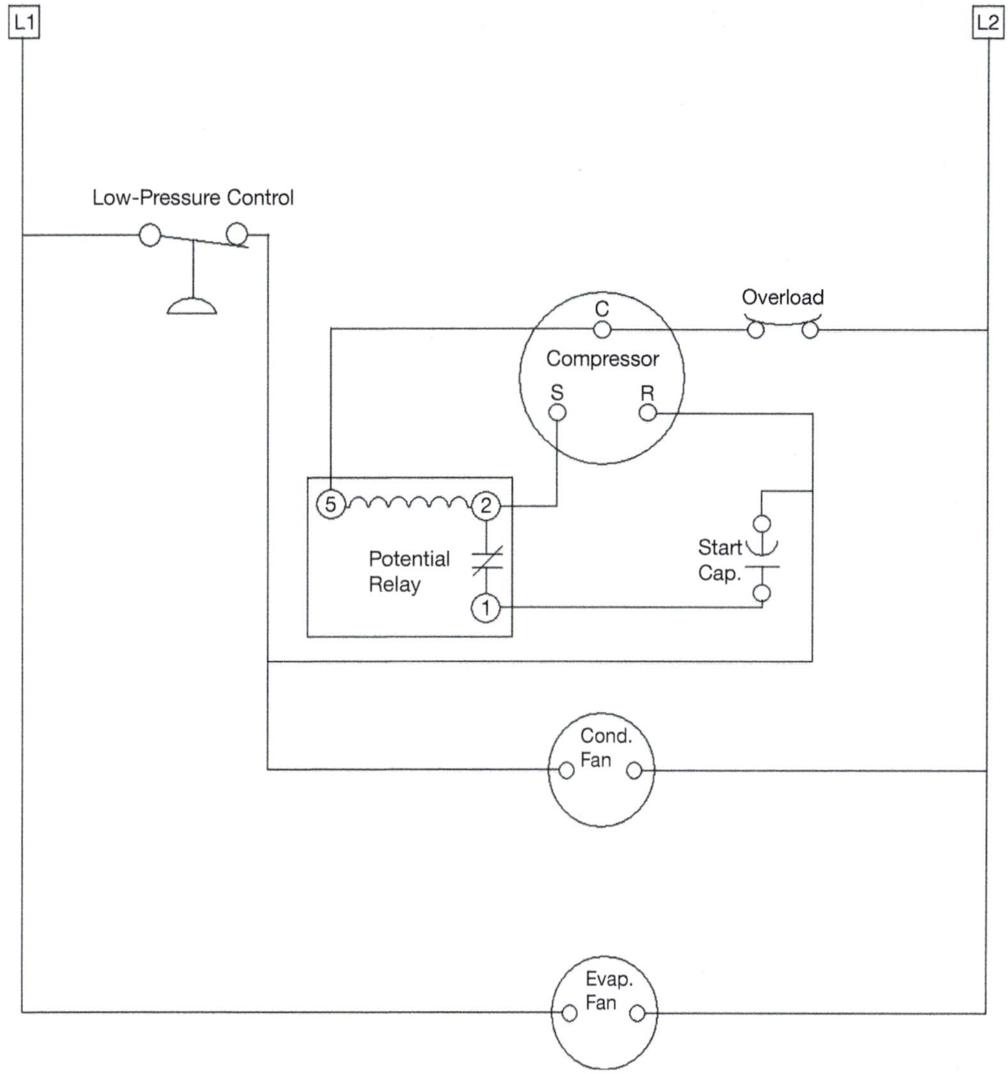

FIGURE 1–9A

Pressure-Controlled Medium-Temperature Cooler with a Capacitor Start Compressor Using a Potential Relay

Typically, mild frost may build up on the evaporator coil during the cooling cycle but will be removed during the off cycle of the compressor. As the cooling thermostat or the N.C. defrost timer contacts open, the compressor shuts down while the evaporator fan stays energized, moving 35 to 38°F air over the evaporator coil. Since this air is above freezing, accumulated frost on the evaporator will melt and the coil will be kept clear. This is considered a time shutdown defrost system with an automatic pump-down cycle.

Sequence of Operation Figure 1–13A–B Diagrams

With the low-temperature electric defrost automatic pump-down freezer, an automatic pump-down cycle is added to the electrical control system. This is typically found when condensing units are located outdoors and outdoor ambient temperatures could drop below cooler box temperatures. This type of control system prevents liquid refrigerant migration back to the compressor during the system's off cycle by pumping all of the system's refrigerant into the condenser and receiver during shutdown.

In operation, the cooling thermostat controls a liquid line solenoid and the low-pressure switch controls the compressor. The cooling thermostat may be set to a −10°F cut-out point with a −5°F cut-in point. The low-pressure switch may have a cut-out point of 2 psig with a cut-in point of 20 psig. During operation, the cooling thermostat is closed, keeping the liquid line solenoid energized, which allows refrigerant to flow through the liquid line and to the expansion valve and from the expansion valve to the evaporator. This keeps low-side/evaporator pressure above the cut-out point of the low-pressure switch. The low-pressure switch stays

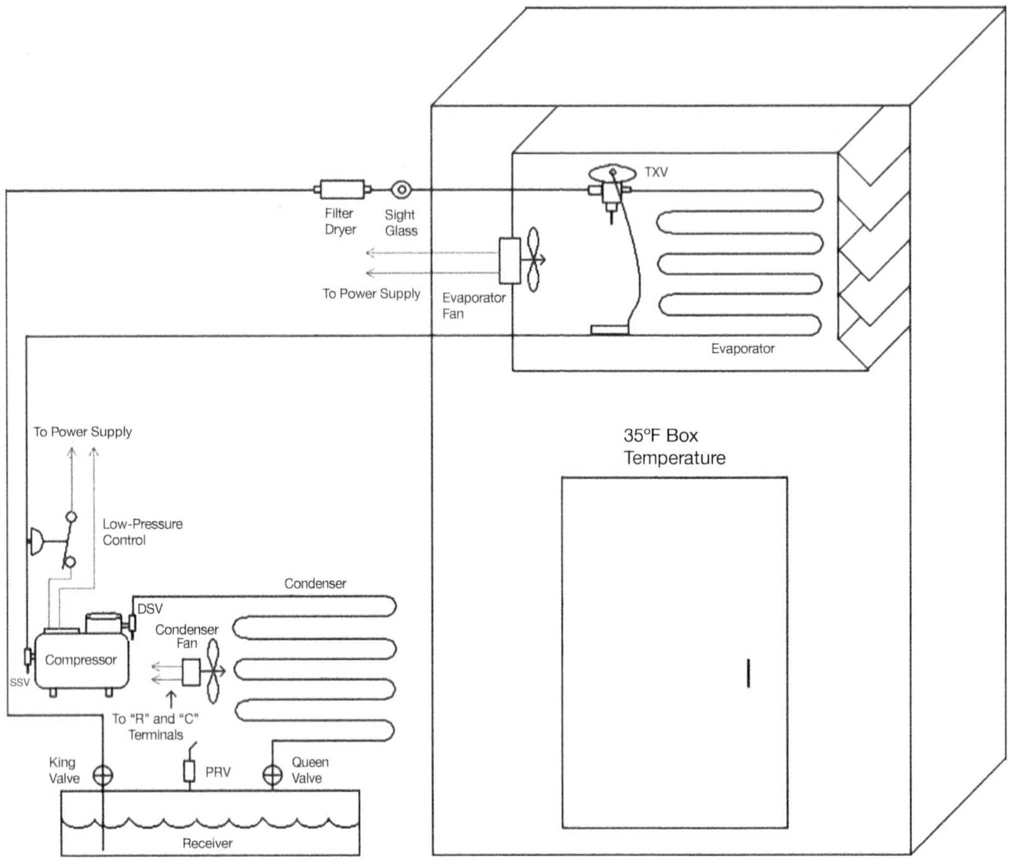

FIGURE 1–9B
Pressure-Controlled Medium-Temperature Cooler

closed and keeps the compressor energized. During shutdown, the box temperature reaches −10°F, which is the cooling thermostat's cut-out temperature; this causes the cooling thermostat to open the electrical circuit to the liquid line solenoid. At this time, the liquid line solenoid closes and stops the flow of liquid to the expansion valve and evaporator coil. All of the refrigerant is pumped out of the evaporator, suction line, and part of the liquid line and stored in the condenser and receiver. The compressor remains in operation until the low-side/evaporator pressure reaches the 2-psig cut-out set point on the low-pressure control. At this time, the compressor shuts down and the evaporator fan continues to operate, circulating freezer box air.

During the off cycle, the freezer box temperature will eventually increase to the cut-in temperature of the thermostat, which is −5°F. At this time, the thermostat closes and energizes the liquid line solenoid. This will send liquid refrigerant to the expansion valve and the evaporator coil, increasing low-side pressure. The low-pressure switch will close when the low-side pressure reaches the cut-in set point of 20 psig. At this time, the compressor will start and the cooling cycle will begin.

As the pressure switch closes, due to an increase of low-side pressure, the single-phase compressor is energized by a potential relay and a start capacitor. The condenser fan, being wired in parallel with the compressor, also energizes, thus moving air over the air-cooled condenser coil.

A defrost timer is added to the system with a supplementary electric defrost heating element and a termination thermostat. Due to extremely low box temperatures, freezers are not capable of removing frost buildup on the evaporator coil by the circulation of air, as are medium-temperature systems. Supplementary heat must be added to the evaporator coil, which increases its temperature well above the melting point of the frost.

The defrost timer motor operates 24 hours per day and controls one normally open and one normally closed set of contacts by the use of trip pins that are preadjusted by the refrigeration technician. The normally closed contacts are wired in series with the cooling thermostat, liquid line solenoid, and the evaporator fan motor. These contacts stay closed during the cooling cycle but open during the defrost cycle. An electric defrost occurs every 6 to 8 hours for about 20 to 40 minutes. When the timer switches from the cooling mode to the defrost mode, the N.C. contacts open and the circuit is broken to the evaporator fan, cooling thermostat, and liquid line solenoid and the system goes into an automatic pump-down. This happens during every defrost mode. The N.O. contacts close and energize an electric defrost heating element that is wired in series with the termination thermostat. The termination thermostat is typically a bimetal switch

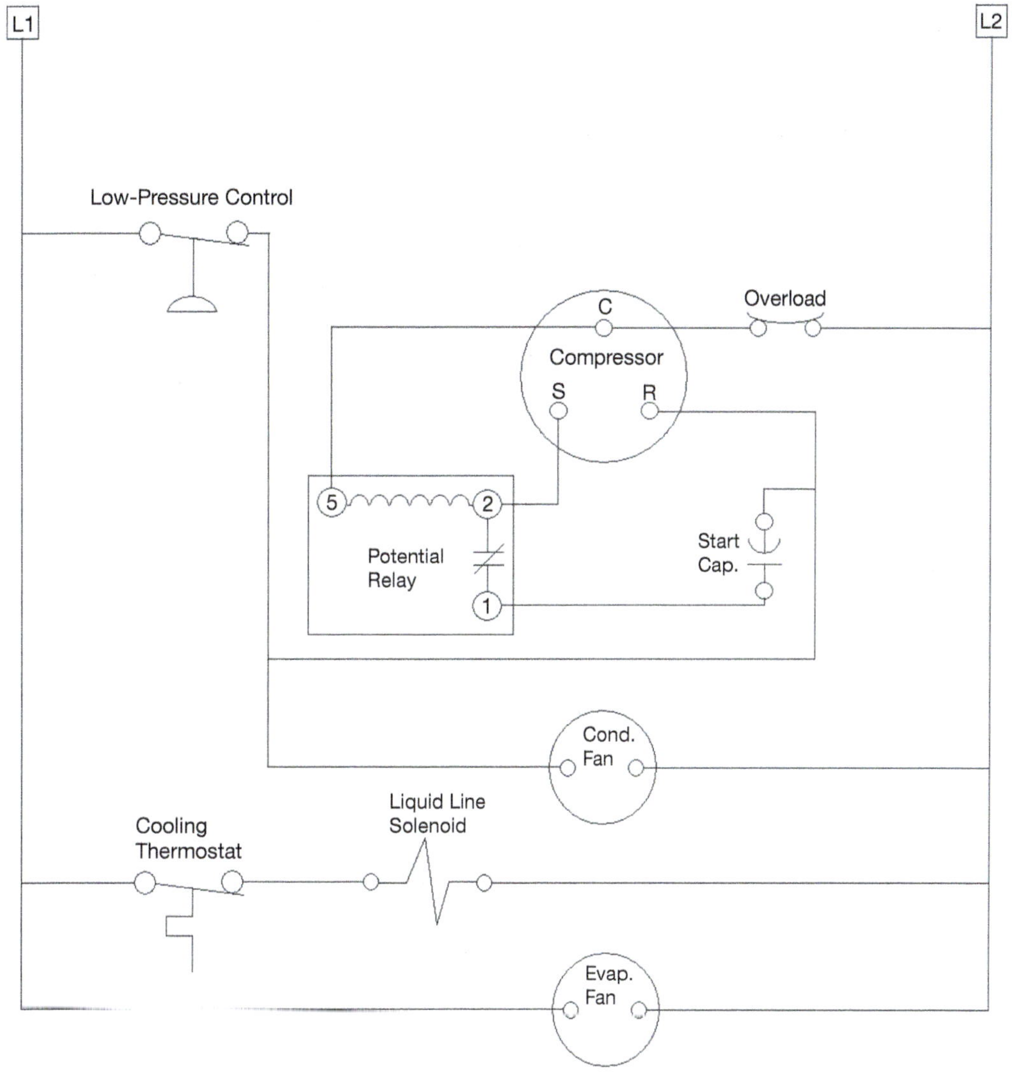

FIGURE 1–10A

Medium-Temperature Automatic Pump-Down System with a Capacitor Start Com-pressor Using a Potential Relay

that is mounted to the evaporator coil, sensing evaporator temperature. The termination thermostat may have a cut-out or opening temperature of 40°F and a cut-in or closing temperature of 20°F. When energized, the electric heating element adds heat to the evaporator coil and melts any frost that may have accumulated on the coil during the freeze cycle. If the timer is set for a 30-minute defrost and the temperature of the coil rises to the termination thermostat's cut-out point before the timer switches back into the cooling mode, the termination thermostat opens and de-energizes the electric heating element, preventing excessive overheating of the freezer.

When the defrost timer switches back into the cooling mode, the N.C. contacts close, which energizes the evaporator fan, cooling thermostat, and liquid line solenoid. This will start the cooling cycle as long as the thermostat is closed. The N.O. contacts open, preventing the operation of the defrost heater circuit. When the evaporator coil temperature drops to the cut-in or closing temperature of the termination thermostat, the switch closes and is ready for another defrost cycle. This is considered a low-temperature electric defrost system with an automatic pump-down control.

Sequence of Operation Figure 1–14A–B Diagrams

With the low-temperature hot gas defrost automatic pump-down freezer, an automatic pump-down cycle is added to the electrical control system. This is typically found when condensing units are located outdoors and outdoor ambient temperatures could drop below cooler box temperatures. This type of control system prevents liquid refrigerant migration back to the compressor during the system's off cycle by pumping all of the system's refrigerant into the condenser and receiver during shutdown.

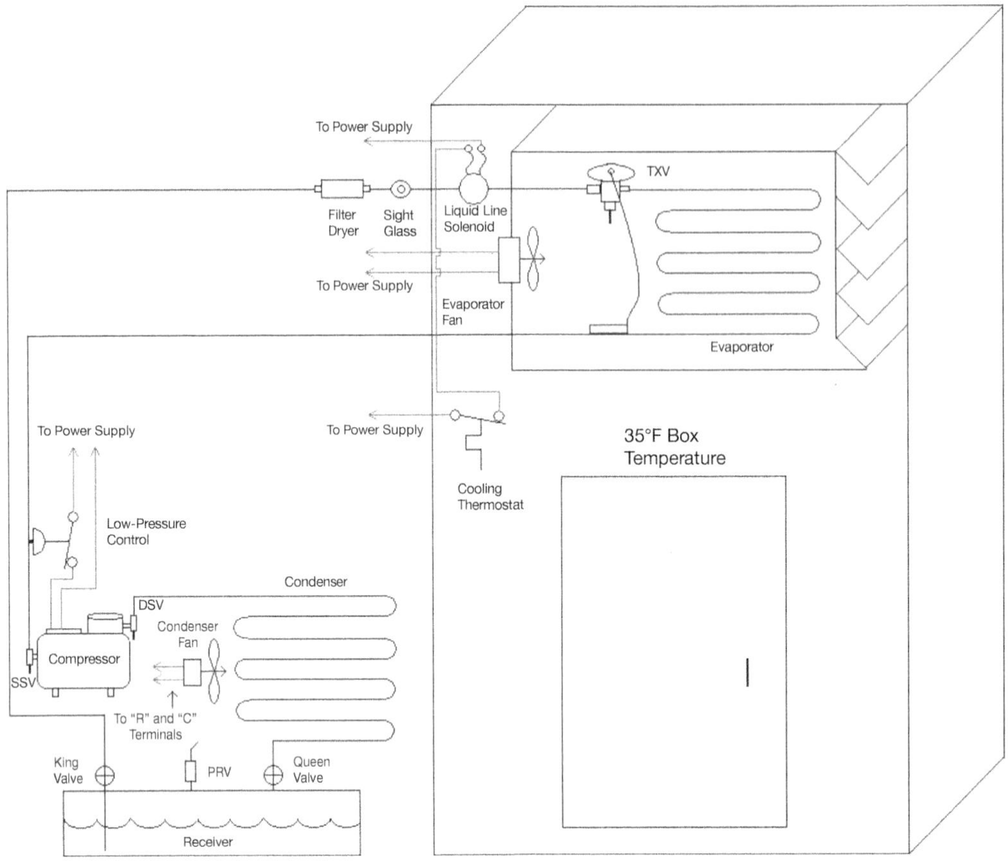

FIGURE 1–10B

Medium-Temperature Automatic Pump-Down System

In operation, the cooling thermostat controls a liquid line solenoid and the low-pressure switch controls the compressor. The cooling thermostat may be set to a −10°F cut-out point with a −5°F cut-in point. The low-pressure switch may have a cut-out point of 2 psig with a cut-in point of 20 psig. During operation, the cooling thermostat is closed, keeping the liquid line solenoid energized, which allows refrigerant to flow through the liquid line and to the expansion valve and from the expansion valve to the evaporator. This keeps low-side/evaporator pressure above the cut-out point of the low-pressure switch. The low-pressure switch stays closed and keeps the compressor energized. During shutdown, the box temperature reaches −10°F, which is the cooling thermostat's cut-out temperature; this causes the cooling thermostat to open the electrical circuit to the liquid line solenoid. At this time, the liquid line solenoid closes and stops the flow of liquid to the expansion valve and evaporator coil. All of the refrigerant is pumped out of the evaporator, suction line, and part of the liquid line and stored in the condenser and receiver. The compressor remains in operation until the low-side/evaporator pressure reaches the 2-psig cut-out set point on the low-pressure control. At this time, the compressor shuts down and the evaporator fan continues to operate, circulating freezer box air.

During the off cycle, the freezer box temperature will eventually increase to the cut-in temperature of the thermostat, which is −5°F. At this time, the thermostat closes and energizes the liquid line solenoid. This will send liquid refrigerant to the expansion valve and the evaporator coil, increasing low-side pressure. The low-pressure switch will close when the low-side pressure reaches the cut-in set point of 20 psig. At this time, the compressor will start and the cooling cycle will begin.

As the pressure switch closes, due to an increase of low-side pressure, the single-phase compressor is energized by a current relay and a start capacitor. The condenser fan, being wired in parallel with the compressor, also energizes, thus moving air over the air-cooled condenser coil.

A defrost timer is added to the system with a supplementary hot gas solenoid valve and a termination thermostat. Due to extremely low box temperatures, freezers are not capable of removing frost buildup on the evaporator coil by the circulation of air, as are medium-temperature systems. Supplementary heat must be added to the evaporator coil, which increases its temperature well above the melting point of the frost.

A hot gas defrost solenoid and a hot gas piping system connects the hot gas or discharge line, leaving the compressor to the inlet to

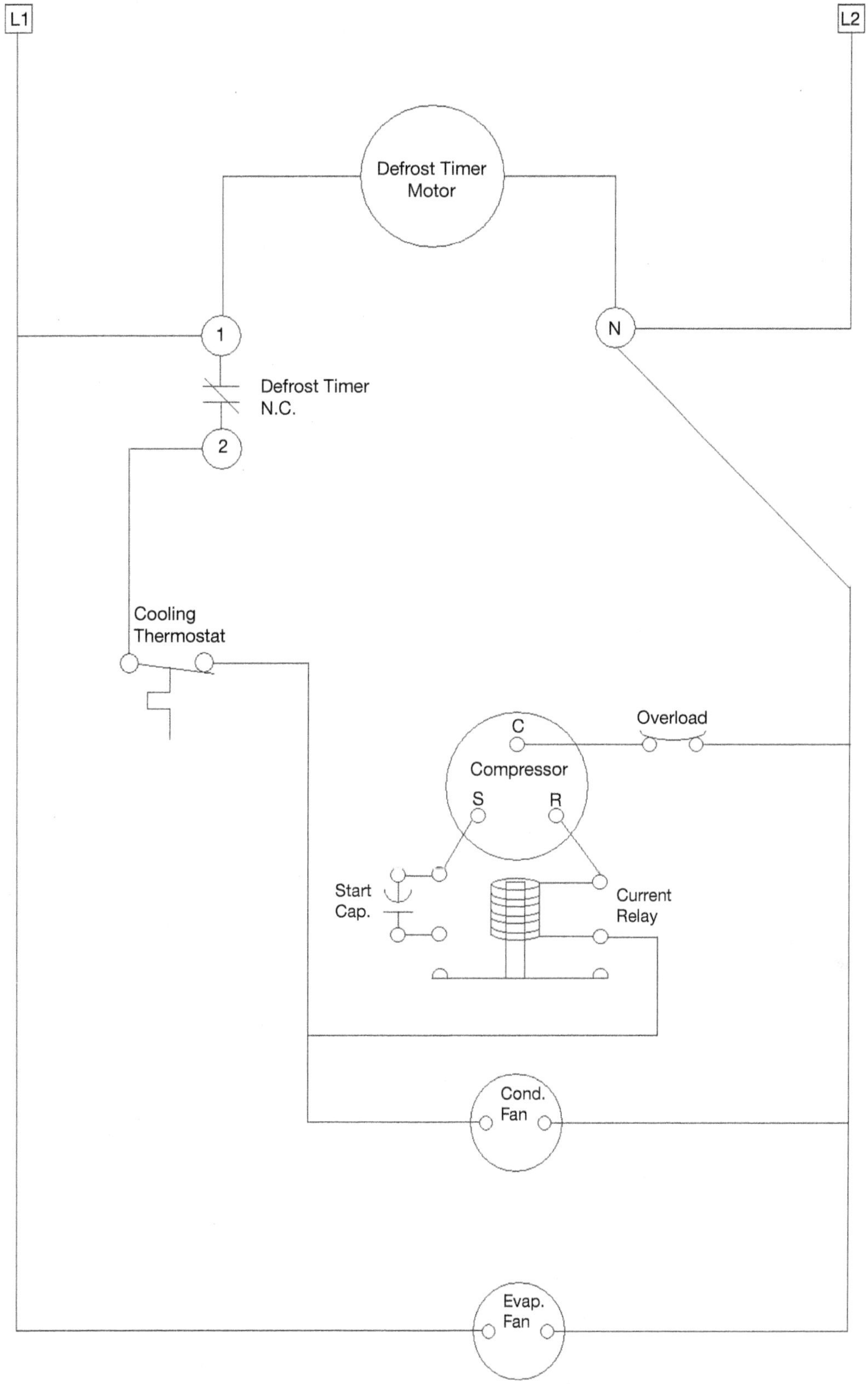

FIGURE 1–11A

Time Shutdown Defrost Medium-Temperature Cooler with a Capacitor Start Com-pressor Using a Current Relay

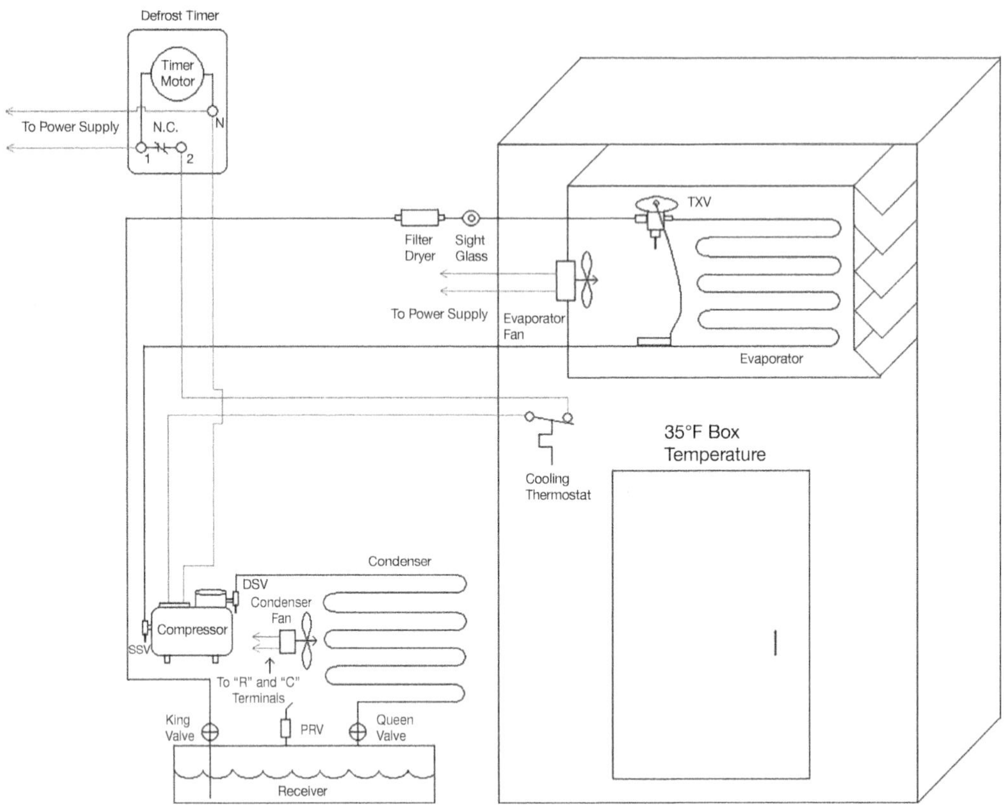

FIGURE 1–11B

Time Shutdown Defrost Medium-Temperature Cooler

the evaporator coil. During the defrost cycle, the compressor remains in operation and the normally closed hot gas defrost solenoid is energized. The valve opens and hot gas from the discharge line bypasses into the inlet side of the evaporator coil and raises its pressure and temperature above freezing. This causes any excessive frost buildup on the evaporator coil to melt and flow down the drain away from the coil. A suction line accumulator must be installed when using hot gas defrost systems to prevent liquid floodback to the compressor during or following a defrost cycle.

The defrost timer motor operates 24 hours per day and controls one normally open and one normally closed set of contacts by the use of trip pins that are preadjusted by the refrigeration technician. The normally closed contacts are wired in series with the cooling thermostat, liquid line solenoid, and the evaporator fan motor. These contacts stay closed during the cooling cycle but open during the defrost cycle. An electric defrost occurs every 6 to 8 hours for about 20 to 40 minutes. When the timer switches from the cooling mode to the defrost mode, the N.C. contacts open and the circuit is broken to the evaporator fan, cooling thermostat, and liquid line solenoid. The N.O. contacts close and energize the hot gas solenoid valve that is wired in series with the termination thermostat. The termination thermostat is typically a bimetal switch that is mounted to the evaporator coil, sensing evaporator temperature. The termination thermostat may have a cut-out or opening temperature of 40°F and a cut-in or closing temperature of 20°F. When energized, the hot gas solenoid valve opens and raises the pressure and temperature of the evaporator coil and melts any frost that may have accumulated on the coil during the freeze cycle. If the timer is set for a 30-minute defrost and the temperature of the coil raises to the termination thermostat's cut-out point before the timer switches back into the cooling mode, the termination thermostat opens and de-energizes the hot gas solenoid valve, avoiding excessive overheating of the freezer. At this point, since the electrical circuit to the liquid line solenoid valve is open, the compressor will pump the system down and remain de-energized by the low-pressure switch until the defrost timer completes its defrost cycle.

When the defrost timer switches back into the cooling mode, the N.C. contacts close, which energizes the evaporator fan, cooling thermostat, and liquid line solenoid. This will start the cooling cycle as long as the thermostat is closed. The N.O. contacts open, preventing the operation of the hot gas defrost valve circuit. When the evaporator coil temperature drops to the cut-in or closing temperature of the termination thermostat, the switch closes and is ready for another defrost cycle. This is considered a low-temperature hot gas defrost system with an automatic pump-down control.

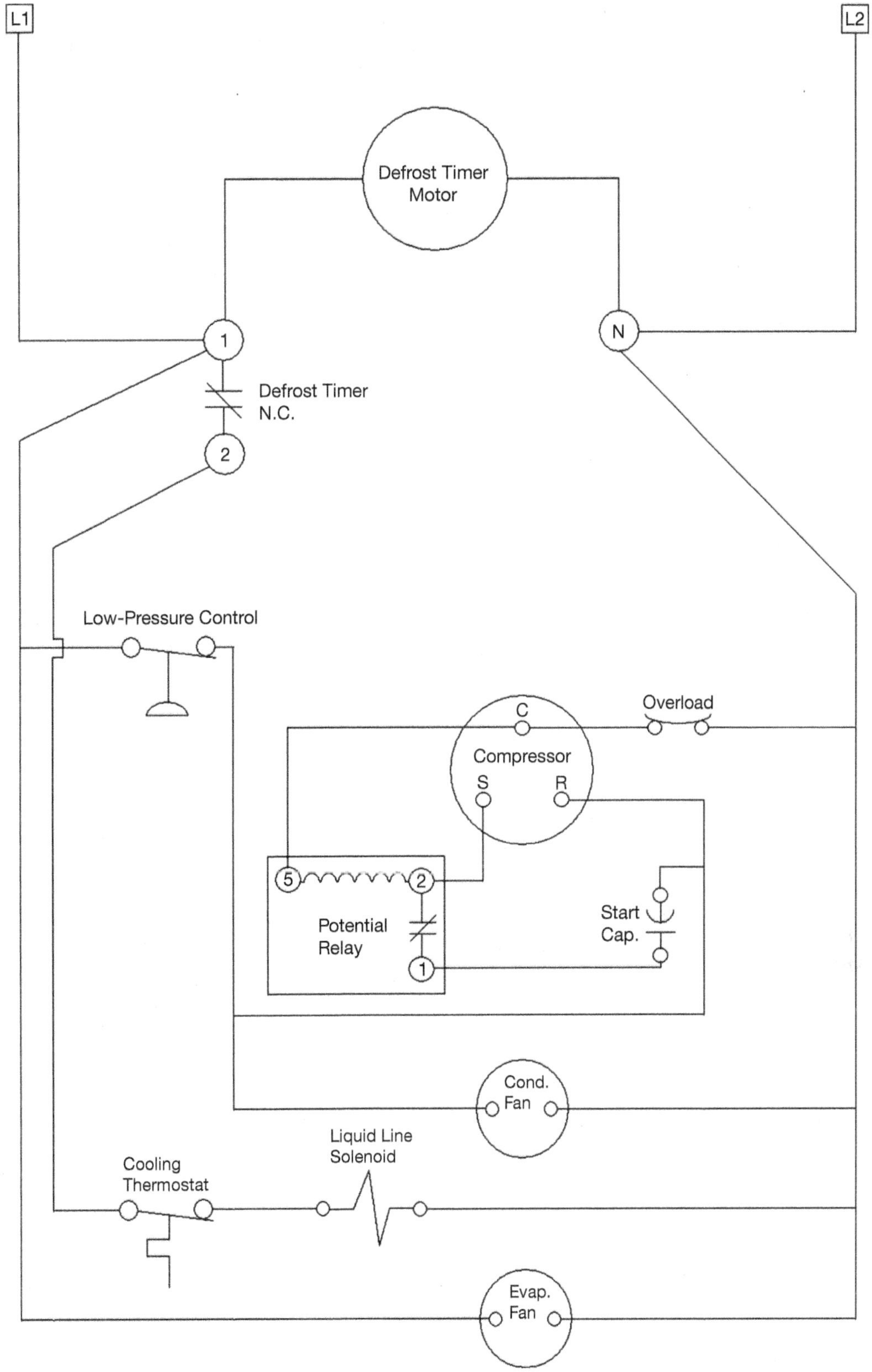

FIGURE 1–12A

Time Shutdown Defrost Medium-Temperature Cooler with Automatic Pump-Down Control

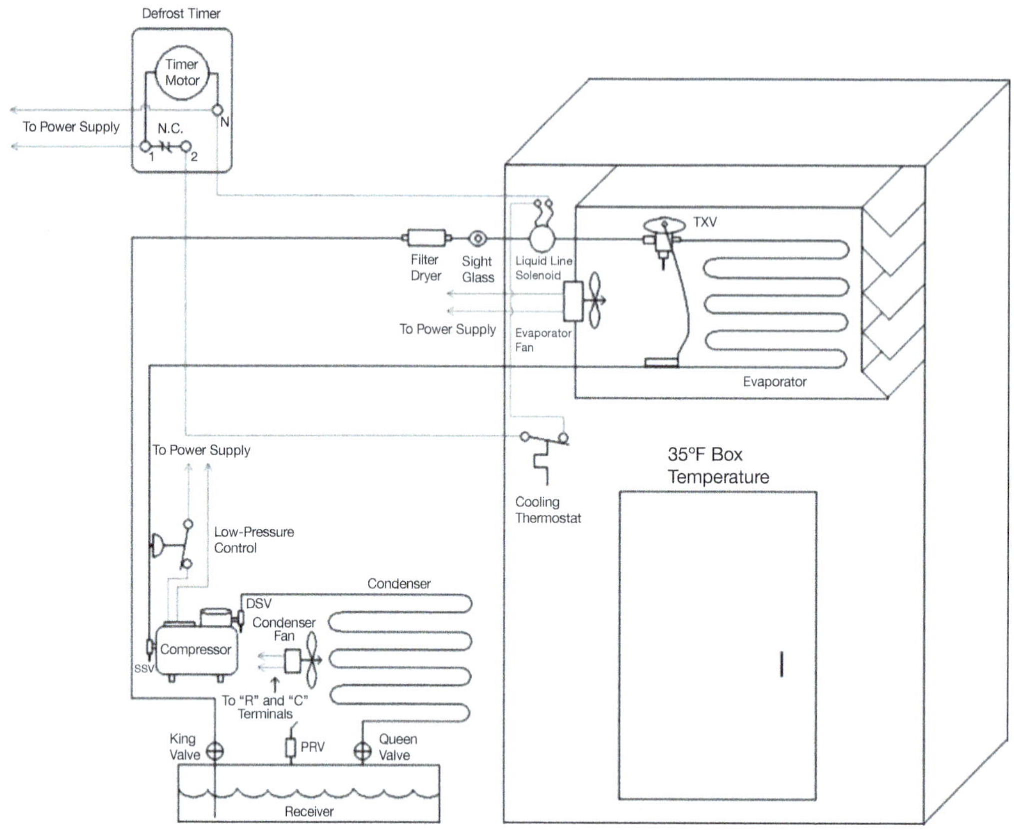

FIGURE 1–12B

Time Shutdown Defrost Medium-Temperature Cooler with Automatic Pump-Down Control

Sequence of Operation Figure 1–15A–B Diagrams

With the low-temperature electric defrost system using a timer release solenoid and an automatic pump-down system, an automatic pump-down cycle is added to the electrical control system. This is typically found when condensing units are located outdoors and outdoor ambient temperatures could drop below cooler box temperatures. This type of control system prevents liquid refrigerant migration back to the compressor during the system's off cycle by pumping the entire refrigerant charge into the condenser and receiver during shutdown.

In operation, the cooling thermostat controls a liquid line solenoid and the low-pressure switch controls the compressor. The cooling thermostat may be set to a −10°F cut-out point with a −5°F cut-in point. The low-pressure switch may have a cut-out point of 2 psig with a cut-in point of 20 psig. During operation, the cooling thermostat is closed, keeping the liquid line solenoid energized, which allows refrigerant to flow through the liquid line and to the expansion valve and from the expansion valve to the evaporator. This keeps low-side/evaporator pressure above the cut-out point of the low-pressure switch. The low-pressure switch stays closed and keeps the compressor energized. During shutdown, the box temperature reaches −10°F, which is the cooling thermostat's cut-out temperature; this causes the cooling thermostat to open the electrical circuit to the liquid line solenoid. At this time, the liquid line solenoid closes and stops the flow of liquid to the expansion valve and evaporator coil. The entire refrigerant charge is pumped out of the evaporator, suction line, and part of the liquid line and stored in the condenser and receiver. The compressor remains in operation until the low-side/evaporator pressure reaches the 2-psig cut-out set point on the low-pressure control. At this time, the compressor shuts down and the evaporator fan continues to operate, circulating freezer box air.

During the off cycle, the freezer box temperature will eventually increase to the cut-in temperature of the thermostat, which is −5°F. At this time, the thermostat closes and energizes the liquid line solenoid. This will send liquid refrigerant to the expansion valve and the evaporator coil, increasing low-side pressure. The low-pressure switch will close when the low-side pressure reaches the cut-in set point of 20 psig. At this time, the compressor will start and the cooling cycle will begin.

As the pressure switch closes, due to an increase of low-side pressure, the single-phase compressor is energized by a potential relay and a start capacitor. The condenser fan, being wired in parallel with the compressor, also energizes, thus moving air over the air-cooled condenser coil.

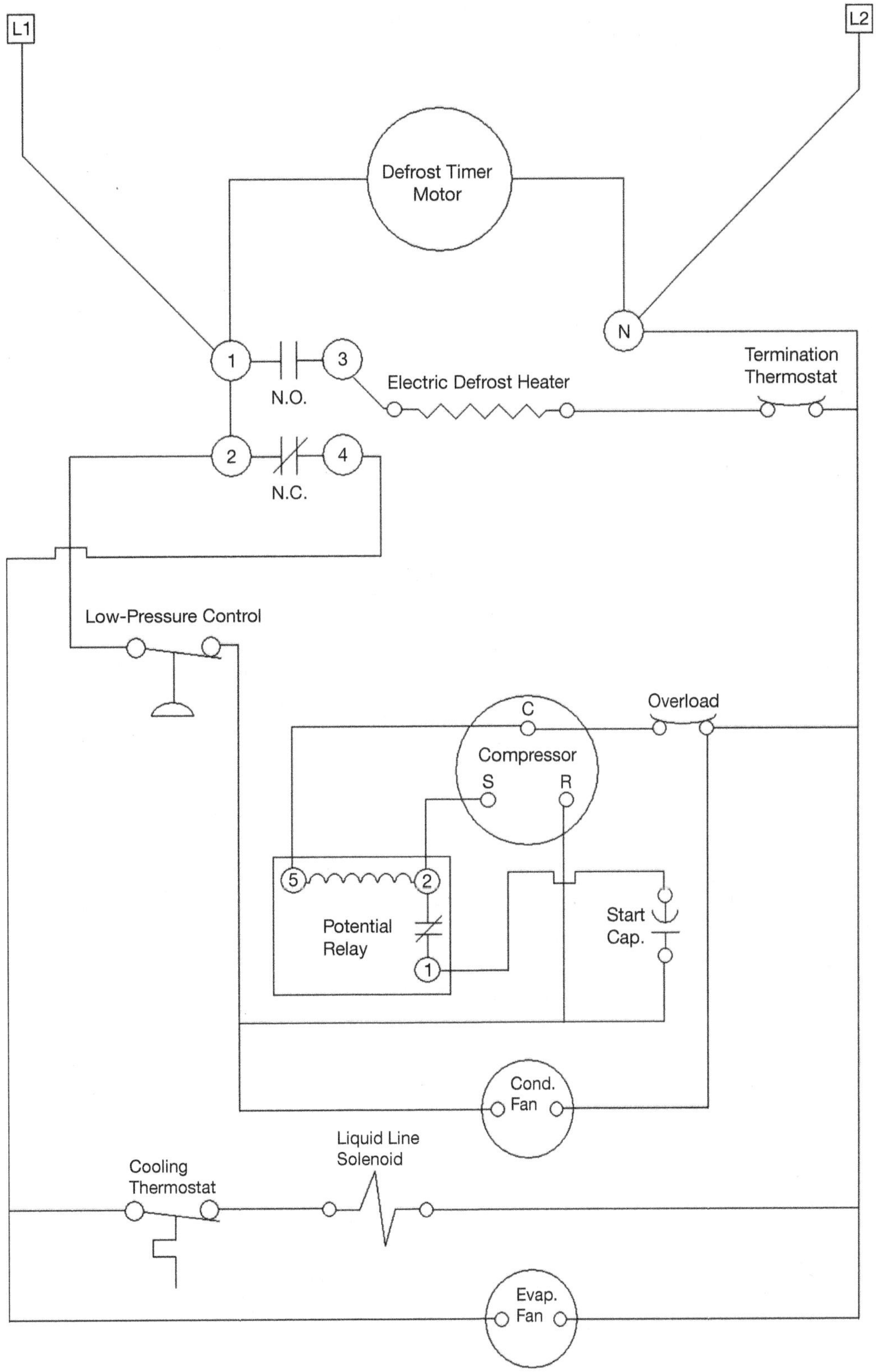

FIGURE 1–13A

Low-Temperature Refrigeration System with Electric Defrost and Automatic Pump-Down

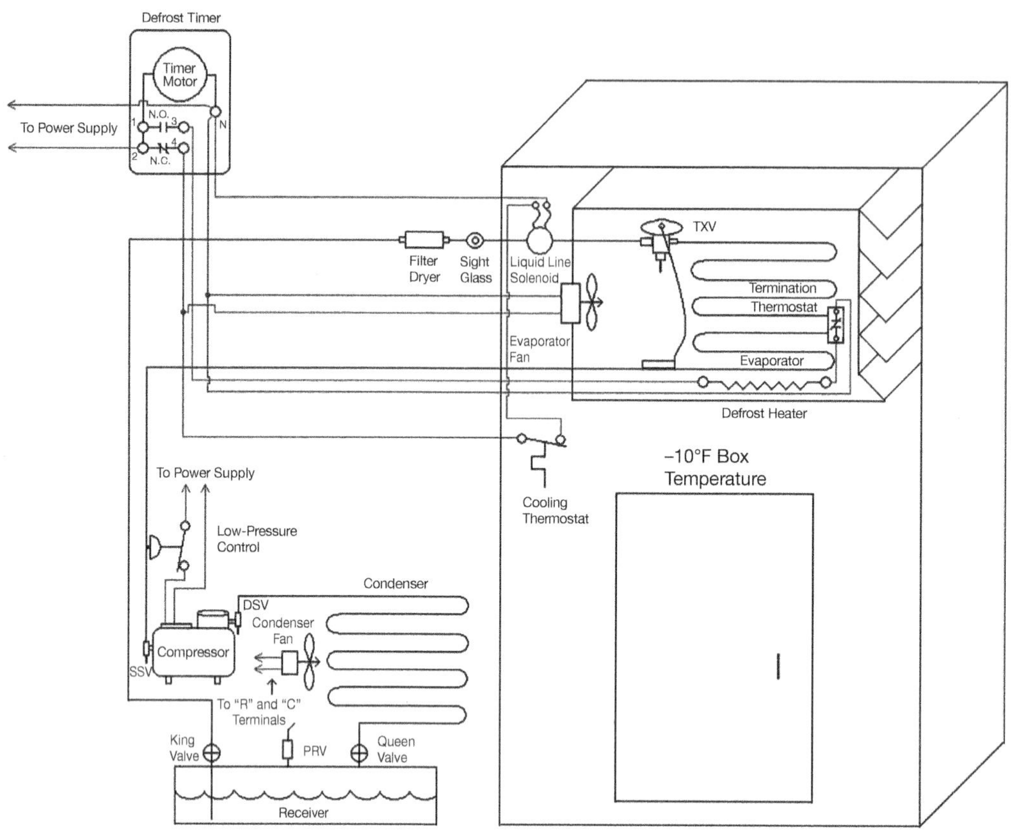

FIGURE 1–13B

Low-Temperature Refrigeration System with Electric Defrost and Automatic Pump-Down

A defrost timer that uses a timer release solenoid is added to the system with a supplementary electric defrost heating element and a single pole, double throw termination thermostat. Due to extremely low box temperatures, freezers are not capable of removing frost buildup on the evaporator coil by the circulation of air, as are medium-temperature systems. Supplementary heat must be added to the evaporator coil, which increases its temperature well above the melting point of the frost.

The defrost timer motor operates 24 hours per day and controls one normally open and one normally closed set of contacts by the use of trip pins that are preadjusted by the refrigeration technician. The normally closed contacts are wired in series with the cooling thermostat, liquid line solenoid, and the evaporator fan motor. These contacts stay closed during the cooling cycle but open during the defrost cycle. An electric defrost occurs every 6 to 8 hours for about 20 to 40 minutes. When the timer switches from the cooling mode to the defrost mode, the N.C. contacts open and the circuit is broken to the evaporator fan, cooling thermostat, and liquid line solenoid and the system goes into an automatic pump-down. This happens during every defrost mode.

The defrost timer's N.O. contacts close and energize an electric defrost heating element that is wired in parallel with the termination thermostat's "close on rise" contacts.

The termination thermostat is typically a bimetal switch that is mounted to the evaporator coil, sensing evaporator temperature. This single pole, double throw termination thermostat has two sets of contacts that are connected to a common terminal point.

The "close on rise" contacts close as the evaporator temperature increases. The other contact is a "close on fall" of evaporator temperature. The termination thermostat's "close on rise" contacts may close at a temperature of 40°F and open at a temperature of 20°F. The "close on fall" closes at 20°F and opens at 40°F.

In the defrost mode, the electric heating element is energized through the N.O. defrost timer contacts, which adds heat to the evaporator coil and melts any frost that may have accumulated on the coil during the freeze cycle. The N.O. defrost timer contacts also complete an electrical path through the timer release solenoid up to the termination thermostat "close on rise" contacts. These "close on rise" contacts should be in the open position at this time since the evaporator coil typically has frost accumulation on the coil.

If the timer is set for a 30-minute defrost and the temperature of the coil rises to the termination thermostat's "close on rise" cut-in point, the timer release solenoid is energized and the defrost timer automatically switches back into the cooling mode and de-energizes the electric defrost heaters. The timer release solenoid and the "close on rise" contacts of the termination thermostat override the timer's defrost cycle.

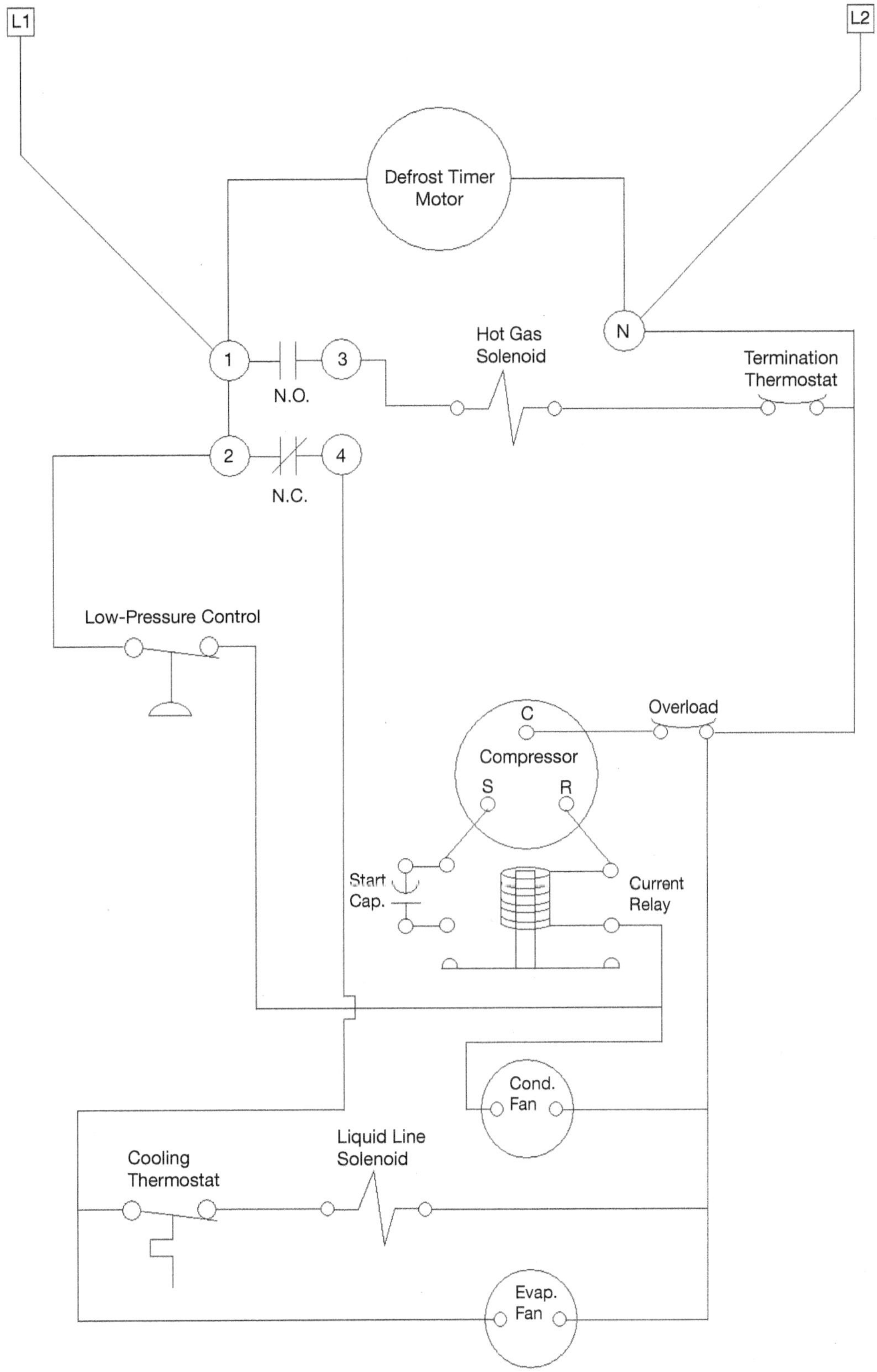

FIGURE 1–14A

Low-Temperature Refrigeration System with Hot Gas Defrost and Automatic Pump-Down

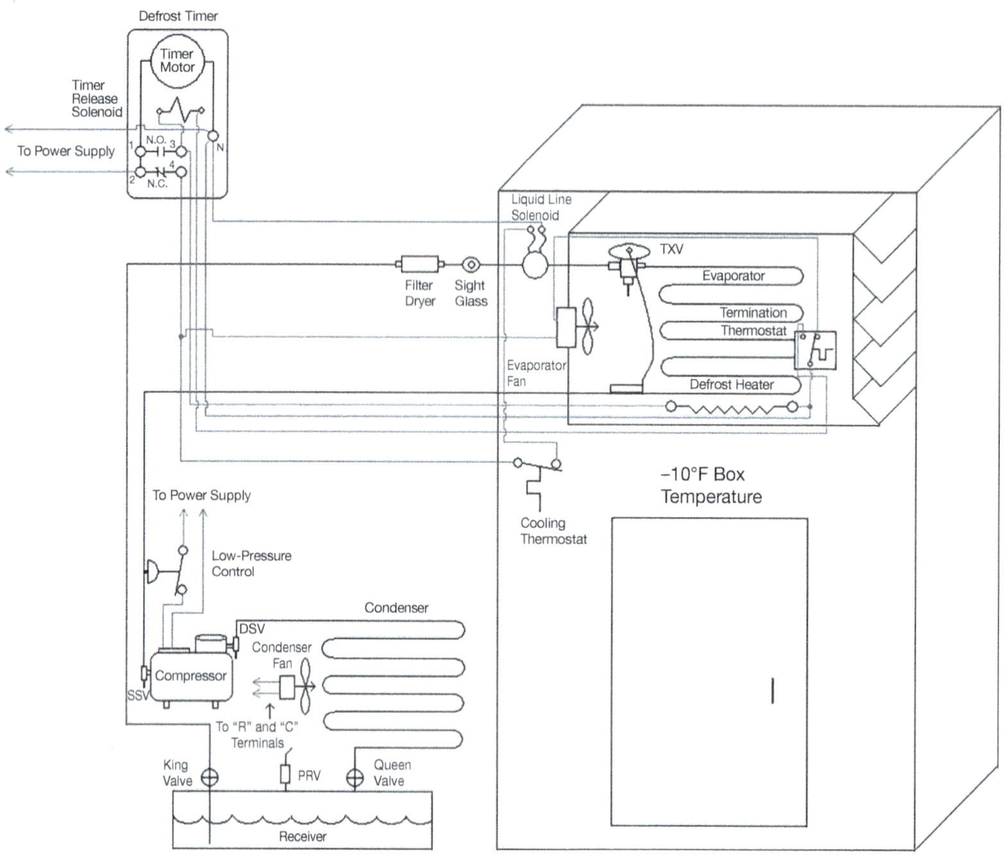

FIGURE 1–14B

Low-Temperature Refrigeration System with Hot Gas Defrost and Automatic Pump-Down

The defrost timer N.C. contacts energize the cooling thermostat and liquid line solenoid. This will start the cooling cycle as long as the thermostat is closed. When the evaporator coil temperature drops to the cut-in or closing temperature of the termination thermostat, the switch closes and is ready for another defrost cycle.

When the defrost timer switches back into the cooling mode, the N.C. contacts close, which completes an electrical path to the "close on fall" contacts of the termination thermostat. The "close on fall" contacts of the termination thermostat are in series with the evaporator fan motor common wire. At this point, the contacts would be in the open position, which would keep the evaporator fan de-energized, preventing the circulation of excessive heat given off by the electric heating element, and this would also allow for the drainage of melted frost from the evaporator coil.

As the refrigeration system drops the temperature of the evaporator coil, the "close on fall" contacts close and energize the evaporator fan motor.

This is considered a low-temperature electric defrost system with an automatic pump-down control.

AIR CONDITIONING COMPONENTS

An air conditioning system consists of various components (see Figure 1–16). Each component helps keep the system operating properly. The technician is responsible for knowing about every system component in order to correctly diagnose malfunctions.

The common **components** that are used in air conditioning systems are as follows:

1. The **compressor** changes the refrigerant from a low-pressure/low-temperature gas into a high-pressure/high-temperature gas.
2. An **oil separator** removes up to 98% of the oil that circulates in a refrigeration system. It is located in the discharge line.
3. The **condenser** changes a high-pressure/high-temperature gas into a high-pressure/high-temperature liquid to remove heat from the system.

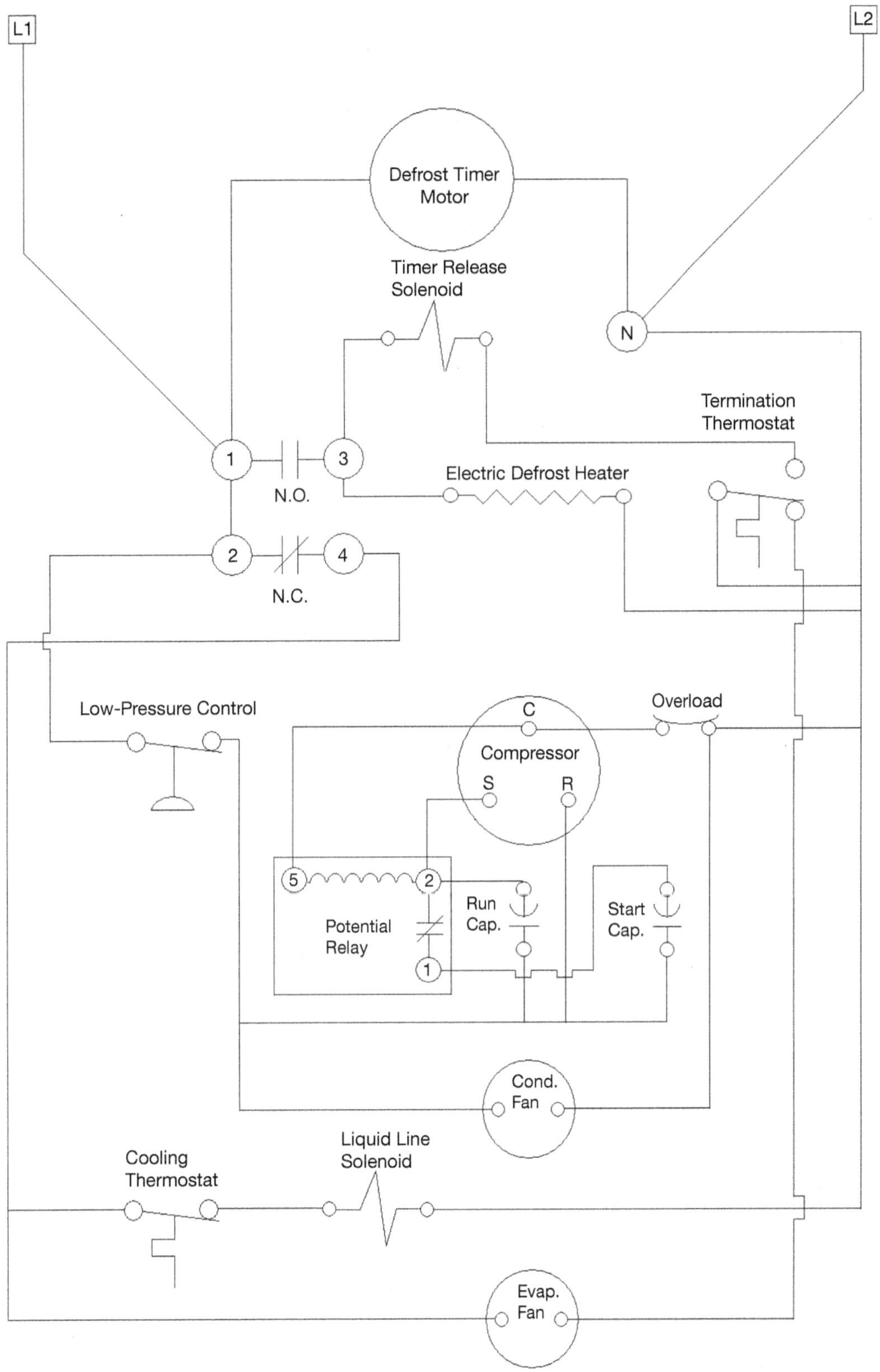

FIGURE 1–15A

Low-Temperature Refrigeration System with Electric Defrost Using a Termination Thermostat and a Timer Release Solenoid with an Automatic Pump-Down Control

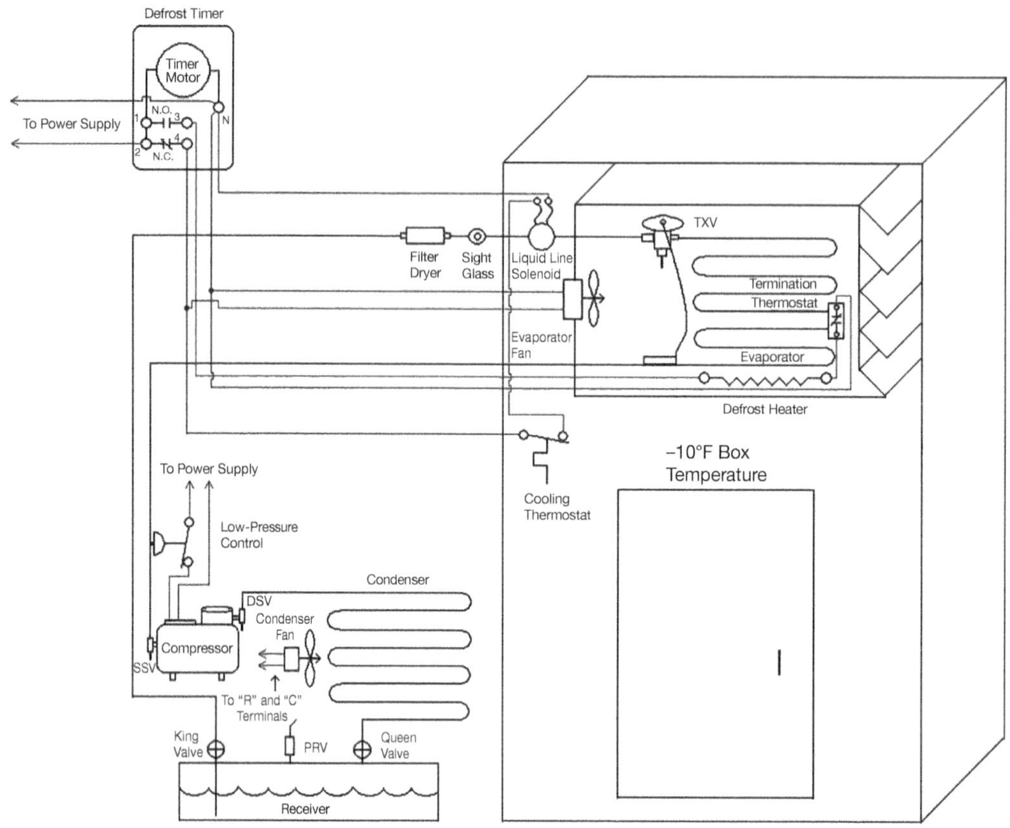

FIGURE 1–15B

Low-Temperature Refrigeration System with Electric Defrost Using a Termination Thermostat and a Timer Release Solenoid with an Automatic Pump-Down Control

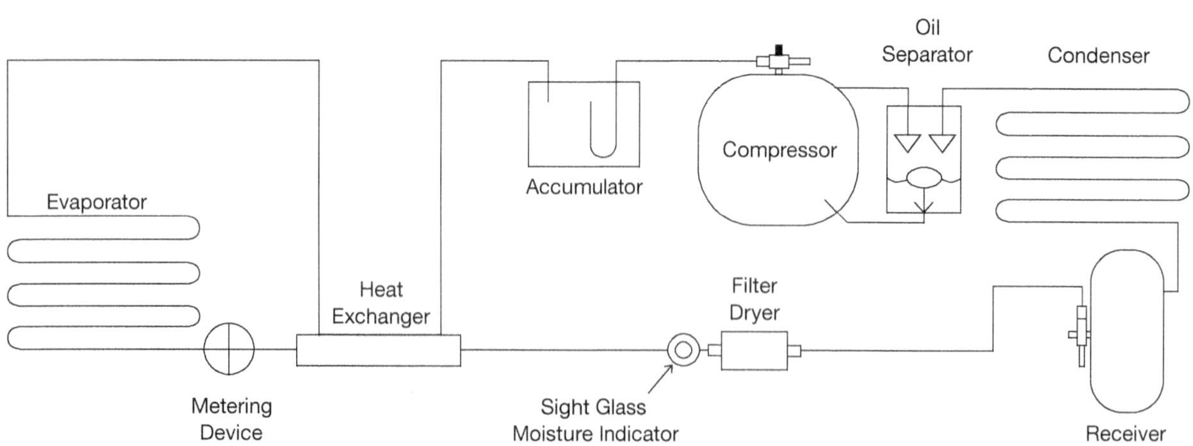

FIGURE 1–16

4. The **receiver** stores liquid refrigerant until it is needed in the evaporator. It is located directly downstream from the condenser.
5. The **filter dryer** traps particles and small amounts of moisture in the system. It is located in the liquid line.
   a. The **sight glass** aids in charging the refrigeration system and is located in the liquid line.
   b. A **moisture indicator** is commonly located in the sight glass. It senses moisture that may be present in the system's refrigerant.
6. The **metering device** changes a high-pressure/high-temperature liquid into a low-pressure/low-temperature liquid.
7. The **evaporator** changes a low-pressure/low-temperature liquid into a low-pressure/low-temperature gas. It absorbs heat into the system.
8. An **accumulator** traps liquid refrigerant and prevents it from entering the compressor. It is located in the suction line.
9. A **heat exchanger** subcools the liquid refrigerant in the liquid line and superheats the refrigerant gas in the suction line. It connects the liquid line to the suction line to increase the efficiency of the system.

As the temperature in a **sealed system** increases, the pressure also increases, and as the temperature decreases, the pressure decreases.

A **split air conditioning system** separates the compressor, known as the *condensing unit,* and the condenser from the metering device and evaporator coil (see Figure 1–17). The condensing unit is connected to the evaporator coil with refrigerant lines.

In a **residential split system,** the evaporator coil and metering device are located in the plenum of the furnace, while the compressor and condenser are located on the outside of the home. The liquid and suction line connect the outdoor and indoor units.

Outdoor condensing units should be mounted on a level concrete slab. Being level will allow any accumulation of water to run evenly out of the cabinet.

A fill of gravel under the **concrete slab** is recommended to prevent settling of the ground. Settling can cause stress and even break the refrigerant lines.

There are several factors to consider for the **location of the condensing unit,** including available space, length of line set, visual appearance, and noise factor (near bedroom windows, patios, etc.).

The manufacturer's **clearance recommendations** must always be followed when installing a condensing unit.

Residential split-system outdoor **condensing units** are wired with a 208-volt/240-volt single-phase power supply.

A **fused disconnect** is located near the system to break the power source supply to the unit during service.

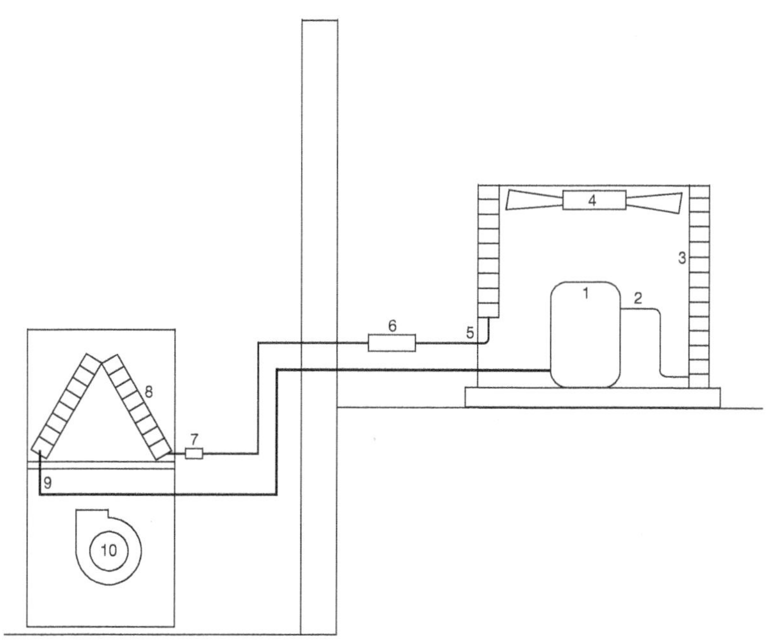

1. Compressor
2. Discharge Line
3. Condenser
4. Condenser Fan
5. Liquid Line
6. Filter Dryer
7. Metering Device (Restrictor)
8. Evaporator
9. Suction Line
10. Indoor Blower

FIGURE 1–17

Technicians must perform a **visual check** to ensure all electrical connections are secure and tight before starting a newly installed system.

A **residential furnace** has a 115-volt line supplied to the system through a disconnect switch. This switch is used to energize the primary side of the control transformer and blower motor.

The **control transformer** of a residential furnace supplies 24 volts to the heating and cooling thermostat. This thermostat controls a fan relay coil, a contactor coil, and the heating control circuit.

A **room thermostat** should be installed on an inside wall away from any heat source. It should be 5′ from the floor and in an area with air motion but away from a supply air register.

On the heating and cooling **thermostat,** the "R" terminal is the power source from the 24-volt transformer.

On the heating and cooling **thermostat,** the "G" terminal energizes the blower relay coil.

On the heating and cooling **thermostat,** the "Y" terminal energizes the outdoor unit's contactor coil.

On the heating and cooling **thermostat,** the "W" terminal energizes the heating control circuit on the furnace.

Four mechanical connections are used to connect the refrigeration line set to the evaporator and the condensing unit. The four connections are quick-connect fittings, compression fittings, flare fittings, and sweat connections.

There are two lines to a **refrigeration line set:** the liquid line and the suction line. The liquid line is the smaller line and the suction line is the larger line. The **suction** line must be insulated to decrease refrigerant temperature rise and exterior moisture condensation.

Refrigeration tubing must be protected by PVC if a line set is installed through a masonry wall. This prevents corrosion and possible refrigerant leaks.

A **suction line** must be pitched toward the flow of vapor to ensure that oil returns to the compressor.

If **refrigerant lines** are coiled, they must be installed horizontally. The vapor must flow from top to bottom to ensure that oil returns to the compressor.

Quick-connect fittings are used on precharged refrigeration line sets. These fittings seal the refrigerant in the tubing until the connections are tightened.

When using **precharged line sets,** excess refrigeration lines must be coiled horizontally to prevent trapping the circulated oil.

If any alteration is made to the length of a **refrigeration line set,** the system must be recharged to match the new length.

Compression fittings use a brass coupling type of nut with an O-ring and a threaded male adapter. The nut with the O-ring is tightened onto the male adapter, compressing the O-ring for a leakless joint.

Flare fittings use a flare and a flare nut connected to the male flare fitting to create a leakless joint.

Sweat (soldered) connections use couplings into which the refrigeration tubing fits. Silver braze is used to seal the tubing together.

When **brazing** a joint, capillary action causes the brazing material to flow throughout the entire joint area.

Nonprecharged line sets and evaporator coils are charged with nitrogen and plugged until installation.

Condensing units are precharged with refrigerant for a specific length of refrigeration line set. If additional line is needed, the refrigerant charge must be increased according to the manufacturer's specifications.

OD stands for the outer diameter of copper tubing. When working with refrigeration tubing, the OD is always taken into consideration.

Soft copper is used on refrigeration line sets up to 3″ OD. Refrigeration line sets of 1″ OD and above are commonly made of **hard drawn copper** tubing.

The common **metering devices** used for split systems below a $7\frac{1}{2}$-ton capacity are the capillary and fixed orifice types.

Thermostatic expansion valve **(TXV)** metering devices are used with split systems that have a capacity of $7\frac{1}{2}$ tons or higher and that have increased to a 12 SEER (Seasonal Energy Efficiency Ratio) or greater.

Attic-space mounted **split systems** use either a horizontal-type furnace or a horizontal air-handling system.

All **ductwork** must be insulated when used in an unconditioned space, such as an attic or crawl space.

An **auxiliary drain pan** must be installed on air conditioning systems that are located in an attic space.

The drain pan must be $1\frac{1}{2}$″ deep and not less than 3″ larger than the air-handling unit.

A normally closed **float switch** controls the cooling circuit and is commonly used to prevent damage to ceiling drywall if an overflow occurs.

A **secondary drain line** should extend from the auxiliary drain pan to a conspicuous place. This serves as an alarm indicating that the primary drain is restricted.

Condensate drains must be **trapped,** as required by the equipment manufacturer.

HEAT PUMPS

In mild winter climates and in the air conditioning season, heat pumps are very popular. It is pertinent that service technicians be knowledgeable of the various types of heat pumps. The following will aid technicians in building on their knowledge.

Cooling Mode

During the cooling mode of the heat pump, high-pressure/high-temperature refrigerant gas flows from the compressor, through the reversing valve, to the outdoor coil. Sensible and latent heat is removed and the refrigerant is changed into a high-pressure/high-temperature liquid. The liquid flows from the outdoor coil, through the check valve, to the de-ice subcooler coil, where the refrigerant

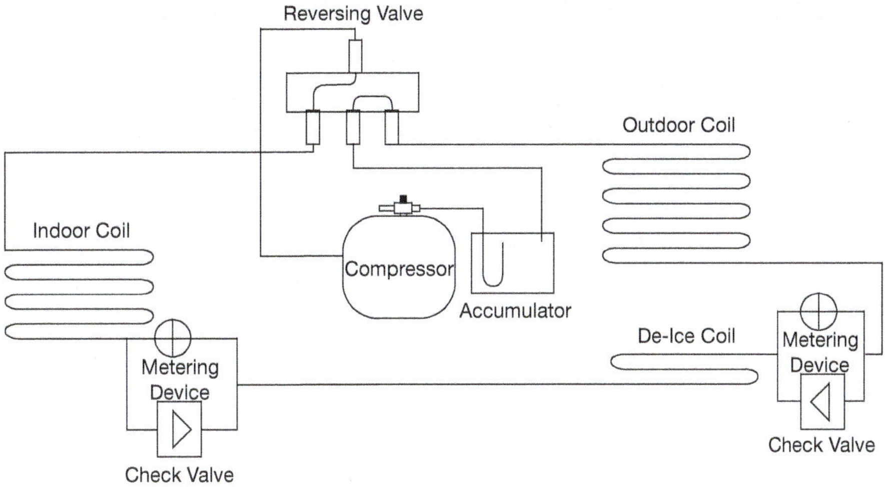

FIGURE 1–18
Heat Pump Diagram Cooling Mode

temperature is lowered to increase the system's capacity. The high-pressure/high-temperature liquid refrigerant flows from the de-ice coil to the metering device, where the refrigerant is changed into a low-pressure/low-temperature saturated liquid. The check valve located at the indoor coil is in the closed position at this time. The refrigerant flows through the indoor coil, absorbing heat, causing the refrigerant to change from a saturated liquid into a superheated gas. From the indoor coil, the refrigerant flows through the reversing valve, to the accumulator and back into the compressor. For an example, see Figure 1–18.

Heating Mode

During the heating mode of the heat pump, high-pressure/high-temperature refrigerant gas flows from the compressor, through the reversing valve, to the indoor coil. Sensible and latent heat is removed and the refrigerant is changed to a high-pressure/high-temperature liquid. The liquid flows from the indoor coil, through the check valve, to the de-ice coil (subcooler), where the refrigerant temperature is lowered to increase the system's capacity. This de-ice coil also prevents the formation of ice buildup on the bottom of the outdoor coil during the heating mode. The high-pressure/high-temperature liquid refrigerant flows from the de-ice coil to the metering device, where the refrigerant is changed to a low-pressure/low-temperature saturated liquid. The refrigerant flows through the outdoor coil. As it flows, it absorbs heat, causing the refrigerant to change from a saturated liquid to a superheated gas. From the outdoor coil, the refrigerant flows through the reversing valve to the accumulator and back into the compressor. For an example, see Figure 1–19.

The **on cycles** of a heat pump during the heating mode are much longer than a standard fossil fuel furnace.

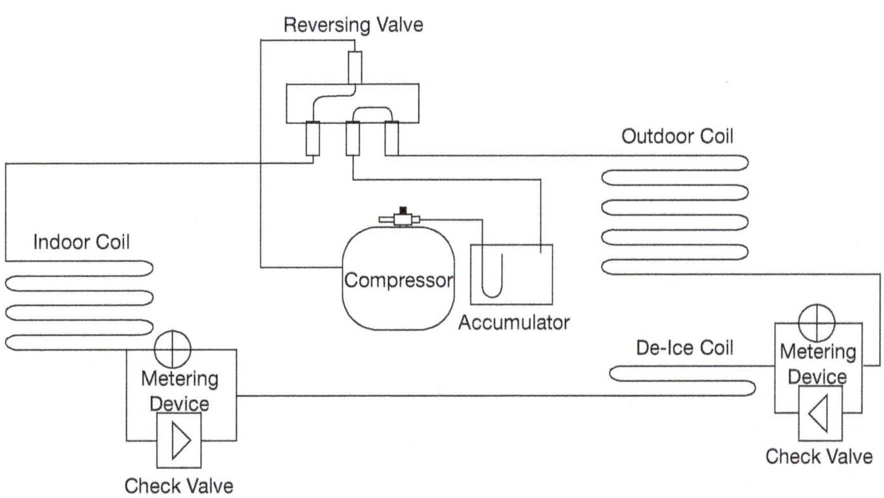

FIGURE 1–19
Heat Pump Diagram Heating Mode

During the heating mode, if the heat pump system needs to be recharged, the best method of charging the system is to weigh in the correct amount of **refrigerant charge.**

The **reversing valve** is used in a heat pump to change the refrigerant flow. During the cooling mode, it causes the outdoor coil to be a condenser and the indoor coil to be an evaporator. During the heating mode, the reversing valve causes the outdoor coil to be an evaporator and the indoor coil to be a condenser. For an example, see Figure 1–20.

The **reversing valve solenoid coil** is designed to be de-energized during the winter in northern climates and during the summer in southern climates. This will extend the life of the valve and conserve electrical energy.

The **reversing valve** has four tubes connected to the system. The top tube is connected to the discharge of the compressor. The bottom center tube is connected to the suction side of the compressor.

The bottom left and right tubes can connect to either the outdoor or indoor coil. In warmer climates, the **left tube** connects to the indoor coil and the **right tube** connects to the outdoor coil. In colder climates, the **left tube** connects to the outdoor coil and the **right tube** connects to the indoor coil.

A heat pump system must have a pressure difference of 75 psig or higher between the high and low side of the system for the **reversing valve** to operate properly.

A **leaking reversing valve** can be the cause of high suction pressures during the heating mode.

To check if a **reversing valve leaks,** a temperature measurement at all four pipe connections should be performed.

When the manifold gauges read **lower than normal head pressure** and **higher than normal suction pressure,** the cause may be that the reversing valve is stuck in the halfway position.

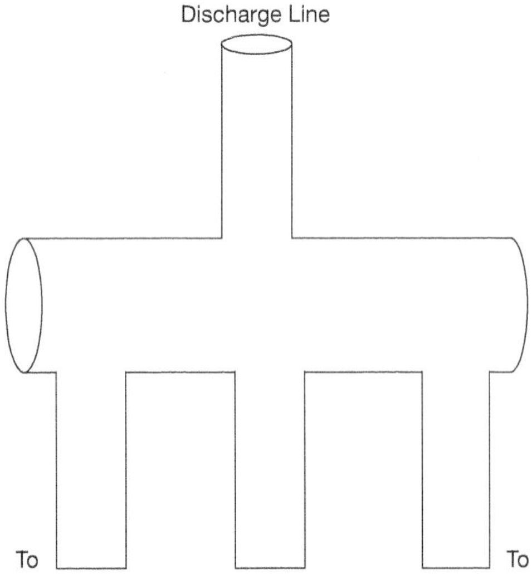

FIGURE 1–20
Reversing Valve

The **vapor line** is the larger line that connects the outdoor coil to the indoor coil.

During the cooling cycle, the **vapor line** carries a low-pressure/low-temperature refrigerant gas from the indoor coil to the outdoor compressor.

During the heating cycle, the **vapor line** carries a high-pressure/high-temperature gas from the compressor to the indoor coil.

The **liquid line** is the smaller line that connects the outdoor coil to the indoor coil.

During the cooling cycle, the **liquid line** carries a high-pressure/high-temperature liquid refrigerant from the outdoor coil to the indoor coil.

During the heating cycle, the **liquid line** carries a high-pressure/high-temperature liquid refrigerant from the indoor coil to the outdoor coil.

The indoor and outdoor coils have one **check valve** and one metering device. The check valves are connected in parallel with one metering device and in series with the other.

A **ball-type check valve** must be installed vertically with its seat located on the bottom. The "seat" refers to where the ball covers the valve opening. When flow stops, the ball will rest over the opening, therefore closing the valve. When soldering a **ball-type check valve** into the system, use a heat sink to prevent warping.

If **check valves** are stuck in the open position, the heat pump will not operate properly—the refrigerant will bypass the metering device.

An oversized piston orifice metering device at the indoor coil will cause **low head pressure** and **high suction pressure** in the cooling mode.

An **accumulator** is commonly located between the bottom center tube on the reversing valve and the inlet to the compressor. The accumulator will prevent liquid from entering the compressor.

When the system is returned to the heating mode, the **liquid refrigerant** is typically found in the vapor line immediately after the defrost cycle.

A **defrost relay** is used in conjunction with an electrical defrost system to achieve the defrost function.

A proper **defrost function** must have the following qualifications:

1. The **reversing valve** must be positioned for the cooling mode.
2. The outdoor coil's **fan** must be de-energized.
3. The system must remain in defrost until the **outdoor coil** has reached a predetermined temperature.
4. **Auxiliary strip heaters** must be energized to increase supply air temperature.

The heat pump commonly develops frost during the heating cycle. **Frost builds up** on the outdoor coil.

A **defrost termination control** is used to de-energize the defrost cycle when the outdoor coil is free of frost.

The **defrost termination thermostat** is commonly located on the outdoor coil and opens at a preset temperature of 55°F.

Four common types of **defrost systems** are temperature initiation–temperature termination; time initiation–temperature termination; static pressure initiation–temperature termination; and static pressure/time initiation–temperature/time termination.

A **temperature initiation–temperature termination** defrost system senses the temperature difference between the outdoor ambient air and the outdoor coil.

A **time initiation–temperature termination** defrost system uses a defrost timer to initiate the defrost cycle and a temperature sensor located on the outdoor coil to terminate the defrost cycle.

A **pressure initiation–temperature termination** defrost system uses a pressure switch that measures a pressure difference across the outdoor coil to initiate the defrost system and a temperature sensor located on the outdoor coil to terminate the defrost cycle.

The **pressure switch** commonly closes at a preset pressure difference of .5″ to .65″ WC.

A **pressure/time initiation–temperature/time termination** defrost system uses a defrost timer and a pressure switch to initiate the defrost system and a temperature sensor located on the outdoor coil to terminate the defrost cycle.

Temperature split is the temperature difference between the air entering a coil and the temperature of the coil.

An **outdoor ambient thermostat** (OAT) is used to energize the auxiliary heat operation if the outdoor ambient temperature drops below a predetermined temperature.

A **hold-back thermostat** is located on the outdoor coil of a heat pump.

A **hold-back thermostat** is used to prevent ice buildup on an outdoor coil during the heating cycle.

A **hold-back thermostat** senses the outdoor ambient temperature. It is used to keep the system energized during a defrost cycle if the room thermostat becomes satisfied.

A **discharge temperature thermostat** is located on the indoor coil of an air-to-air heat pump. It is used on a system that incorporates fossil fuel as a means of auxiliary heat.

A **discharge temperature thermostat** de-energizes the fossil fuel system when the discharge temperature reaches 95°F during the defrost cycle.

A heat pump must always be **sized** for the cooling load of the building.

The proper amount of **auxiliary heat** must be installed to handle the total heating requirement of the building.

Auxiliary strip heaters are found in heat pumps to aid in the heating process during low outdoor ambient temperatures.

Auxiliary strip heaters are used during the defrost cycle to increase the temperature of the supply air.

The **auxiliary strip heaters** are located at the discharge side of the blower and downstream from the indoor coil on an air-to-air heat pump. They are electrically controlled by the second stage of a two-stage thermostat.

The **de-ice coil** is the lower section of the outdoor coil. It is used during the cooling mode to increase the subcooling of the refrigerant.

The **de-ice coil** prevents the formation of ice buildup on the bottom of the outdoor coil during the heating mode.

The **coefficient of performance (COP)** is a heat pump's rating in the heating mode. It is determined by comparing the total amount of heat produced (heat output) to the amount of electrical energy used (electrical input).

When installing an **outdoor coil** on an air-to-air heat pump, the unit should be located as close as possible to the indoor coil. The length of the vapor line is critical when the system is in the heating mode. Longer lines are less efficient, while shorter lines are more efficient.

An **outdoor unit** must not be located under an overhang. This could cause recirculation of air and decrease the efficiency of the system.

When installing the **outdoor coil,** a gravel foundation should be placed below the pad, or base, the unit will rest on.

Refrigeration lines must be kept as short as possible to prevent pressure drop, heat loss, or heat gain through the line set.

The **vapor line** must be insulated to decrease refrigerant temperature rise and exterior moisture condensation during the cooling mode. It will also prevent refrigerant temperature drop during the heating mode.

Common **static pressures** in the design of a heat pump duct system are .15″ WC for the supply duct and .05″ WC for the return duct.

A common **temperature rise** over the indoor coil during the heating mode is 35°F.

For service accessibility, a minimum of a **20″ clearance** is commonly required for the indoor unit.

A **1″ airspace** is required between the supply air duct and any combustible material.

When installing an indoor downflow-type unit, a **noncombustible floor base** is required between the floor and the system.

On horizontal heat pump installations, an **auxiliary drain pan** must be located under the entire unit.

When an indoor coil is installed on the negative side of the blower, air is pulled over the coil. A **"P" trap** is mandatory on the drain line to prevent condensation from entering the duct system.

A **secondary drain line** should extend from the auxiliary drain pan to a conspicuous place. This serves as an alarm, indicating that the primary drain is restricted.

An **electrical disconnect** must be installed at the indoor unit if circuit breakers are not a part of the unit.

A standard heating and cooling **thermostat** cannot be used for a heat pump system.

An air-to-air heat pump uses an **8-strand, 18-gauge** wire for the control circuit.

The **"Y1" terminal** on a heat pump thermostat controls the reversing valve.

The **"W1" terminal** on a heat pump thermostat controls the compressor motor operation.

The **"W2" terminal** on a heat pump thermostat controls the auxiliary heat circuit.

The **"R" terminal** on the heat pump thermostat is the power supply from the transformer.

The **"C" terminal** on the heat pump thermostat is the 24-volt common from the transformer.

The **"G" terminal** on the heat pump thermostat controls the fan relay coil.

The **"X" terminal** on the heat pump thermostat controls the trouble light at the thermostat.

The **"E" terminal** on the heat pump thermostat controls the emergency heat light at the thermostat.

Figure 1–21 is a typical air-to-air heat pump control wiring diagram.

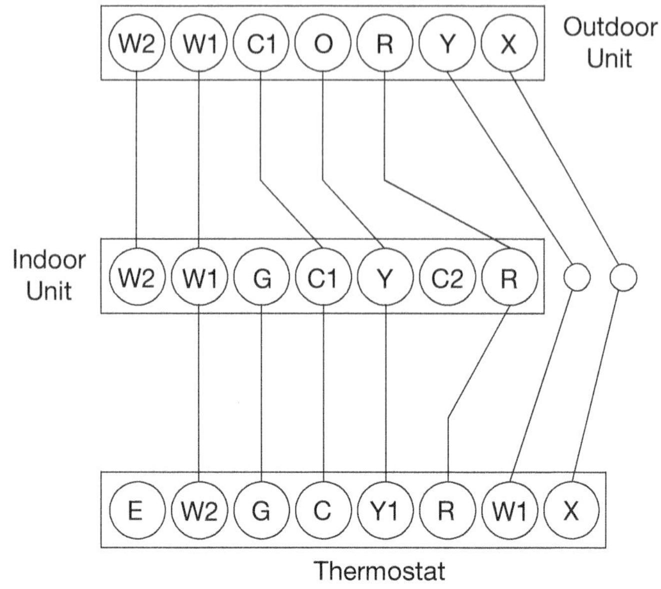

FIGURE 1–21

SECTION 2

Gas and Oil Forced-Air and Hot Water Heating Systems

FUNDAMENTALS OF GAS COMBUSTION

It is of vital importance that technicians have a knowledge of the fundamentals of gas combustion. The following are basic related terms.

Two common **types of gases** used today are natural (methane) and LPG (liquified petroleum gas).

Air has a **specific gravity** of 1.0.

Natural gas has a **specific gravity** of 0.6. It is lighter than air.

LPG has a **specific gravity** of 1.5. It is heavier than air.

Natural gas has a heating value of approximately 1,000 Btu per cubic foot.

LPG has a heating value of approximately 2,500 Btu per cubic foot.

Combustion is a chemical reaction, or the rapid oxidation of fuel that results in the release of heat and light.

The requirements for proper combustion are **fuel, oxygen,** and **ignition.**

An **oxygen molecule** is 2 oxygen atoms (O_2).

For **complete combustion,** 1 part methane (CH_4) is combined with 2 parts oxygen ($2O_2$).

A **carbon dioxide molecule** is 1 carbon atom combined with 2 oxygen atoms (CO_2).

A **water vapor** molecule is 2 hydrogen atoms combined with 1 oxygen atom (H_2O).

During **complete combustion,** (CH_4) + ($2O_2$) will change to (CO_2) + 2(H_2O) + HEAT.

The percentage of CO_2 in the flue gases for natural gas should be from 7 to 9%.

Air is approximately 21% oxygen and 79% nitrogen.

The proper air and fuel mixture for **complete combustion** is 1 part fuel and 10 parts air. An additional 5 parts of **excess air** is added to the combustion air to ensure complete combustion.

Carbon monoxide (CO) is a deadly gas caused by incomplete combustion and flame impingement on a cold surface of a heat exchanger.

No more than 400 ppm of **CO** is allowed in furnace flue gases air-free measurement and 225 ppm with excess air included.

Air-free measurements are flue gases minus excess air.

The maximum allowable concentration for continuous 24-hour exposure of CO according to ASHRAE is .009 ppm.

CO measurements should be taken in the vent pipe leaving the furnace, around the burner area and around the draft diverter.

A **natural gas** (methane) molecule is 1 carbon atom combined with 4 hydrogen atoms (CH_4).

Heat transfer is accomplished in three ways: conduction, convection, and radiation.

Thermodynamics is a science that deals with the relationship of heat and mechanical energy.

GAS FURNACE CONTROLS AND COMPONENTS

Heating service technicians must know the function of numerous controls and components. Technicians must also understand the ignition systems that contain many of these components.

A **standing pilot furnace** uses a continuous pilot designed to ignite the gas at the main burners of the furnace.

A **standing pilot furnace** uses a thermocouple or a thermopile to prove the existence of a pilot.

A **standing pilot ignition system** consists of a main gas valve, a pressure regulator, a pilot assembly, and a thermocouple.

A **thermocouple** is made of two dissimilar metals and is part of the pilot assembly.

The **thermopile** is a pilot generator rated at a higher millivolt rating than a thermocouple.

The **pilot flame** should engulf the thermocouple tip by $\frac{3}{8}''$ to $\frac{1}{2}''$.

On a standing pilot furnace, the pilot is proven by a **30-millivolt** signal from the thermocouple to the gas valve.

A **thermopile** millivolt system will range from 550 to 750 millivolts.

An **intermittent pilot furnace** uses an ignition system that ignites the pilot on every cycle of the furnace. The pilot ignition system is commonly a spark or a hot surface type.

A **spark-type ignition system** uses an ignition transformer, a cable, and an electrode to produce the spark for the pilot.

A **hot-surface-type ignition system** controls current through a silicon carbide ignitor that increases the temperature above the ignition point of the fuel. Its operation is much quieter than a spark ignition system.

In a **hot-surface ignition system,** the components consist of an ignition control module, a combination gas valve, a flame sensor, and a hot surface igniter.

Most **hot surface igniters** operate on 120 volts. Igniters in some ignition systems have recently started operating on 24 volts.

In a **direct ignition system,** the components consist of a combination gas valve, ignition control module, and an igniter/sensor.

A **direct ignition furnace** uses an ignition system that ignites the main burners on every cycle of the furnace. This furnace has no pilot.

The **direct ignition system** is commonly a spark or a hot-surface type.

The **voltage** created by an ignition control module commonly ranges from 7,000 to 15,000 volts.

A **spark gap,** used on a spark ignition system, is commonly $\frac{1}{8}''$.

The **pilot** or **main burner flame** should engulf the tip of the rod on an igniter/sensor by $\frac{3}{8}''$ to $\frac{1}{2}''$.

All electrical connections to any heating appliance should be done in compliance with the **National Electrical Code.**

All furnaces must have their own separate **120-volt circuit.**

The four **furnace designs** are downflow/counterflow, upflow, horizontal, and multipositional.

The **combination gas valve** consists of a pilot solenoid, a main valve solenoid, a pressure regulator, a main shutoff, and a pilot adjustment.

The **pilot solenoid** controls the gas flow to the pilot.

The **main gas valve solenoid** controls gas flow to the main burners.

WC refers to water column. It is an instrument in which water is placed. The column has increments in inches. It is used for measuring low pressure.

The **pressure regulator** is adjustable and controls the amount of gas pressure to the main burners. It commonly reduces the gas pressure of 7″ WC to 3.5″ WC for natural gas. If the regulator is turned clockwise, the pressure will increase.

A **burner orifice** is a drilled, threaded port located in the gas manifold. It directs a stream of gas to the burners.

Primary air is air that is needed for complete combustion. It is introduced and mixed with the gas before the point of ignition.

Too much **primary air** causes the flame to lift off the burner ports.

Not enough **primary air** causes insufficient combustion and produces carbon monoxide.

Secondary air is air that ensures complete combustion. It surrounds the flame at the point of ignition.

A **burner** is used to mix combustion air with fuel. Then it directs the mixture to the combustion chamber or heat exchanger for ignition. The burner must be installed properly or a noisy ignition could result.

The **venturi** is part of the burner in which air and fuel is mixed.

A **draft diverter** allows the air within the furnace location to blend with the products of combustion.

Ported, slotted, ribbon, or inshot are the most common types of **gas burner designs.**

A **fan control** starts and stops the blower during the furnace operation. It commonly delays on start and stop.

There are **three common types of fan controls:** a thermal-activated type, a 24-volt timed-delay type, and a solid-state circuit-board type.

The **thermal-activated fan control** uses a bimetal to sense the temperature of the heat exchanger.

The **timed-delay fan control** uses an electric heater and a bimetal fan switch to cycle the blower motor.

A **defective fan control** stuck open prevents the indoor blower from operating, which causes the heat exchanger to overheat.

The **circuit-board fan control** uses a timed-delay feature that is built into the solid-state circuit board.

A **limit switch** de-energizes the gas valve if the heat exchanger exceeds a predetermined temperature, commonly 180 to 200°F.

A **pressure switch** proves the operation of the induced draft motor.

A **door interlock switch** is used to disconnect all electrical power supply to the furnace heating and cooling circuits when the blower door is moved.

A normally open **float switch** senses the level of water in a condensate tank and starts a condensate pump.

A **flame roll-out switch,** or fusible link, de-energizes the gas valve if the flame from the burners is pushed out of the heat exchanger. This is commonly caused by a plugged heat exchanger.

A **heat exchanger** safely transfers the heat created by combustion to air. The air is moved over the heat exchanger by the blower.

A primary heat exchanger is commonly constructed of **cold rolled** or **stainless steel.**

The joints of a heat exchanger are usually **welded** or **mechanically crimped.**

In a condensing furnace, the products of combustion from a primary heat exchanger are directed to a **secondary heat exchanger.** The gases are condensed to increase the efficiency of the furnace. The secondary heat exchanger is located directly in front of the blower outlet.

An **induced draft blower** is used to move products of combustion out of a heat exchanger and into a vent. It is located at the outlet of a heat exchanger.

An **induced draft blower** creates a positive pressure in the venting system of a condensing furnace and a negative pressure in the venting system of a noncondensing furnace.

An **auxiliary limit control** is used as a backup to the main limit control. It is most commonly found on downflow and multipositional furnaces.

A **vent safety control** is used to de-energize the gas valve if the vent temperature increases above a predetermined temperature.

A **flame proving device** proves the presence of the pilot or main burner flame.

A **thermocouple** proves a pilot flame by changing heat energy to electrical energy. It operates at 30 millivolts.

A **flame rod sensor** uses a flame rectification system to prove the presence of the pilot or main burner flame. Common microamp signals range from .2 to 6 μA DC.

A **bimetal flame sensor** proves the presence of a pilot flame. It does so by using the heat from the pilot to switch a set of contacts and energize the main gas valve.

A **mercury flame sensor** proves the presence of a pilot flame. It does so by using the heat from the pilot to switch a set of contacts and energize the main gas valve.

A **gas valve** controls the flow of gas to the main burners.

A **two-stage** gas valve operates in low fire during mild outdoor temperatures and high fire during colder outdoor temperatures.

A **two-stage** gas valve typically uses a two-stage thermostat: W1 controls low fire and W2 controls high fire.

The first-stage heat on a **two-stage** gas valve produces a lower capacity, which will result in a reduction of Btu output with longer burner cycles.

A high-efficiency condensing furnace's **heat exchanger** can be tested by blocking off the intake and exhaust air openings and setting the thermostats fan selector switch to "On." A manometer will then be connected to the hose leading to the pressure switch and if a positive reading is obtained, the furnace has a leak in the heat exchanger.

The Operational Sequence of a High-Efficiency Indirect Ignition Furnace Is as Follows:

On a call for heat, the thermostat closes between the "R" and "W" terminals. The induced draft motor is energized, and this causes the normally open pressure switch to close. If the limit control is in the closed position, the ignition module is energized. The ignition system either sparks or energizes a hot surface igniter. The pilot valve is energized and the flame is proved by the pilot safety circuit. The main gas valve is then energized and the main burners are ignited. After approximately 30 to 60 s, the indoor blower is energized by the fan control circuit. The system will operate until the thermostat is satisfied and the "R" and "W" contacts open, which terminates the need for heat. The induced draft motor, main burner, and pilot circuits are de-energized. The indoor blower will remain in operation until a predetermined time or temperature. After this period, the indoor blower will de-energize.

COMBUSTION AIR

All fossil fuel equipment needs air for the combustion process to occur. The quantity of combustion air varies with the furnace size and type. The following are terms and information related to combustion air.

Combustion air is the air that is used for the combustion, dilution, and ventilation process of a furnace.

Dilution air is the air surrounding a furnace location. When this air is induced into the diverter, it becomes diluted with the products of combustion (exhaust gases) leaving the chimney.

Infiltration is outdoor air that enters into a home around loose-fitting doors and windows.

Infiltration is measured in air changes per hour.

Infiltration rates:

Tight construction = .3 to .6 air changes per hour

Average construction = .6 to 1.0 air changes per hour

Loose construction = 1.0 to 2.5 air changes per hour

Gas-fired appliances use a **15:1 ratio** of gas and combustion air mixture. Every 1 cu ft of gas requires 15 cu ft of combustion air.

Natural draft appliances use a **14:1 ratio** of gas and dilution air mixture. Every 1 cu ft of gas requires 14 cu ft of dilution air.

Fan-assisted draft furnaces do not use dilution air.

Combustion air needs to be free from household chemicals, or corrosion of the heat exchanger and vent system could occur.

Gas appliances can be located in unconfined or confined spaces.

An **unconfined space** is a space with a volume that is *greater than 50 cu ft per 1,000 Btu/h* of the total input of all gas-fired appliances.

Example: If a 150,000 Btu/h furnace is located in a basement with a 40,000 Btu/h water heater, the total input is 150,000 + 40,000 = 190,000 Btu/h. To determine if the appliances are in an unconfined area, use this formula: (190,000 Btu/h)/(1,000 Btu/h) × (50 cu ft) = 9,500 cu ft. If the volume of the basement is greater than 9,500 cu ft, then it is considered an unconfined space.

Combustion air can be taken from the unconfined space if the construction of the home is loose.

Combustion air must be provided from the outdoors to the unconfined space in tight-constructed homes.

When introducing **outdoor combustion air** into an unconfined space, two equal openings must be provided—one at or within 12″ of the ceiling and the other at or within 12″ of the floor.

Each **opening** used to introduce combustion air into an *unconfined space* must have the total free area of at least *1 sq in. per 4,000 Btu/h input* and *not less than 100 sq in.*

Example: An 80,000-Btu/h furnace and a 40,000-Btu/h water heater are installed in an unconfined space of a loose-constructed home. The total input is (120,000 Btu/h)/(4,000 Btu/h) = 30 sq in. per opening. Because the minimum opening is 100 sq in., each opening must be at least 100 sq in.

When **horizontal ducts** are used to introduce outdoor air to the unconfined space, the openings must be *1 sq in. for every 2,000 Btu/h* input and *not less than 100 sq in.*

Louvers and grills reduce the percentage of free area in a combustion air opening. This must be taken into consideration when introducing combustion air into a confined or unconfined space.

Example: A 10″ × 10″ grill with the free area of 70% is used for combustion air. 10″ × 10″ = (100 sq in. × .7). The total free area of the opening is reduced to 70 sq in.

A **confined space** is a space with a volume *less than 50 cu ft per 1,000 Btu/h* of the total input of all gas-fired appliances.

Combustion air must be provided from the outdoors to the confined space in tight-constructed homes.

When introducing **outdoor combustion** air into a confined space, two equal openings must be provided—one at or within 12″ of the ceiling and the other at or within 12″ of the floor.

Each **opening** used to introduce combustion air into a *confined space* must have the total free area of at least *1 sq in. per 1,000 Btu/h* total input and not *less than 100 sq in.*

> ***Example:*** An 80,000-Btu/h furnace and a 40,000-Btu/h water heater are installed in a confined space of a loose-constructed home. The total input is $(120{,}000\ \text{Btu/h})/(1{,}000\ \text{Btu/h}) = 120$ sq in. Each opening must be at least 120 sq in.

When **horizontal ducts** are used to introduce outdoor air to the *confined space*, the openings must be *1 sq in. for every 2,000 Btu/h* total input and *not less than 100 sq in.*

The minimum dimension of a **horizontal duct** must not be less than 3″.

When *a* **vertical duct** is used to introduce outdoor combustion air into a *confined space,* the opening must be *1 sq in. for every 4,000 Btu/h* total input and *not less than 100 sq in.*

When a **vertical duct** is used to introduce combustion air into a confined space, it must terminate 1′ above the floor.

It is extremely important to completely separate the furnace **supply and return air** from the **combustion air**, or the negative pressure from the blower could cause products of combustion to mix within the living space of the home.

Direct vent, high-efficiency condensing furnaces must be installed according to the manufacturer's recommendation.

Many **direct vent,** high-efficiency condensing furnaces allow for a vertical or horizontal venting system. The vent pipe is commonly schedule 40 PVC.

When installing combustion air for a **direct vent,** high-efficiency condensing furnace, support the pipe every 5′, insulate the pipe if necessary, and have an airtight seal.

The **combustion air** and **vent pipes** used on a direct vent, high-efficiency condensing furnace must be located near the same area to prevent the system from operating under different air-pressure zones. If they are too close to each other, it could cause the furnace operation to cycle on the low-pressure switch.

VENTS AND CHIMNEYS

Vents and chimneys are the passageways for combustion gases. It is important for the service technician and installer to be knowledgeable of the types of vents and chimneys and their related terms.

Venting is the removal of combustion gases through a vent system.

A **vent** is the passageway that transports combustion gases to the atmosphere.

A **vent connector** is the horizontal or lateral section of pipe that connects the appliance to a vertical vent or chimney.

Combustion gases are commonly composed of carbon dioxide (CO_2), hydrogen (H), oxygen (O_2), and nitrogen (N).

Incomplete combustion and flame impingement on a cold surface create **carbon monoxide** as a product of combustion.

Buoyancy is the moving of gases through a vent system due to high temperatures.

Buoyancy increases as combustion gas temperature increases.

The temperature difference between the ambient air and the flue gases creates a **density difference**. This difference helps move flue gases through the vent system on a natural draft furnace.

The vent height and the temperature difference between the flue gases and the ambient air create a negative **static pressure** in a category 1 vent system.

Heat loss through a vent minimizes the **static pressure** required for proper venting.

Category 1 operates at a negative pressure, at least 140°F above flue gas dewpoint temperature, and depends on a natural draft to remove the flue gases.

Category 2 operates at a negative pressure, less than 140°F above flue gas dewpoint temperature, and depends on a natural draft to remove the flue gases.

Category 3 operates at a positive pressure, at least 140°F above flue gas dewpoint temperature, and requires a fan-powered airtight vent to remove the flue gases.

Category 4 operates at a positive pressure, below 140°F, and requires a fan-powered airtight vent to remove the flue gases.

When induced draft noncondensing and natural-draft-type furnaces are used, a **category 1** venting system is most common.

The **vent system** that is used for an induced draft condensing furnace is a category-4-type vent system. Schedule 40 PVC is a common material used for high-efficiency furnace venting. Always use the manufacturer's recommendation.

"Wet time" is a period at which vent gases may reach a dewpoint temperature and condense within the vent system.

"Wet time" refers to the time that it takes a vent system to warm up. This happens at the beginning of a furnace cycle. This time should be less than 2 min.

In a natural draft furnace, **dilution air** decreases the dewpoint temperature of the flue gases, which decreases the possibility of condensation.

Because there is no **dilution air** used with an induced draft noncondensing furnace, it has more problems with condensation.

A double wall **type "B" vent** reduces the heat loss through a vent system and aids in keeping flue gases from reaching their dewpoint temperature.

To resist corrosion, the inner wall of a **type "B" vent** is made of aluminum.

The outer wall of a **type "B" vent** is made of galvanized steel.

An airspace located between the inner and outer wall on a **type "B" vent** will decrease heat loss through the vent by 50% when compared to standard single wall vent pipe.

A **type "B" vent** must be used when passing through floors, ceilings, or an unconditioned space.

A **firestop**, made of noncombustible material, must be used when vent pipe passes through floors.

A **type "B" vent** must have a 1″ to 3″ clearance from combustibles.

Standard **single wall vent** pipes must have a 6″ clearance from combustibles.

A **vent connector** with a diameter of 5″ or less must be at least 26-gauge galvanized metal.

A **vent connector** with a diameter from 6″ to 9″ must be at least 24-gauge galvanized metal.

A **vent connector** with a diameter from 10″ to 16″ must be at least 22-gauge galvanized metal.

A **vent connector** with a diameter that is greater than 16″ must be at least 16-gauge galvanized metal.

All **vent connectors** entering a masonry chimney must be cemented to the chimney with an approved cement.

A **vent connector** must have a vertical rise of at least $\frac{1}{4}''$ per 1′ of horizontal run.

> ***Example:*** 10′ of pipe must have a 2.5″ vertical pitch. (.25″ × 10′ = 2.5″)

A single wall uninsulated **horizontal vent connector** must not exceed $1\frac{1}{2}'$ per inch diameter of vent pipe.

One **90-degree elbow** has the resistance of 20 diameters in length of the vent being used.

> ***Example:*** One 6″ elbow has the same resistance as
>
> 1. 6″ × 20 = 120″
> 2. 120″/12″ = 10′

The following problems cause **condensation** in an induced draft noncondensing furnace:

1. Too low of a temperature rise across the heat exchanger caused by gas pressure adjustment set too low or blower speed adjustment set too high.
2. Too large of a venting system.
3. An oversized furnace or heat anticipator set too low, which causes short cycling.
4. Too long of a single wall vent horizontal run.
5. Too many 90-degree elbows.

The following problems cause **pressurization** in an induced draft noncondensing furnace:

1. Too small of a vent pipe.
2. Too high of a gas pressure.
3. Too long of a single wall vent horizontal run.
4. Too many 90-degree elbows.

When installing a **main vent,** a straight pipe or two 45-degree elbows are acceptable, but never use two 90-degree elbows.

A **main vent** must terminate a minimum of 5′ above the nearest connected appliance vent hood.

The minimum height of a **main vent** is 5′ from the nearest connected appliance vent hood.

The height of the **main vent** depends on the pitch of the roof.

If a main vent is located near a vertical wall, the vent height should be not less than 2′ above the wall within 10′ away from the wall.

When installing the **vent pipe** of a direct vent high-efficiency condensing furnace, the pipe should be pitched toward the furnace $\frac{1}{4}$ of an inch per linear foot of vent pipe used.

The **vent pipe** of a direct vent high-efficiency condensing furnace should have an air-tight seal and be supported every 5′ of pipe length.

All chimneys or vents must have a **minimum area** equal to the area of the furnace connection and a **maximum area** of no more than two sizes greater than the vent diameter.

A **masonry chimney** must be built to building code standards.

A **masonry chimney** commonly uses clay tile liners and brick with a refractory mortar.

A **masonry chimney** that is designed for a gas appliance must use a clay tile liner.

A **masonry chimney** must extend 2′ minimum height and not less than 2′ above any part of a building. It must be within 10′ of a chimney.

A **chimney liner** must be used when installing an induced draft noncondensing furnace.

The space surrounding a **chimney liner** in a masonry chimney cannot be used for venting.

Vent tables (see Figure 2–1) are used to properly vent a category 1 heating appliance. Category 2, 3, and 4 appliances must be vented to the manufacturer's specifications.

A **vent table** will help a technician to determine

1. the type of vent connector that can be used (single wall or double wall).
2. the size of vent pipe from the appliance to the main vent.
3. the size that the main vent should be to properly vent the appliances that are being used.

Single wall vent connectors are commonly used with lower-efficiency furnaces that use dilution air.

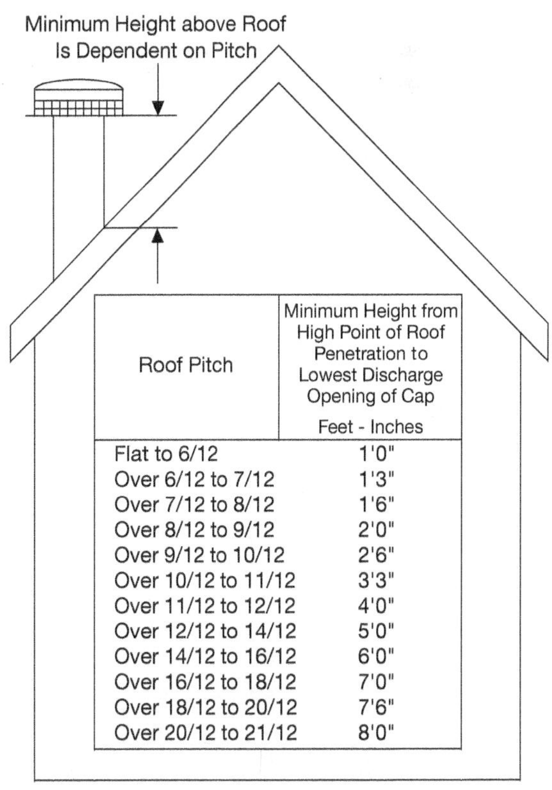

Roof Pitch	Minimum Height from High Point of Roof Penetration to Lowest Discharge Opening of Cap Feet - Inches
Flat to 6/12	1'0"
Over 6/12 to 7/12	1'3"
Over 7/12 to 8/12	1'6"
Over 8/12 to 9/12	2'0"
Over 9/12 to 10/12	2'6"
Over 10/12 to 11/12	3'3"
Over 11/12 to 12/12	4'0"
Over 12/12 to 14/12	5'0"
Over 14/12 to 16/12	6'0"
Over 16/12 to 18/12	7'0"
Over 18/12 to 20/12	7'6"
Over 20/12 to 21/12	8'0"

FIGURE 2–1

Double wall type "B" vent pipe is used as the main vent between ceilings, floors, and unconditioned spaces.

Double wall type "B" vent pipe is commonly used as the vent connector in induced draft noncondensing furnaces.

The following are some typical venting applications:

A type "B" double wall vent with a type "B" double wall vent connector serving one category 1 appliance.

A type "B" double wall vent with a single wall vent connector serving one category 1 appliance.

A type "B" double wall vent with a type "B" double wall vent connector serving two or more category 1 appliances.

A type "B" double wall vent with a single wall vent connector serving two or more category 1 appliances.

To use a **vent table,** determine the table that matches the type of vent system that is being installed. When vent sizing a single appliance to a single vent system, the procedures are different from common venting.

Use the following procedures to size a vent system using a single appliance with a single vent system:

1. Determine the table that is to be used. (Is the vent connector going to be single wall or double wall? Is a vent or a chimney going to be used?)
2. Determine the height of the vent or chimney being used.
3. Determine the lateral distance the appliance will be from the vent.
4. Using the proper vent table, determine the vent connector type and size.

Example: An induced draft noncondensing 80,000-Btu/h furnace is to be installed in a home that uses a 15′ double wall type "B" vent. The furnace will be 5′ from the vent.

1. Use Table 2–1. This table shows the capacity of type "B" double wall vents with single wall metal connectors serving a single catagory 1 appliance.
2. Follow the chart and find the 15′ *vent height* that is being used.
3. Since the furnace is going to be 5′ *from the vent,* find the 5′ column for the lateral vent connector.
4. In a fan-assisted draft furnace, follow the 5′ *lateral* to the point where the 80 (80,000 Btu/h) falls between the MIN/MAX numbers.
5. Follow this column up to the top of the chart to determine the proper size for the vent.
6. Since a 4″ *single wall vent* is found on the chart, it can be used.

Common venting is a term that is used when connecting two or more appliances to one vent system.

When **common venting** two appliances:

1. The smaller appliance with lower Btu rating should connect as high as possible above the larger appliance with greater Btu rating.
2. The smaller appliance should have a shorter horizontal run than the larger appliance.
3. The smaller appliance must never have more elbows than the larger appliance.

TABLE 2–1 Capacity of Type "B" Double Wall Vents with Single Wall Metal Connectors Serving a Single Category 1 Appliance

VENT AND CONNECTOR DIAMETER IN INCHES

Height	Lateral	3″			4″			5″			6″		
H (ft)	L (ft)	FAN MIN	FAN MAX	NAT MAX	FAN MIN	FAN MAX	NAT MAX	FAN MIN	FAN MAX	NAT MAX	FAN MIN	FAN MAX	NAT MAX
6	0	38	77	45	59	151	85	85	249	140	126	373	204
	2	39	51	36	60	96	66	85	156	104	123	231	156
	4	NR	NR	33	74	92	63	102	152	102	146	225	152
	6	NR	NR	31	83	89	60	114	147	99	163	220	148
8	0	37	83	50	58	164	93	83	273	154	123	412	234
	2	39	56	39	59	108	75	83	176	119	121	261	179
	5	NR	NR	37	77	102	69	107	168	114	151	252	171
	8	NR	NR	33	90	95	64	122	161	107	175	243	163
10	0	37	87	53	57	174	99	82	293	165	120	444	254
	2	39	61	41	59	117	80	82	193	128	119	287	194
	5	52	56	39	76	111	76	105	185	122	148	277	186
	10	NR	NR	34	97	100	68	132	171	112	188	261	171
15	0	36	93	57	56	190	111	80	325	186	116	499	283
	2	38	69	47	57	136	93	80	225	149	115	337	224
	5	51	63	44	75	128	86	102	216	140	144	326	217
	10	NR	NR	39	95	116	79	128	201	131	182	308	203
	15	NR	NR	NR	NR	NR	72	158	186	124	220	290	192

Appliance Input Rating in Thousands of Btu/h

Use the following procedures to size a vent system using two or more appliances with a single vent:

1. Determine the table that is to be used. (Is the vent connector single wall or double wall? Will a vent or a chimney be used?)
2. Determine the height of the vent or chimney being used.
3. Determine the type of venting that each common vented appliance will use. The options are natural or fan assisted.

NAT = Natural Draft

FAN = Fan Assisted

There are three different combinations: FAN + FAN, FAN + NAT, NAT + NAT.

1. Determine the height of the vent connector rise for each individual appliance. (Common vent tables assume that the connector length is no more than $1\frac{1}{2}'$ per inch of vent diameter.)
2. Using the vent connector capacity chart, determine the vent connector type and size for each appliance.
3. Using the common vent capacity chart, determine the main vent size.

Example: An induced draft noncondensing 85,000-Btu/h furnace with a 3′ vent connector rise is to be common vented with a 40,000-Btu/h natural draft water heater with a 2′ vent connector rise. The home has an existing 5″ double wall type "B" main vent that is 20′ high.

1. Use the *vent connector capacity chart* (Table 2–2). This table shows the capacity of a type "B" double wall vent with a type "B" double wall vent connector serving two or more category 1 appliances.
2. To size the vent connector for the 40,000-Btu/h water heater, use the *vent connector capacity chart* and follow the column on the chart that uses a *20′ vent height.* Using a *2′ connector rise,* follow the column over until it reads 66 or 66,000 Btu/h. By following the column up, one sees that a natural draft appliance that is less than 66,000 Btu/h can use a *4″ vent connector.*
3. To size the vent connector for the 85,000-Btu/h induced draft noncondensing furnace, use the vent connector capacity chart and find the column on the chart that uses a *20′* vent height. Using a *3′ connector rise,* follow the column over until it reads a MIN of 35 or 35,000 Btu/h and a MAX of 110 or 110,000 Btu/h. By following the column up, one sees that an induced draft appliance that is 85,000 Btu/h can use a *4′ vent connector.*
4. To size the main vent, first add the total input of all appliances that are to be common vented. In this case, the total input is 125,000 Btu/h. Use the *common vent capacity chart* (Table 2–3) and find the column on the chart that uses a *20′ vent height.* Since one appliance is a fan-assisted induced draft and the other appliance is a natural draft appliance, the *FAN + NAT column* will be selected. Follow the column over until 183 or 183,000 Btu/h is found. This is the maximum input that the main vent can handle. Follow this column up and one sees that the existing *5″* double wall "B" type vent can be used.

Terms for Table

NAT = Natural Draft

FAN = Fan Assisted

NR = Not Recommended

H = Height

L = Lateral

TABLE 2–2 Capacity of Type "B" Double Wall Vents with Type "B" Double Wall Connectors Serving Two or More Category 1 Appliances

VENT CONNECTOR CAPACITY CHART

VENT CONNECTOR DIAMETER IN INCHES

Connector Height	Rise	3″			4″			5″			6″		
H (ft)	L (ft)	FAN MIN	FAN MAX	NAT MAX	FAN MIN	FAN MAX	NAT MAX	FAN MIN	FAN MAX	NAT MAX	FAN MIN	FAN MAX	NAT MAX
8	1	22	40	27	35	72	48	49	114	76	64	176	109
	2	23	44	32	36	80	57	51	128	90	66	195	129
	3	24	47	36	37	87	64	53	139	101	67	210	145
10	1	22	43	28	34	78	50	49	123	78	65	189	113
	2	23	47	33	36	86	59	51	136	93	67	206	134
	3	24	50	37	37	92	67	52	146	104	69	220	150
15	1	21	50	30	33	89	53	47	142	83	64	220	120
	2	22	53	35	35	96	63	49	153	99	66	235	142
	3	24	55	40	36	102	71	51	163	111	68	248	160
20	1	21	54	31	33	99	56	46	157	87	62	246	125
	2	22	57	37	34	105	66	48	167	104	64	259	149
	3	23	60	42	35	110	74	50	176	116	66	271	168

Appliance Input Rating in Thousands of Btu/h

TABLE 2–3 Common Vent Diameter in Inches

COMMON VENT CAPACITY CHART

	4″			5″			6″		
Vent Height (ft)	FAN +FAN	FAN +NAT	NAT +NAT	FAN +FAN	FAN +NAT	NAT +NAT	FAN +FAN	FAN +NAT	NAT +NAT
8	101	90	73	155	129	114	224	178	163
10	110	97	79	169	141	124	243	194	178
15	125	112	91	195	164	144	283	228	206
20	136	123	102	215	183	160	314	255	229

GAS PIPING

Gas piping carries fuel gases to an appliance. It is very important for the installer to know basic terms and methods involved in gas piping. The following will help the technician/installer to better understand gas piping.

A **drip leg** should be installed on a vertical gas line to catch contaminants that may flow with the gas.

A permit is required by most municipalities to install or alter **fuel gas piping.**

A **manual gas cock valve** shutoff should be installed externally to the furnace before the union, the drip leg, and the appliance.

All **manual gas cock valves** must be approved for gas fuel only.

Gas shutoff valves should be installed within the appliance.

When gas piping supplies gas to more than one building in one area, a **separate shutoff valve** is needed for each building.

A **system shutoff valve** should be installed near the liquefied petroleum gas tank point of connection.

A **gas union** should be installed between the manual gas cock valve, the combination gas valve, and the gas valve control.

The gas supply line to a **mobile home** should be a minimum of $\frac{3}{4}''$.

The gas piping used to supply **mobile homes** should be installed at least 18″ below grade (below ground level).

Gas pipe installations should be kept at least 6″ above the grade or the structure. Examples of structures are porches, steps, breezeways, and so on.

After an installation, **gas piping** must be pressure tested to ensure that there are no leaks present.

Most **gas piping** is constructed of standard-weight wrought iron steel. It may be galvanized or black.

Gas piping that is not protected can be installed above grade in recesses or channels that are approved by a mechanical inspector.

Plastic gas piping should be at least 18″ below grade.

Leaks in gas piping can be located with a soapy bubble solution.

Never use an **open flame** to locate a gas leak.

When **liquefied petroleum gas (LPG)** is used, it must maintain a pressure of 11″ WC from the regulator at the tank to the appliance.

When using natural gas, a **gas pressure** of 7″ WC is required from the meter to the appliance.

When using natural gas, a **gas pressure** of 3.5″ WC is commonly required from the furnace to the burner.

Specific gravity refers to the relative weight of an amount of gas as compared to the weight of an equal amount of air.

The **specific gravity** of natural gas is approximately 0.65 sp. gr.

Supports for hanging gas piping should be constructed of metal straps or hooks. They should be placed at approved intervals.

Underground gas piping that is ferrous and metallic should be at least 12″ below grade.

Underground piping should be supported by compacted soil or sand.

To Properly Size a Gas Pipe System, Use the Following Procedures:

1. Sketch a diagram of the gas piping layout. Include the following:
   A. The appliance Btu/h input ratings.
   B. The distance of each appliance from the main supply line.
   C. The length of the main line and the measurements between each tee.
2. Convert the Btu/h input rating to Btu/h per cu ft. If it is natural gas, divide the Btu/h input rating by 1,000 Btu/h per cu ft.

   Example: 35,000 Btu/h input/1,000 Btu/h per cu ft = 35 cu ft.
3. Determine the distance between the gas meter and the farthest appliance. Use this distance for sizing all of the sections on the chart.
4. Starting at the farthest appliance, determine the amount of gas each section of pipe will have to carry, and size accordingly.

   Example: To solve this problem, use Figure 2–2 and Table 2–4.

   Maximum Capacity of Pipe in Cubic Feet per Hour

1. The distance between the meter and the farthest appliance (the water heater) is 55′. Because there is no entry for 55′ in Table 2–4, 60′ length will be used.
2. Total the Btu/h input ratings on all appliances in the system. 150,000 Btu/h + 75,000 Btu/h + 50,000 btu/h = 275,000 Btu/h.
3. Convert the Btu/h input rating to Btu/h per cu ft by dividing 275,000 Btu/h by 1,000 Btu/h per cu ft = 275 cu ft.

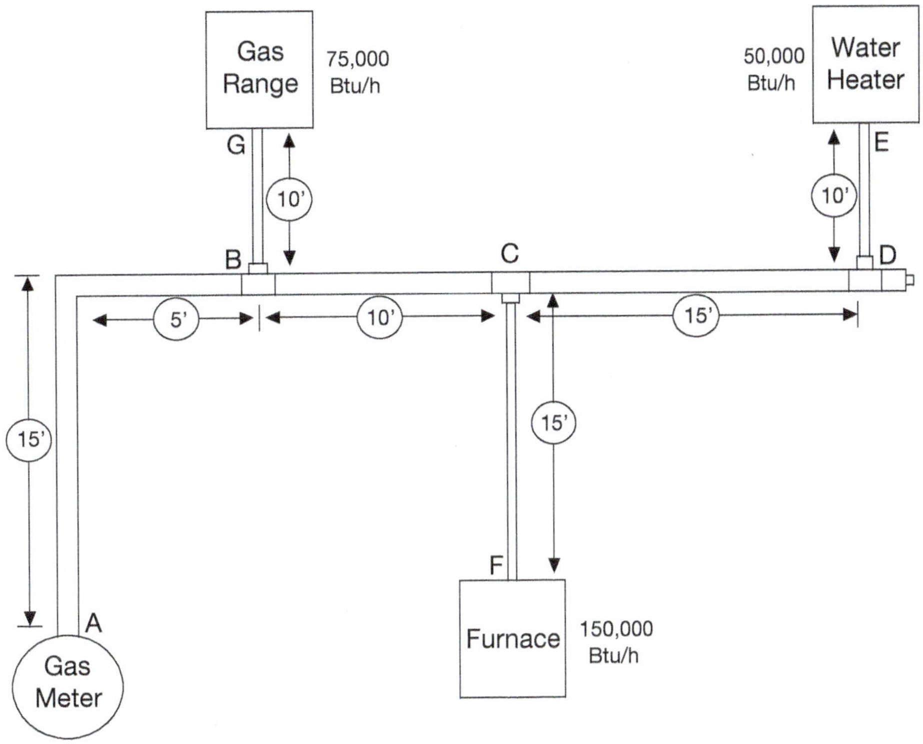

FIGURE 2–2

4. On the chart provided, find the column that uses 60′ and follow it down until it reads 275 cu ft, or the column succeeding 275 cu ft. This shows that the main gas line from points "A" to "B" will be $1\frac{1}{4}''$. The function of the line is to carry all the gas necessary to supply all appliances that will branch off from it. This line has to be larger than the other pipes in the system, because it is carrying all the initial gas from the meter.
5. Going to the farthest appliance (the water heater at 50 cu ft) and staying at the 60′ length, follow the column to 50 cu ft. Table 2–4 shows that the gas pipe section from "D" to "E" and "C" to "D" will be $\frac{1}{2}''$ in size. This occurs because only 50 cu ft will be carried by this section of pipe.

TABLE 2–4 Pipe Equivalent Chart

LENGTH OF PIPE IN FEET

Nominal Iron Pipe Size in Inches	10′	20′	30′	40′	50′	60′
$\frac{1}{4}''$	32	22	18	15	14	12
$\frac{3}{8}''$	72	49	40	34	30	27
$\frac{1}{2}''$	132	92	73	63	56	50
$\frac{3}{4}''$	278	190	152	130	115	105
1″	520	350	285	245	215	195
$1\frac{1}{4}''$	1,050	730	590	500	440	400
$1\frac{1}{2}''$	1,600	1,100	890	760	670	610

6. The next appliance is the furnace that uses 150 cu ft of gas. Staying on the 60′ length, follow Table 2–4 until it reads 150 cu ft (or the column that succeeds 150 cu ft). This column is 195 cu ft. The section of pipe from "C" to "F" will be 1″ in size.
7. Since section "B" to "C" carries gas to the furnace and the water heater, the quantity of gas from both appliances will have to be totaled to size the section. 150 cu ft + 50 cu ft = 200 cu ft. Staying on the 60′ length, follow the table down until it reads 200 cu ft (or the column that succeeds 200 cu ft). This column is 400 cu ft. The section from "B" to "C" will be $1\frac{1}{4}''$ in size.
8. The last appliance is the gas range. It uses 75 cu ft of gas. Staying on the 60′ length, follow the table down until it reads 75 cu ft (or the column that succeeds 75 cu ft). This column is 105 cu ft.

The section of pipe from "B" to "G" will be $\frac{3}{4}''$ in size.

The following is the list of gas pipe sections and their sizes.

"A" to "B" is 114″

"B" to "C" is 114″

"C" to "D" is 12″

"D" to "E" is 12″

"C" to "F" is 1″

"B" to "G" is $\frac{3}{4}''$

Carbon is what gives black iron pipe its color.

Black iron pipe is commonly used in gas piping and where it will not be affected by corrosion.

Black iron pipe is measured in inches by its **nominal size.**

Nominal size is an approximation of the inside diameter for pipe up to 12 inches.

The two (NPT) **American National Standard Pipe Threads** are tapered and straight.

Pipe joint compound applied to threads is used to provide a seal and as a lubricant when assembling.

Each joint has approximately seven perfect threads and two or more imperfect threads.

ESCO INSTITUTE-HVAC EXCELLENCE CARBON MONOXIDE SAFETY CERTIFICATION EMPLOYMENT READY (ER)

The abbreviation for **carbon monoxide** is CO.

CO is typically measured in **ppm (parts per million).**

Low concentrations of CO can compound preexisting health conditions.

When **CO** is inhaled, it bonds O_2 (oxygen) in the blood.

When CO is inhaled, it also produces **carboxyhemoglobin (COHb)** in the bloodstream.

The three symptoms of CO poisoning are headaches, sinus head pressure, and constant stuffiness.

A standard **therapy for CO poisoning** is to administer supplemental O_2 (oxygen), provide ventilator support, and monitor heart rate.

A burner requires fuel to be mixed with a specific amount of **air or O_2 in the air.**

Parts per million (ppm) references the number of CO molecules present in a million molecules of air.

Three components that must be present for combustion are **fuel, air/oxygen, and ignition.**

The **three "Ts" of combustion** are time, temperature, and turbulence.

Medical studies have shown that exposure to low levels of **CO causes damage to the endothelium** (membrane, inner lining of blood vessels).

COHb percent is CO measured in the blood by taking a blood sample.

An alternate means to measure COHb is by use of a **breath analyzation module**.

A healthy person exposed to **200 ppm** will experience headaches, dizziness, and nausea.

After 2 to 3 hours of exposure, the concentration of CO will be **life-threatening at 400 ppm**.

The medical symptoms and **health effects of CO poisoning** can vary based on the age, state of health, and weight of an individual.

The responder that responds to a CO alarm must administer a **breath analysis of the occupants** to verify the presence of CO.

The main cause of accidental CO poisoning in North America is from back drafting appliances or appliances that do not draft all of the time.

A cracked heat exchanger is the least common cause of CO poisoning but must always be investigated.

The **installation of a CO detector is recommended** when any hydrocarbon appliance is installed.

The installation of **CO alarms designed for current standard criteria** may not provide adequate protection for children, older people, or people with deteriorating health issues.

Levels of CO_2 in the outdoors are typically **350 to 450 ppm;** amounts of CO_2 in indoor levels that will cause complaints of stuffiness and odors are 600 to 1000 ppm.

An **HVACR technician tests for CO levels in a home** and from HC appliances to prevent the production of CO.

If a CO alarm is activated, the homeowner must **vacate the premises**.

CO test equipment must be calibrated per the manufacturer's recommendations.

A **false CO reading** from a CO meter may be caused by a high concentration of CO outside the building.

Carefully enter the building using a **self-contained breathing apparatus (SCBA)** and have the CO test instruments in operation when a CO alarm has been activated and nobody is outside the building.

A **CO breath analyzer** can confirm that a person has been exposed to CO even though the building has been ventilated.

Once confirmed that CO is within a building, the building should be condemned until the source or sources are found.

The **two circumstances that make CO a hazard** are a strong enough concentration that can harm and can get into breathable air.

Too much fuel or too little oxygen during combustion will produce **incomplete combustion and carbon monoxide.**

Use **soap bubbles** or another approved leak detection method for gas leaks and never an open flame.

Measuring the pressures of the home as well as HVAC equipment will ensure that all gas-fired equipment will vent flue gasses properly.

The air pressure at sea level is **14.7 psi**.

Air on a **supply side** of a blower is under **positive pressure**.

Pressure testing within a building is done to compare the building pressure to the outside pressure.

In a **building stack effect**, air leaks in at holes near the base of the structure and leaks out at holes near the top of the structure.

Positive pressure is on the windward side of a structure.

By closing off a zone that has return air grills, the zone will **not be able to return air back to the furnace**.

The key to understanding **back drafting** is to understand the driving force that creates a negative pressure.

The **tightness of a building** will increase pressure differences within the building.

The definition of pressure is **force per unit of area $P = F/A$.**

OIL FURNACES

An oil heating service technician needs to understand fuel fundamentals as well as the controls and components the system uses. The following information is needed when servicing oil furnaces.

No. 2 fuel oil has a heating value of approximately 140,000 Btu per gallon.

High-pressure gun burners commonly operate at 100 psig. (Some models operate between 140 and 200 psig.)

The primary voltage of an **ignition transformer** is 120 volts. The secondary voltage is 10,000 volts. An ignition transformer is used to create the spark for the ignition of the oil on a high-pressure gun burner.

Electrodes are metal rods wrapped in a ceramic insulator that carry the high voltage from the ignition transformer to the point of ignition.

A common **spark gap** between the electrode tips is $\frac{1}{8}''$.

A **bus bar** is a brass or copper plate that connects electrodes to the ignition transformer.

An **oil nozzle** is used to atomize the fuel oil before ignition. Spray angles range from 30 to 90 degrees.

Nozzles have three different spray patterns: solid, hollow, and semihollow.

Nozzles are rated in gallons per hour (gph).

A **squirrel cage combustion blower** is used to supply air for combustion.

The **burner motor** is a split-phase motor used to turn the oil pump and the combustion air blower.

The speed of a **burner motor** can be 1,725 rpm or 3,450 rpm.

There are **three locations** for oil storage tanks: indoor, aboveground, and underground.

An **indoor tank** must be installed within 7′ of the burner. The fill pipe must be a 2″ iron pipe size and the air vent must be a $1\frac{1}{4}''$ iron pipe size. An indoor tank is commonly installed as a single pipe system.

Indoor tanks commonly have a 275-gallon capacity.

An **indoor tank globe valve** must be of the fuse type to prevent oil leakage during a fire.

A 275-gallon fuel tank is **constructed** of 14-gauge steel.

Fuel oil tanks must be approved by Underwriter Laboratories (UL).

An **underground tank** commonly has a 500-, 1,000-, or 1,500-gallon capacity. The fill pipe must be a 2″ iron pipe size, the air vent must be a $1\frac{1}{4}''$ iron pipe size, and both must be installed as a two-pipe system.

A **decrease of primary air** increases CO_2 percentage and increases combustion gas temperatures, decreases stack temperatures, and increases efficiency.

An **increase of primary air** decreases CO_2 percentage, decreases combustion gas temperatures, increases stack temperatures, and lowers efficiency.

Viscosity is the oil's resistance to flow. The higher the viscosity, the thicker the oil.

The **pour point** is the temperature at which oil will start to flow.

Cloud point refers to the temperature at which wax crystals begin to form in oil.

Single-stage fuel pumps are used with aboveground fuel tank applications.

Two-stage fuel pumps are used with underground fuel tank applications.

The **cad-cell** is a separate component from the cad-cell relay primary control. The cad-cell is wired to the primary control. It receives and sends messages to the cad-cell's primary control relay.

A **cad-cell** is a photo-type cell that has a high resistance to darkness. It is made of cadmium sulfide.

The **cad-cell** should be 1,500 ohms or less when sensing light.

The **cad-cell** should be around 100,000 ohms or more when sensing darkness.

A **cad-cell relay** is a control device used to sequence and control the burner motor and ignition transformer while sensing the **light** of the flame.

The **cad-cell relay** uses a black wire to supply line voltage to the relay, a white wire to supply the common wire to the relay, and an orange wire to feed line voltage to the burner motor and ignition transformer.

A **stack mount relay** is a control device used to sequence and control the burner motor and ignition transformer while sensing the **heat** of the flame.

The **stack mount relay** uses terminal number 1 to supply line voltage to the relay, terminal number 2 to supply the common wire to the relay, and terminal number 3 to feed line voltage to the burner motor and ignition transformer.

A **coupler** is a device that is used to connect the shaft of the fuel pump and the shaft of the burner motor.

An **air shutter** is used to adjust the primary air on a high-pressure gun burner.

A **barometric damper** is a device that is installed in the flue of an oil-fired furnace. It is used to control draft through the chimney.

There are three **fuel filters** in the fuel oil system: one on the inlet line, one in the fuel pump, and another at the nozzle.

Turbulent flow refers to airflow created by spiral vanes within the air blast tube.

Most residential oil furnaces use an **oil supply line** that is a $\frac{3}{8}''$ OD copper tubing.

The **lubrication** fluid used in burner motors is a number-10-weight nondetergent motor oil.

A **delayed oil valve** is used to allow the oil pump to build up pressure on the startup of a high-pressure gun burner. It is wired in parallel with the oil burner motor.

In an oil furnace, the **products of combustion** are:

- Carbon dioxide (CO_2)
- Carbon monoxide (CO)
- Oxygen (O)
- Nitrogen (N)
- Water vapor (H_2O)

High CO_2 readings indicate a hot fire.

A **flue gas analyzer** is commonly used to measure CO_2.

Common CO_2 **readings** are from 10 to 12%.

Common CO_2 **readings** for natural gas furnaces should be from 8.5 to 9.5%.

A **draft gauge** is used to measure the amount of draft created in a furnace.

A common reading **over the fire draft** for an oil furnace ranges from −.01″ to −.02″ WC. Over-the-fire-draft reading refers to the exhaust draft, a draft reading after a fire (combustion process).

A common **breech draft** reading for an oil furnace ranges from −.03″ to −.06″ WC.

Breech draft reading refers to the draft within the chimney piping.

A draft for an oil-fired furnace can be adjusted using the **barometric damper** located on the stack. Opening the damper decreases the draft, and closing the damper increases the draft.

A **smoke tester** is used on an oil furnace to measure the amount of smoke that is created in the combustion process.

Common **smoke readings** for an oil-fired furnace are from #1 to #2 as seen on the smoke sample card. The smoke card is part of a combustion analysis kit.

A **high smoke reading** indicates that there is not enough combustion air.

A **low smoke reading** indicates that there is too much combustion air.

A **stack thermometer** commonly measures temperatures that range from 200 to 1,000°F.

To obtain the **net stack temperatures** of flue gases, measure the gases exiting the flue and subtract the ambient air temperature.

The **gross stack temperature** is room air temperature plus the net stack temperatures, or the reading of the stack thermometer when placed in the furnace flue.

Common **gross stack temperatures** are from 500 to 550°F.

If **combustion air** must be supplied to an oil furnace, there must not be less than 1 sq in. per 5,000 Btu.

HYDRONIC HEATING THEORY

Hydronic heating systems, also known as hot water heating systems, are widely used in the HVAC/R industry. There are many different types of hydronic heating systems in use today, such as low- and high-pressure steam boilers as well as radiant panel heating systems. However, this section covers the hot water boiler, in which water pumps and baseboard radiation are used to transfer heat from the boiler water to the zones that are to be heated.

Heat is transferred from one object to another in three general ways: conduction, convection, and radiation.

Conduction is the transfer of heat from one molecule to another when two objects are in contact with each other. The objects must have a difference in temperature.

Heat flows from the warmer object to the cooler object.

Convection is the transfer of heat through the movement of a fluid. The fluid is most commonly air or water.

Radiation is the transfer of heat by the use of waves through space. When a heat source gives off heat waves, they travel through space and heat the object that they come in contact with. Dark, dull objects absorb more heat than shiny, light objects.

Heat energy is absorbed into a fluid when it changes from a liquid to a gas; this is known as **evaporation.**

A **Btu** (British thermal unit) is a measurement of heat energy.

1 **Btu** is the amount of heat needed to change 1 lb of water 1°F.

1 gallon of water weighs **8.33 lb.**

There are approximately 500 lb of water pumped in 1 hour if the **flow rate** through a system is 1 gal/min. That is, 1 gal/min × 8.33 lb/gal × 60 min/h = 500 lb/h.

The **temperature difference** between the water leaving the hot water boiler (supply) and the water entering the boiler (return) is 20°F.

A **2.31-ft water column** exerts a pressure equal to 1 psi.

MECHANICAL CONTROLS

Many types of controls are needed on the hot water heating system, some electrical and some that are either automatic or manual controls. This section covers the nonelectric mechanical controls.

The **pressure** of a hot water boiler should not exceed 30 psi or a temperature of 250°F.

Air trapped within a hot water boiler tends to move to the highest point of the system and prevents the circulation of water.

The **cold start** pressure of a hot water boiler is 12 psi.

Common **operating pressures** of a hot water boiler are 18 to 25 psi.

Common **operating temperatures** of a hot water boiler range from 180 to 200°F.

When sizing a hot water boiler, the **net IBR rating** should be equal to or slightly higher than the heat loss of the building. **IBR** stands for the Institute of Boiler and Radiator Manufacturers. The **IBR net rating** is the amount of heat supplied to the home after deducting the piping heat loss and cold startup of the system.

AGA (American Gas Association) is an agency responsible for testing and certifying gas appliances.

ASME (American Society of Mechanical Engineers) is responsible for the design and construction features of hot water boilers.

The **DOE** (Department of Energy) sets the standards and tests procedures to be followed in which boiler and appliance manufacturers certify the efficiency levels of their heating equipment.

AFUE (Annual Fuel Utilization Efficiency) is the efficiency rating of the appliance; it is calculated as 100% minus stack losses during normal system operation.

The **main shutoff valve** is the first component of the boiler water system. Its purpose is to shut off the supply water to the boiler when the piping system is being serviced. The globe valve has the greatest resistance to flow when compared to the ball-and-gate valve.

The **backflow preventor** is the next component down stream from the main shutoff.

The **backflow preventor** allows water to flow in one direction. The **backflow preventor** allows city water into the boiler but does not allow boiler water into the city water supply.

A **pressure-reducing valve** is located at the inlet to a hot water heating system downstream from the main shutoff and the backflow preventor.

The main purpose of a **pressure-reducing valve** is to reduce the water pressure entering the boiler down to 12 psig.

Air vents are used to purge air from a hot water heating system.

Manual air vents, also known as coin vents, are located at the highest sections of the radiation units within a home. The valves are opened to release trapped air out of the system.

The **automatic-float-type air vent** uses a float assembly and chamber to trap air within a hot water heating system. The vent is open to the atmosphere when the chamber is full of air; air is pushed out of the vent until the chamber fills with water, then the vent closes.

A **purge valve** is commonly located in the return piping side of the circuit. A butterfly valve is used to direct the flow of water to a globe valve. When the globe valve is opened, trapped air within the circuit purges out to the atmosphere.

An **air scoop** is used to direct air within the boiler system to the automatic air vent. Its purpose is to aid in the removal of unwanted air within a boiler system.

An **expansion tank** is used to allow for water expansion on a hot water heating system. There are two types of expansion tanks. One type is a compression tank that is located high above the hot water boiler, usually in the boiler room. Water enters at the bottom of the compression tank, and air is located above the water at the top. The air at the top gives the water the room needed for its expansion. Another type of expansion tank is the diaphragm type of expansion tank. A rubber diaphragm is located between the system water and an air pocket within the tank. The diaphragm is generally precharged with air to 12 psig. As the water is heated, it expands and compresses the air on the other side of the bladder.

A **pressure-relief valve** is a safety device that is located on a hot water boiler.

The **pressure-relief valve** is listed with ASME (American Society of Mechanical Engineers).

If the pressure of the boiler happens to reach unsafe limits, the **pressure-relief valve** opens and vents the boiler water to the atmosphere, preventing an explosion.

A hot water boiler **pressure-relief valve** opens at 30 psig.

The Btu/h rating on a **pressure-relief valve** must never be less than the gross Btu output of the boiler.

The **pressure-relief valve** should be terminated within 6″ from the floor.

The diameter of the discharge piping of the **pressure-relief valve** must never be less than the outlet diameter of the valve.

The **tridicator** is the pressure gauge located on the boiler. It is used to measure boiler temperature, pressure, and altitude.

Common hot water **boiler temperatures** during winter months are from 180 to 220°F.

ELECTRICAL CONTROLS AND MECHANICAL COMPONENTS

Understanding the operation of boiler electrical controls is essential for service technicians to properly troubleshoot and repair a hot water heating system. The following are electrical controls that are used on hot water heating systems.

The zone or room **thermostat** is used to directly energize the zone valve motor.

A zone valve circuit that uses a **40-VA transformer** should use no more than three zone valves.

Electric zone valves are either 4-wire or 5-wire type. The **4-wire zone valve** is power open and spring close.

The purpose of a **zone valve** is to allow water to flow to the heated zone during system operation.

An **end switch** is a normally open switch that is built into the zone valve and is used to prove when the valve is in the open position. The **end switch** closes when the zone valve is fully open.

There are commonly two red wires on the 4-wire zone valve that are connected to the **end switch.** When the zone valve is fully opened, the end switch must be in the closed position. Two other wires, commonly yellow, are used to energize the zone valve motor. A single pole, single throw thermostat is wired in series with the zone valve motor.

The **5-wire zone valve** is power open and power close. Two wires are used for an end switch and three wires are used for the zone valve motor. A single pole, double throw thermostat must be used to control the **5-wire zone valve**.

The lever on the **zone valve** is used to manually open the zone valve.

Zone valves can be located on either the supply or the return side of the piping system.

A **zone valve** that uses a resistance heater type of motor has a heater element and a wax substance inside the valve. When the heater is energized, the wax expands and opens the valve.

Nonelectric zone valves are located at the inlet of the baseboard section of each room. They are a modulating type of control that senses room temperature at the valve.

Nonelectric zone valves do not use end switches, so the circulator pump must be in operation at all times.

A **heating thermostat** opens on a temperature rise and closes on a temperature drop.

A heating thermostat has a **heat anticipator** wired in series with the heating control circuit. The heat anticipator prevents system overshoot.

The zone thermostat's **heat anticipator** must be set to the zone valve motor control amperage.

The **heating control amperage** can be measured by placing an in-line ammeter across terminals "R" and "W." It can also be measured by placing a 10-loop multiplier wire across terminals "R" and "W" and measuring the amperage with a clamp-on ammeter.

A **high-limit control** is used with a hot water boiler. Its purpose is to control the maximum temperature of the water within the boiler.

If the **high-limit control** reaches its cut-out setting, it shuts down the burners to the boiler, allowing the temperature of the boiler to drop. Commonly, a 10°F differential is preset within the control.

If the cut-out on a **high-limit control** was set to 200°F, then the differential of 10°F would cause the burners to fire at 190°F.

A **low-limit control** performs two functions simultaneously: It keeps the temperature of the boiler water at a minimum and controls the operation of the circulator pump.

When the boiler water temperature drops below the set point of the **low-limit control,** the circulator pump shuts off and the heating control circuit is energized. This prevents the circulator pump from moving cool water to the zones.

The "R" and "B" terminals on the **low-limit control** operate the heating circuit.

The "R" and "W" terminals on the **low-limit control** operate the circulator pump circuit.

An **aquastat boiler controller** is used to control the burners and the circulator pump. The high-limit control is built into the **aquastat boiler controller.**

The **operational sequence** for a system that uses a zone valve and a boiler controller is as follows: The zone thermostat calls for heat energizing the zone valve motor; this opens the zone valve, allowing water to flow to the zone, and closes the end switch within the zone valve. The end switch is connected to the aquastat boiler controller. Closing of the end switch energizes the aquastat boiler controller, starts the circulator pump, and energizes the burner system, which heats the water within the boiler.

The **reverse-acting low limit** prevents the circulator pump operation until a minimum boiler temperature is reached, commonly 160°F.

A **low-water cut-out** is used to sense the level of water within a boiler; if the water level is too low, it will de-energize the burner system of the boiler.

A **vent safety control** is used to de-energize the gas valve if the vent temperature increases above a predetermined temperature. Its primary purpose is to shut down the gas valve control circuit if the venting system were to become restricted and cause flue gas spillage.

A **flame roll-out** switch or fusible link de-energizes the burner circuit if the flame from the burners is pushed out of the heat exchanger. This is commonly caused by a plugged heat exchanger.

A **flame proving** device proves the presence of the pilot or main burner flame.

A **thermocouple** proves a pilot flame by changing heat energy to electrical energy; it operates at 30 millivolts.

A **flame rod sensor** uses a flame rectification system to prove the presence of the pilot or main burner flame; common microamp signals range from .2 to 6 μA.

Both a **bimetal flame sensor** and a **mercury flame sensor** prove the presence of a pilot flame by using the heat from the pilot to switch a set of contacts and energize the main gas valve.

A **pressure switch** proves the operation of the induced draft motor.

The **circulator pump** is commonly located on the return side of the boiler's piping system. It is used to force water through the system. It is rated in gallons per minute (gpm).

A **circulator pump** uses an impeller to move the water through the piping system.

The inlet to the circulator pump is known as the **volute.**

Water enters the volute as the centrifugal pump is spinning and a **pressure difference** is created due to the centrifugal force of the impeller.

The pressure difference across the circulator pump is known as the **pump head.**

Pump head is measured in feet of water column.

The greater the pump head, the lower the **pump capacity.**

Centrifugal pumps are rated in **horsepower.**

One **horsepower** is equivalent to 746 watts of power.

Direct drive pumps often have **flexible couplings** that connect the shaft of the pump motor to the shaft of the circulator pump. This is necessary due to motor/pump alignment changes when the temperature of the water changes.

Two types of **bearings** are used on centrifugal pumps: sleeve bearings and ball bearings.

Sleeve bearings are commonly made of bronze.

Sleeve bearing pumps operate more quietly than ball bearing pumps.

Sleeve bearing pumps must be oiled each season.

Ball bearings handle a larger load and must be greased occasionally.

A **circulator pump** must never be installed between the pressure-reducing valve and the boiler; this will cause the system to be overfilled.

The purpose of a **vent system** on gas appliances is to safely remove the by-products of combustion from the boiler combustion chamber to the outside of the building.

A **vent system** consists of a draft hood assembly, a vertical riser, horizontal connectors, and the chimney (if used).

The purpose of the **draft hood** is to dilute the flue gases with room air, to lower flue gas temperatures, and to divert a downdraft.

The **vent damper** located on the hot water boiler allows the products of combustion to flow to the venting system.

A **motorized vent damper** opens to the venting system during system operation and closes during the off cycle to prevent building heat loss.

During boiler operation, the **vent damper** motor can be either energized or de-energized, depending on the manufacturer.

An **end switch** in the vent damper proves the motor is in the open position and allows the heating system to energize.

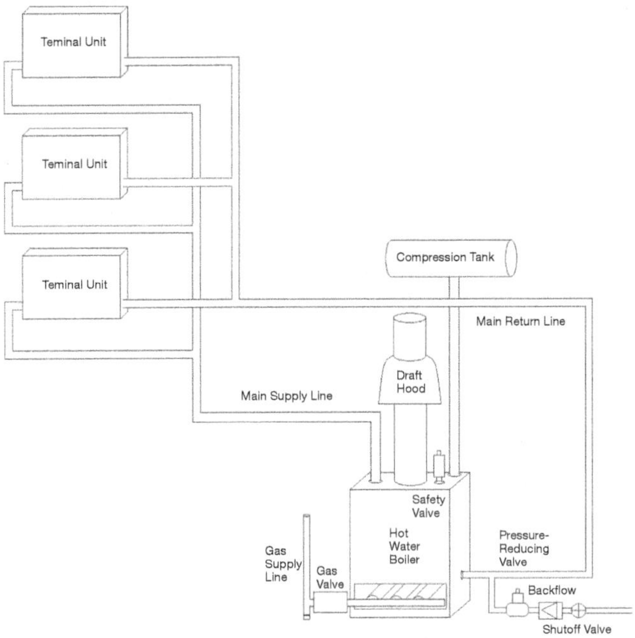

FIGURE 2–3
Direct Return Piping System Diagram

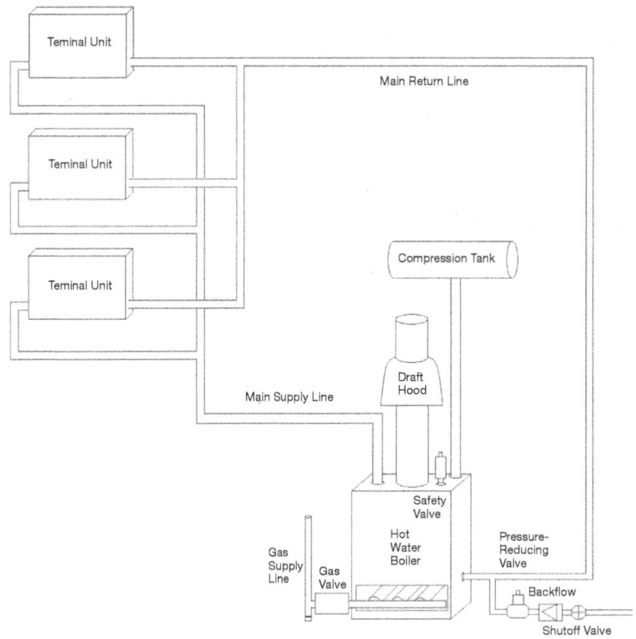

FIGURE 2–4
Reverse Return Piping System Diagram

There are many configurations of piping for the hot water heating system. Understanding water flow through a hot water heating system helps a service technician pinpoint where system problems may exist. The following covers different types of piping arrangements and fittings that are used on hot water heating systems.

A **gravity supply hot water heating system** operates on the principle that the density of water decreases as it is heated.

On a **gravity hot water system,** the boiler is located at the lowest point of a system.

System piping for a **gravity hot water system** is larger in diameter when compared to a forced convection system; this reduces frictional loss through the system.

A gravity hot water system is usually a two-pipe system, either a direct return or a reverse return.

A **direct return** two-pipe system has the first radiation unit supplied, the first to be returned back to the boiler (see Figure 2–3).

A **reverse return** two-pipe system has the first radiation unit supplied, the last to be returned back to the boiler.

The **reverse return** two-pipe system is a self-balancing system (see Figure 2–4).

A **one-pipe system** uses one main pipe from the supply side of the boiler back to the return side.

One-pipe systems use **flow-diversion tees** that are located at the inlet of each terminal unit (radiation).

The **flow-diversion tee** is designed to force water into the terminal unit. Another name for a flow-diversion tee is the monoflow tee.

Mixing-tees may be located at the outlet of a terminal unit on a one-pipe supply loop heating system. The **mixing-tee** mixes the water leaving a terminal unit with the water in the one-pipe system (see Figure 2–5).

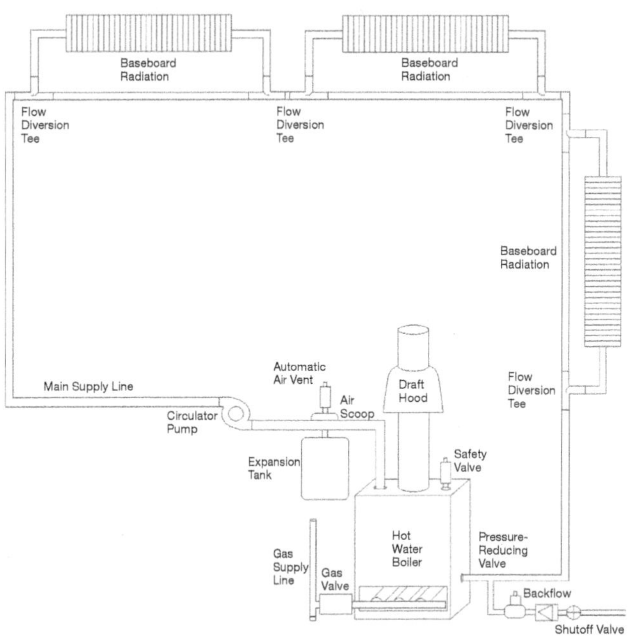

FIGURE 2–5
Single Pipe System with Flow Diversion Tee Diagram

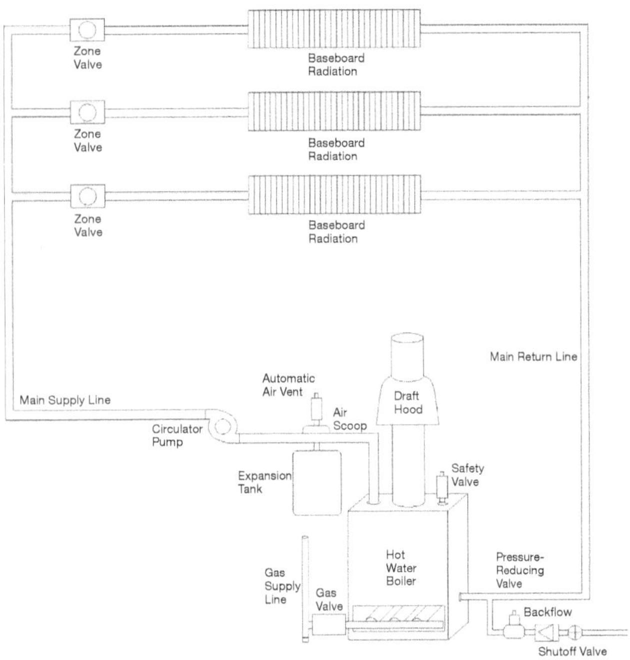

FIGURE 2–6
Multiple-Zone-Valve System

A **two-pipe system** uses one main supply pipe to carry heated water to individual zones and a return pipe to carry the return water back to the boiler. Figure 2–6 shows a multiple-zone-valve system.

A **series loop** piping system has one continuous loop of piping from the supply pipe, leaving the boiler through each section of radiation throughout the building, then it is returned back to the boiler (see Figure 2–8).

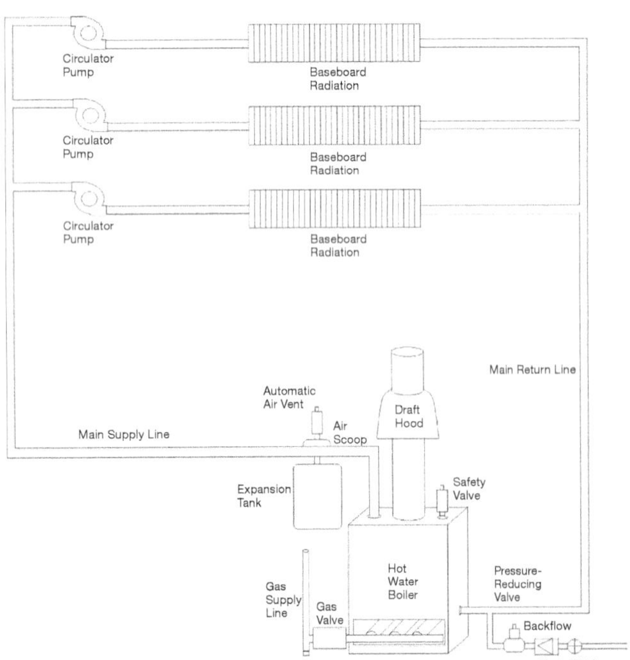

FIGURE 2–7
Multiple-Pump System Diagram

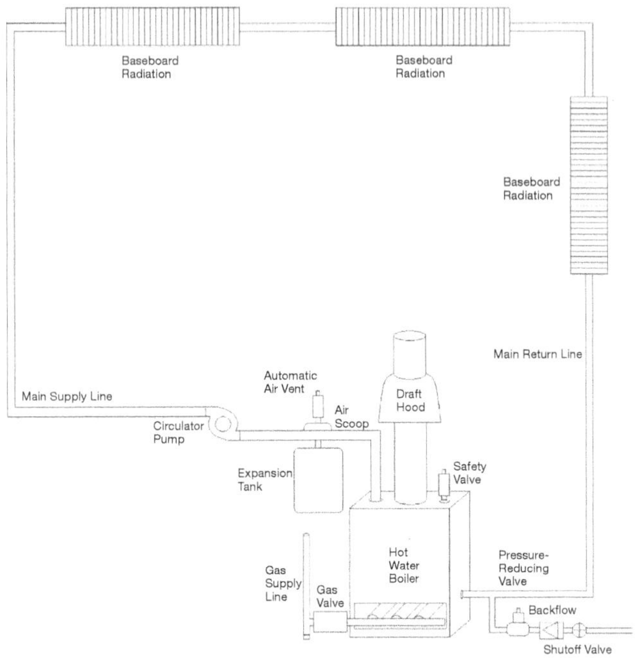

FIGURE 2–8
Series Loop Piping System

A **radiant panel** heating system supplies hot water through a piping system that is located in ceilings, walls, or floors (see Figure 2–9).

A **flow control valve/flow check valve** is used to prevent thermosiphoning through the piping system when the circulator pump is de-energized.

A **flow control/flow check valve** is recommended when a system uses separate circulators for each zone.

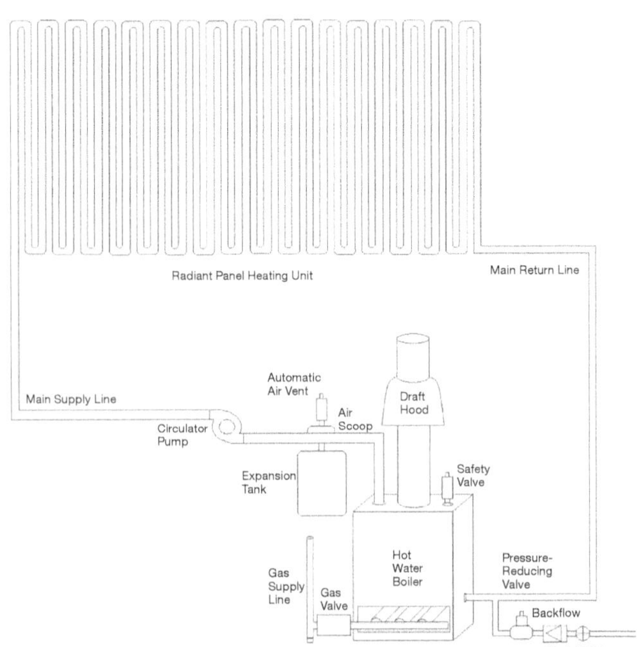

FIGURE 2–9
Radiant Panel Heating System

Three types of **check valves** are swing check, ball check, and magnetic check.

The **swing check valve** is a critically mounted valve. It must be mounted horizontally. If it is mounted in a vertical direction with a downward flow, it will not close. It must be mounted so the flow is upward if it is used vertically. The weight of the water will close the valve.

A **balancing valve/flow adjustment valve** is a butterfly-type valve that restricts the flow of water within a zone. It is used to balance the proper amount of water supplied to each zone.

A **bypass valve** is a valve that is placed in parallel with the pressure-reducing valve. It is used to quickly fill a boiler using city water pressure.

The **standard rating** per linear foot of $\frac{3}{4}''$ radiation for a hot water system operating at approximately 200°F is 600 Btu.

Baseboard **radiation ratings** for hydronic heating systems are expressed in Btu per linear foot per hour.

Friction loss is the loss of pressure through a piping system due to the restriction of flow through the piping and the fittings.

Friction loss is measured in foot of head.

TROUBLESHOOTING HYDRONIC HEATING SYSTEMS: ELECTRICAL DIAGRAMS

Sequence of Operation of the Multiple-Zone-Valve System Diagram (Figure 2–10)

With the disconnect switch in the closed position, 120 volts energizes the primary side of the transformer and the boiler aquastat control; 24 volts is present at the secondary side of the 40-VA transformer. The transformer is used to operate the zone valve motor circuits, which are controlled by the zone thermostats. If all thermostats are in the open position, there will be no action from the control circuit. If any of the thermostats closes, the zone valve motor controlled by that thermostat will energize, allowing water to flow to the zone. When the zone valve is fully open, the end switch will close, jumping out

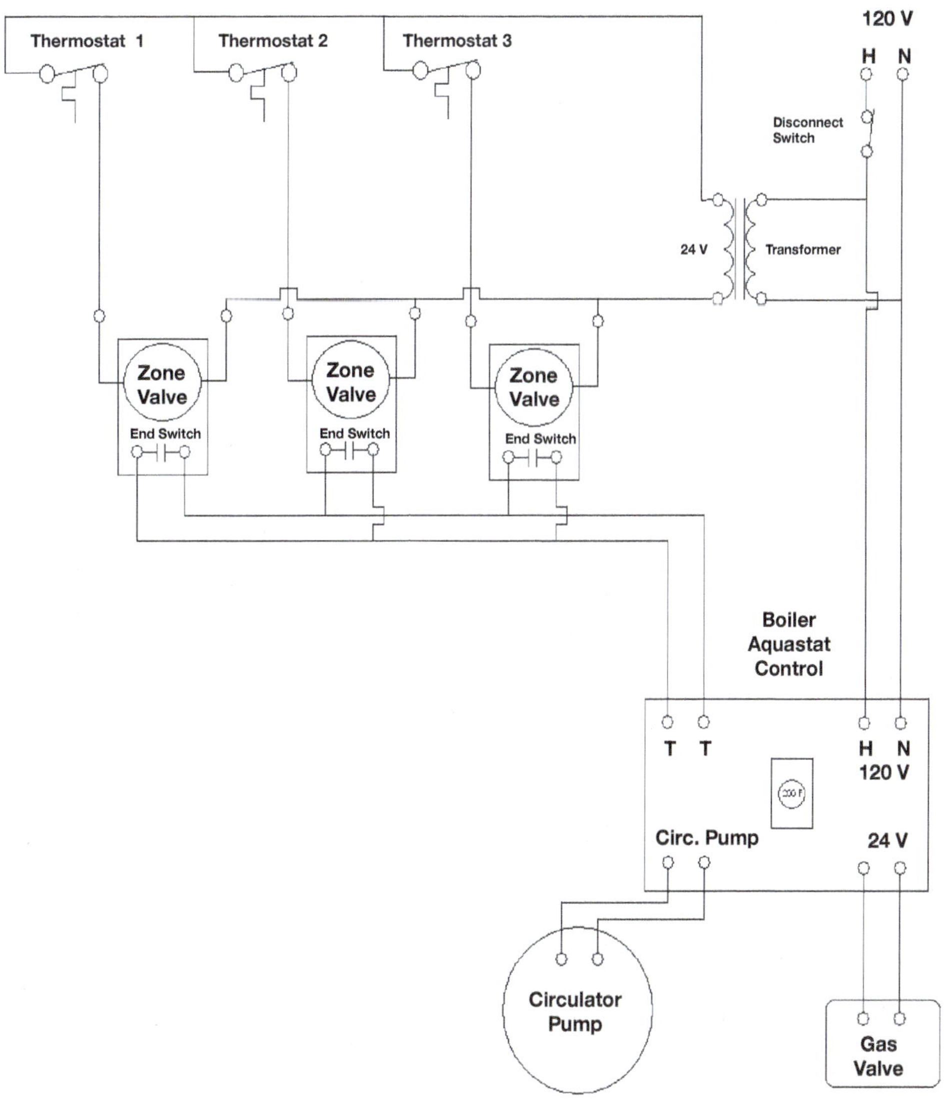

FIGURE 2–10

terminals "TT" on the aquastat boiler controller. A 120-volt/24-volt control transformer is built into the boiler controller, which energizes a relay circuit that causes the circulator pump and gas valve to energize. A limit control is also built into the aquastat boiler control. Its purpose is to cycle the burners, preventing the boiler water from reaching a predetermined temperature, usually 180 to 200°F.

Sequence of Operation of the Oil-Fired Multiple-Zone-Valve System Diagram (Figure 2–11)

With the disconnect switch in the closed position, 120 volts energizes the primary side of the transformer and the boiler aquastat control; 24 volts is present at the secondary side of the 40-VA transformer. The transformer is used to operate the zone valve motor circuit, which is controlled by the zone thermostats. If all thermostats are in the open position, there will be no action from the control circuit. If any of the thermostats closes, the zone valve motor controlled by that thermostat will energize, allowing water to flow to the zone. When the zone valve is fully open, the end switch will close, jumping out terminals "TT" on the aquastat boiler controller. A 120-volt/24-volt control transformer is built into the boiler controller, which energizes a relay circuit that causes the circulator pump to energize, and 120 volts is sent to the oil-fired primary control. A limit control is also built into the aquastat boiler control. Its purpose is to cycle the burners, preventing the boiler water from reaching a predetermined temperature, usually 180 to 200°F. When 120 volts energizes the oil-fired primary control, the ignition transformer and the burner motor are energized. The burner motor turns an oil pump, which moves oil to the nozzle at 100 psig. The nozzle atomizes the oil as it sprays it into the combustion chamber. The ignition transformer sends 10,000 volts to the electrode assembly, which ignites the atomized oil. The cad-cell senses the light from the flame after combustion and allows the system to operate. If a flame is not sensed for 45 s, a safety circuit in the primary control shuts down the burner motor and ignition transformer. The resistance of the cad-cell decreases in the light and increases in the dark.

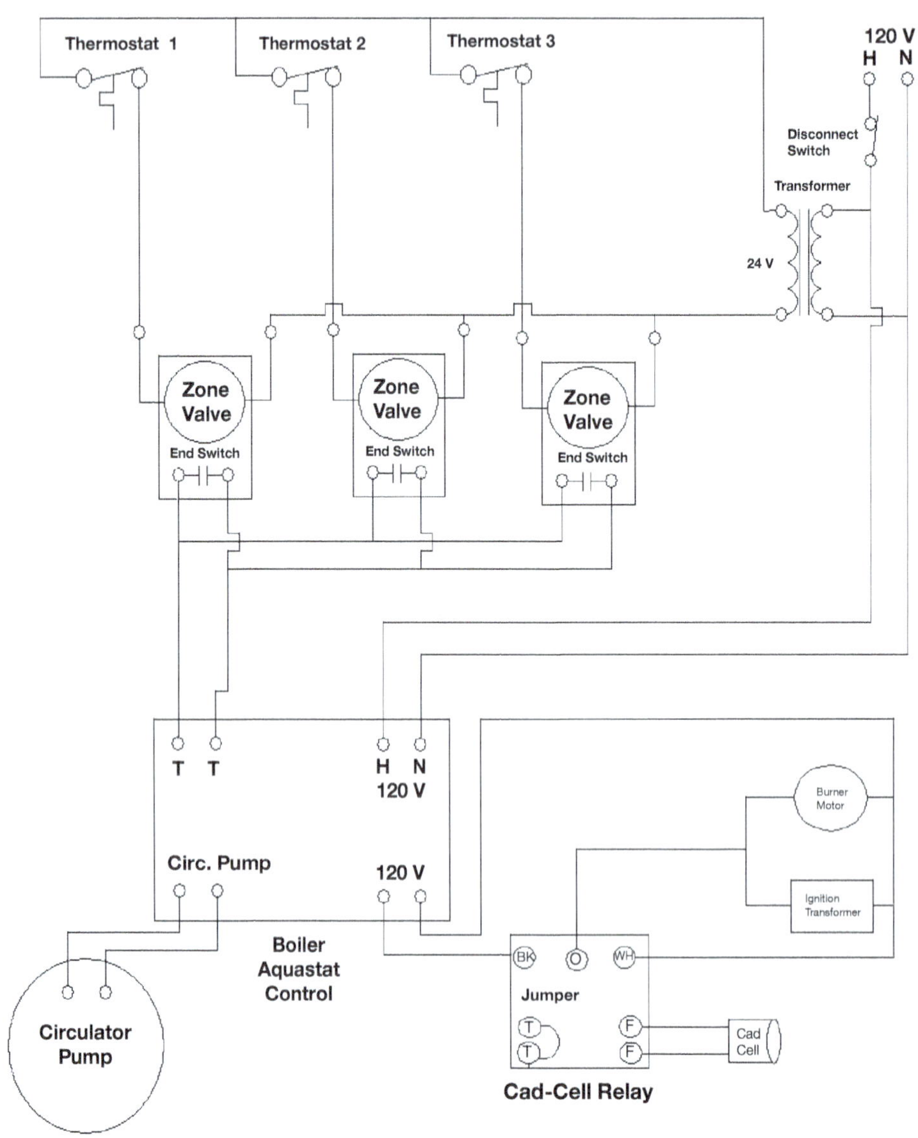

FIGURE 2–11

Sequence of Operation of the Multiple-Circulator Pump System Diagram (Figure 2–12)

With the disconnect switch and the high-limit aquastat in the closed position, 120 volts energizes the primary side of the 120-volt/24-volt 40-VA transformer. A 24-volt output energizes the gas valve circuit, and the boiler water is heated. This system will keep the boiler water temperature in a range from 180 to 200°F. When a zone thermostat closes, it causes the switching relay to close a set of contacts to its circulator pump circuit. The switching relay is constructed of a 120-volt/24-volt transformer and a relay; 120 volts is applied to terminals 1 and 2 on the switching relay and the internal transformer is energized; 24 volts is sent through the relay coil and to the "TT" terminals. When the thermostat closes to the "TT" terminals, a set of normally open (N.O.) contacts closes between terminals 1 and 3, which sends 120 volts to the circulator pump. Because the boiler water is preheated, the zone calling will receive heat immediately.

Sequence of Operation of the Single-Pump Relay-Controlled System Diagram (Figure 2–13)

With the disconnect switch in the closed position, 120 volts energizes the primary side of the transformer, and 24 volts will be present at the secondary side of the 40-VA transformer; 24 volts will be brought to terminals 1 and 4 at the vent damper. On a call for heat, the thermostat closes between the "R" and "G" terminals on the relay center. The relay coil is energized with 24 volts and N.O. contacts "R1" and "R2" close. "R1" completes the circuit to the "Y" terminal on the relay center. From the "Y" terminal, current will flow through the high limit to terminal 2 on the vent damper. The vent damper motor is energized, opening to the venting system and closing the end switch, completing the path to terminal 3. This will feed current out of the vent damper through the flame roll-out sensor and spill switch to the gas valve. The burners will ignite and be cycled by the high-limit aquastat, keeping the boiler water temperature at about 200°F.

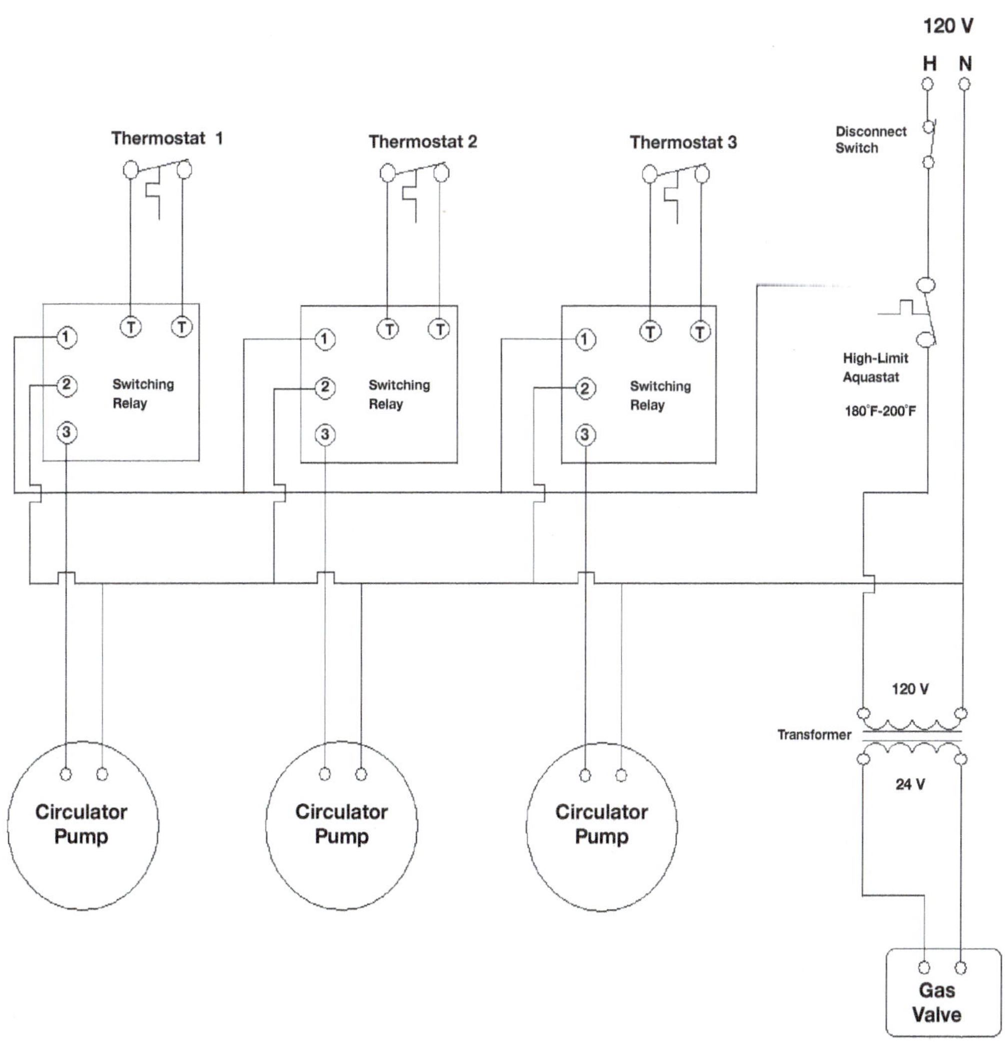

FIGURE 2–12

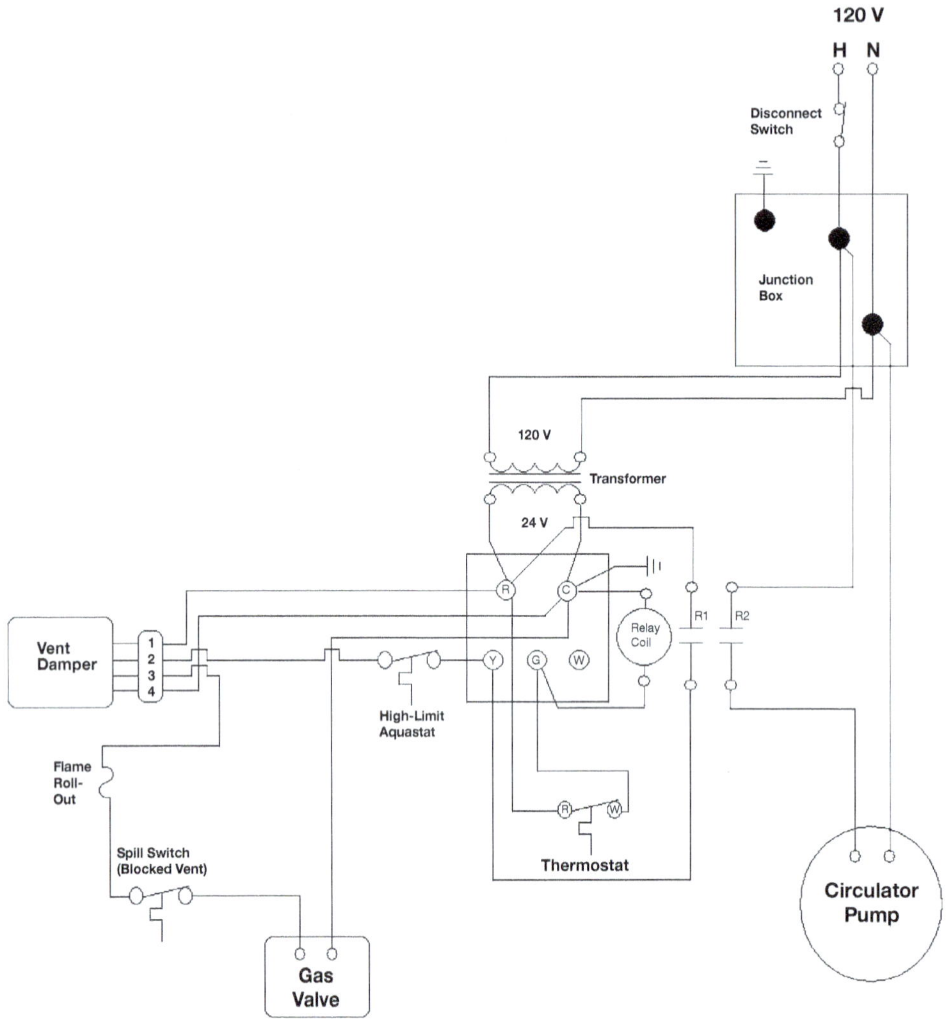

FIGURE 2–13

Sequence of Operation of the Single-Pump Relay-Controlled System with Ignition Control Module Diagram (Figure 2–14)

With the disconnect switch in the closed position, 120 volts energizes the primary side of the transformer, and 24 volts will be present at the secondary side of the 40-VA transformer; 24 volts will be brought to terminals 1 and 4 at the vent damper. On a call for heat, the thermostat closes between the "R" and "G" terminals on the relay center. The relay coil is energized with 24 volts and N.O. contacts "R1" and "R2" close. "R1" completes the circuit to the "Y" terminal on the relay center. From the "Y" terminal, current will flow through the high limit to terminal 2 on the vent damper. The vent damper motor is energized, opening to the venting system and closing the end switch, completing the path to terminal 3. This will feed current out of the vent damper through the flame roll-out sensor and spill switch to the ignition control module. The pilot gas valve and the ignition transformer are energized, the flame is sensed by the igniter/sensor, the main gas valve is energized, and the burners are ignited by the pilot flame. As long as the pilot flame is sensed, the main gas valve and the pilot gas valve stay open. The ignition control module will be cycled by the high-limit aquastat, keeping the boiler water temperature at about 200°F. See Figure 2–15.

Sequence of Operation of the Multiple-Zone-Valve System Ignition Control Module and Vent Damper Diagram (Figure 2–15)

With the disconnect switch in the closed position, 120 volts energizes the primary side of the transformer and the boiler aquastat control; 24 volts is present at the secondary side of the 40-VA transformer. The transformer is used to operate the zone valve motor circuits, which are controlled by the zone thermostats. If all thermostats are in the open position, there will be no action from the control circuit. If any of the thermostats closes, the zone valve motor controlled by that thermostat will energize, allowing water to flow to the zone. When the zone valve is fully open, the end switch will close, jumping out terminals "TT" on the aquastat boiler controller. A 120-volt/24-volt control transformer is built into the boiler controller, which opens the vent damper and a relay circuit that causes the circulator pump and ignition control module to energize. The pilot gas valve and the ignition transformer are

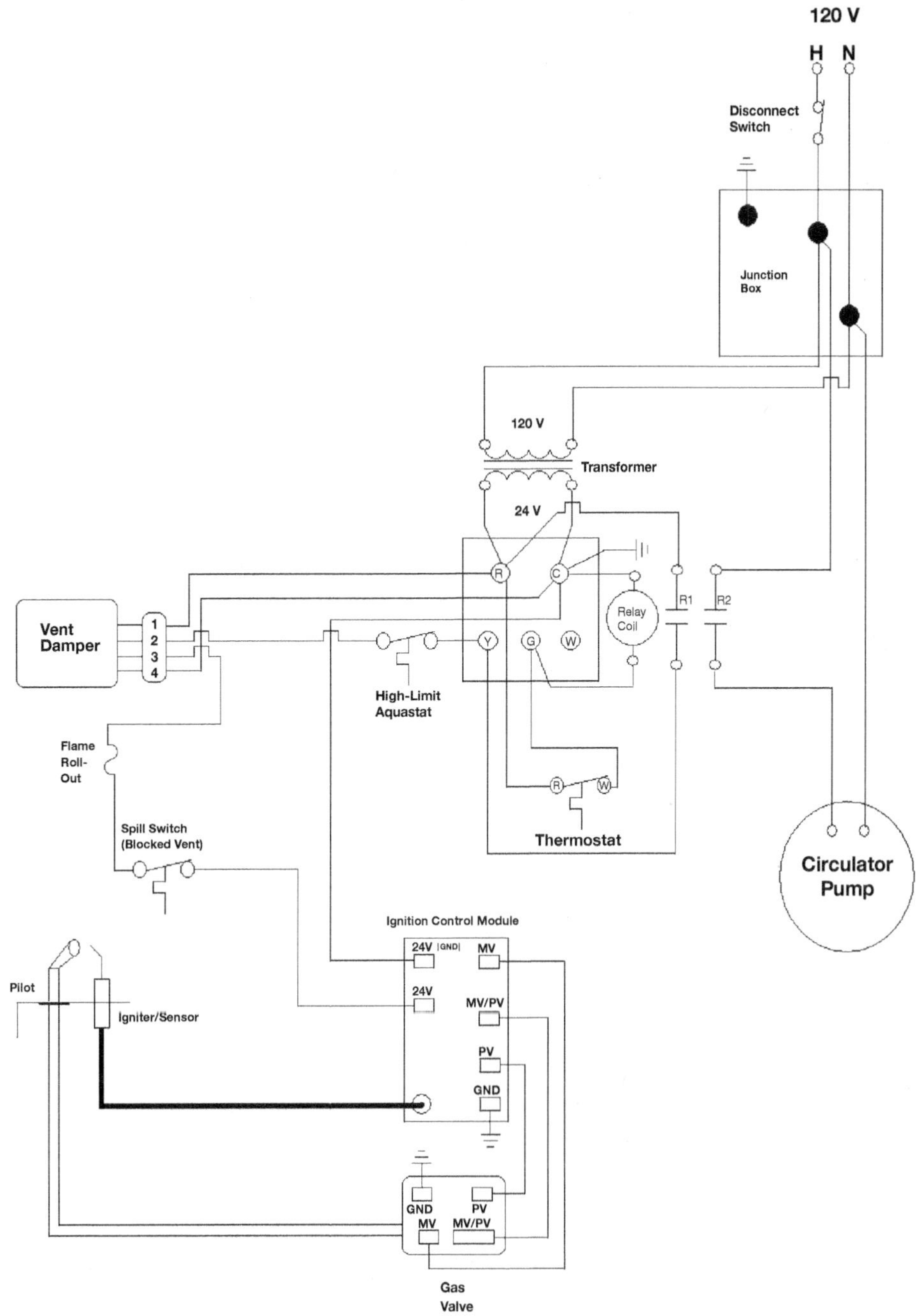
120 V
H N
Disconnect Switch
Junction Box
120 V
Transformer
24 V
R
C
Relay Coil
R1
R2
Vent Damper
1
2
3
4
Y
G
W
High-Limit Aquastat
Flame Roll-Out
R
W
Thermostat
Circulator Pump
Spill Switch (Blocked Vent)
Ignition Control Module
24V
MV
24V
MV/PV
PV
GND
Pilot
Igniter/Sensor
GND
PV
MV
MV/PV
Gas Valve

FIGURE 2–14

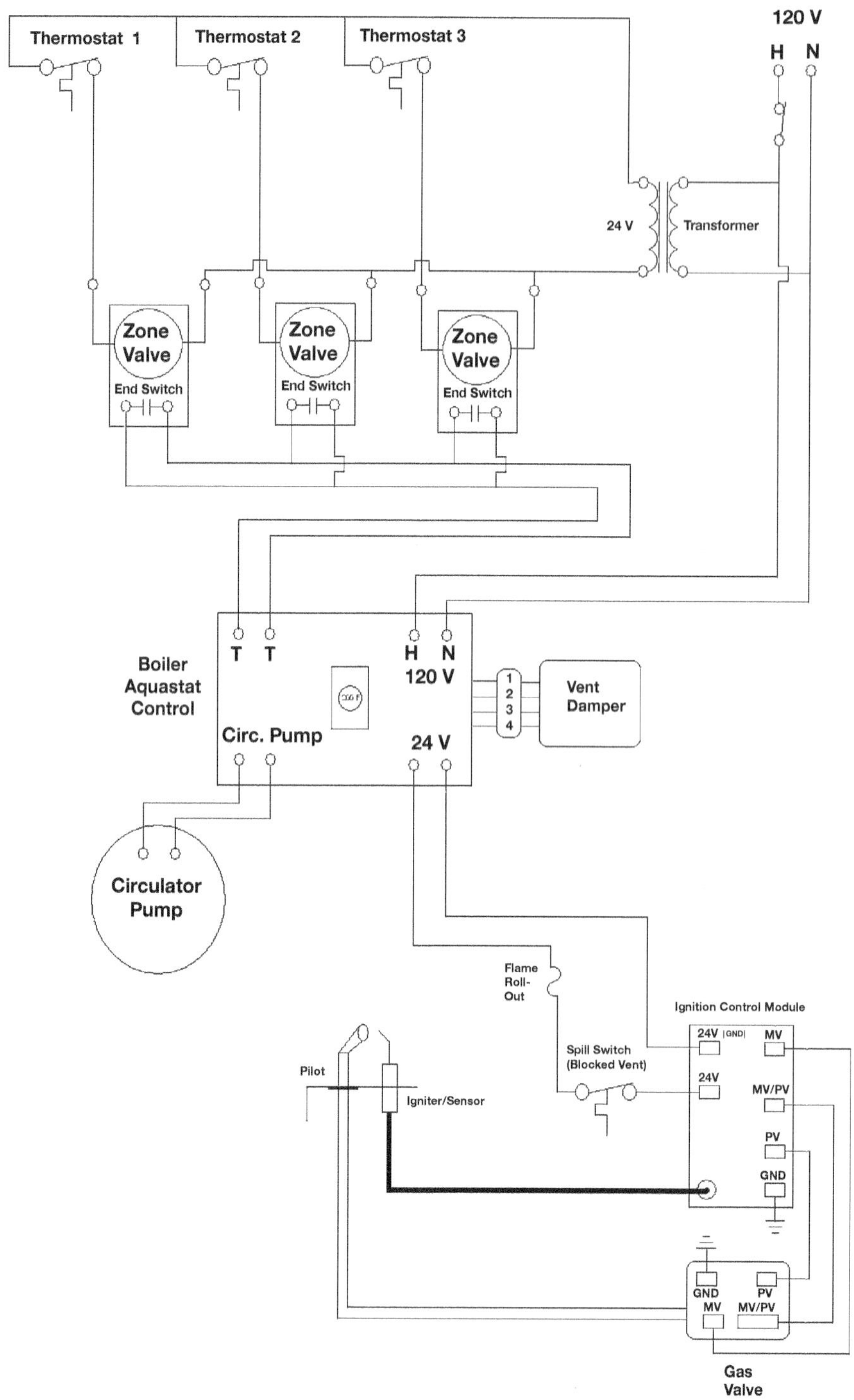
Thermostat 1
Thermostat 2
Thermostat 3
120 V
H N
24 V
Transformer
Zone Valve
End Switch
Zone Valve
End Switch
Zone Valve
End Switch
Boiler Aquastat Control
T T
H N
120 V
Circ. Pump
24 V
1
2
3
4
Vent Damper
Circulator Pump
Flame Roll-Out
Ignition Control Module
Spill Switch (Blocked Vent)
Pilot
Igniter/Sensor
24V
MV
24V
MV/PV
PV
GND
GND
PV
MV
MV/PV
Gas Valve

FIGURE 2–15

energized, the flame is sensed by the igniter/sensor, the main gas valve is energized, and the burners are ignited by the pilot flame. As long as the pilot flame is sensed, the main gas valve and the pilot gas valve stay open. A limit control is built into the aquastat boiler control. Its purpose is to cycle the burners, preventing the boiler water from reaching a predetermined temperature, usually 180 to 200°F. See Figure 2–16.

Sequence of Operation of the Multiple-Pump System with Boiler Controller, Ignition Control Module, and Vent Damper Diagram (Figure 2–16)

With the disconnect switch in the closed position, 120 volts is applied to the boiler aquastat control and to terminals 1 and 2 on the switching relay. Each switching relay has an internal transformer that is energized at this time; 24 volts from the switching relay is sent through the relay coil and to the "TT" terminals. When a thermostat closes to the "TT" terminals, a set of normally open contacts closes between terminals 1 and 3, which sends 120 volts to the circulator pump. A normally open set of contacts at the "XX" terminals closes at this time, jumping out terminals "TT" on the aquastat boiler controller. A 120-volt/24-volt control transformer is built into the boiler controller, which opens the vent damper and a relay circuit that causes the ignition control module to energize. The pilot gas valve and the ignition transformer are energized, the flame is sensed by the igniter/sensor, the main gas valve is energized, and the burners are ignited by the pilot flame. As long as the pilot flame is sensed, the main gas valve and the pilot gas valve stay open. A limit control is built into the aquastat boiler control. Its purpose is to cycle the burners, preventing the boiler water from reaching a predetermined temperature, usually 180 to 200°F.

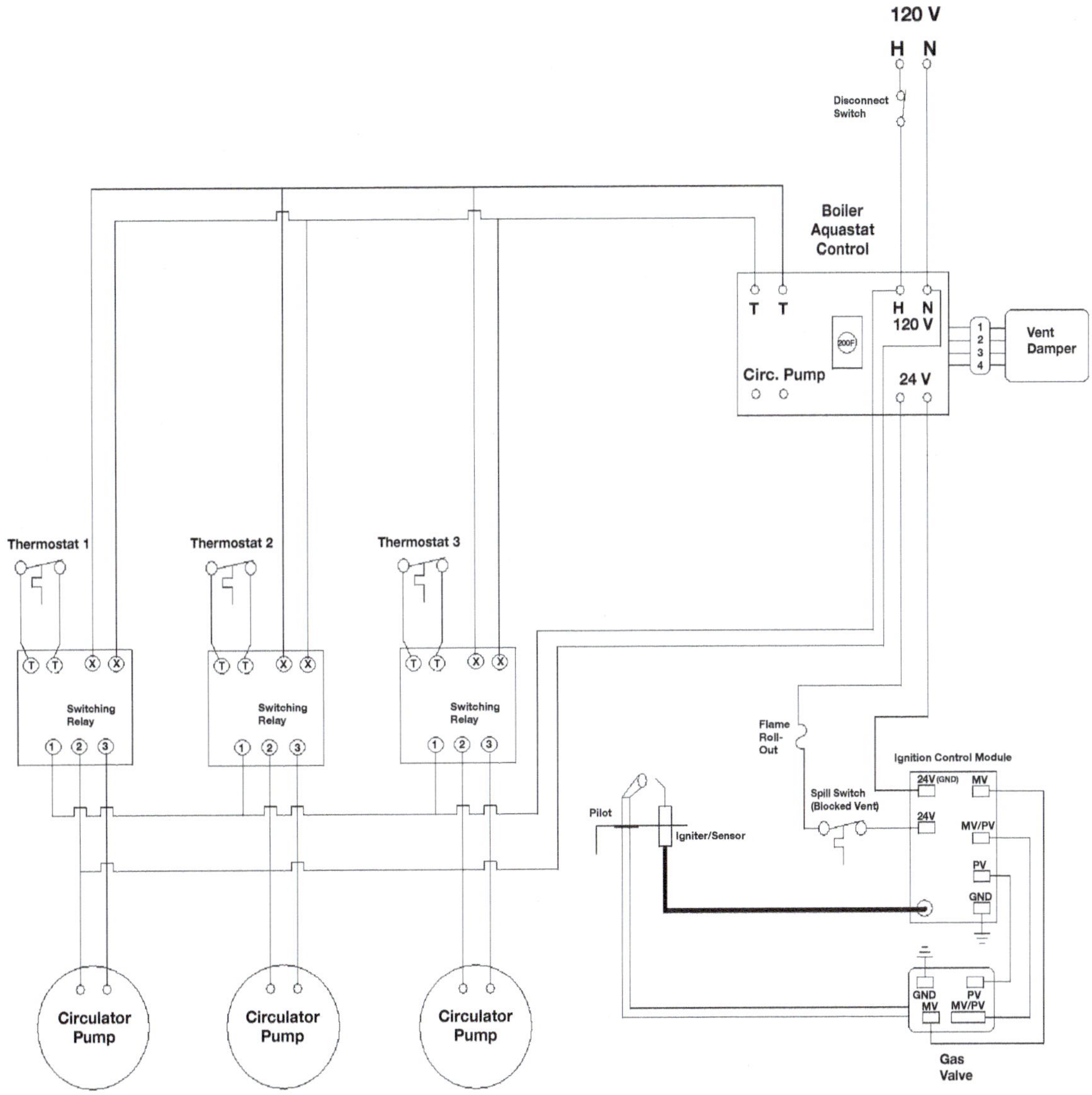

FIGURE 2–16

SECTION 3

Electrical Theory and Application

HVAC/R technicians must have a thorough understanding of electricity and electrical controls. Theory, application, meters usage, and troubleshooting are covered in this section.

ELECTRICAL SAFETY

Service work consists of testing and handling electrical wiring. The technician must be extremely careful when working with electrical circuits. If caution is not taken, serious injury or death could occur. The following is information to help the technician to work in a safe manner. Always respect electricity.

After installing or servicing a system, **inspect all electrical wiring** for loose connections, improper grounding, and defective conductor insulation.

Always **observe precautions, literature, and labels** regarding electrical equipment.

All electrical installations and repairs made must meet the **National Electrical Code** (NEC).

Always be aware of the **electrical rating** of the equipment that is being serviced.

To prevent possible electrocution when repairing or replacing electrical components, always place the **electrical disconnect switch in the "off" position.**

Always use a voltmeter before servicing a system to determine if **voltage** is present.

A **power lock-out tag-out** should be placed on electrical disconnects to prevent the power from being turned on while the equipment is being serviced.

An **electrical shock** is when current flows through a body.

An electric shock across both arms of a technician can cause **heart damage.**

If skin is moist, it is more susceptible to an **electric burn.**

Electrocution from as low as +**25 ma** (milliamps) can kill or cause a serious injury.

To prevent equipment damage or a personal injury, never leave an **electrical safety control** bypassed.

To prevent electrocution, always properly connect equipment **ground wires**.

A **ground fault circuit interrupter (GFCI)** is used to prevent electrocution due to an equipment ground fault.

When using power tools, use an extension cord protected by a GFCI outlet to prevent electrocution.

Never remove the bottom prong of a **three-prong plug.** This will cause the equipment to be improperly grounded.

Always use a **ground adapter** for a two-prong plug to properly ground the electrical equipment.

Never wear **jewelry** when servicing electrical equipment.

Never place both hands near electrical equipment when troubleshooting. This can cause a serious electrocution.

Never work in **damp areas** when servicing electrical equipment.

To prevent a burn or a possible electrocution, always **discharge capacitors** with a 20,000-ohm 2-watt resistor before touching their terminals.

To prevent a possible fire, **never exceed the rated amperage** of the equipment when replacing a fuse or circuit breaker. Always use a timed-delay fuse on motor circuits.

To prevent a possible electrocution, use an insulated **fuse puller** to remove fuses from electrical disconnects.

The technician should know the functions of all electrical components that are in the system's electrical diagram.

To prevent electrocution, inspect hand tools for **adequate insulation** when performing an electrical repair.

Properly secure all hanging wires after an installation or during a service procedure to prevent electrocution or system damage.

Make sure electrical power tools are properly **grounded** before use to prevent electrocution.

The **electrical panels** must be tightly secured on all equipment after service has been performed.

All moving parts should be inspected by the technician to ensure that safety guards are in their proper position.

Technicians should keep their **test instrumentation** properly calibrated.

Technicians should carry a **first aid kit** in their work vehicle.

To prevent a personal injury, always use **adequate lighting** when servicing HVAC/R equipment.

When working on electrical panels, a technician should have thick **rubber-soled shoes** to prevent grounding and possible electrocution.

To prevent a personal injury, **inspect and tighten** all fan motors and blades.

To prevent a personal injury, keep hands and fingers away from **moving fan belts**.

To prevent a personal injury, never wear **loose clothing** when working on HVAC/R equipment.

To prevent a personal injury, always keep **long hair** pulled back or tucked in a cap when working on mechanical equipment.

ELECTRICAL THEORY

Electrical theory is the basis for understanding electricity and electrical controls. Understanding the following electrical terms will help the technician with certification exams as well as with troubleshooting electrical circuits.

An **atom** consists of electrons, protons, and neutrons.

An **electron** has a negative charge.

A **proton** has a positive charge.

A **neutron** has a neutral charge.

An atom that has a **negative electrical charge** is one that has an excess of electrons.

An atom that has a **positive electrical charge** is one that has an excess of protons.

An atom that has a **neutral charge** has the same number of protons as electrons.

Like-charged particles repel each other. Unlike-charged particles attract each other.

Conductors are materials that easily pass electrons. Some examples of conductors are gold, silver, and copper.

Insulators are materials that do not easily pass electrons. Some examples of insulators are rubber, plastic, and glass.

A power supply is created by moving a conductor through a magnetic field. This induces a **current** in the conductor.

Electrical pressure is measured in **voltage.** "E" is the symbol for voltage.

Electrical pressure, also known as **voltage** (or electromotive force), is the potential difference that pushes the electron through an electrical circuit.

A **millivolt** is one thousandth (1/1,000) of a volt. Millivolts should be measured with a millivoltmeter.

A **coulomb** is a quantity measurement of electrons. One coulomb contains 6.25×10^{18} electrons.

Electrons that flow in an electrical circuit are known as **current.** "I" is the symbol for current.

Current is measured in **amperes** or **amps.**

One amp is defined as one coulomb per second.

A **milliamp** is one thousandth (1/1,000) of an amp and should be measured with a milliammeter.

A **microamp** is one millionth (1/1,000,000) of an amp and should be measured with a microammeter.

The opposition to current flow in an electrical circuit is known as **resistance.** "R" is the symbol for resistance.

Resistance is measured in **ohms.** The symbol for ohms is the Greek symbol omega (Ω).

The resistance of a wire will be twice as much if its length is **doubled.**

A **megaohm** is one million (1,000,000) ohms and should be measured with an ohmmeter.

When electrons flow through an electrical circuit, electrical **work** is being accomplished.

Power is the rate at which electrons do work in an electrical circuit. "P" is its symbol.

A **watt** is the measurement of the power in an electrical circuit. "W" is its symbol.

1 watt = 3.412 Btu

1,000 watts = 1 kilowatt

1 kilowatt = 3,412 Btu

1 horsepower = 746 watts

A **wattmeter** measures the true electrical work done in an electrical circuit.

To convert kW to Btu/h, you must use the equation **watts × 3.413**.

When working with an electric heater, the Btu/h conversion would be **3413 Btu/kilowatt.**

MAGNETISM

Many HVAC/R electrical controls and loads operate using the principle of magnetism (electromagnetism). The following terms and concepts are used to aid the technician in gaining a better understanding of magnetism theory.

All **magnets** have both a north pole and a south pole.

Unlike **poles** attract each other and like **poles** repel each other.

Magnetic lines of force flow between the north and south poles of a magnet.

Magnetic lines of force are known as **flux.**

A **permanent magnet** is a piece of material that holds its magnetic force after passing through a magnetic field.

An **electromagnet** is produced by coiling a wire around an iron core and applying current to the wire.

Magnetism is used in inductive loads such as relay coils and motor windings.

Magnetic flux lines are created when current flows through a conductor. When the conductor is coiled, a magnetic field is created.

To **intensify the magnetic field** in an electromagnet, add more turns to the conductor or increase the current flow through the conductor.

An **iron core** tends to center itself in the coil when there is current flow through the conductor.

DIRECT CURRENT

Direct current flows in one direction, from negative to positive. Understanding direct current will help the service technician become more proficient in diagnosing systems problems.

With **direct current (DC),** electrons flow in one direction. When using electron theory, the electrons flow from the negative pole to the positive pole of a DC power source.

A simple circuit consists of three components, **a power source, a path,** and **a load.**

The **power source** is the voltage or potential used to push the electrons. Higher voltage creates a greater push. On a DC power source, red is positive and black is negative.

The **electrical path** is the wire (conductor) that carries the electrons through the circuit. Larger wires have less resistance and greater capability for an increase of current flow. Longer wires increase the resistance.

The **electrical load** is a current-consuming device that converts electron flow into heat, light, or magnetism.

Switches are placed in an electrical circuit to control the electron flow. A closed switch allows current to flow in the circuit. An open switch breaks the circuit, disrupting the current flow.

Switches can be operated manually or automatically. Examples of manually operated switches are push-button, toggle, or lever types. Examples of automatically operated switches are thermostats, pressure switches, magnetic relays, and timed-delay relays.

Ohm's law is used for DC (direct current) or AC (alternating current) circuits that are purely resistive.

The following is the formula for Ohm's law:

E = voltage

I = amperage

R = resistance

$$E = I \times R \text{ or } I = E/R \text{ or } R = E/I$$

Watt's law is used to find the power in an electrical circuit.

The following is the formula for Watt's law:

E = voltage

I = amperage

R = resistance

P = power

$$P = E \times I \text{ or } E = P/I \text{ or } I = P/E$$

$$P = I^2 \times R$$

$$P = E^2/R$$

A **power factor** (Pf) is used to calculate the power rating of an inductive circuit. Power in an inductive circuit is lost due to the out-of-phase relationship that the current has with the voltage.

The calculation of power in an AC inductive circuit is $P = E \times I \times Pf$.

SERIES CIRCUITS

A common series circuit is a switching device that is wired in series with a load for the purpose of control. The following is information on series circuits.

A series circuit is a circuit that has **one path** for current to flow through.

Current stays the same throughout a series circuit.

Voltage drops across loads that are wired in series.

Current decreases as loads are added in series.

Total resistance (RT) increases as loads are added in series.

The calculation for resistance in a series circuit is **RT = R1 + R2 + R3K**

PARALLEL CIRCUITS

Load devices in HVAC/R electrical circuits are usually wired in parallel. The following is information on parallel circuits.

A parallel circuit is a circuit that has **two or more paths** for current to flow through.

Voltage stays the same at each load that is wired in parallel.

Current divides across the individual branches in a parallel circuit.

Current increases as loads are added in parallel.

Total resistance decreases as loads are added in parallel.

The calculation for resistance in a parallel circuit is

$$RT = (R1 \times R2)/(R1 + R2) \text{ or}$$

$$1/RT = 1/R1 + 1/R2 + 1/R3K$$

ALTERNATING CURRENT

Most circuits in the HVAC/R industry operate on an alternating current. It is important for the service technician to understand the principles of alternating current. The following is information that will be helpful to technicians.

An **alternating current (AC)** is a current that reverses the electron flow in a circuit.

Alternating current is generated by moving a conductor through a magnetic field.

Frequency is the number of complete cycles that occur in an AC circuit in 1 s.

Frequency is measured in **hertz (Hz).**

In a **60-Hz** electrical circuit, electrons change direction 120 times in 1 s.

An **AC sine wave** is a graphic illustration that is used to represent the position of the windings of a generator. Generators are used to produce alternating current.

One complete cycle **sine wave** represents one complete cycle of the generator.

An **AC sine wave** contains a total of 30 electrical degrees. It reaches its peak positive voltage at 90 degrees, returns to a value of 0 volts at 180 degrees, increases to its maximum negative voltage at 270 degrees, and returns to 0 volts at 360 degrees. Each complete wave is called a cycle.

The resistance to the passing of magnetic flux lines through a substance is called **reluctance.**

Capacitance is the ability to store electrons in an electrical circuit.

The resistance in an AC circuit due to the change of current flow is called **reactance.**

Capacitive reactance is the opposition to change in voltage caused by the capacitor in an AC capacitive circuit. Current leads voltage in a capacitive circuit.

Inductance is the generation of a voltage by the variance of the current flow rate in an AC circuit.

Inductive reactance is the opposition to current flow in an AC inductive circuit caused by magnetic lines of flux. Voltage leads current in an inductive circuit.

Impedance is the total resistance of an AC circuit, caused by inductive reactance, resistance, and capacitive reactance.

A **single-phase AC** power supply uses two wires to bring voltage to an electrical load. A 120-volt circuit uses one leg of power and a common (neutral) wire. A 240-volt circuit uses two legs of power and does not use a common (neutral) wire in the electrical circuit.

The **hot leg** is the wire used in a power supply that feeds a voltage source to an electrical circuit. In a 120-volt power supply, there is only one leg and it must be fused. In a 240-volt single-phase power supply, both legs of power must be fused.

The hot leg on a 120-volt power supply and both hot legs on a 240-volt power supply must be **fused.** The neutral leg does not need to be fused.

A **three-phase system** has three hot legs that are 120 degrees out of phase with each other.

A **three-phase alternating current** uses three wires to bring voltage to a load. All three legs on a three-phase system must be fused.

The **voltage** between any two hot legs on a 240-volt three-phase power supply is 240 volts and on a 480-volt power supply is 480 volts.

Polarity is the condition that indicates the direction of electron flow. In a 120-volt circuit, black is hot, white is neutral, and green or bare copper is ground.

A **ground** wire connects from the frame of an electrical load to the earth ground. It protects against electrocution during a short-to-ground situation. Its color code is green.

POWER DISTRIBUTION

It is important to understand how utility companies distribute power to customers. The following is information on voltage phases and distribution configurations that utility companies use to distribute electrical power.

Power supplied to substations is **120,000 volts** or higher. The substation reduces the voltage to **4,800 volts** for residential areas and **34,000 volts** for commercial areas.

A **step-down transformer** will reduce the voltage from 4,800 volts to a common low-voltage residential application of 240 volts single phase.

Light commercial applications use either a **delta-** or a **wye-**configured transformer (Figures 3–1 and 3–2). A delta configuration steps the voltage down to 240 volts, three phase (60 Hz). A wye configuration steps the voltage down to 208 volts, three phase (60 Hz).

A **208-volt three-phase wye transformer** is considered a balanced system. Across any two hot legs, a reading of 208 volts should be obtained. From any hot leg to ground, the voltage reading should be 120 volts.

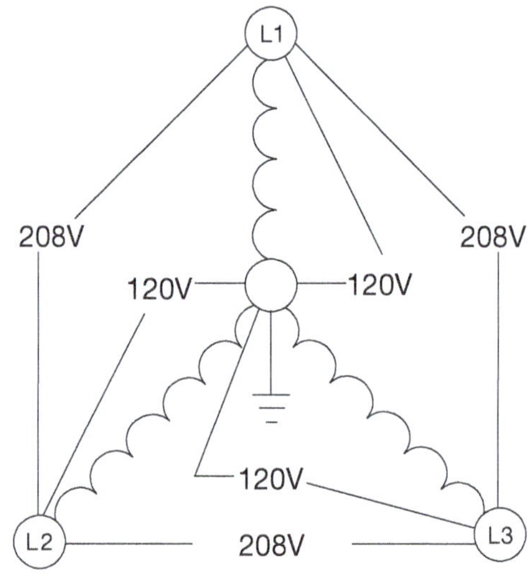

FIGURE 3–1

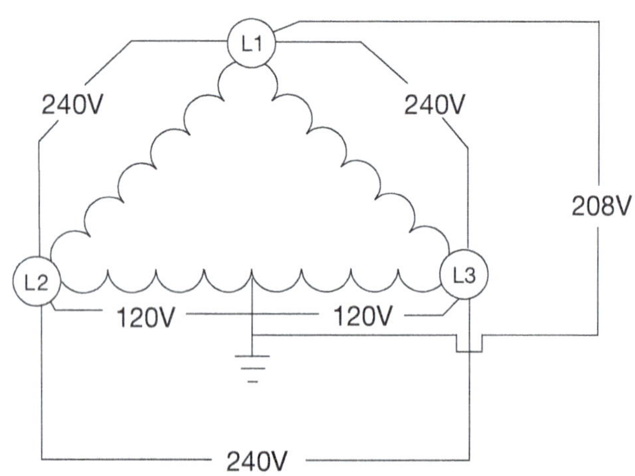

FIGURE 3–2

A **240-volt three-phase delta transformer** is an unbalanced system. Across any two hot legs, a reading of 240 volts should be obtained. From two hot legs to ground, the voltage reading should be 120 volts. From one hot leg to ground, the voltage reading should be 208 volts. The leg that reads 208 volts to ground is considered the wild leg.

Larger commercial and industrial applications will step voltage down to 460 volts, three phase (60 Hz), using a delta-configured transformer. When a wye-configured transformer is used, the voltage will be 480 volts, three phase (60 Hz).

When placing a voltmeter across any two legs on a **460-volt three-phase delta system,** a reading of 460 volts should be obtained.

When placing a voltmeter across any two legs on a **480-volt three-phase wye system,** a reading of 480 volts should be obtained.

When testing a **480-volt three-phase wye system,** any hot leg to ground should measure 277 volts.

The voltage that is supplied to an electrical load must be within **10%** of the designed voltage. An electrical load that is designed for 120 volts can operate on a voltage from 108 to 132 volts. An electrical load that is designed for 240 volts can operate on a voltage from 216 to 264 volts.

AC MOTORS

All service technicians need to understand AC motor principles. The following information covers types of motors, motor controls, capacitors, starting relays, and related terms. This section will help the technician with competency exams as well as in the field.

Induction motors use induced magnetism to convert electrical energy to mechanical energy.

Motors are designed to operate with a supply voltage that is **+ or − 10%** of the voltage that is labeled on the nameplate.

The **nameplate** is a plate located on the motor that lists the design characteristics of the motor.

The **rotor** is the rotating iron core that turns in the magnetic field. The rotor is connected to the shaft of the motor.

The **stators** are the stationary electromagnets. They are also known as poles.

The **stators** have opposite magnetic polarities.

Motor speeds depend on the number of stationary poles in a motor.

A **2-pole** motor is 3,600 rpm.

A **4-pole** motor is 1,800 rpm.

A **6-pole** motor is 1,200 rpm.

Motors are rated in **horsepower.** One horsepower is the amount of power required to lift 550 lb in 1 s.

746 watts equals 1 horsepower.

External motors are **cooled** by air motion. Hermetically sealed compressors are cooled by refrigerant gas and air motion.

Single-phase motors use a **run winding** for motor operation and a **start winding** that gives a motor the starting torque.

The **run winding** is less resistant than the start winding.

The **run winding** is in the circuit during the full operation of the motor.

The **start winding** has a higher resistance than the run winding and is in the circuit on motor startup.

The **start winding** is taken out of the circuit after the motor has reached 75% of its running speed.

The **start winding** is 90 degrees out of phase with the run winding.

The **run and start windings** are wired in parallel to each other.

When **testing** the resistance of motor windings with an ohmmeter, always use the R × 1 ohm scale.

When **measuring the resistance** of the motor windings, run to start is the highest resistance, common to start is the next lowest resistance, and common to run is the lowest resistance.

When measuring resistance, an **open** run winding will read infinite resistance between run and common. An open start winding will read infinite resistance between start and common.

When measuring resistance, a **shorted** run winding will read zero or a low resistance between run and common. A shorted start winding will read zero or a low resistance between start and common.

When testing for a **short to ground motor winding,** always use the highest ohm scale.

If a **motor winding** is shorted to ground or the internal windings are shorted together, it commonly trips a circuit breaker or blows a fuse.

When testing for a **grounded motor winding,** a measurable resistance from any motor terminal to ground shows a short-to-ground condition.

The **contacts** of a start relay are wired in series with the start winding.

Start relays used on refrigeration compressors must be external due to the arcing of the contacts. This could cause acids to form in a refrigeration system.

A **centrifugal switch** is a normally closed switch that takes the start winding out of the circuit.

A **centrifugal switch** uses the centrifugal force of the rotor to open a set of normally closed contacts that are in series with the start winding.

A **current relay** uses the operating current of the run winding to control a set of contacts.

A **current relay** has a normally open set of contacts in series with the start winding.

A **potential relay** uses the back electromotive force (back voltage) of the start winding to operate a set of contacts.

A **potential relay** has a set of normally closed contacts in series with the start winding.

On a **potential relay,** terminals 1 and 2 are the normally closed contacts and terminals 2 and 5 are the coil.

On a **potential relay,** terminal 1 connects to the start capacitor, terminal 2 connects to the start winding, and terminal 5 connects to the common terminal located on the compressor.

A **solid-state starting relay** uses a positive temperature coefficient (PTC) thermistor in series with the start winding.

On a **solid-state relay,** as the temperature increases, the resistance increases.

A **hard start kit** consists of a start capacitor and start relay.

FLA, or full load amps, is the operating or running amperage of a motor.

LRA, or locked rotor amps, is the starting amperage of the motor.

The **LRA** is commonly five times higher than normal running amperage.

Overload protectors are commonly a bimetal snap disc or a thermistor.

Overload protectors open a set of normally closed contacts by sensing excessive heat, excessive current, or both.

Overload protectors can be external or internal types.

External overloads are located outside a motor.

Internal overloads are located in the windings of a motor.

Overload protectors can be used in a pilot duty or line break circuit.

Pilot duty overload protectors use a normally closed set of contacts in series with a contactor or starter coil that controls the power supply to the motor being protected. When testing an open internal overload protector with an ohmmeter, use the R × 1 scale.

Line break overload protectors use a normally closed set of contacts in series with the common side of a motor or compressor.

When testing for an **open internal overload protector,** the ohmmeter will read measurable resistance across run and start, infinite resistance across run and common, and infinite resistance across start and common.

When testing for an **open internal overload protector** using an ammeter and a voltmeter, the ammeter will read zero while the voltmeter will read source voltage across the motor.

Capacitors are used to give a motor starting torque or to aid a motor in its running operation.

Capacitors are rated in **microfarads** (mfd or µf).

A capacitor is designed with a specific **VAC (volts AC)** rating. The voltage to a capacitor must never exceed this rating or the capacitor could be damaged.

The **start capacitor** is used on Capacitor Start (CS) and Capacitor Start/Capacitor Run (CSR) motors. It is a dry-type capacitor and uses a 20,000-ohm 2-watt bleed resistor to discharge the capacitor during the off cycle.

Start capacitors are designed to energize the start windings of a motor until its rotor reaches about 75% of its running speed.

The **start capacitor** is de-energized by a start relay. The most common start relays are the current relay, the potential relay, the solid-state relay, and the centrifugal switch.

The **run capacitor** is used on the Permanent Split Capacitor (PSC) and Capacitor Start/Capacitor Run (CSR) motors. It is an oil-filled capacitor, and the oil dissipates heat during operation.

Run capacitors are used to give the motor a small starting torque. They are left in the circuit to increase the running efficiency of the motor.

A **dual capacitor** is two run capacitors set in one container.

Dual capacitors are usually found on residential and window air conditioning systems. They are connected to a PSC-type compressor and condenser fan motor.

Capacitors can be wired in series or in parallel to increase or decrease the capacitance in a circuit. The following are calculations that are used to figure the circuit's capacitance:

C1 = Capacitor number one

C2 = Capacitor number two

TC = Total capacitance

When two **capacitors** are wired in parallel, the calculation for the capacitance is

$$C1 + C2 = TC.$$

When two **capacitors** are wired in series, the calculation for the capacitance is

$$\frac{C1 \times C2}{C1 + C2} = TC.$$

When replacing a run or start capacitor, the **voltage rating** must always stay the same.

When **testing** a start capacitor with an ohmmeter, the bleed resistor must be temporarily removed.

When **testing** a capacitor with an ohmmeter, a good capacitor will move from infinite resistance to zero resistance then back to infinite. A shorted capacitor will stay at zero or a readable resistance. A capacitor that is internally burned open will stay on infinite resistance.

A **capacitor analyzer** is used to measure the microfarad, the power factor, and the condition (shorted or open) of the capacitor.

SHADED-POLE MOTORS

Shaded-pole motors are used for small condenser and evaporator fans and small, low-Btu-output furnaces.

Shaded-pole motors have copper rings located on the stator of the shaded pole. A current is induced in the shaded pole from the main winding. The shaded pole produces a magnetic field that is out of phase with the main winding, thus giving the motor its starting torque.

Reversing the direction of a shaded-pole motor can be done by disassembling the motor, removing the rotor, turning the rotor 180 degrees, and reinstalling it.

Shaded-pole motors can have many different **speeds,** but most have two or three speeds.

SPLIT-PHASE MOTORS

A **split-phase-type compressor motor** is used on refrigeration systems that incorporate a capillary tube metering device. Split-phase motors are also used for oil burners, exhaust fans, and small water pumps.

A **split-phase motor** uses an inductive run winding to operate the motor and a resistive start winding to give the motor its starting torque.

A **split-phase motor** does not use a run or start capacitor.

A **split-phase motor** is commonly a fractional horsepower type of motor.

A **split-phase motor** commonly uses a centrifugal switch, a current relay, or a solid-state relay to de-energize the start winding after the motor has reached 75% of the running speed.

Reversing a split-phase motor is done by reversing the leads on the start winding.

CAPACITOR START INDUCTION RUN (CSIR) MOTORS

A **capacitor start induction run (CSIR) motor** produces a high starting torque. The CSIR motor is a split-phase motor with a start capacitor added in series with the start winding.

When a CSIR motor reaches **75% of its running speed,** the start winding and the start capacitor are de-energized by a centrifugal switch or a starting relay. The motor run winding stays energized to keep the motor in operation.

A **CSIR motor** is used on compressors up to 5 hp, larger water pumps, large belt drive fans, and with compressors where the refrigeration system uses a TXV-type metering device.

If a start capacitor has too low of a **microfarad rating** for the CSIR motor, the motor will not start.

PERMANENT SPLIT CAPACITOR (PSC) MOTOR

The **permanent split capacitor (PSC) motor** is designed to have a low starting torque with a high running torque.

A **run capacitor** is wired in series with the start winding. The start and run windings are in the circuit during the full operation of the motor.

The **run capacitor** limits the electron flow through the start winding so it will assist with the running torque of the motor.

A **PSC** motor does not need a start relay or a start capacitor.

PSC compressors are used with systems that equalize during the off cycle.

Multispeed PSC motors with direct drive-blower motors are very common.

If a run capacitor has too low of a **microfarad rating** for the PSC motor, the motor might not start. It could also increase the amperage of the motor during the running cycle.

CAPACITOR START/CAPACITOR RUN (CSR) MOTOR

The **capacitor start/capacitor run (CSR) motor** produces high starting and running torque. It is actually a CSIR motor with a run capacitor added to the start winding.

When the motor reaches **75% of its running capacity,** the start capacitor is de-energized by a starting relay. The run capacitor controls a small amount of current through the start winding during the running operation of the motor. This causes the motor to operate with an increase of running torque.

A **CSR motor** is used where high starting and running torque is required, such as large hermetic and semihermetic compressors.

A **CSR motor** commonly uses a potential-type starting relay.

If a *start* capacitor has too low of a **microfarad rating** for the CSR motor, the motor will not start.

If a *run* capacitor has too low of a **microfarad rating** for the CSR motor, the motor may not start or it could increase the amperage of the motor during the running cycle.

THREE-PHASE MOTORS

Three-phase motors are large, rugged motors. They are much stronger than single-phase motors.

Three-phase motors use three windings that are 120 electrical degrees out of phase. This causes a large starting torque and an excellent running efficiency.

Three-phase motors do not have start windings, capacitors, or starting relays as do single-phase motors.

Common **voltages** for three-phase motors are 208/240 and 440/480 volts.

They have two basic types of **winding configurations,** the wye and the delta.

To **reverse the direction** of a three-phase motor, reverse any two power leads connected to the motor.

Three-phase motors are used for large compressor motors, large blower motors, and large pumps normally above 5 hp.

An **integral controlled motor (ICM)** is a variable speed motor controlled by the output of a circuit board.

ELECTRONIC COMMUTATED MOTORS (ECM) VS. PERMANENT SPLIT CAPACITOR MOTORS

Permanent split capacitor (PSC) motors are single-phase induction motors.

PSC motors have a run and start winding.

PSC motors use a **run capacitor** that is wired in series with the start winding.

PSC motors have **2 to 5 speeds** that are manually adjustable.

PSC motors range from **40 to 60%** efficiency.

PSC motors have sleeve bearings or ball bearings.

Sleeve bearings will not lubricate below 500 rpm and will overheat.

Typical speed ranges for PSC motors are 800 to 1100 rpm; the most common is 1,075 rpm and is affected by the external static pressure of the HVACR system.

$$\textbf{RPM} = \frac{60 \text{ Hz} \times 120 \text{ cycles per seconds}}{\text{Number of poles}}$$

$$\textbf{\textit{Example:}}\ 1{,}200 \text{ rpm} = \frac{60 \text{ Hz} \times 120 \text{ cycles per second}}{6 \text{ poles}}$$

Average constant fan power consumption for **PSC motors** is approximately 515 watts.

Variable speed and constant torque motors use a motor stator winding, an electronic motor control, and a permanent magnet for the rotor.

Variable speed and constant torque (ECM) motors are **80%** efficient.

Variable speed and constant torque **motor controllers** convert single-phase AC power to three-phase DC power.

The **new energy code** in 2019 requires HVAC equipment to have ECM motors installed during manufacturing.

Variable speed motors (VSMs) work at different speeds according to **static pressure** of the system.

VSM operates at a **low constant airflow** that helps prevent dust collection, prevents stagnation of the air in a building, and increases optimum humidity control.

Variable speed (VS) is a **brushless three-phase DC**-type motor.

VS motors can show the **actual cfm.**

VS motors' power consumption is approximately **83 watts.**

The VS motor **starts unloaded and operates slowly**, which reduces the locked rotor amps and places less stress on the blower wheel.

VS motors can be set to a specific cfm using a **16-pin harness** and adjust using system static pressure.

Variable speed and constant torque motors have a **5-pin power supply** connection.

Speed ranges for VS motors are 200 to 1,300 rpm.

Constant torque motors use a **4-pin harness** and operate on either 120 V or 230 V.

A constant torque motor (CTM) is a **brushless three-phase DC-type motor.**

Output ranges for single-phase PSC, CT, and VS motors are $\frac{1}{3}$, $\frac{1}{2}$, $\frac{3}{4}$, and 1 hp.

Input single-phase voltages for PSC, CT, and VS motors are 120 V or 240 V.

Speed ranges for CTM are 600 to 1,100 rpm.

CTM 2–5 torques are preprogrammed.

VS motors are programmed with adjustable variable airflow or torque.

CTM and VS motors have **ball bearings.**

Sleeve bearings will not lubricate below 500 rpm.

CTM power consumption is approximately **200 watts.**

The **VS motor** operates quieter than the PSC and CT motors.

ELECTRICAL CONTROLS

There are numerous electrical controls in the HVAC/R industry. The following information will give the technician a better understanding of electrical control operation.

Fuses and circuit breakers are considered overcurrent protectors. They are used to open a circuit if the current reaches a value that will cause excessive or dangerous temperatures in the conductor or conductor insulator.

A **fuse** (Figure 3–3) is rated for a maximum amperage and a maximum voltage.

To test a fuse with an ohmmeter, disconnect the power supply, remove the fuse from the circuit, set the ohmmeter to the R × 1 scale, and place the leads across the fuse. If the fuse is in good condition, there will be zero resistance. If it is burnt open, there will be infinite resistance.

To test a circuit breaker with a clamp-on ammeter, clamp the ammeter around one wire that leads to the circuit breaker. Set the circuit breaker to the "on" position and check the amperage reading of the circuit. If the amperage exceeds the rating of the circuit breaker, test the circuit for an electrical short or a grounded condition. If the amperage does not exceed the rating of the circuit breaker and the circuit breaker trips the circuit off, the breaker is defective.

A heating and cooling **thermostat** is used to sense space temperature. Many different types of sensing elements are used in thermostats, such as liquid- or gas-charged sensing bulbs, bimetal elements, and thermistors.

A **bimetal sensing element** is two different metals bonded together, commonly brass and invar, to cause a warping action. This warping action will either open or close a set of contacts.

A **bimetal snap disc** sensor is used on thermostats, limit controls, and fan controls.

A liquid-, gas-, or cross-charged (gas and liquid) **sensing bulb** is commonly used to operate a bulb-type temperature control.

The **liquid-charged** sensing bulb is extremely sensitive. It is completely full of liquid and has no vapor present.

A **gas-charged** sensing bulb contains all gas. It has no liquid in the bulb at any time. For proper operation, it must be installed so that the bulb is the coldest part of the control.

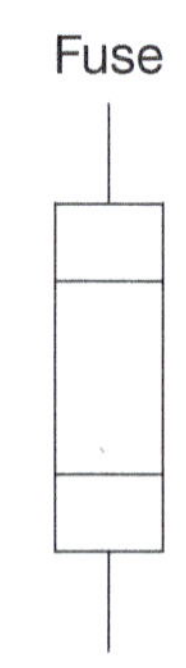

FIGURE 3–3

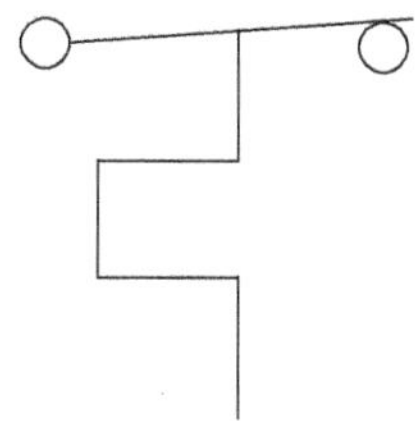

FIGURE 3–4

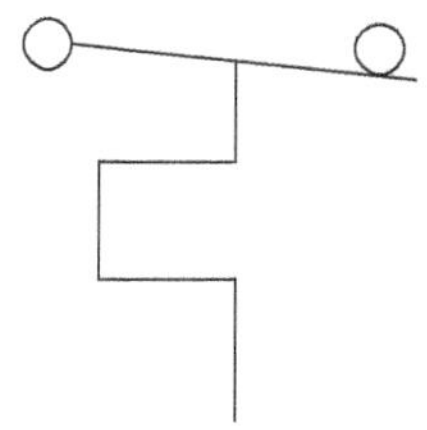

FIGURE 3–5

A **cross-charged** sensing bulb always has liquid and vapor present. The temperature control can be installed where the bulb is at a warmer or colder location than the rest of the control.

A **positive temperature coefficient thermistor** is used on electronic thermostats as the sensing element. Its resistance increases as the temperature increases.

A **negative temperature coefficient thermistor** is used on electronic thermostats as the sensing element. Its resistance increases as the temperature decreases.

A **heating thermostat** (Figure 3–4) opens on a temperature rise and closes on a temperature drop.

The heat anticipator is used to maintain stable **temperatures** within a space. It will prevent system overshoot (too long of an on cycle) and system lag (too short of an off cycle).

The **heat anticipator** must be set to the heating control amperage.

The **heating control amperage** can be measured by placing an in-line ammeter across terminals "R" and "W." It also can be measured by placing a 10-loop multiplier wire across terminals "R" and "W" and measuring the amperage with a clamp-on ammeter.

Mercury is commonly used in a sealed glass bulb filled with an inert gas to close switch contacts on heating and cooling thermostats.

Single-stage heating and cooling thermostats commonly use **four-strand wire** for control. The red wire is power from the transformer, the white wire is to the heating control circuit, the green wire is for the fan relay circuit, and the yellow wire is for the cooling circuit.

Two-stage thermostats will have two extra wires for the second-stage cooling circuit and the second-stage heating circuit.

Automatic changeover is commonly incorporated in commercial thermostats. This allows for the system to change from heating to cooling automatically.

Deadband is the point at which an automatic changeover thermostat is balanced between the heating and cooling cycle.

Outdoor thermostats are commonly used on electric heating systems to sense the outdoor temperatures.

When the outdoor temperatures drop below a predetermined set point, the thermostat energizes a bank of electric strip heaters.

A **cooling thermostat** (Figure 3–5) opens on a temperature drop and closes on a temperature rise.

A **cooling anticipator** is wired in parallel with the cooling circuit. It prevents system lag.

The **cooling anticipator** is commonly a nonadjustable resistor placed across the cooling thermostat.

An **outdoor ambient thermostat** is used on cooling systems to control condenser fan speeds if the temperature drops below a predetermined point. During low ambient conditions, the fan speed is decreased.

An **outdoor ambient thermostat** is used on cooling systems to de-energize the compressor circuit if the temperature drops below a predetermined point. During low ambient conditions, the compressor is locked out to prevent compressor damage.

A **programmable thermostat** is used to control an HVAC system with more accuracy than a standard bimetal thermostat.

A programmable thermostat uses a **thermistor** instead of a bimetal to sense room temperature.

A programmable thermostat's **cooling anticipator** has a fixed setting.

A programmable thermostat's **heat anticipator** can be adjustable or have a fixed setting.

LED lights are commonly used on programmable thermostats as indicator lights.

An **emergency heat light** is used with a programmable heat pump thermostat to indicate the operation of the emergency heat cycle.

A **24-volt power supply** is commonly brought to a programmable thermostat to energize the LED lights located in the thermostat.

A **two-stage programmable** heat pump thermostat uses the first stage to energize the compressor and the second stage to energize the auxiliary strip heaters.

A **battery** is used in a programmable thermostat to save the program and maintain clock time if there is a power failure.

Automatic changeover is commonly used in a programmable thermostat. This allows for the system to change from heating to cooling automatically.

Programmable thermostats are available with a **5- or 7-day** programming capability.

Temperature set back is the temperature at which a thermostat is programmed during an unoccupied time for the purpose of energy conservation.

Set point is the setting of the thermostat during an occupied time.

Computed or intelligent recovery is commonly used in programmable thermostats. This feature determines the rate that the space temperature deviates from the system's set-point temperature to the system's set-back temperature. The thermostat then starts the system to allow the space temperature to increase gradually to the predetermined set point at a preset time.

The **override** feature in a programmable thermostat is used to temporarily bypass the system's programmed schedule. This can be accomplished by pressing the override or hold button.

A **limit switch** (Figure 3–6) is used to de-energize a gas valve if the temperature of the furnace rises above a predetermined temperature.

A limit **switch** commonly uses a bimetal as the sensing element and can be an automatic or a manual reset type.

A **low-pressure switch** (Figure 3–7) is used in an air conditioning system to sense the low-side pressure on a system. The low-pressure switch opens when there is a drop in pressure and closes when there is a rise in pressure.

The low-pressure switch can be used as a safety control to sense a drop in system pressure due to a loss of refrigerant charge. This will protect the compressor from both overheating and loss of lubrication.

The low-pressure switch can also be used as an operating control. As the space temperature decreases, the system's operating pressure decreases. This fluctuation in system pressure can cause a low-pressure switch to open and close the circuit to the compressor.

A **high-pressure switch** (Figure 3–8) is used in an air conditioning system to sense a rise in high-side pressure. If the pressure increases above a predetermined point, the high-pressure switch will open and de-energize the compressor.

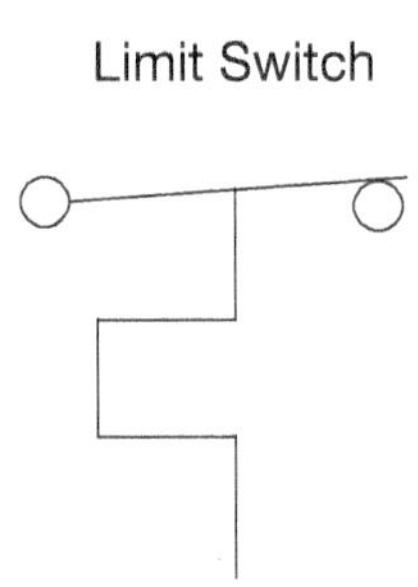

FIGURE 3–6

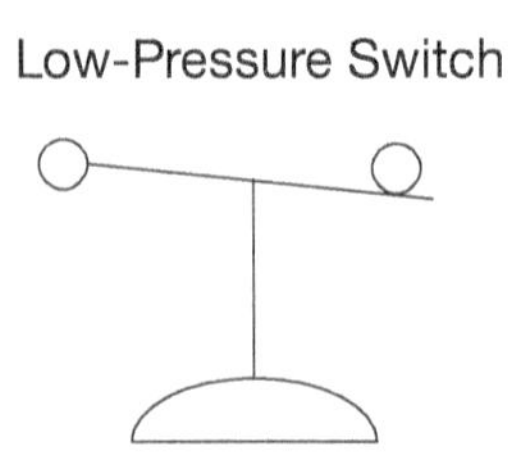

FIGURE 3–7

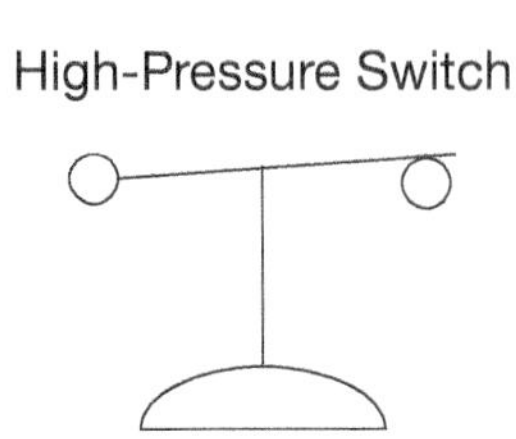

FIGURE 3–8

The **high-pressure switch** is used as a safety control. It is either an automatic or a manual reset type.

A **reverse-acting high-pressure switch** is used during low ambient conditions to control condenser fans on an air-cooled condenser. By cycling the condenser fans, the control of high-side pressure is accomplished.

A **relay** (Figure 3–9) is used to energize or de-energize a load circuit by using the signal of a control circuit. When the coil of a relay is energized, the contacts in the relay change their positions. A normally open contact will close and a normally closed contact will open. Relays commonly energize motors, heaters, lights, contactors, or other relay coils.

The **contacts** of a relay must never exceed 20 amps.

When using an ohmmeter, a relay coil should have a **measurable resistance** across the coil.

When **control voltage** is applied across the coil, the switch contacts should change positions.

The **inrush current** on the startup of a relay is higher than the holding current during relay operation.

If a **relay coil** is shorted or burnt open, the switch contacts will not change positions.

When measuring the resistance of a relay coil and an **infinite resistance** is obtained, the coil is burnt open.

When measuring the resistance of a relay coil and a **zero resistance** is obtained, the coil is shorted.

A **shorted relay coil** commonly damages the control transformer.

When **replacing a relay,** it is important to know the relay's switch configuration and the coil voltage.

A **contactor** (Figure 3–10) is used to energize or de-energize high-voltage loads. When the coil of a contactor is energized, the contacts in the contactor change their positions. A normally open contact will close and a normally closed contact will open. Contactors commonly energize motors and electric heaters. The coil voltage of a contactor can be 24, 120, 240, or 480 volts.

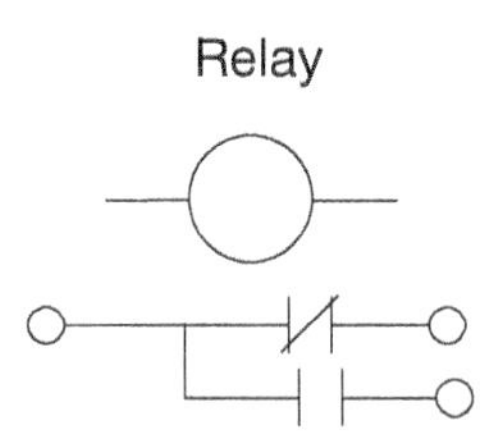

FIGURE 3–9

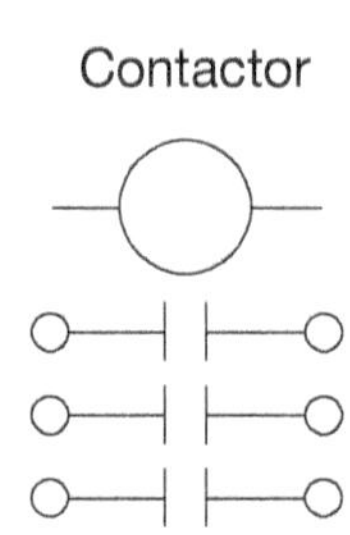

FIGURE 3–10

When using an ohmmeter, a contactor coil should have a **measurable resistance** across the coil.

When **control voltage** is applied across the coil, the switch contacts should change positions.

If a **contactor coil** is shorted or burnt open, the switch contacts will not change positions.

When measuring the resistance of a contactor coil and an **infinite resistance** is obtained, the coil is burnt open.

When measuring the resistance of a contactor coil and a **zero resistance** is obtained, the coil is shorted.

A **shorted contactor coil** commonly damages the control transformer.

When **replacing a contactor**, it is important to know the contactor's switch configuration, coil voltage, and contact rating.

The difference between a **relay and a contactor** is the amount of current that the contacts can handle. A relay can handle under 20 amps and a contactor can handle over 20 amps.

A **starter** (Figure 3–11) is composed of contacts, a solenoid coil, and a set of overloads. When control voltage is applied across the coil of the starter, the contacts change positions. A normally open contact will close and a normally closed contact will open. Starters commonly control the operation of motors. Overload protectors are built into the starter to de-energize the coil if an overload occurs.

When using an ohmmeter, a starter coil should have a **measurable resistance** across the coil.

If a **starter coil** is shorted or burnt open, the switch contacts will not change positions.

When measuring the resistance of a starter coil and an **infinite resistance** is obtained, the coil is burnt open.

When measuring the resistance of a starter coil and a **zero resistance** is obtained, the coil is shorted.

A **shorted starter coil** commonly damages the control transformer or main line fuse.

When **replacing a starter,** it is important to know the starter's switch configuration, coil voltage, and contact rating.

The difference between a **contactor** and a **starter** is the overload protectors that are built into the starter.

There are **three types of overload protectors** used in starters. They are the bimetal relay, the thermal relay, and the molten-alloy relay.

The **bimetal relay** uses two dissimilar metals bonded together. When current increases, more heat is produced. This causes the bimetal to warp, breaking the circuit to the starter coil.

The **thermal overload** uses a wire that is in series with the load being controlled. On an overload condition, the wire overheats and opens a set of contacts that are in series with the starter coil. **Overload condition** refers to excessive electrical current in the form of heat flowing through the wire.

The **molten-metal relay** places the molten alloy in series with the load being controlled. If an overload condition occurs, the metal melts. This will cause a ratchet wheel to open a set of contacts that are in series with the starter coil.

A **timed-delay relay** (TDR) (Figure 3–12) is used to delay the starting or stopping of a load. TDRs are used on furnaces to delay the blower operation or on air conditioning systems as an antishort cycle device for the compressor. The length of the delay can be preset or adjustable.

A timed-delay relay is a control switch used to transfer an action in the control circuit to the operation of a higher-voltage load. The timed-delay relay has a heater or a solid-state circuit controlling a set of normally open (N.O.) or normally closed (N.C.) contacts.

The timed-delay relays can either be a **delay on make** or **delay on break** type. The delay circuit is energized by the control circuit.

A **delay on make relay** (Figures 3–13 and 3–14) delays the opening or closing of contacts for a predetermined amount of time.

The **delay on make relay** is commonly used in air conditioning circuits to delay the starting of the compressor circuit.

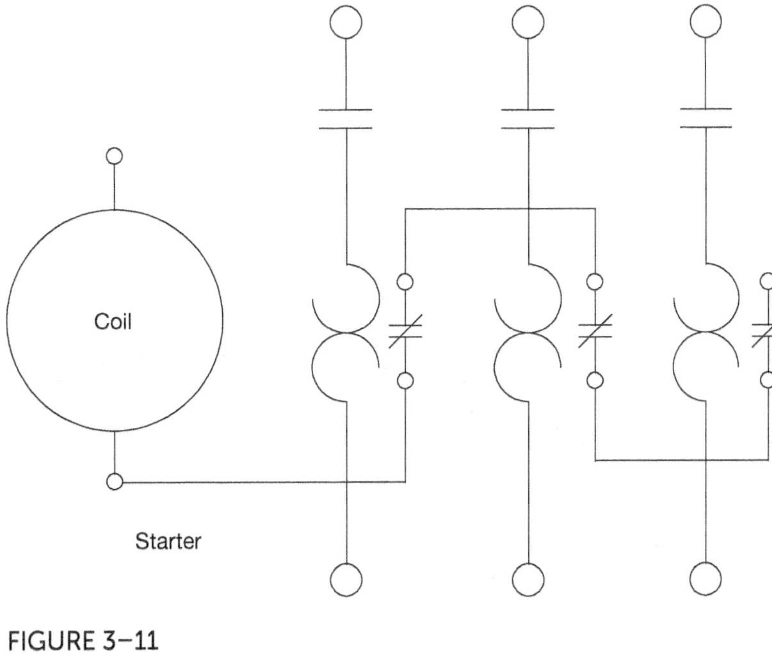

FIGURE 3–11

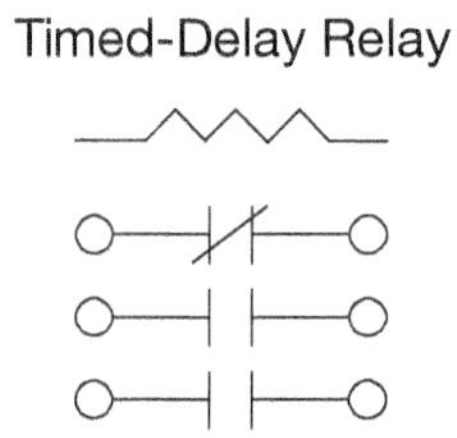

FIGURE 3–12

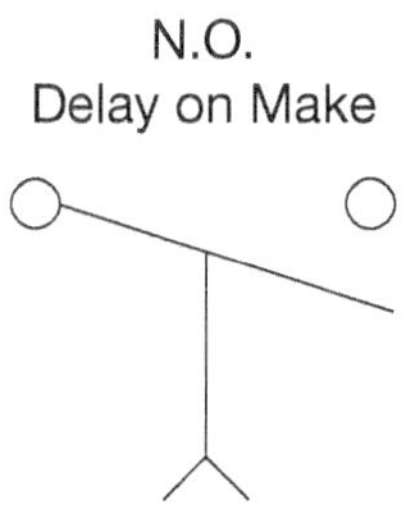

FIGURE 3–13

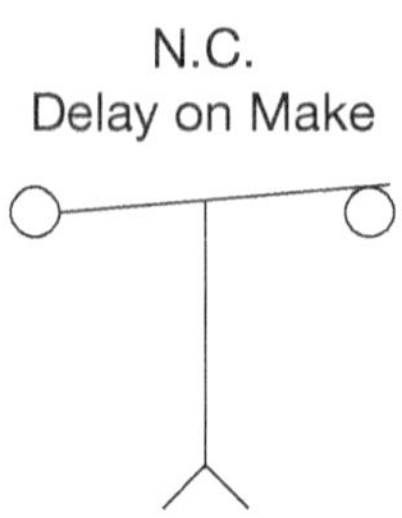

FIGURE 3–14

A **delay on break relay** (Figures 3–15 and 3–16) delays the opening or closing of contacts for a predetermined amount of time.

The **delay on break relay** is commonly used in heating circuits to delay the stopping of the indoor blower after the heating cycle.

A **sequencer** (Figure 3–17) is a series of switches that sequentially controls loads in an HVAC system.

There are many different methods of **sequential operating controls,** such as thermal-operated timed-delay relays, solid-state circuit boards, and cam-operated timers.

A **sequencer** is commonly used in an electric heating system. The sequencer energizes the heaters, delaying current flow to each heater element. This will decrease the inrush current on the startup of the system.

Microprocessors are commonly used as sequencers. They are preprogrammed circuit boards that have many different switching combinations.

A **defrost cycle** is commonly controlled by an electric defrost timer or a solid-state circuit board.

An electric **defrost timer** (Figure 3–18) uses a synchronous motor to control a cam, which in turn controls switch contacts. Defrost timers are commonly used to control a defrost cycle located in a refrigeration or heat pump system.

Microprocessors can be used as defrost timers. They can be programmed to energize a defrost circuit at a specified period in time or when the need for defrost is detected.

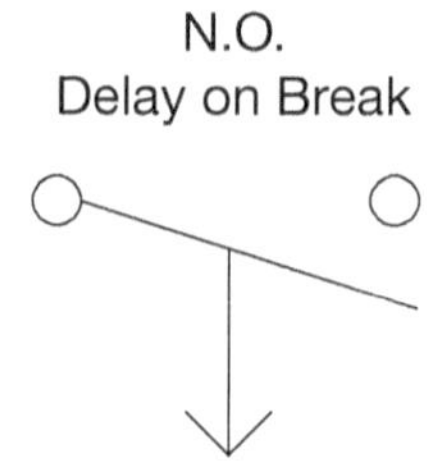

FIGURE 3–15

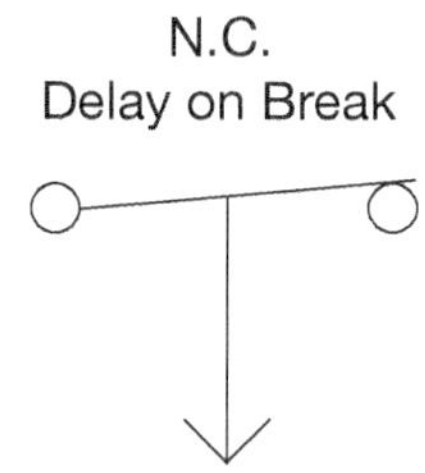

FIGURE 3–16

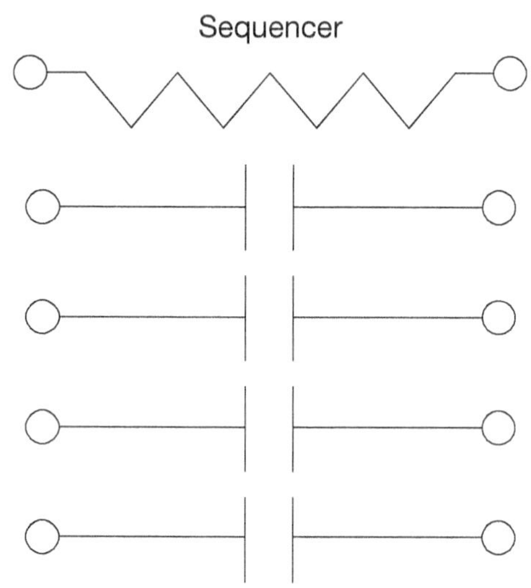

FIGURE 3–17

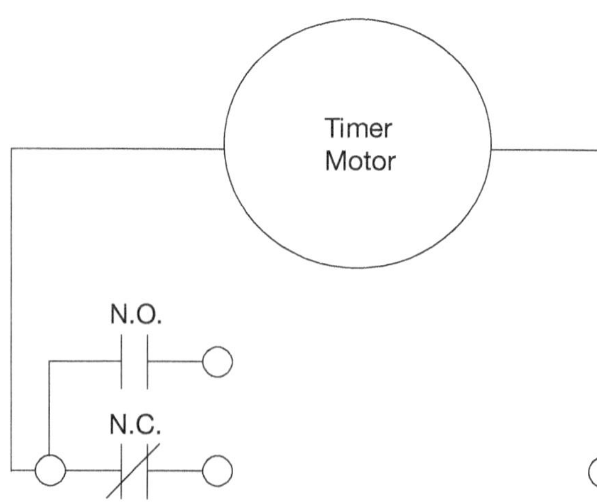

FIGURE 3–18

A **solenoid valve** (Figure 3–19) is used to control the flow of refrigerant in an air conditioning or refrigeration system. When energized, a normally closed solenoid valve will open and a normally open solenoid valve will close.

On an **automatic pump-down system,** a normally closed solenoid valve is placed in the liquid line. A thermostat directly controls the solenoid valve and a low-pressure switch controls the compressor. When the thermostat closes, the solenoid valve opens. This will allow refrigerant to flow to the evaporator coil. When refrigerant enters the evaporator, the pressure increases. This will close the pressure switch and energize the compressor. When the thermostat is satisfied, it will open and de-energize the solenoid valve. This will cause the valve to close and it will block the flow of refrigerant to the evaporator coil. The compressor will continue operating until the refrigerant is pumped out of the evaporator. This will cause the pressure switch to open and de-energize the compressor.

During a defrost on refrigeration systems, a **hot gas solenoid valve** is used to direct hot gas from the discharge line to the inlet of the evaporator.

A **reversing valve solenoid** is used to reverse the flow of refrigerant in a heat pump system.

Hot gas bypass solenoid valves are used for capacity control on commercial air conditioning systems.

Crankcase heaters are placed in or around the shell of a compressor. They are used to prevent migration of refrigerant into the crankcase during low ambient temperatures.

Crankcase heaters are commonly energized during the system's off cycle.

Electric heater elements are used in electric furnaces as the primary heat source.

Electric heaters (Figure 3–20) are rated in watts. One watt is equal to 3.412 Btu.

Auxiliary strip heaters are used in heat pumps to aid in the heating and defrost cycle during low outdoor ambient temperatures.

The **auxiliary strip heaters** are located downstream of the indoor coil on a heat pump. They are electrically controlled by the second stage of a two-stage thermostat.

Fusible links are safety controls used to break a circuit on an increase of temperature.

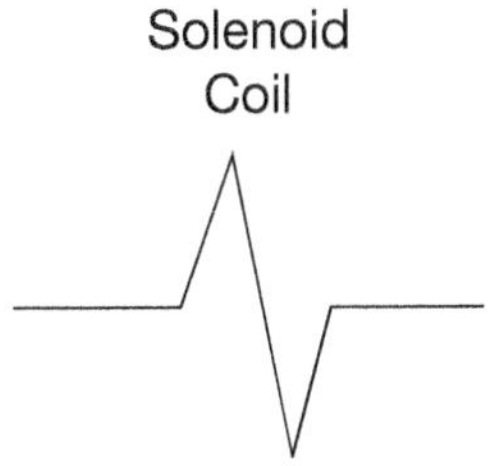

FIGURE 3–19

FIGURE 3–20

On auxiliary strip heaters, **fusible links** (Figure 3–21) are wired in series with the limit control and the electric heater. This secondary safety control is used to prevent overheating of the heating element.

Fusible links are also used as flame roll-out sensors on many gas-fired furnaces.

The **step-up transformer** (Figure 3–22) takes a lower voltage and increases it to a higher voltage. They are used to apply a spark across an electrode for the ignition of a heating cycle. An ignition transformer steps voltage up from 240, 120, or 24 volts to 7,000 volts to 10,000 volts.

The **step-down transformer** takes a higher voltage and decreases it to a lower voltage. They are commonly used as the control voltage for relays, contactors, and other low-voltage controls in an HVAC/R system. A step-down transformer commonly steps voltage down from 120 or 240 volts to 24 volts.

The **turn ratio** on the secondary winding compared to the primary winding of a transformer determines whether it is a step-up or a step-down type. If there are more turns on the secondary winding than on the primary winding, it is a step-up transformer. If there are more turns on the primary winding than on the secondary winding, it is a step-down transformer.

If a **transformer** uses a 240-volt primary winding that has 400 turns and the secondary has 200 turns, the output voltage will be 120 volts.

When replacing a step-down transformer, the **VA rating** is important. This is the power or wattage that is being consumed in a control circuit. If the VA rating of the transformer is lower than the power rating of the control circuit, the transformer will burn out.

To calculate the **VA rating,** measure the amperage of the control circuit and multiply it by the voltage of the control circuit.

Example: The amperage of a system's control circuit is 1.6 amps and the voltage being used is 24 volts.

$E \times I = W : 24 \text{ volts} \times 1.6 \text{ amps} = 38.4 \text{ watts.}$

FIGURE 3–21

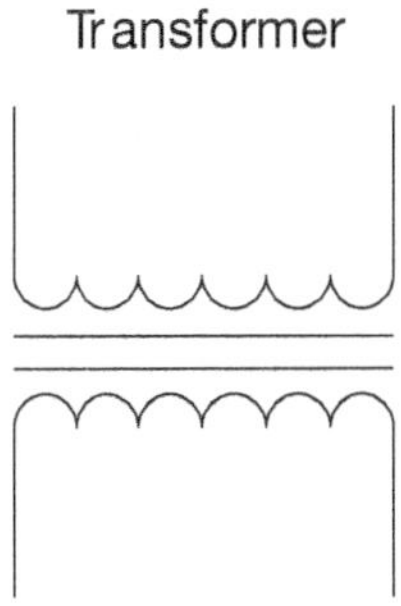

FIGURE 3–22

The transformer that can be used on this circuit must have a minimum VA rating of 38.4. The transformer that would be used for this circuit would commonly be a 40-VA-rated transformer.

When testing a transformer **using a voltmeter**, source voltage at the primary and secondary side of the transformer shows that the transformer is functional.

If there is **voltage** at the primary side but not at the secondary side of the transformer, the problem could be a bad primary or secondary winding.

When testing a transformer using an **ohmmeter**, measurable resistance should be across the secondary and the primary windings.

An infinite resistance across the **secondary winding** of a transformer shows an open secondary winding.

A zero or low resistance across the **secondary winding** of a transformer shows a shorted secondary winding.

An infinite resistance across the **primary winding** of a transformer shows an open primary winding.

A zero or low resistance across the **primary winding** of a transformer shows a shorted primary winding.

SOLID-STATE ELECTRONICS

New HVAC/R equipment uses solid-state components. The service technician must be able to understand the components of solid-state controls in order to be capable of troubleshooting system malfunctions. The following are terms related to solid-state electronics.

Solid-state circuit boards are becoming more common in residential and light commercial HVAC/R. They are being used to replace electrical and electromechanical controls that are used for safety and operation.

Never use an ohmmeter when testing a circuit board. The voltage from the battery located in the meter can damage the solid-state circuitry.

Semiconductors conduct electricity in a controlled manner. There are N-type (negatively charged) and P-type (positively charged) semiconductors.

Diodes allow current to flow in one direction. One end of a diode is an anode and the other end is a cathode.

Transistors are made of N-type and P-type semiconductor material. They have three terminals, a base, a collector, and an emitter. Transistors can be used as a solid-state relay or an amplifier.

Thermistors are semiconductors that are used to sense temperature. There are two types of thermistors: a PTC or positive temperature coefficient type and an NTC or negative temperature coefficient type.

A **PTC thermistor's** resistance increases when the temperature increases and an NTC thermistor's resistance decreases when the temperature increases.

A **capacitor** is used to store electrons. In a DC circuit, a capacitor blocks current flow, and in an AC circuit, a capacitor aids current flow.

A **resistor** is used to drop the voltage in a solid-state circuit.

A **rectifier** is used to change AC to DC.

WIRING LAYOUTS AND ELECTRICAL DIAGRAMS

It is imperative that the technician be well versed in interpreting electrical diagrams and layouts. Frequently, the technician has to refer to schematic diagrams to become familiar with a system's components and system operations. The following are terms and information on wiring layouts and electrical diagrams.

A residential **furnace electrical** circuit is sized to provide 120 volts, single phase, 60 Hz through a properly sized circuit breaker.

An electrical **disconnect switch** must be located at the furnace to disconnect the power supply during service.

An electrical **disconnect** must be installed near an air conditioning condensing unit to disconnect the power supply during service.

Common 120-volt **loads** that operate in an HVAC system are the indoor blower motor, the induced draft blower, the transformer's primary winding, an electronic air cleaner (if used), and a humidifier transformer (if used).

Any electrical **pigtail** wires that are twisted together must be connected with the proper-sized wirenut.

A 240-volt, 60-Hz, single-phase **power supply** is used for the condensing unit on residential packaged or split systems.

The **loads** that are operated by the 240-volt circuit on a package HVAC unit are the compressor, condenser fan, transformer's primary winding, the supply air blower, the induced draft blower, and crankcase heater.

The input of a **control transformer** for a residential split air conditioning system is 120 volts, 60 Hz, single phase, and the output is 24 volts.

The input of a **control transformer** for a packaged HVAC system is 208/240 volts, 60 Hz, single phase, and the output is 24 volts.

The **24-volt control circuit** directly energizes relay coils, contactor coils, gas valves, and control modules in an HVAC system.

A **schematic** is a group of lines and electrical symbols arranged in a ladder form to show the operational sequence of an electrical system. It is used when troubleshooting an HVAC/R system.

In a **schematic,** the vertical lines represent the power supply and the horizontal lines represent the individual circuits.

A **legend** is used with a schematic diagram to identify the label of electrical circuit components.

The **top portion** of a schematic is usually the line-voltage side of the electrical circuit.

The **bottom portion** of a schematic is usually the low-voltage side of the electrical circuit.

A **transformer** is commonly located between the top portion and the bottom portion of the schematic diagram.

A **pictorial diagram** shows the actual factory-installed wiring and location of system components.

A **pictorial diagram** is also known as a label or line diagram.

A **factual diagram** consists of an electrical schematic and a pictorial diagram.

An **installation diagram** is an aid for the system installer. It displays the system's fuse size, wire size, and its color code.

An **installation diagram** commonly shows the placement of the thermostat and power supply wiring for the system.

Factory-installed wiring is the electrical wiring that is color coded, sized, and installed in a system at the factory.

Field-installed wiring is the wiring that is brought to an HVAC system by an installation technician as the system is being installed.

A **heavy dashed line** on an installation wiring diagram indicates field-installed line voltage wire.

A **heavy solid line** on an installation wiring diagram indicates factory-installed line voltage wire.

A **light dashed line** on a wiring diagram indicates field-installed low-voltage wire.

A **light solid line** on a wiring diagram indicates factory-installed low-voltage wire.

A **connection diagram** shows the actual wiring of electrical components along with their proper color-coded wire.

Figure 3–23 is a schematic of a split air conditioning system. The operational sequence is as follows: On a call for cooling from the cooling thermostat **(CTH)**, terminals **R** and **Y** close, energizing the contactor coil **(C)** and the fan relay coil **(FR)** in the 24-volt circuit. The **C1** and **C2** contacts change position and energize the compressor **(COMP)** and condenser fan **(CF)** in the 240-volt circuit. The **FR1** and **FR2** contacts change positions and energize the indoor blower motor **(IBM)** in the 120-volt circuit. If the low-pressure

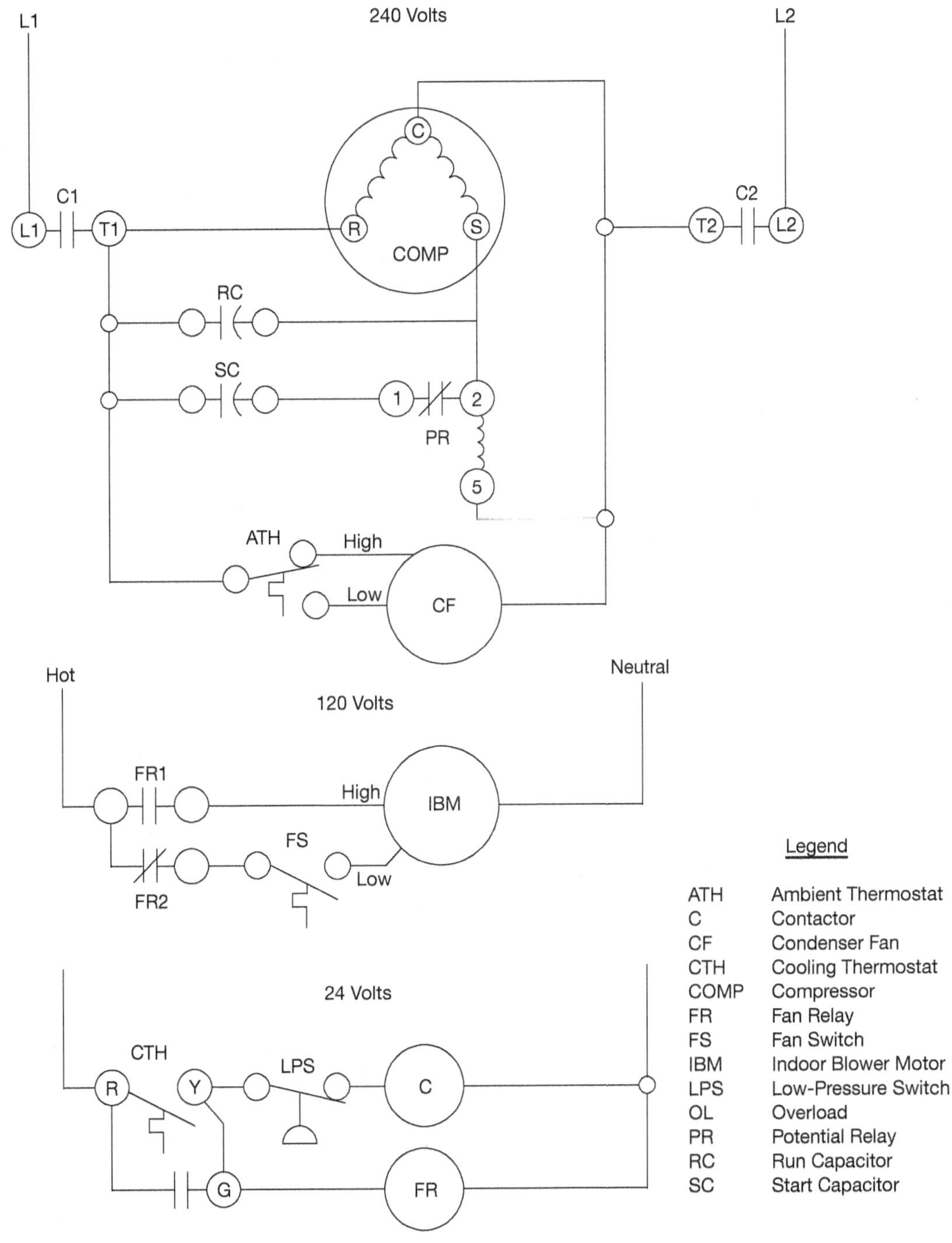

FIGURE 3–23

switch **(LPS)** opens, the contactor coil **(C)** will de-energize, and contacts **C1** and **C2** will open. This will break the circuit to the compressor **(COMP)** and condenser fan **(CF)**.

Figure 3–24 is a schematic of a forced-air gas furnace. The operational sequence is as follows: With the door switch **(DS)** in the closed position, the transformer **(TR)** is energized by 120 volts; 24 volts will then be available on the control side of the transformer. On a call for heat, the heating thermostat closes terminals **R** and **W** and energizes the induced draft relay **(IDR)** with 24 volts. The normally opened **IDR** contacts close, energizing the induced-draft motor **(IDM)** with 120 volts. After the motor is energized, the pressure switch **(PS)** closes, which proves the operation of the induced-draft motor. The **PS** directs current to the high limit **(HL)** and secondary limit **(SL)**. If both limit controls are in the closed position, the ignition control module is energized across terminals **TH** and **TR**. This starts a sequence of operation by energizing the pilot valve **(PV)** and the spark ignition system. When the pilot lights, the flame is proven by the flame sensor. Then the main valve **(MV)** is energized and the main burners are ignited. The heat from the heat exchanger causes the fan switch **(FS)** to close, energizing the indoor blower motor **(IBM)** with 120 volts. If the thermostat's "auto/on" switch is set to the "on" position, the thermostat will close from terminals **R** to **G**. This will energize the fan relay **(FR)** coil, causing **FR1** and **FR2** contacts to change positions. The normally closed **FR2** contacts will open, de-energizing the low-speed circuit, and the normally open **FR1** contacts will close, energizing the high speed on the indoor blower motor **(IBM)** with 120 volts.

Figure 3–25 is a schematic of a circuit board common to many forced-air furnaces. The operational sequence is as follows: On a call for heat, the heating thermostat will close from **R** to **W** and bring 24 volts to the **W** terminal on the circuit board. Once the **W** terminal has been powered by 24 volts, this will start the sequence of operation by energizing the **GAS1** and **GAS2** terminals with 24 volts. This will energize the pick coil on the gas valve through the normally closed set of contacts on the bimetal pilot sensor. At the same time, it will energize the pilot igniter and the hold coil on the gas valve. With the pick coil and pilot igniter energized, gas will flow to the pilot and be ignited by a spark from the pilot igniter. The pilot flame will heat the bimetal pilot sensor and a set of normally closed contacts will change to the open position after approximately 45 s. This will de-energize the pick coil and the pilot igniter. The hold coil will stay energized, since it is directly controlled from the **GAS1** terminal and is not controlled by the bimetal flame sensor. The hold coil will keep the pilot valve open, which will keep the pilot lit and ready for the ignition of the main gas burners. The bimetal pilot sensor also has a set of normally open contacts that will close once the pilot flame is sensed, approximately 45 s after pilot ignition. This proves the existence of the pilot and energizes the main gas valve. The main gas valve will feed gas to the main burners, and the gas will be ignited by the pilot. When voltage is sent to terminal **GAS3**, *simultaneously*, voltage will also be sent to the solid-state timed-delay circuit. It is this voltage that will start the energizing of the solid-state timed delay. After 45 s, the fan will energize on the heating speed. When the thermostat is satisfied, terminals **R** and **W** will open. This will de-energize the (hold) pilot valve and the main gas valve. The blower will operate for 45 s, and then it will shut down, completing the cycle.

Figure 3–26 is a schematic of an oil-fired forced-air furnace. The operational sequence is as follows: With the disconnect switch **(DISC)** and the limit switch **(LIM)** in the closed position, 120 volts is brought to the black and white wires. This will energize the

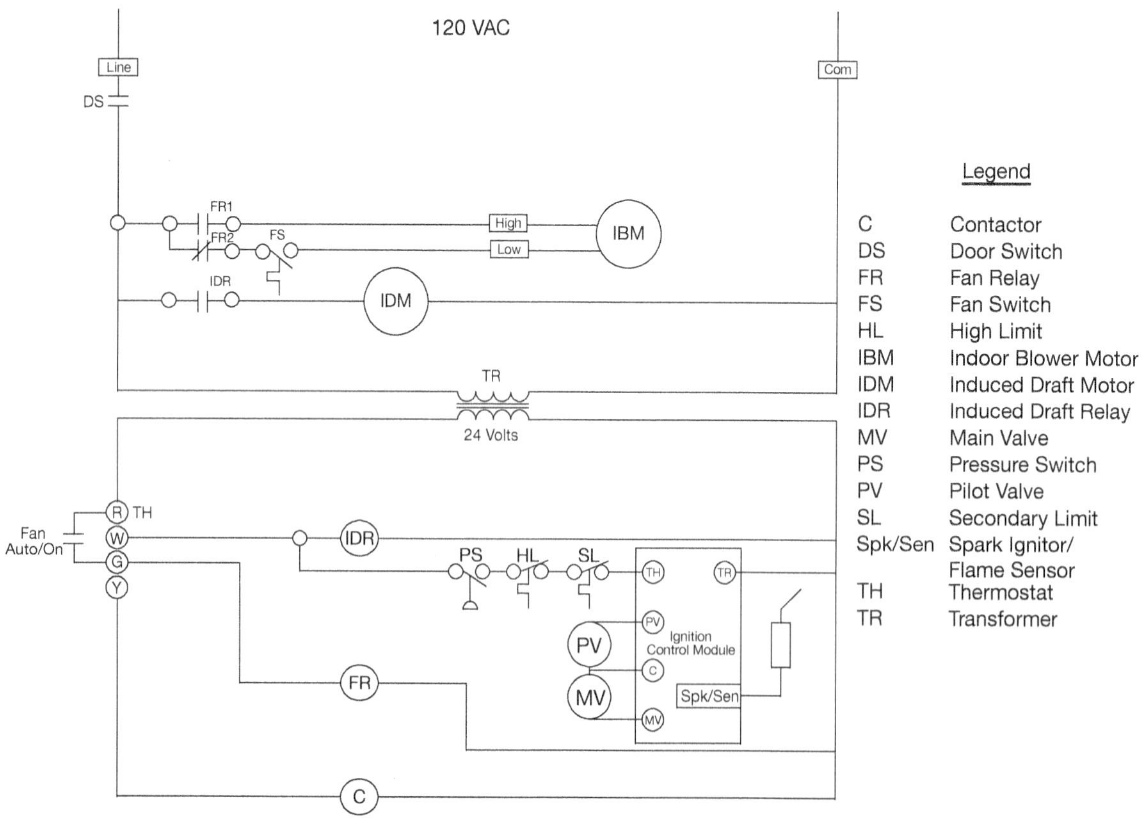

FIGURE 3–24

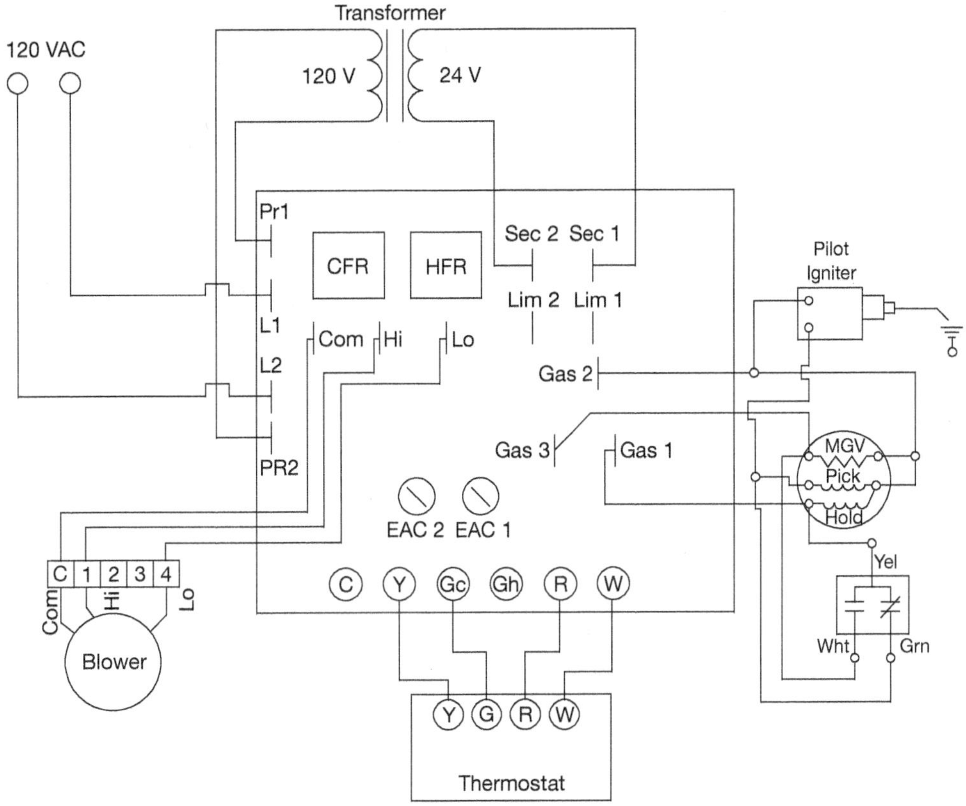

FIGURE 3–25

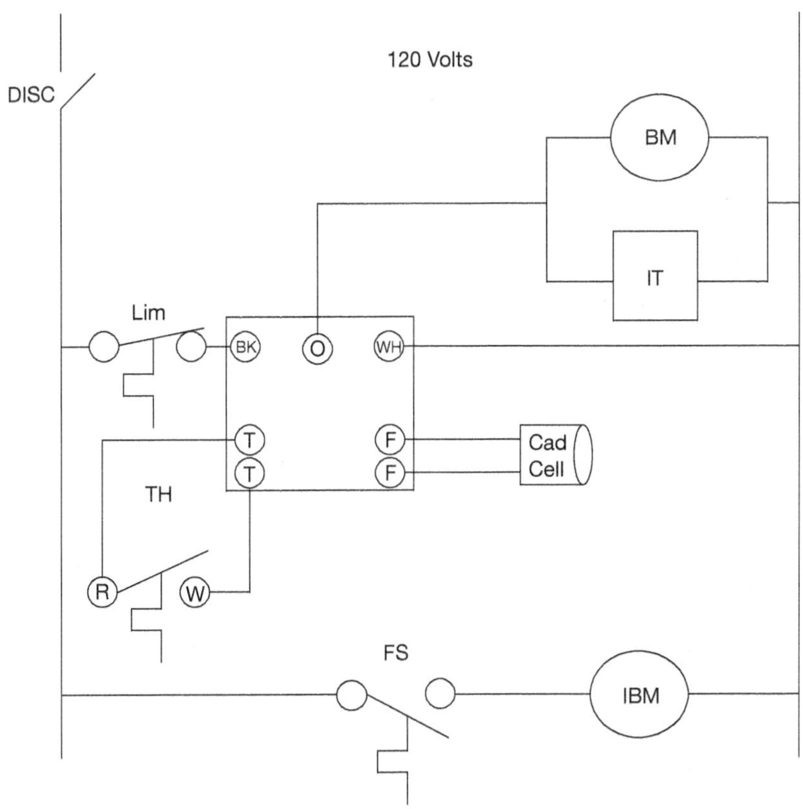

Legend

BM	Burner Motor
DISC	Disconnect
FS	Fan Switch
IBM	Indoor Blower Motor
IT	Ignition Transformer
LIM	Limit
TH	Thermostat

FIGURE 3–26

cad-cell control. On a call for heat, terminals **R** and **W** close on the thermostat **(TH)**. This will cause the orange wire on the control to energize the burner motor **(BM)** and the ignition transformer **(IT)** with 120 volts. The burner motor drives the fan and the oil pump, which supplies the fuel and the air for combustion. The ignition transformer supplies the spark for ignition. The heat from combustion transfers into the heat exchanger and increases the temperature, causing the fan switch **(FS)** to close and energize the indoor blower motor **(IBM)** with 120 volts. The cad-cell proves the flame by sensing its light. This will keep the system energized until the thermostat is satisfied. If the cad-cell does not sense the light that is produced by the flame, the cad-cell controller will de-energize the burner motor and the ignition transformer after approximately 45 s. If the cad-cell senses light before the system is energized, the system will not start.

Figure 3–27 is a schematic of an air-to-air heat pump. The operational sequence is as follows in the next four sections.

Cooling Cycle

When the thermostat is set to the cooling cycle and the fan switch is set in the auto position, the control relay **(CR),** the fan relay **(FR),** and the compressor contactor **(C)** are energized. The **CR** energizes the reversing valve solenoid, which directs refrigerant so that the outdoor unit is used as the condenser and the indoor unit is used as the evaporator. The **FR** energizes the indoor fan on high speed, and the **C** energizes the compressor and condenser fan.

Heating Cycle

When the thermostat is set to the heating cycle and the fan switch is set to the auto position, the compressor contactor **(C)** and the fan relay **(FR)** are energized. The reversing valve solenoid is in the

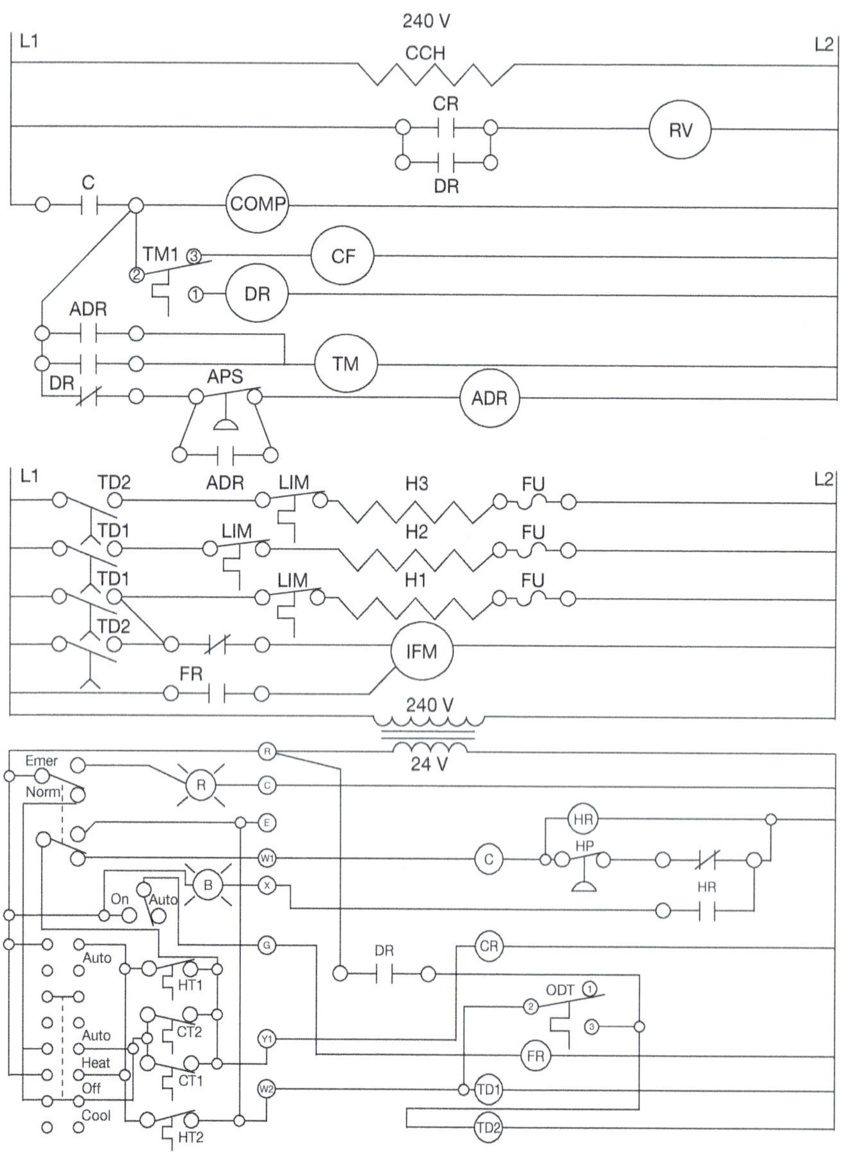

Legend	
ADR	Auxiliary Defrost Relay
APS	Air Pressure Switch
C	Contractor
CCH	Crankcase Heater
CF	Condenser Fan
COMP	Compressor
CR	Control Relay
CT1	Cooling Thermostat 1st Stage
CT2	Cooling Thermostat 2nd Stage
DR	Defrost Relay
FR	Fan Relay
FU	Fusible Link
H1	Auxiliary Strip Heater #1
H2	Auxiliary Strip Heater #2
H3	Auxiliary Strip Heater #3
HP	High-Pressure Switch
HR	Holding Relay
HT1	Heating Thermostat 1st Stage
HT2	Heating Thermostat 2nd Stage
IFM	Indoor Fan Motor
LIM	Limit Switch
ODT	Outdoor Thermostat
RV	Reversing Valve
TD1	Timed Delay Relay #1
TD2	Timed Delay Relay #2
TM	Timer Motor

FIGURE 3–27

de-energized position. This directs the refrigerant so the outdoor unit is used as the evaporator and the indoor unit is used as the condenser. The **FR** energizes the indoor fan on high speed and the **C** energizes the compressor and condenser fan. If the room temperature drops two degrees below the setting of the first-stage thermostat, the second-stage heating circuit is energized. The second-stage thermostat **(H2)** closes, energizing the timed-delay relay #1 **(TD1)**. This sequentially closes two timed-delay relay contacts, which in turn energizes two auxiliary strip heaters, **H1** and **H2.** If the outdoor temperature is below the predetermined temperature on the outdoor thermostat **(ODT),** the timed-delay relay #2 **(TD2)** is energized. This energizes strip heater #3 **(H3).**

Emergency Heat Cycle

During the emergency heat cycle, the normal/emergency switch is placed in the emergency position. The outdoor unit, containing the compressor and condenser fan, is de-energized. The fan relay **(FR)** energizes the fan motor, while the timed-delay relay #1 **(TD1)** energizes auxiliary strip heaters **H1** and **H2.** The red light **(R)** at the thermostat indicates that the system is operating on the emergency heat cycle.

Defrost Cycle

During the defrost cycle, the system temporarily changes from the heating to the cooling cycle. The initiation of the defrost cycle is accomplished by the air pressure switch **(APS),** the outdoor coil temperature, and defrost time **(TM1).** If the pressure difference across the outdoor coil reaches a predetermined point, the **APS** closes, energizing the auxiliary defrost relay **(ADR).** This will close one set of contacts across the pressure switch, forming a holding circuit on the ADR coil. A second set of **ADR** contacts will close, energizing the timer motor. This will cause the timer motor's cam to turn and aid in closing the **TM1** contacts. The **TM1** contacts will close if the temperature of the coil is below a predetermined point and the timer's cam turns to a specific location. This will break the contacts to the condenser fan and make the contacts to the defrost relay **(DR)** coil. Energizing the DR coil will close a set of contacts to the reversing valve, close a set of contacts to the timer motor, open a set of contacts to the **ADR,** and close a set of contacts to the timed delay relay #2 **(TD2),** which will energize the second-stage heaters **H3.** After the system is defrosted, the outdoor coil temperature increases, reaching a predetermined temperature, and the timer's cam turns to a specific location. The system returns back to the heating cycle.

ELECTRICAL METERS

Understanding how to apply electrical meters when troubleshooting is extremely important to the HVAC/R technician. This section covers the use of voltmeters, ohmmeters, and ammeters. It will also help technicians with competency exams on electrical meter usage.

An **analog meter** uses the magnetic field in an electrical circuit to cause a deflection of a needle. The needle deflection used with a scale provides an analog readout. The most accurate reading of an **analog meter** is from the middle to the upper two-thirds of the scale.

Digital meters use solid-state circuitry to create a digital readout. They have no moving parts and are more rugged and accurate than analog meters.

Many times, digital meters have **autoranging.** This will automatically set the volt, amp, or ohm scale.

A **voltmeter** measures the operating and source voltage to an electrical load.

A **voltmeter** should always be set to the highest scale when working with an unknown voltage.

A **voltmeter** should be placed in parallel with the load being tested.

When **testing for voltage** across a set of opened contacts, a voltmeter should record source voltage.

When **testing for voltage** across a set of closed contacts, a voltmeter should read 0 volts.

A **millivolt** is one thousandth of a volt. When measuring millivolts, a millivoltmeter must be used.

Analog ammeters use a needle type of readout and usually have a hold button to lock in the reading.

Digital ammeters have a digital readout and are more accurate than analog meters.

When using an **ammeter,** the circuit that is being tested must be energized.

An **ammeter** measures the startup and operating amperage of a load.

An **ammeter** should always be set to the highest scale when working with an unknown amperage.

An **in-line ammeter** should be placed in series with the load that is being tested.

Always clamp around one wire of a load when checking amperage using a **clamp-on ammeter.**

A **milliamp** is one thousandth (1/1000) of an amp. When measuring milliamps, a milliammeter must be used.

A **10-loop multiplier** and a clamp-on ammeter can be used to measure milliamps. If a wire is coiled 10 times, the amp reading will be multiplied by 10.

VOMS are used to measure volts, ohms, and milliamps of an electrical circuit.

An **ohmmeter** measures the resistance of a load and the continuity of a switch or circuit.

An **ohmmeter** uses a battery for its power source.

The source voltage should always be disconnected from the circuit when using an **ohmmeter.**

When testing a component with an **ohmmeter,** the component must be isolated from the circuit.

When testing across a set of normally closed contacts on a relay, an **ohmmeter** should read zero ohms of resistance.

When testing across a set of normally open contacts on a relay, an **ohmmeter** should read infinite ohms of resistance.

When using an **ohmmeter,** an open circuit will read infinite resistance and a short circuit will read zero resistance.

A **wattmeter** is used to read the power output rating of a load. When measuring the output rating of a motor, the power factor must be taken into consideration.

Example: ($P = E \times I \times Pf$) If a motor has a power factor (Pf) of 85% or .85 and the voltage (E) of 240 volts with the amperage (I) at 5 amps, the power rating (P) would be: ($P = 240 \text{ volts} \times 5 \text{ amps} \times .85$)$P = 1020$ watts

ELECTRICAL TROUBLESHOOTING

Troubleshooting is an everyday occurrence in the work of a service technician. The following are terms and information that will help the technician be proficient in this work and on certification exams.

Using a Voltmeter

A **voltmeter** should always be placed in parallel with the power supply being tested.

A **voltage test** is one of the first tests that should be made on an inoperable electrical circuit.

Energizing a 24-volt relay or contactor coil should change the positions of contacts. If a contact is normally open, it should close, and if a contact is normally closed, it should open.

Closed contacts in an operating system should read zero volts. If a measurable amount of voltage is read, the contacts may be burned.

Open contacts in an operating system should read "source voltage."

If **source voltage** is read to a load such as a motor, a heater, or a relay coil, the load should operate. If the load does not operate, it is usually burnt open.

Hopscotch is a method of using a voltmeter to test an electrical circuit. To use this method, one of the voltmeter leads is placed on the common side of a load. The other lead is placed on the line side of the load to test for voltage. If there is no voltage to the load, the lead located on the line side is moved toward the power supply, across wires and switches, until source voltage is found.

Using an Ohmmeter

Always disconnect the power supply to the system when using an **ohmmeter.**

Always isolate the component that is being tested when using an **ohmmeter.**

Fuses are generally tested using the ohmmeter. A good fuse will measure zero ohms and a bad fuse will measure infinite ohms.

Closed contacts in a nonoperating system should read zero ohms. If a measurable amount of resistance is obtained, the contacts may be burned.

Open contacts in a nonoperating system should read "infinite resistance." If zero ohms are found, the contacts are stuck in the closed position.

When **measuring the resistance** of a load, always start on the lowest ($R \times 1$) scale.

Measurable resistance should be obtained when using an ohmmeter to test an electrical load such as a motor, a heater, or a relay coil.

Continuity means a continuous or complete circuit.

It can either be zero ohms or a measurable resistance.

A **shorted** electrical load will read zero or very low ohms of resistance.

An open electrical load will read "infinite" ohms of resistance.

To test a load for a **short-to-ground condition,** isolate the load from the circuit, use the highest ohm scale, and place one of the meter leads to the line side of the load. Place the other lead to the uninsulated shell or ground terminal of the load.

Using an Ammeter

The system must be in operation to obtain amperage when **using an ammeter.**

To read amperage on a 24-volt control circuit using a clamp-on ammeter, a multiplier must be used to amplify the current reading.

A **multiplier** is a wire that is coiled 10 times and placed in series from the power supply. This will amplify the current by 10 to obtain the proper current. The reading must be divided by 10.

Example: Using a 10-loop multiplier, an amperage of 8 amps is obtained. $8/10 = .8$ amps total.

An in-line milliammeter can be used to test for proper amperage on a 24-volt control circuit.

A **short circuit** will have an excessive current draw. This will blow fuses or trip breakers in a 120-volt or 220-volt circuit. In a 24-volt control circuit, a short will overheat and burn out a control transformer when using the cooling circuit and will burn out the heat anticipator when using the heating circuit.

An **open circuit** does not allow current to flow and will read zero amps.

The **VA rating** is the power output of a transformer. To test for an underrated transformer, divide the VA rating of the transformer by the control voltage. This will give the maximum allowable amperage of the circuit: $VA/V = A$.

Example: 40-VA transformer/24-volt control voltage = 1.66 amps maximum.

SECTION 4

Airflow Components and Duct Fabrication

This section consists of airflow components and types of duct fabrication. It is very important for a technician to have a good understanding of airflow test equipment, fans/blowers, types of duct systems, materials used to fabricate ductwork, types of air-side components, air filters, humidifiers, and temperature/humidity measurements. The following will aid the technician when studying for competency exams.

AIRFLOW TOOLS AND MEASUREMENTS

There exist many types of airflow test equipment that are used daily. The technician needs to be knowledgeable in the use of airflow tools and where and when to obtain measurements. The following is related information that will assist the technician.

A **tachometer** is used to measure the speed of a rotating object such as a motor or a blower. A common type of tachometer uses a strobe light and measures rpms. This is known as a strobe tachometer.

A **Pitot-tube** is a tool that is used to test air velocity pressure, static pressure, and total pressure on a duct system.

When using a **Pitot-tube,** it must be pointed into the air stream to measure the velocity pressure.

The small hole in the end of a **Pitot-tube** is used to measure total pressure. The holes on the side of the tube are used to measure the static pressure in the duct system.

Total pressure minus static pressure is the **velocity pressure.**

The velocity and static pressures are added and the sum is the **total pressure.**

The **speed of the air** traveling in a duct is referred to as velocity. It is measured in inches of water column (WC) or feet per minute (fpm).

Extremely small amounts of pressure are measured using a water column gauge. **Water column** is the pressure that it takes to push a column of water up vertically. A half-inch of water column is the amount of pressure it would take to push a column of water in a tube half an inch.

A **Pitot-tube traverse** is an airflow testing method used to take an average air velocity pressure reading in an air distribution system. This is important when determining air volume within the system.

A **Pitot-tube traverse** can be done by taking air velocity pressure readings in the main supply and return duct and/or at individual branch runs.

A **velometer** measures the speed of the air in a duct system or at a diffuser. Meter readouts can be in analog or digital form.

A **diffuser probe** used with a velometer will measure air speed at the outlet of a diffuser.

A **Pitot-tube** used with a velometer will measure airspeed in a duct system.

Vane anemometers can have a digital or an analog readout. The digital is more accurate than the analog. The anemometer measures the velocity of air in feet per minute (fpm).

A **rotating vane anemometer** uses a propeller-type wheel that measures air velocity in feet per minute (fpm).

Thermal anemometers use a hot wire that senses the velocity of the air in fpm when placed in an airstream. They are designed to sense low velocities.

The **static pressure** of a duct system is the outward air pressure that pushes on the duct. It is measured in inches of WC.

A **manometer** used with a Pitot-tube will test a system's velocity pressure, static pressure, and total pressure.

Static Pressure + Velocity Pressure = Total Pressure

(**SP + VP = TP**)

A **U-tube manometer** uses a U-shaped tube and a liquid, usually water, to measure air velocity pressure, static pressure, or total pressure in a duct system.

A **U-tube manometer** is used to test pressures above 1.0″ WC.

An **incline manometer** is a transparent plastic measuring device, filled with a liquid, that measures pressure in a duct system.

The **incline manometer's** scale is on a slope and measures pressures below 1.0″ WC.

A **calibrated** or **differential pressure gauge** uses a high- and a low-pressure port to measure the pressure difference between two different pressure locations.

Volume of air in a duct system is the quantity that is flowing through the system. It is measured in cubic feet per minute (cfm).

An **airflow hood** is a tool that is placed over a diffuser to determine the volume of air in cfm.

When sizing a duct system, the **equal friction loss method** is one of the best sizing methods used.

AIRFLOW CALCULATIONS

Airflow through a Duct System

V = velocity in fpm (feet per minute)

A = area of a duct in sq ft (square feet)

Q = quantity of air in cfm (cubic feet per minute)

Q = AV or (cfm = sqft × pfm)

V = QA or (fpm = cfm/sqft)

A = Q/V or (sqft = cfm/fpm)

Calculation for Furnace Blower Output

cfm = Output Btu/h/1.08 × temperature difference

Calculation for Weight of Air

Weight of air = W Volume = V Density =

W = DV or W/D = V or W/V = D

Duct sizing calculators are used to help design duct systems. Figure 4–1 is an example of a duct sizing calculator.

To use the duct calculator, first find the cfm of the equipment, then determine the velocity or allowable friction loss per 100 feet.

Example:

1. Assume the required volume of air is 1,600 cfm and the velocity is 1,100 fpm.
2. Set the 1,600 cfm opposite to the velocity, 1,100 fpm. The friction scale will be set at .10″ per 100′ of duct.
3. The round duct diameter scale shows that a 16″ round duct must be used.
4. The equivalent rectangular duct chart shows that the area of a 16″ duct is the same as 8″ × 30″.

A **friction loss chart** is used to size duct systems if a duct calculator is not used. A friction loss chart has the same features as a duct calculator, except in the way the round duct is converted to rectangular duct. Figure 4–2 is an example of a friction loss chart.

The **horizontal lines** on the chart show the cubic feet per minute (cfm) of air.

The **vertical lines** show the friction loss of water column per 100′.

The **diagonal lines** from right to left show the air velocity in fpm.

The **diagonal lines** from left to right show the round duct diameter sizes.

Two variables are needed to use the **friction loss chart.**

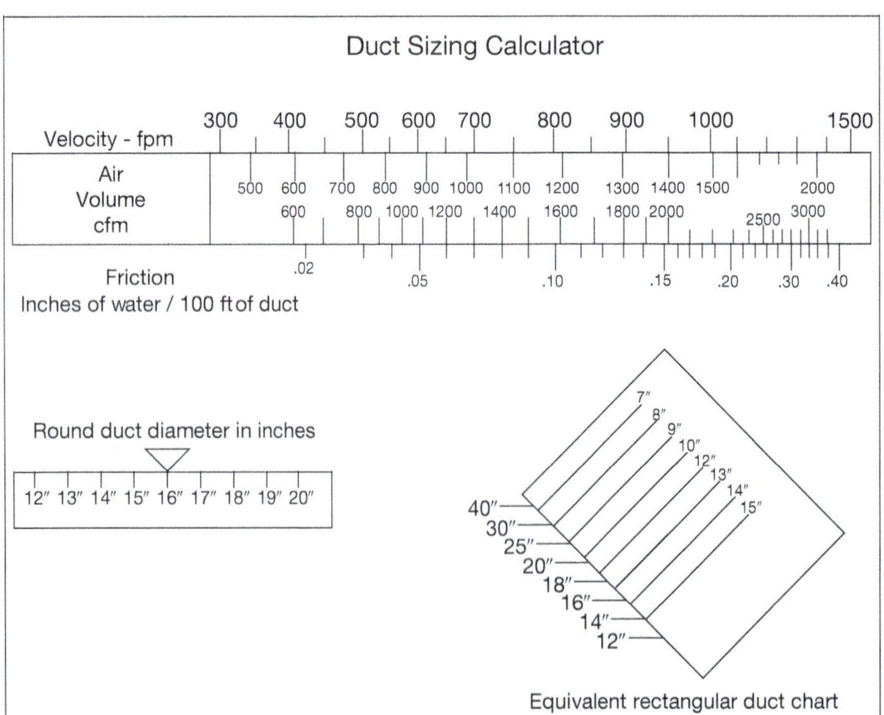

FIGURE 4–1

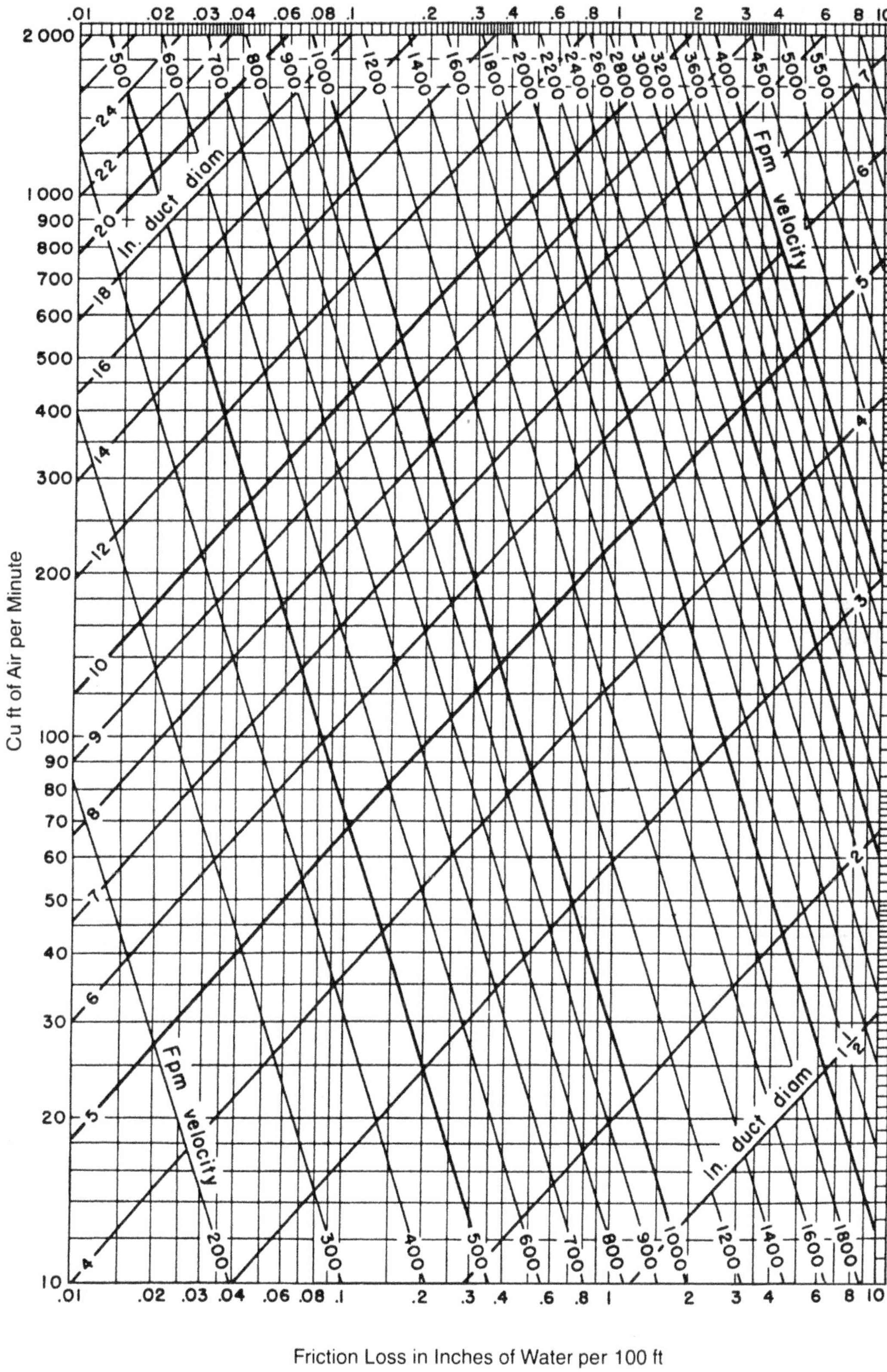
Cu ft of Air per Minute
Friction Loss in Inches of Water per 100 ft
In. duct diam
Fpm velocity
.01 .02 .03 .04 .06 .08 .1 .2 .3 .4 .6 .8 1 2 3 4 6 8 10
2 000
1 000
900
800
700
600
500
400
300
200
100
90
80
70
60
50
40
30
20
10
500 600 700 800 900 1000 1200 1400 1600 1800 2000 2200 2400 2600 2800 3000 3200 3600 4000 4500 5000 5500
200 300 400 500 600 700 800 900 1000 1200 1400 1600 1800
24 22 20 18 16 14 12 10 9 8 7 6 5 4
7 6 5 4 3 2 1 1/2

FIGURE 4–2

Follow the chart using these examples.

Example 1 If the air quantity of a system is 70 cfm and the round duct size is 5″, the friction loss of the duct is .1″ WC per 100″.

The velocity is 500 fpm.

Example 2 If the velocity of the air is 900 fpm and the friction loss per 100′ is .15″ WC, the quantity of the air is 400 cfm.

The duct size of the system is a 9″ round.

BLOWERS AND FANS

A large part of troubleshooting HVAC/R equipment is directly related to blowers and fans. The technician must know about air volume, bearing types, lubrication, motor amperage, motor speeds, fan blades, and so on. The following is information that will aid the technician in this area.

Dry bearings cause excessive amp draw on a motor.

A **bushing** or **sleeve bearing** is a bronze sleeve around a motor shaft.

A **sleeve bearing** can be permanently oiled and sealed at the factory, or it can require field oiling.

Oil-type motors have oil ports and must be oiled with 3 to 6 drops of nondetergent oil every 6 months. Overoiling can cause an accumulation of dirt to build up in the stator windings.

Ball-bearing-type motors are more rugged and can handle more of a load than sleeve-bearing-type motors.

Sealed-ball-bearing motors should be disassembled and regreased every 3 to 5 years, depending on use.

Larger-bearing-type motors, commonly over 1 horsepower, use zerk fittings to allow the motor to be greased approximately every 3 months.

Most outdoor **condenser fans** use an axial or propeller-type blade designed to move the proper quantity of air across the condenser coil.

Air volume over a fan will change if any of the following factors are changed:

Speed of motor rotation

Number of fan blades

Blade pitch

Blade length

Blade width and shape

The **motor amperage** will increase on a fan motor if the air volume is increased and will decrease if the air volume is decreased.

Fan motors are weatherproofed by using rain shields to protect the motor from water.

A **condenser grill** shields the operating fan blade to protect against injury during fan operation.

Direct-drive condenser fans lower than 5 horsepower are commonly a permanent split capacitor (PSC) type of motor.

During **low ambient conditions,** condenser fans must be cycled to prevent the high-side pressure from dropping too low. When the high-side pressure is too low, it causes the evaporator coil to become starved of refrigerant liquid. This decreases the efficiency of the system.

Condenser fans that have a two-speed motor often use a reverse-acting pressure control to cycle the fan depending on the high-side pressure or outdoor ambient temperature.

A **solid-state motor control** varies the speed of the condenser fan so a constant high-side pressure is maintained. It commonly uses a thermistor to sense the liquid line temperature. It will lower the speed of the condenser fan on a decrease of liquid line temperature and raise the speed on an increase of liquid line temperature.

Indoor blowers use a centrifugal or squirrel-cage fan design to move the proper quantity of air over the heat exchanger and evaporator coil.

A **blower door switch** is used to de-energize the blower if the panel near the blower is removed. This will prevent injury during blower operation.

Direct-drive indoor blowers lower than 5 horsepower commonly use a PSC motor.

Belt-drive indoor blowers lower than 5 horsepower commonly use a capacitor start induction run type of motor.

Single-phase direct drive blower motors are commonly multi-speed. A three-speed direct drive 120-volt blower uses different color-coded wiretaps for the different speeds.

The following are color-coded wires that are commonly used for different **speeds:**

Red—low speed

Black—high speed

White—common speed

Other **speed taps,** such as medium low, medium, and medium high, have a range of various color codes, depending on the manufacturer. Color codes that are used for the midrange speeds are blue, yellow, and brown.

When installing a blower motor, the wires controlling the **extra speeds** that are not being used must be capped with a wirenut and taped or tie strapped to prevent an electrical short.

To **change the speeds** on a direct drive blower motor, the colored wiretaps must be changed. The white or orange common wire must remain in the circuit.

Fans and **blower motors** that are 5 horsepower or above are commonly three-phase motors.

Fans and blowers that have **bent blades** will decrease the airflow performance and cause excessive vibration.

Single-phase motors often use run and start capacitors to aid in motor starting torque and operation. Three-phase motors do not use capacitors.

FLA is the full load amperage of a motor. It can also be stated **RLA,** which stands for running load amperage.

LRA is the locked rotor amperage, or starting amperage, of the motor.

The **fan speed calculation** for a belt-drive blower is:

Fan rpm = diameter of motor pulley/ diameter of fan pulley × rpm

Example: A motor has a 3″-diameter pulley and it turns a blower with a 6″-diameter pulley. The motor has a 1,050 rpm.
3″/6″ × 1,050 rpm = 525 rpm

Air volume is measured in cubic feet per minute (cfm). When using an air conditioning system, a common volume of air per ton of refrigeration effect is **400 cfm.**

A higher **cfm** of air is used during a cooling cycle and a lower cfm is used during a heating cycle.

The common external **static pressure** for residential and light commercial blowers is .5 WC.

As the **static pressure** in a duct system increases, the amperage on the blower motor increases. As the static pressure decreases, the amperage decreases.

Nameplate information, including voltage, amperage, rotation, rpm, and horsepower, is important when replacing a fan motor.

AIR DUCT SYSTEMS AND FABRICATION

The air duct system contains and directs air from the blower unit to the conditioned space. Air duct systems and duct fabrication are important for a technician in the HVAC/R field to understand. The following are many of the terms that are related to air duct systems and fabrication.

Horizontal rectangular air ducts up to 60″ wide must be **supported** with 1″-wide galvanized steel hangers at intervals not exceeding 10′.

Joints, seams, and **fittings** in air duct systems must be made airtight by the use of tapes, mastics, gaskets, and other means.

A **round metal duct** crimped joint must have a contact lap of at least $1\frac{1}{2}''$.

A **round metal duct** must be connected with at least three screws that are spaced evenly apart.

The **flex connectors or vibration isolators** are used to eliminate vibrations and noises within the duct system.

Vibration isolators that are installed between metal ducts and mechanical equipment must not exceed 10″ in length.

A **metal duct system** must be installed at least 4″ above the earth's surface.

When a **metal duct system** is installed under a concrete slab, the duct must be encased in at least 2″ of concrete.

Circular bands that are used to connect round duct must be at least 1″ wide and must use the same gauge metal as the duct.

The minimum total area of a **supply and return air duct** used on a blower-type warm-air furnace must be at least 2 sq in. per 1,000 Btu/h approved output rating.

The minimum total area of a **supply and return air duct** used on a heat pump must be at least 6 sq in. per 1,000 Btu/h approved output rating.

The following must be printed on the outside of external **duct insulation:** the name of the manufacturer, the "R" value at the installed thickness, and the flame spread index.

An air duct system that is installed in an unconditioned space, such as a crawl space, attic area, or rooftop, must be **insulated and sealed with an approved vapor barrier.**

When a duct system is located on the exterior of a roof, it must be **insulated** with at least 3″ of a mineral fiber blanket with an approved weatherproof barrier.

When a duct system is located in an attic garage or crawl space, it must be **insulated** with at least 1″ of a mineral fiberglass blanket with an approved vapor barrier.

In a single-dwelling unit, exposed **rectangular ducts** or **plenums** 14″ or less must be fabricated with a minimum of 28-gauge galvanized steel.

In a single-dwelling unit, exposed **rectangular ducts** or **plenums** over 14″ must be fabricated with a minimum of 26-gauge galvanized steel.

A **cross break** on a section of duct increases its rigidity and decreases the noise caused by expansion and contraction.

Low-pressure duct systems are designed for air pressure up to 2″ WC.

When air ducts are run through floors, a **firestop** around the duct is required. A common type of firestop is fiberglass insulation.

When installing **air ducts** in a basement, the supply and return air trunks should run perpendicular to the floor joists.

In a home with a **slab floor construction,** the air duct system is commonly installed in the attic area.

In a home with a **crawl space construction,** the air duct system is installed either in the attic or in the crawl space area.

The **plenum** is a sheet-metal-like box that is installed on an air handler used to discharge the air into the duct system.

An **extended plenum** duct system is one where the furnace is in a central location of the home and the main duct supply is branched from two sides of the furnace.

Each main duct supply on an **extended plenum system** should not exceed 24′ in length without reducing the duct area.

When using a **reducing extended plenum,** the furnace is located at one end of the house and the trunk size is reduced as it increases in distance from the furnace.

The decrease in size of a **reducing extended plenum** causes an increase in air velocity.

A **duct transition** is a section of duct that connects two different sizes of duct.

A **crimping tool** is commonly used to decrease the diameter of the male fitting on a round duct.

The **compass snip** is used to cut circles on sheet metal.

Branch runs are the supply air ducts that distribute air to each room in the house. They should never be connected to the end of any plenum system, as this will cause imbalanced airflow.

Balancing dampers must be provided in branch runs as close to the supply air plenum as possible.

When adjusting a **balancing damper,** if the adjusting handle is in line with the duct, the damper is 100% open. If the adjusting damper is perpendicular to the duct, the damper is 100% closed.

A **main supply air plenum** must be located at least 1″ from a combustible surface.

A **register** is adjustable and is used on the supply air side of a system. A **grill** is nonadjustable and is used on the return air side of a system.

Supply air registers used with first-floor upflow and attic furnace installations commonly use a ceiling-type mount.

Supply air registers used with basement, crawl space, or first-floor downflow installations commonly use a floor-type mount.

Bathrooms, kitchens, and utility rooms use **side wall** or **toe kick** supply air registers. The reason is to prevent water damage to the furnace and duct system.

A **perimeter loop** duct design is used for homes with a slab floor construction. Before the concrete slab is poured, a duct loop is placed around the perimeter of the home. Supply air openings in the duct are made, and four branch feeder ducts are run to feed the supply air from the furnace to the perimeter loop. The concrete slab is then poured around the duct system.

Perimeter radial or **central plenum** duct systems are found when furnaces are installed on the first floor of a home. The duct system will either be run in an attic or a crawl space area. The furnace is centrally located in the home and the plenum is extended above or below the furnace. When it is extended below the furnace, the plenum is located in a crawl space and branch supplies are connected from the plenum to the outside of the perimeter of the home in the floor. If it is extended above the furnace, the plenum is located in the attic and branch supplies are connected from the plenum to the outside of the perimeter of the home in the ceiling.

Return air systems can use an extended plenum, or a reduced extended plenum, when the furnace is located in the basement.

Return air systems use floor joists, wall studs, and gypsum board as channels for the return of conditioned space air back to the furnace.

In basement installations, **return air trunks** are mounted tight to the floor joists and are commonly made of 30-gauge sheet metal.

Central return systems are common with attic and first-floor furnace installations.

The common locations for **return air grills** are high or low on the inside wall.

Return air grills that are located high on the wall are used in a region where the cooling season is more predominant.

Return air grills that are located low on the wall are used in a region where the heating season is more predominant.

Figures 4–3 through 4–9 are diagrams of common metal duct connectors.

"S" slips are commonly connected horizontally on the duct sections.

Drive slips are commonly connected vertically on the duct sections.

When **drive cleats** are used, make sure to bend the end tabs to prevent movement of the cleat.

The **Pittsburgh lock seam** is one of the most common metal duct connections used in the HVAC/R industry. Its purpose is to connect sections of metal duct together along its length to make an air-tight seam.

The **Pittsburgh lock seam** can be made by hand using a sheet-metal brake and hand seamers or it can be constructed by the use of a Pittsburgh machine. See **Figure 4–10.**

Duct construction must conform to the requirements of **SMACNA Standards.**

Static pressure must be taken into consideration when determining the thickness of duct material.

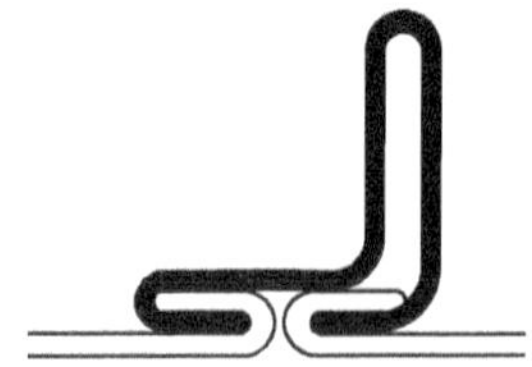

Standing Drive Slip

FIGURE 4–3

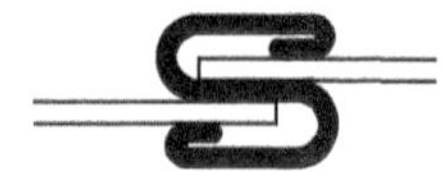

Hemmed "S" Slip

FIGURE 4–4

Double "S" Slip

FIGURE 4–5

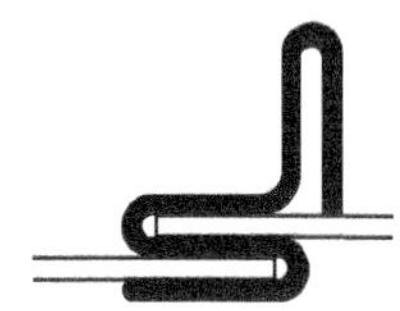

Standing "S" Slip

FIGURE 4–6

Inside Slip Joint

FIGURE 4–7

Drive Slip

FIGURE 4–8

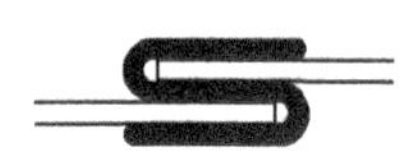

Plain "S" Slip

FIGURE 4–9

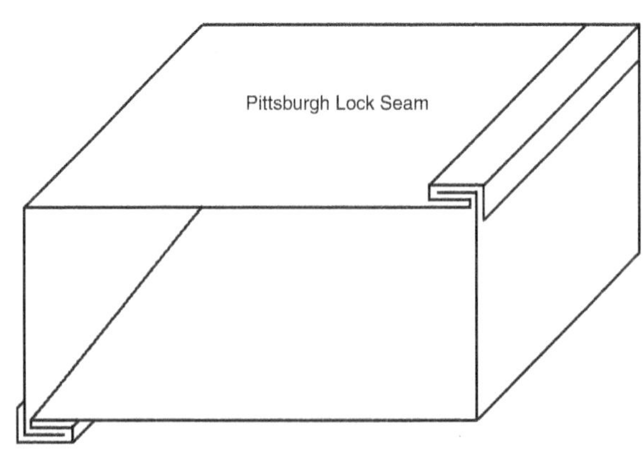

FIGURE 4–10

Tests for **fire performance** characteristics used on duct systems must conform to **UL 181.**

Nonmetallic and metallic duct systems must be **tested and classified** in accordance with **UL 181.**

The maximum discharge temperature permitted by industry standards for warm-air heating system is 250°F.

When gypsum board is used as an air shaft, the **air temperature** of the air must not exceed 125°F.

FLEXIBLE (FLEX) DUCT

A flexible air duct is generally used in approved areas where it is more appropriate to use than a standard sheet-metal duct. The following information pertains to flex duct.

When installing **flexible duct,** the section being installed must be fully extended to prevent the reduction of airflow.

The **radius** of a flexible duct must not be less than the diameter of the duct being used.

Sheet metal collars that are connected to a flexible duct must be a minimum of 2″ in length.

Sheet metal sleeves used for joining two sections of flexible duct must be a minimum of 4″ in length.

A horizontal flexible duct must be **supported** every 4′.

A vertical flexible duct must be **supported** every 6′.

A **saddle material** must be used when supporting flexible duct to prevent any restriction of the duct diameter.

The **minimum width** of saddle material used to support a flexible duct is $1\frac{1}{2}''$.

The **maximum sag** for a flexible duct is $\frac{1}{2}''$ per foot between the supports.

FIBERGLASS DUCTBOARD

Many HVAC/R companies prefer to use fiberglass ductboard instead of sheet metal. The following information is related to ductboard and its applications.

Fiberglass ductboard is commonly 1″ or $1\frac{1}{2}''$ thick.

The **maximum static pressure** that can be obtained in a fiberglass ductboard system is 2″ WC negative pressure and 2″ WC positive pressure.

The **maximum air velocity** in a fiberglass ductboard system is 2,400 fpm.

The **maximum inside temperature** for a fiberglass ductboard system is 250°F.

The **maximum outside surface temperature** for a fiberglass ductboard system is 150°F.

When **screws** are used to fasten ductboard to a metal surface, a #10 sheet-metal screw with a $2\frac{1}{2}''$ square washer must be used.

Screws used to fasten ductboard material must be a minimum of $\frac{1}{2}''$ longer than the board thickness.

The following materials must be used when **sealing a fiberglass ductboard:** pressure-sensitive aluminum foil tape, heat-activated aluminum foil tape, mastic, and glass fabric tape.

A shiplap joint flap must be **stapled** every 2″ on center.

Aluminum foil tape is used over a stapled shiplap joint flap.

Crosstabs are used to hold seams in position when staples are not used.

A **crosstab** is made of aluminum foil tape with a minimum length of 8″. It is placed across a taped joint every 12″.

The **minimum width** of pressure-sensitive aluminum foil tape that is used on fiberglass ductboard is $2\frac{1}{2}''$.

Aluminum foil tape must **overlap** each side of a fiberglass ductboard joint by a minimum of 1″.

If aluminum foil tape is stored in an area less than 50°F, the tape must be **preheated.**

When using **heat-activated tape,** the tape must be heated from 550 to 600°F after being applied to the fiberglass ductboard.

The **minimum width** of heat-activated tape used on fiberglass ductboard is 3″.

Mastic with a glass fabric system can be used to seal the seams of a fiberglass ductboard system.

A **tie rod** or **channel reinforcement** must be used on larger ductboard systems to increase duct rigidity.

A **tie rod** is commonly placed vertically in a fiberglass ductboard system.

A **tie rod** used for ductboard reinforcement is made of 12-gauge galvanized steel wire.

When a transverse joint is used in a ductboard system, the **tie rod** must be placed 4″ from the end of the female joint.

When a butt joint is used in the ductboard system, the **tie rod** must be placed 3″ from either side of the joint.

There are generally three types of seams used with **fiberglass ductboard:** shiplap joints, butt joints, and V-grooves.

Air must flow from the male to the female end of a shiplap joint.

V-grooves can be used on single flat board sections to create a rectangular duct.

Square cut or **butt joints** are not permitted on corners of ductboard sections.

Figures 4–11 through 4–17 are diagrams of common fiberglass ductboard closures.

B = Square Cut
BF = Square Cut with Flap
M = Male Shiplap
S = Female Shiplap
SC = Shiplap Corner Fold
SF = Female Shiplap with Flap
V = V – Groove
VC = V Corner Groove

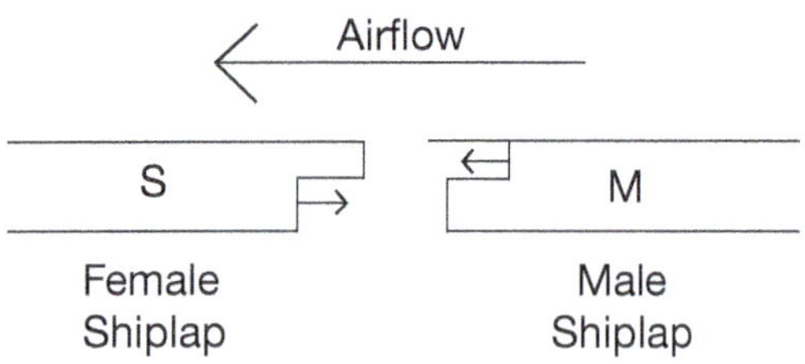

FIGURE 4–11

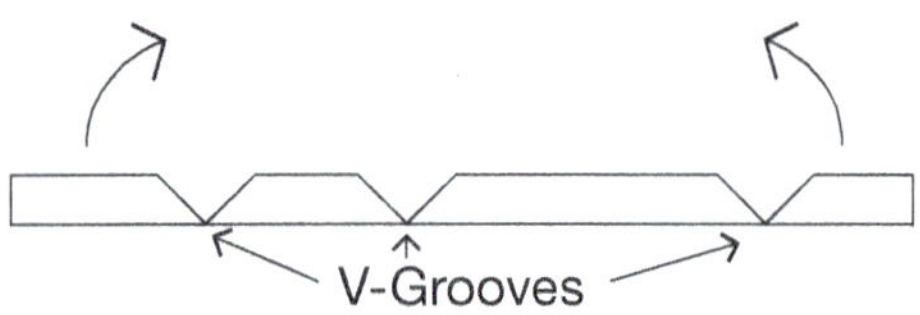

FIGURE 4–12

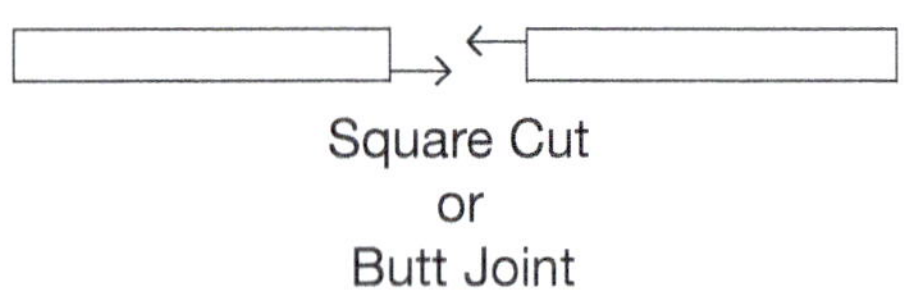

FIGURE 4–13

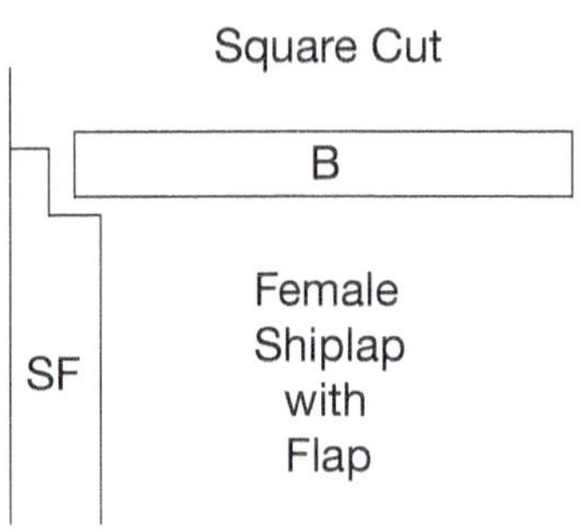

FIGURE 4–14

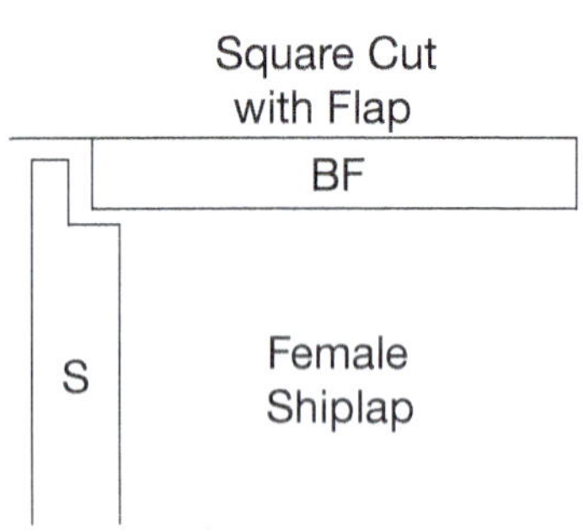

FIGURE 4–15

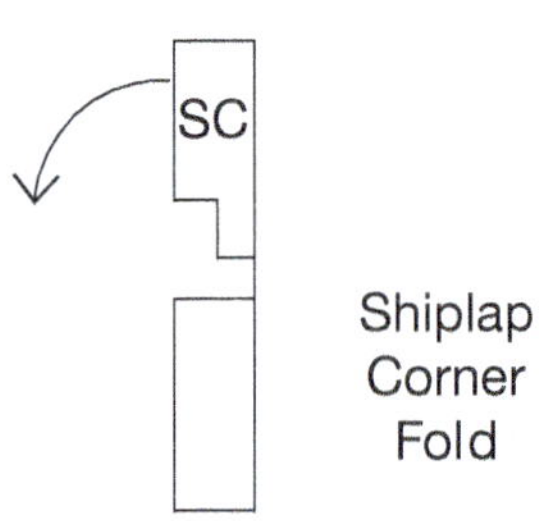

FIGURE 4–16

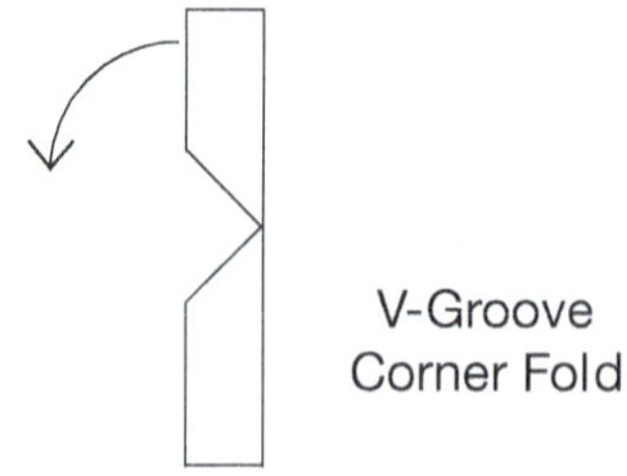

FIGURE 4–17

AIR-SIDE COMPONENTS

Technicians must understand component functions and terms that pertain to air-side components, which might include types of air dampers and damper systems.

There are two general types of **damper** blades: the **parallel blade** and the **opposing blade.**

The blades of the **parallel-blade damper** open together uniformly. The blades of the **opposing-blade damper** are constructed so that each blade opens and closes in the opposite direction.

The **opposing-blade damper** gives more uniform airflow through the system.

Mixed air control mixes return air with outside air for ventilation and free cooling. This system is controlled by an economizer circuit.

The **economizer circuit** uses a damper motor along with enthalpy sensors, discharge air sensors, and/or mixed air sensors to determine the amount of outside air to bring in for free cooling.

Exhaust or relief dampers are used in systems to prevent the pressurization of a building.

In a constant air volume system, **face** and **bypass dampers** are used together to control the amount of air over a heating or cooling coil. This type of control is found on larger commercial systems.

A **variable air volume (VAV) system** controls a quantity of air to a conditioned space depending on the load. During cooling, air volume increases as load increases. During heating, air volume decreases as heat load increases.

The air **temperature difference** between conditioned air and ventilation air must not exceed 10°F.

Dampers balance airflow by opening or closing certain ducts (zone control).

Some **dampers** are located in the diffuser, grill, or right in the duct.

Dampers should fit tight for accurate air control and minimum leakage.

Commercial and industrial buildings must have **automatic fire dampers** installed where the duct sections are run between rooms.

Fire dampers are usually held open by a fusible link that will melt with heat and cause the damper to close.

Smoke dampers use a photoelectric device to detect smoke. This causes the damper to close.

Proper ventilation is needed to increase oxygen and decrease carbon dioxide levels in the conditioned space.

An **Energy Recovery Ventilator (ERV)** is used to add fresh ventilation air and remove stale exhaust air in a home.

A technician inspecting a **motorized economizer** should check the condition of the vanes against binding or damage.

AIR FILTERS

Various types of filters and filtering equipment are used to help clean the air we breathe. The following are filter types and terms that relate to air filters.

Air filters should always be placed in the return air system upstream from a heat exchanger or a coil.

Air filters must be located so that they can be readily removed.

Return air boots must have turning vanes to evenly distribute the air across the filter media.

Ionizing means to charge a particle with a negative or positive charge.

Electronic air cleaners (EAC) are used to remove small particles of dust from an air distribution system. There are three common types: an ionized plate type, a charged-media nonionizing type, and a charged-media ionizing type.

The **ionized plate type** generates positive ions on wires that come in contact with dust particles. The wires pass the positively charged ions to the dust that is sent through collector plates. The collector plates have an electrical field that attracts and holds the dust particles. This system uses a high DC voltage of 12,000 volts.

The **charged-media nonionizing type** uses a dielectric medium consisting of fiberglass and cellulose mats supported by an electrically charged grid. The electrical charge attracts the dust particles into the filter media.

The **charged-media ionizing type** uses a combination of ionizing dust particles and electrically charging the filter media for an effective filtration system.

Most **electrostatic air cleaners** have a screen to trap large particles.

An **electronic air cleaner** removes particles as small as 0.0001 micron.

An **electronic air cleaner** may be equipped with a pressure drop indicator and an automatic shutoff control.

Cells of an **electronic air cleaner** should be cleaned every 2 or 3 months using chemicals or detergents.

The framework (casing) of the **electronic air cleaner** is connected with sheet-metal screws to the return-air side of the furnace.

Portable electronic air cleaners can be installed in ceiling applications. Most electronic air cleaners use a **prefilter,** installed before the electronic element, to catch large dust particles.

The **electrostatic filter** puts a static electrical charge on all particles passing through it, which then is drawn to collector plates with an opposite electrical charge.

Because of high voltage, the **electrostatic filter** is designed to shut off automatically when the service door is open.

A **manometer** is used to measure pressure drop across an air filter. The pressure drop is measured in inches of WC.

Disposable filters should be changed at least twice a year.

An **adhesive filter** is coated with an adhesive liquid or oil to trap particles in an air distribution system.

Adhesive filters remove up to 90% of dust particles.

HUMIDIFIERS

There are various models and types of humidifiers installed on heating appliances. A technician needs to know how to troubleshoot and install them correctly. The following information will help the technician when dealing with humidifiers.

Humidifiers are added to a forced-air heating system to increase humidity levels in the home during the winter months.

Insufficient **humidity** affects human comfort.

The three general **types of humidifiers** are evaporative, atomizing, and vaporizing.

The five common **evaporative-type humidifiers** are plate type, rotating-plate type, rotating-drum type, fan-powered type, and flow-through type.

A **rotating-drum humidifier** commonly uses a bypass duct to circulate supply air through the humidifier and back into the return air. This increases the efficiency of the humidifier and, with the use of a damper, allows the humidifier to be closed off during the summer months.

The media of a **plate-type humidifier** revolves in a pan full of water.

A **fan-powered humidifier** uses a fan to induce air over the humidifier media.

The **flow-through humidifier** uses a stationary media and a solenoid valve for water control. When the solenoid is energized, water flows over the top and through the media, allowing the warm air from the furnace to evaporate the water.

A **flow-through humidifier** commonly uses a **solenoid valve** for water control.

A **flow-through humidifier** commonly uses a bypass duct to circulate supply air through the humidifier back into the return air. This increases the efficiency of the humidifier and, with the use of a damper, allows the humidifier to be closed down during the summer months.

An **atomizing humidifier** commonly uses a rotating disc to atomize the water.

An **atomizing humidifier** commonly uses a fan to move water vapor.

The **vaporizing humidifier** uses an electrical heating element at the bottom of a water pan to evaporate the water.

The **vaporizing humidifier** can be installed in the supply or return-air section of the heating system.

The plate type, rotating-plate type, rotating-drum type, and fan-powered humidifiers are most commonly installed on the **supply air duct** of a forced-air heating system.

A **float valve** is used to regulate water into the pan of plate type, rotating-plate type, rotating-drum type, and fan-powered humidifiers.

The **humidifier media** is part of the humidifier that holds the water until it is evaporated into a heating system.

A **plate type** humidifier uses porous plate-type media located in a rack inside the humidifier.

The media of a **rotating-plate** humidifier revolves in a pan full of water.

The media of a **rotating-drum humidifier** revolves on a drum located in a pan full of water.

The average recommended relative humidity for a home during the winter months is **35 to 45%.**

If the proper **humidity level** is used in a home during the winter months, it will decrease the rate of moisture evaporation off of the skin. This will improve the comfort level.

Humidifiers commonly have their own 24-volt, 20-VA control transformer.

The **humidistat** is a control that senses humidity in a home. When the humidity level decreases, the humidistat closes, energizing the humidifier motor or solenoid valve.

The **humidifier circuit** is commonly energized when the indoor blower is energized during the heating cycle.

A **current sensing relay** is commonly used to energize the primary side of the 120-volt humidifier control transformer or the 24-volt humidifier circuit. It senses the flow of current through the heating speed of the blower and closes a set of normally opened contacts.

Many modern furnace controls use terminals marked **H** or **HUM** for the humidifier circuit. These terminals can energize the 120-volt primary side of a control transformer or a 24-volt humidifier circuit.

Long heat cycles and an **intermittent fan** motor operation are best suited for good humidification.

A humidifier that has its bypass damper closed will not provide enough moisture throughout the living structure, resulting in too much **static electricity.**

INDOOR AIR QUALITY (IAQ) AND SAFETY

This section provides information on indoor air quality, personal safety, refrigerant safety, soldering/brazing safety, and electrical safety. It also consists of a study guide for EPA Certification exams. Technicians must always be aware of the conditions and situations they are called to work under. Technicians must always work in a safe manner, so that they will not invite injury to themselves or to others. The following section will aid the technician in understanding indoor air quality (IAQ) and safety when working with electrical and mechanical systems.

INDOOR AIR QUALITY (IAQ)

Many types of contaminants are carried by airflow through supply registers and return-air ducts of a heating and cooling system. Contaminants in the airflow can result in poor health conditions for human life. Mild to severe health conditions may occur, depending on the type of contaminant. The following information will provide the technician with a better understanding of indoor air quality.

The difference between a **contaminant** and a **pollutant** is that a contaminant may or may not be a health risk and a pollutant is always a health risk.

Respirable particles that are breathed deep into the lungs are commonly 10 microns or smaller.

Asbestos is a name used to describe a number of fibers that are made from a mineral called **chrysolite.** If asbestos is breathed into the lungs, it could cause scarring of the lung tissue, lung cancer, and damage to the abdomen.

The first thing a technician should do to **improve air quality** is to change a system's air filters.

An inexpensive **disposable filter** is only about 7% efficient.

A **pleated filter** is commonly about 30% efficient.

Carbon-type filters are commonly used to efficiently remove tobacco smoke particles.

Stratification refers to air that settles in different temperature layers. To eliminate stratification, air must be kept in motion.

Air contaminants are classified into four groups: gases/vapors, inert particles, microorganisms, and pollen.

Synthetic fibers, dust, and cigarette smoke are considered **inert particles.**

Inert particles, such as dust, are the main cause of poor air distribution in HVAC/R equipment.

Microorganisms include fungi, mold, and bacteria.

Gases and vapors originate from the human body, hair sprays, paints, adhesives, solvents, radon, and ozone.

Pollen comes from grass, weeds, seasonal plants, and trees.

Sick building syndrome (SBS) commonly causes the following symptoms: eye, nose, and throat irritation; dizziness; headaches; drowsiness; and fatigue.

A high content of **carbon dioxide** depletes the oxygen level in a building. A satisfactory CO_2 measurement in an indoor environment is a maximum of 1,000 ppm.

Indoor air contaminants are commonly two to five times higher than outdoor pollutants.

Legionnaires' disease is an illness caused by the *Legionella pneumophila* bacteria. The bacteria are commonly caused by improperly treating cooling tower water.

A solution of **90% water** and **10% bleach** is commonly used to clean nonporous materials and prevent mold growth.

Some **air leakage** into a building will help decrease the percentage of indoor air contaminants.

A **high relative humidity** within a building can increase the growth of mold.

A well-balanced **ventilation** system will exhaust stale indoor air and introduce the same quantity of fresh outside air.

Common types of **ventilation systems** are passive, exhaust only, nonheat recovery, balanced mechanical, and heat recovery types.

For proper **ventilation,** a building must have an outside air supply and an exhaust system.

Having a slight positive pressure within a residency will help to **minimize infiltration** and can improve air quality.

A **VOC (volatile organic compound)** is a highly evaporative chemical that gives off gases.

VOCs can cause a person to have rashes, nausea, sinus congestion, drowsiness, headaches, and respiratory problems.

Radon is a radioactive gas that comes from the earth's crust. It enters the home through cracks, joints, nontrapped drains, sump pumps, water supplies, underground service openings, and crawl spaces.

Radon commonly stays near ground level because it's nine times heavier than air.

Radon is not considered harmful to humans until it decays.

Radon levels inside should be considered harmful when the average annual concentration in the normal living areas exceeds 800 becquerels (Bq) per cubic meter. The becquerel is a measurement of radon levels.

Polynuclear aromatic hydrocarbons (PAHs) are produced when materials containing carbon and hydrogen are burned.

PAHs outside the home are commonly caused by wood-burning appliances, coal-burning equipment, and internal combustion engines.

PAHs inside the home are commonly caused by cigarette smoke, the burning of food, and fireplaces.

Ozone (O_3) is an unstable oxygen molecule made up of 3 oxygen atoms.

Ozone is produced when air is heated by an electrical arc during an electrical storm.

The level of **ozone** inside the home is directly correlated with the level outside the home.

Nitric oxide (NO) is a toxic gas that is produced during high-temperature combustion.

Nitric oxide and **nitrogen dioxide** can be produced inside a home when fuel-burning appliances are not vented. Another indoor source is cigarette smoke.

Once nitric oxide is released into the air, it quickly changes to **nitrogen dioxide.**

Common indoor levels of **nitrogen oxide** are 0.25 ppm (parts per million) or less.

Carbon monoxide (CO) is a poisonous gas that is produced during incomplete combustion of fossil fuels.

Common indoor sources of **carbon monoxide** are cracked heat exchangers, cracked chimneys, unvented appliances, and cigarette smoke.

Common indoor levels of **carbon monoxide** range between 0.5 and 5 ppm.

If a home has an excessive negative pressure, the combustion gases of an appliance could be drawn back into the home and produce **carbon monoxide.**

Formaldehyde is classed as a volatile organic compound (VOC). It is in many building materials such as medium-density fiberboard (MDF) and particle board.

An increase of temperature and humidity levels in a home is directly proportional to the rate that a **formaldehyde gas** is released.

TEMPERATURE, HUMIDITY, AND PSYCHROMETRICS

Technicians need to know various types of temperature sensors. They must also be capable of performing humidity calculations and understand the importance of humidity in a system. The following section covers temperature and humidity.

A liquid column **thermometer** is a glass thermometer that is filled with mercury or red alcohol. Its common range is from −30 to +120°F.

A **dial thermometer** uses a bimetal sensor when it incorporates an analog readout, and a thermistor when using a digital readout. Its range is from −40 to +160°F.

Electronic thermometers use a thermistor-type sensor to measure temperatures. A thermistor's resistance changes when its temperature changes. An electronic thermometer is much more accurate than a dial or bimetal thermometer.

Infrared thermometers use an infrared beam to check the surface-area temperature of an object. Infrared thermometers are usually used to check temperatures of objects that are difficult to reach with standard thermometers.

Recording thermometers are used to check and record the temperatures of a system over a period of time. Recording thermometers are usually used on a system from 1 to 7 days.

The dry-bulb delta-T is the dry-bulb temperature difference between the conditioned (inside space) compared to the unconditioned (outside space).

The wet-bulb delta-T is the wet-bulb temperature difference between the conditioned (inside space) compared to the unconditioned (outside space).

Enthalpy is a property of substances that is a measure of their heat content.

Humidity ratio is the weight of water vapor in each pound of water vapor.

Specific volume is the cubic feet of the mixture per pound of dry air.

Sensible heat ratio refers to the ratio of the sensible heat to total heat within a process.

A **sling psychrometer** is used to measure the wet-bulb and dry-bulb temperature of air to determine the relative humidity.

A **sling psychrometer** uses liquid-filled glass tube thermometers that are extremely accurate.

The **dry-bulb temperature** of air is the sensible temperature measured with a standard thermometer.

The **wet-bulb temperature** of air is measured by placing a wick that is saturated with water around a thermometer bulb and rapidly moving air over the wick. The dryer the air, the lower the wet-bulb temperature will be.

Electronic **humidity-measuring equipment** is used to measure the relative humidity and the dewpoint of the air. Some electronic meters use remote sensing probes that are placed in an airstream to measure the relative humidity level.

A **psychrometric chart** is used to determine the properties of air, such as dry-bulb temperatures, wet-bulb temperatures, dewpoint temperatures, moisture content, relative humidity levels, specific volume, and more.

The **vertical lines** are used to measure dry-bulb temperatures (see Figure 4–18).

The **horizontal lines** are used to measure grains or pounds of moisture and dewpoint temperatures (see Figure 4–19).

The **45-degree diagonal lines** are used to measure the wet-bulb and the total heat content (see Figure 4–20).

The **diagonal lines** are used to measure the specific volume.

The **curved lines** are used to measure the percentage of relative humidity (see Figure 4–21).

Complete **air conditioning** means to heat, humidify, cool, dehumidify, ventilate, filter, and circulate conditioned space air.

A typical **comfort goal** is maintaining air at a 75°F dry-bulb temperature at 50% relative humidity.

Insufficient **humidity** levels affect human comfort by causing a dry, uncomfortable feeling.

Too much **humidity** causes a person to feel hot and causes perspiration.

Too high dry-bulb **temperature** decreases the rate of heat transfer from a person to the surrounding air. Too low dry-bulb temperature increases the rate of heat transfer from a person.

Too high **relative humidity** decreases the rate of evaporation of moisture from a person to the surrounding air. Too low relative humidity increases the rate of evaporation from a person.

$$\frac{\text{Heat Loss} - \text{Heat Gain}}{\text{Heat Loss} - \text{Conduction}}$$

Conduction is the transfer of heat from molecule to molecule through a substance by a chain reaction.

Convection is the transfer of heat by the movement of molecules from one place to another.

A fluid such as water or air is used to transfer the heat by **convection.**

Radiation is the transfer of heat by passing from the heat source through space without heating the space in between.

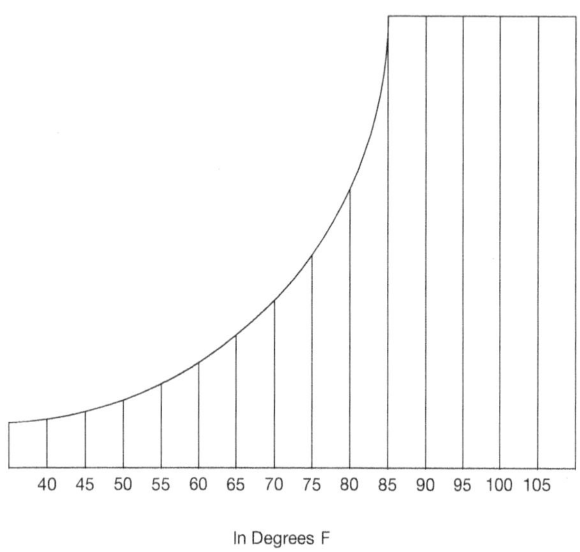

FIGURE 4–18

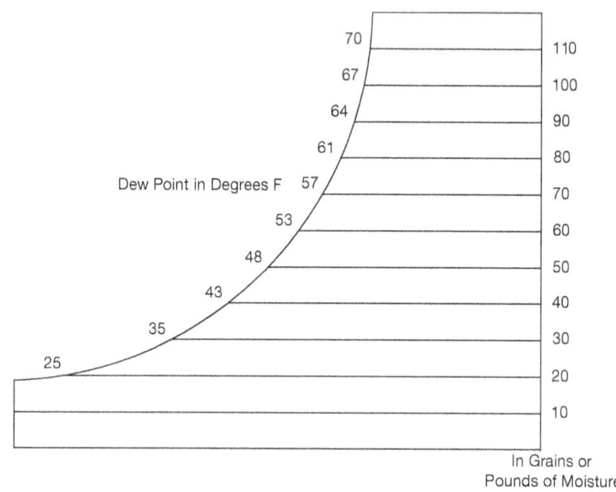

FIGURE 4–19

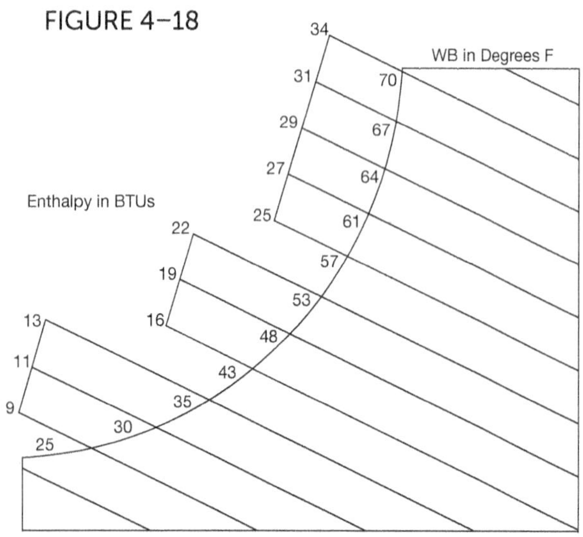

FIGURE 4–20

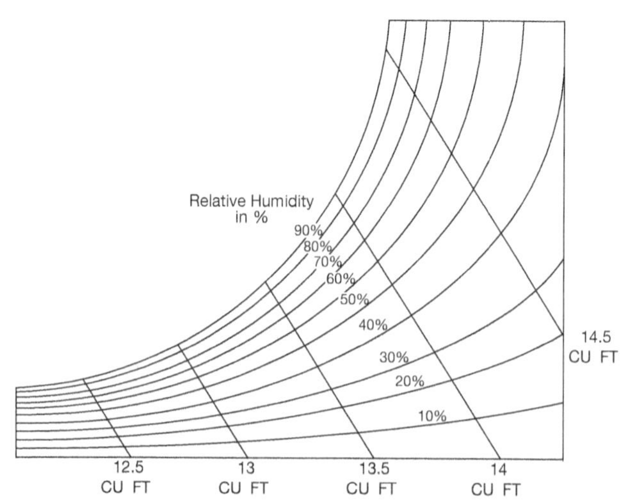

FIGURE 4–21

Heat flow through a wall separating two spaces at different temperatures depends on three factors:

The area of the wall.

The temperature difference between the two spaces.

The heat-conducting properties of the wall.

The larger the **wall area,** the more heat it can conduct.

A 200-ft^2 area will **conduct** twice as much heat as a 100-ft^2 area.

A **greater temperature** difference between the two spaces causes a greater heat transfer.

If the **temperature difference** is 100°F as compared to 50°F, twice as much sensible heat will transfer through the wall. This principle also applies to windows, roofs, and other building surfaces.

The **R-value** of a material is its ability to resist the flow of heat through it.

The higher the **R-value** of a material, the more resistive it is to heat transfer.

To determine heat transfer through a wall using the **R-value,** use the following calculation:

Q = 1/R × A × TD

Q = Heat in Btu/hr

A = Area in ft^2

TD = Temperature difference across the wall

Wall material and thickness affect the **heat transfer** through the wall.

There are two terms used when it comes to the **construction of walls:**

Conductance (C) value is the ability of a material to transfer heat in Btu/hr/ft^2/F C = 1/R(R = resistance to heat flow).

Conductivity (k) value is the conductance of a material per unit of thickness.

C = k/L (L = thickness in inches).

Conductivity is the ability of a material to conduct heat.

Thermal **conductivity** is defined as the heat flow per hour (Btu/hr) through 1 ft^2 of 1″-thick homogeneous material when the temperature difference between the two faces is 1°F.

Conductance is a term used to denote heat flow through nonhomogeneous materials such as glass blocks, hollow clay tiles, and concrete block.

Conductance is the heat flow rate in Btu/hr through 1 ft^2 of nonhomogeneous material of a certain thickness when there is 1°F temperature difference between the two surfaces.

Conductivity and **conductance** are not interchangeable terms.

Conductivity is the heat flow through 1″ of a homogeneous material and conductance is the heat flow through a nonhomogeneous material.

Air-space conductance is another factor that must be considered when calculating the heat loss of a structure.

Air space is defined as heat flow in Btu/hr through 1 ft^2 of air space for a temperature difference of 1°F between the bounding surfaces.

An **air film** at the inside and outside area of the wall will also produce a C value based on air velocity.

Fi = Inside air film for still air

C = 1.6 (R = 1/1.6) = 0.625.

Fo = Outside air film for air at 15 MPH is

C = 6(R = 1/6)0.17.

To determine overall heat transfer of a wall that uses four or five different materials, the **U-factor** is used, which simplifies this calculation.

The **U-factor** works with both homogeneous and nonhomogeneous materials.

The **U-factor** is defined as the heat flow per hour through 1 ft^2 of material when the temperature difference is 1°F between the air on the two sides of the wall or roof.

The **overall heat transfer** formula is:

Q = A × U × td

Q = heat flow in Btu/hr

A = area in ft^2

U = overall heat transfer coefficient in Btu/ft^2/°F (U-factor)

TD = temperature difference in degrees F

HEAT GAIN—CONDUCTION AND RADIANT

To determine the **heat gain** through an exterior wall or door:

1st—Determine the **indoor design** and the **outdoor design** temperatures and wet-bulb or % relative humidity levels.

2nd—Determine the wall materials and **"U" value** of the wall.

3rd—Determine the **time of day** to be considered. Example: 4:00 p.m. (16 p.m. military time)

4th—Determine the **direction** the wall is facing (N, S, E, W, etc.).

5th—Determine the **CLTD** (cooling load temperature difference) using the appropriate ASHRAE charts and use the following formula:

U × A × CLTD = Btu/hr

Solar heat gain happens when the sun shines through the windows and adds a heating load to the building.

Windows gain heat by:

1. Conduction.
2. Solar gain or SHGF (solar heat gain factor).

To determine the heat gain through an **exterior window**:

1st—Find wall location (N, S, E, W, etc.).

2nd—Determine the time of day (4:00 p.m. is commonly used).

3rd—Find SHGF based on the type of window and location (e.g., N-40° latitude or N-32° latitude).

4th—Find shade coefficients 1 and/or 2—SC1 or SC2.

5th—Use the formula:

SHGF × SC1 × SC2 × Area = Btu/hr

To determine the heat gain through a **roof**:

1st—Determine the roof construction and determine the "U" value of the roof.

2nd—Determine the roof heat gain factor based on type of construction and time of day (example: 4″ LW concrete at 4:00 p.m. sun time) using the proper ASHRAE charts.

3rd—Determine the area of the roof in ft^2.

Use the following formula to determine the heat gain in Btu/hr:

Btu/hr = Factor × Area × U

For a **slab floor**, consider 0 Btu/hr heat gain.

If the **floor** is over a crawl space or above-grade basement, heat gain can be determined by using proper ASHRAE tables and determining the area and exposed floor or above- and below-grade and using the following formula:

U × A × CLTD (or TD if not exposed to sunlight) = Btu/hr.

INFILTRATION

Infiltration is heat lost or gained into a building by outdoor air being introduced into the building.

Common **infiltration rates:**

Tight construction = .3 to 6 air changes per hour

Average construction = 6 to 1.0 air changes per hour

Loose construction = 1.0 to 2.5 air changes per hour

It is required that people within a building receive 15 to 60 cfm of **ventilation** air per person.

Infiltration for windows and doors is determined by using the **crack method:** The **length of crack** around the perimeter of a fixed, nonopening window in linear feet multiplied by the cfm factor on a chart when compared to the design outdoor wind velocity for the area. (Opening windows will take into account the crack of the opening section of the window.)

Infiltration due to **door opening** is determined by the outdoor wind velocity (5, 10, 15 mph) and number of people opening the doors (possible customers) × cfm factor/person = cfm.

Exhaust fans are always rated in cfms.

To determine the **building's cfm** of infiltration, take the larger number of the two:

1. The required 15 cfm/number of building occupants
2. Window infiltration + door infiltration + exhaust fan cfm

Example:

(calculated window infiltration) 244 cfm +
(calculated door infiltration) 145 cfm +
(total building exhaust fan) cfm 450 cfm = 839 cfm

Compared to 15 cfm/person × 65 people = 975 cfm.
The greater cfm of 975, known as deliberate cfm required for ventilation, will be used.

To determine heat lost or gained into a building, take the required ventilation of the building of 975 cfm and use the following formulas to determine **sensible and latent heat loss or gain:**

Sensible heat Btu/hr = cfm × 1.08 or (1.1) × TD

Latent heat Btu/hr = cfm × .68 × grains of moisture difference

Ventilation refers to changing the air in a building structure by causing air to flow from one area to another.

Stratification is a condition where there is little to no air movement; air stays layered within a room.

INTERNAL HEAT GAIN

There are many considerations when determining the **internal heat gain** in a building, such as:

Lighting

Appliances

Motors

People

To determine the heat gain due to incandescent and fluorescent **light bulbs**, use the following formulas:

Incandescent bulb wattage × number of bulbs × 3.41 Btu/watt = Btu/hr
Fluorescent wattage × number of bulbs per fixture × number of fixtures × 1.2 = total watts × 3.41 = Btu/hr

To determine the **heat gain due to appliance**, use the proper ASHRAE charts to determine the amount of sensible and latent heat gain by specific appliances.

Example:

According to ASCRAE charts, a 5-gallon electric coffee urn (no hood) produces 3,850 Btu/hr sensible heat and 1,250 Btu/hr latent heat.

An electric hot plate 18″ × 20″ × 13″ produces 5,300 Btu/hr sensible heat and 3,600 Btu/hr latent heat.

To determine **heat gain by motors**, use the following formula:

HP (Horsepower) × 2,545 Btu/hr/hp = Btu/hr

1.5 HP × 2,545 Btu/hr/hp =
Sensible – 3,818 Btu/hr, Latent – 0

People add both latent and sensible heat into buildings depending upon their activity.

To determine the **heat gain due to people**, use the proper ASHRAE charts to determine the amount of sensible and latent heat gain by activity of people.

PART II
Questions

SECTION 1

Refrigeration and Air Conditioning Tools and Components

THE FOLLOWING QUESTIONS MAY BE HELPFUL TO STUDY FOR THESE CERTIFICATION TESTS:
NATE—CR-SV, CR-IN, AC-SV, AC-IN, HP-SV, HP-IN, HG-SV, HG-IN, HO-SV, HO-IN, AD-SV, AD-IN, OL-SV, OL-IN, GS-SV, GS-IN, RC, SV, RC-IN, RL-SV, RL-IN.
ESCO INSTITUTE HVAC EXCELLENCE CERTIFICATIONS

PERSONAL SAFETY

1. What type of safety equipment is commonly used to protect against falling objects?
   A. safety glasses
   B. ear plugs
   C. gloves
   D. hard hats
2. What type of safety equipment is commonly used to protect against flying objects?
   A. safety glasses
   B. respirators
   C. gloves
   D. a safety harness
3. What type of safety equipment is commonly used to protect against scrapes, cuts, and burns?
   A. safety glasses
   B. respirators
   C. gloves
   D. a safety harness
4. What type of safety equipment is commonly used to protect against falling objects and punctures?
   A. ear plugs
   B. respirators
   C. steel-toe boots
   D. a safety harness
5. What type of safety equipment is commonly used for protection in high-noise-level areas?
   A. ear plugs
   B. respirators
   C. steel-toe boots
   D. hard hats
6. What type of safety equipment is commonly used for protection in a highly contaminated area?
   A. ear plugs
   B. respirators
   C. safety glasses
   D. hard hats
7. What type of safety equipment is commonly used for protection against falls when working in high areas?
   A. ear plugs
   B. respirators
   C. steel-toe boots
   D. safety harness
8. A safety harness should be used when working on objects
   A. more than 4′ above the ground.
   B. more than 10′ above the ground.
   C. more than 6′ above the ground.
   D. more than 12′ above the ground.
9. When altering any safety equipment, what should be done after use?
   A. Always change it back to its original working order after use.
   B. Place a note near the equipment explaining the alteration in detail.
   C. If the alteration worked, nothing is needed.
   D. Safety equipment must never be altered in any way.

10. To prevent an injury when using a wrench, it is safer to

A. pull the wrench toward you.

B. push the wrench away from you.

C. pull or push the wrench.

D. use maximum force at all times.

11. When lifting objects,

A. it is best to use your back muscles only.

B. bend your back and use your leg muscles.

C. keep your back straight and use your leg muscles.

D. always use your arm and chest muscles.

12. When lifting heavy objects,

A. always use your back muscles.

B. always use your arm muscles.

C. always use your arm and back muscles.

D. always ask for assistance.

13. To prevent a personal injury when working with power tools, always

A. wear hand protection.

B. wear eye protection.

C. wear long sleeves to protect arms.

D. use AGA-approved power tools.

14. To prevent a personal injury when handling sheet metal, always

A. wear gloves that are a nonslip type.

B. wear safety glasses.

C. ask for assistance when handling large or heavy sections.

D. All of the above

15. When using an extension ladder, you should

A. keep three-point contact with the ladder.

B. extend the ladder all the way.

C. extend the ladder only one half the way.

D. always get help extending the ladder.

16. Never use a hammer and chisel without

A. hand protection.

B. ear protection.

C. eye protection.

D. foot protection.

17. Gasoline should never be used to clean tools because

A. the vapors are flammable.

B. it will dissolve the tool being cleaned.

C. vapors will remain with the equipment being cleaned.

D. it does not cut grease well enough.

18. To prevent a personal injury,

A. never use a hammer with a rounded head.

B. never use a worn chisel.

C. use only well-oiled hand tools.

D. hold a screwdriver only by its handle when walking.

THE FOLLOWING QUESTIONS MAY BE HELPFUL TO STUDY FOR THESE CERTIFICATION TESTS:
NATE—CR-SV, CR-IN
ESCO INSTITUTE HVAC EXCELLENCE CERTIFICATIONS

HAND TOOLS

1. Which of the following tools has a six-point star-shaped head?

A. a Phillips screwdriver

B. a nut driver

C. a spade bit

D. a torx driver

2. Which of the following tools has a hexagon-shaped head and is used to remove a screw with a six-point hex-head shape?

A. a Phillips screwdriver

B. a nut driver

C. a spade bit

D. a torx driver

3. Which tool has a variable open end that is used to remove a nut or a bolt?

A. an Allen wrench

B. a flare-nut wrench

C. an adjustable wrench

D. a box-end wrench

4. Which tool is used to loosen or tighten set screws?

A. an Allen wrench

B. a flare-nut wrench

C. an adjustable wrench

D. a box-end wrench

5. Which tool has a closed end and is used to loosen a nut or bolt?

A. tongue-and-groove pliers

B. a pipe wrench

C. an adjustable wrench

D. a box-end wrench

6. Which tool is used to loosen or tighten a mechanical fitting on a refrigeration or air conditioning system?

A. a flare-nut wrench

B. a pipe wrench

C. an adjustable wrench

D. a service wrench

7. Which tool is used to loosen or tighten a nut or a bolt from the side?

A. a service wrench

B. a box-end wrench

C. a socket wrench

D. an open-end wrench

8. Which is the proper tool to use when assembling or disassembling threaded pipe?

A. a pipe adapting kit

B. a pipe wrench

C. a crescent wrench

D. a torque wrench

9. Which is the correct tool to use when manipulating access valves on a refrigeration or air conditioning system?

A. an adjustable wrench

B. tongue-and-groove pliers

C. a torque wrench

D. a service valve wrench

10. What are the most common size openings for a service wrench used on light commercial refrigeration and air conditioning systems?

A. $\frac{3}{16}''$, $\frac{1}{4}''$, $\frac{5}{16}''$, and $\frac{3}{8}''$

B. $\frac{1}{4}''$, $\frac{1}{2}''$, $\frac{3}{4}''$, and $1''$

C. $\frac{3}{16}''$, $\frac{5}{16}''$, $\frac{7}{16}''$, and $\frac{9}{16}''$

D. $\frac{5}{16}''$, $\frac{3}{8}''$, $\frac{1}{2}''$, and $\frac{5}{8}''$

11. Which tool would most likely be used to fit on top of a $\frac{1}{2}''$ bolt?

A. vise grip pliers

B. socket wrench

C. needle-nose pliers

D. service valve wrench

12. Which tool is used to apply a specific amount of force to a nut or a bolt?

A. a pipe wrench

B. an adjustable wrench

C. a torque wrench

D. a service wrench

13. The tool that is used to ensure that ductwork and other work is level is

A. a plumb bob.

B. a carpenter's square.

C. a tape measure.

D. a level.

14. Which of the following is an L-shaped tool used to measure materials and check for uniformity?

A. a plumb bob

B. a carpenter's square

C. a tape measure

D. a level

15. What type of saw is used to cut wood against the grain?

A. a ripsaw

B. a crosscut saw

C. a hacksaw

D. a keyhole saw

16. What type of saw is used to cut wood with the grain?

A. a ripsaw

B. a crosscut saw

C. a hacksaw

D. a keyhole saw

17. What type of saw is primarily used on metal?

A. a ripsaw

B. a crosscut saw

C. a hacksaw

D. a keyhole saw

18. What type of saw is most commonly used to cut drywall?

A. a ripsaw

B. a crosscut saw

C. a hacksaw

D. a keyhole saw

19. Which file has one row of teeth?

A. a single-cut file

B. a double-cut file

C. a triple-cut file

D. a rasp

20. Which file has two rows of teeth that crisscross each other?

A. a single-cut file

B. a double-cut file

C. a triple-cut file

D. a rasp

21. The three common drive sizes that a drill is available in are

A. $\frac{3}{8}''$, $\frac{1}{2}''$, and $\frac{3}{4}''$.

B. $\frac{1}{4}''$, $\frac{3}{8}''$, and $\frac{3}{4}''$.

C. $\frac{5}{16}''$, $\frac{3}{8}''$, and $\frac{1}{2}''$.

D. $\frac{1}{4}''$, $\frac{3}{8}''$, and $\frac{1}{2}''$.

22. The tool used with a drill to install or remove hex-head sheet-metal screws is called

A. a hex-head adapter.

B. a screw adapter.

C. a torx adapter.

D. a hex-head spade bit.

23. What tool is used to recess holes in wood and other material so that screws can be driven flush or slightly below the surface?

A. a spade bit

B. a countersink

C. a punch

D. a reamer bit

24. Which bit is preferred for drilling holes in wood?

A. a masonry bit

B. a countersink

C. a spade bit

D. a keyhole bit

25. What tool has a corkscrew shape and is used to bore holes into wood?

A. a masonry bit

B. a spade bit

C. a corkscrew bit

D. an auger bit

26. What tool is used to bore holes in brick, block, concrete, stone, or mortar?

A. a masonry bit

B. a countersink

C. a spade bit

D. a keyhole bit

27. The proper tool to use when creating an indentation in sheet metal is

A. a punch.

B. a countersink.

C. a torx driver.

D. an auger bit.

28. What tool is used to cut a threaded hole in metal material?

A. a tap

B. a die

C. a countersink

D. a cutting tool

29. Which of the following would be the proper tool for threading gas pipe?

A. a punch

B. a tap

C. a die

D. a reamer

30. What tool is used to remove the burrs from the inside of tubing?

A. a reamer

B. a die

C. a tube cutter

D. a countersink

31. Which hammer would most likely be used to remove a nail from a 2 × 4 stud?

A. a ball-peen hammer

B. a claw hammer

C. a masonry hammer

D. a sledgehammer

32. Which hammer, designed for metal work, has a flat back and a round front?

A. a ball-peen hammer

B. a claw hammer

C. a masonry hammer

D. a sledgehammer

33. Which hammer has a large head that is designed for pounding block or concrete?

A. a ball-peen hammer

B. a claw hammer

C. a masonry hammer

D. a sledgehammer

34. What care should be taken to prevent corrosion after using hand tools?

A. Hand tools must be stored at 60°F to prevent corrosion and metal fatigue.

B. Hand tools should be wiped down with machine oil.

C. Hand tools should be cleaned with mineral spirits after each use.

D. Hand tools should be wiped down with a saltwater solution.

THE FOLLOWING QUESTIONS MAY BE HELPFUL TO STUDY FOR THESE CERTIFICATION TESTS:
NATE—CR-SV, CR-IN.
ESCO INSTITUTE HVAC EXCELLENCE CERTIFICATIONS

BASIC CONSTRUCTION

1. The two most common roof designs used on homes today are

A. the flat roof and the truss-built roof.

B. the rafter-built roof and the flat roof.

C. the truss-built roof and the rafter-built roof.

D. the gable roof design and the hip roof design.

2. The most common base used under shingles to strengthen the roof by adjoining the rafters or trusses together is

A. $\frac{1}{4}''$ plywood.

B. $\frac{1}{2}''$ OSB sheathing.

C. $\frac{1}{2}''$ particle board.

D. $\frac{1}{2}''$ pine board.

3. Trusses are commonly constructed of
   A. 2″ × 8″ or 2″ × 10″ lumber.
   B. 1″ × 8″ or 1″ × 10″ lumber.
   C. 2″ × 2″ or 2″ × 4″ lumber.
   D. 1″ × 2″ or 2″ × 4″ lumber.
4. What type of roof construction uses boards that connect the ridge of the roof and the exterior wall?
   A. a truss-built roof
   B. a flat roof
   C. a hip roof
   D. a rafter-built roof
5. The material that is used for a standard ceiling is
   A. $\frac{1}{4}$″ drywall.
   B. $\frac{5}{8}$″ drywall.
   C. $\frac{3}{4}$″ drywall.
   D. $\frac{7}{8}$″ drywall.
6. What would the effect of heat transfer be if the R-value on a roof were increased?
   A. It would increase the heat transfer.
   B. It would decrease the heat transfer.
   C. It would have no effect on heat transfer.
   D. It would reduce the thermal efficiency.
7. Common exterior residential walls are constructed of
   A. 2″ × 2″ or 2″ × 4″ studs
   B. 2″ × 4″ or 1″ × 4″ studs.
   C. 2″ × 4″ or 2″ × 6″ studs.
   D. 1″ × 4″ or 1″ × 6″ studs.
8. How far apart are the studs spaced in an interior and exterior wall?
   A. 12″ on center
   B. 14″ on center
   C. 16″ on center
   D. 18″ on center
9. The bottom board located on a wall that the studs are nailed to is known as a
   A. lower rafter.
   B. horizontal stud.
   C. load-bearing joist.
   D. bottom plate.
10. What is known as a load-bearing wall?
    A. This is the wall that uses two studs to return air back to the furnace.
    B. This is the wall that always has a beam resting on top of it.
    C. This is the wall that has the weight of the roof on top of it.
    D. This is the wall that should never have a beam resting on top of it.
11. What is the purpose of a vapor barrier?
    A. to keep moisture from entering and exiting a home
    B. to protect the insulation of an exterior wall
    C. to retain steam created in bathrooms
    D. to remove moisture from a home
12. Name three common exterior wall insulation R-values.
    A. R-13, R-19, R-30
    B. R-5, R-11, R-13
    C. R-19, R-30, R-40
    D. R-11, R-13, R-19
13. What is the purpose of the 1″ × 2″ wood slats that are placed between the floor joists?
    A. They are used to decrease the squeak of a floor.
    B. They are bridging used to stiffen the floor.
    C. They are used to support the floor joists.
    D. They are used to hold supply air ducts in place.
14. A subfloor is commonly made with
    A. $\frac{1}{2}$″ tongue-and-groove plywood.
    B. $\frac{3}{4}$″ OSB tongue-and-groove sheathing.
    C. $\frac{5}{8}$″ OSB sheathing.
    D. $\frac{5}{8}$″ plywood.
15. An underlayment floor is made of
    A. $\frac{1}{4}$″ plywood.
    B. $\frac{1}{4}$″ OSB sheathing.
    C. $\frac{1}{2}$″ plywood.
    D. $\frac{1}{2}$″ OSB sheathing.

THE FOLLOWING QUESTIONS MAY BE HELPFUL TO STUDY FOR THESE CERTIFICATION TESTS:
NATE—CR-SV, CR-IN, AC-SV, AC-IN, HP-SV, HP-IN, RC-SV, RC-IN, RL-SV, RL-IN.
ESCO INSTITUTE HVAC EXCELLENCE AIR CONDITIONING AND COMMERCIAL REFRIGERATION CERTIFICATION

REFRIGERANT SAFETY

1. To prevent an injury when working with refrigerants, what safety precautions are necessary?
   A. Always have a fire extinguisher nearby.
   B. Always have a phosgene sensor near the equipment being serviced.
   C. Always use safety glasses and gloves.
   D. Always use ear protection.
2. What two types of injuries can CFC, HCFC, and HFC refrigerants cause?
   A. lung cancer and heart attacks
   B. heart attacks and frostbite burns
   C. carbon monoxide poisoning and frostbite burns
   D. frostbite burns and suffocation
3. Which of the following may cause suffocation?
   A. using a torch on a refrigeration system under pressure
   B. venting refrigerant in a confined space
   C. using a vacuum pump in a confined space
   D. using a torch while in an enclosed walk-in cooler
4. Which one of the following statements is true?
   A. A system should be evacuated while under pressure.
   B. It is safe to heat a refrigerant cylinder above 125°F.
   C. Refrigerant is lighter than air.
   D. Air is lighter than refrigerant.
5. What is required in equipment rooms that use systems containing large quantities of refrigerant?
   A. phosgene and refrigerant alarms and sensors
   B. refrigerant and carbon monoxide alarms and sensors
   C. carbon monoxide and oxygen alarms and sensors
   D. oxygen and refrigerant alarms and sensors
6. When halogens are in contact with an open flame,
   A. an explosion will occur.
   B. phosgene gas is produced.
   C. it is noticeable from a green color.
   D. it is noticeable from a purple color.
7. What care must be taken when working around toxic fumes?
   A. Open a window, if possible, for ventilation.
   B. Hold your breath as long as possible.
   C. A breathing apparatus must be used.
   D. No extra care must be taken.
8. What must always be available when working with large amounts of refrigerants in equipment rooms?
   A. a self-contained breathing apparatus
   B. a fire extinguisher
   C. a carbon monoxide detector
   D. a smoke detector
9. When leak testing systems using nitrogen, the technician should
   A. not mix refrigerant with the nitrogen.
   B. leak test through the highside only.
   C. always use a pressure regulator and safety relief valve at the nitrogen tank.
   D. leak test through the lowside only.
10. When using nitrogen in a system for leak detection, a technician
    A. must use two separate gauges when connecting the nitrogen and a trace refrigerant to the system at the same time.
    B. should add mostly refrigerant when testing a system for leaks.
    C. does not have to use a pressure regulator on the nitrogen tank.
    D. must never connect nitrogen and refrigerant to the system at the same time.
11. When pressure testing a refrigeration system,
    A. the system must always be in operation.
    B. the pressures must be at least 50 psig.
    C. the pressures must not exceed system nameplate pressures.
    D. the system must always be de-energized.
12. What mixture is the most explosive?
    A. air and refrigerant
    B. air and nitrogen
    C. nitrogen and oil
    D. oxygen and oil
13. To prevent an explosion, never leak test a system using
    A. CO_2.
    B. oxygen.
    C. nitrogen.
    D. R-22.
14. When charging refrigerant into an air conditioning system, the technician should
    A. never charge in a liquid state.
    B. never charge in a vapor state.
    C. always charge by superheat method only.
    D. always charge by the manufacturer's recommendation.

15. To prevent an explosion, the discharge service valve on a compressor must never be

A. in the midseated position during operation.

B. in the back-seated position during operation.

C. in the front-seated position during operation.

D. in the back-seated position during a system pump-down.

16. Which refrigerant lines might cause a serious burn?

A. the discharge line

B. the suction line

C. the liquid line

D. the inlet line

17. When handling large air conditioning fin and tube coils,

A. always use the proper safety harness.

B. always evacuate the coil.

C. always use hand protection and ask for assistance.

D. always remove the coil from the package system.

18. Acidic oil from compressors is harmful because

A. it causes skin burns.

B. it contains oil.

C. it contains oil and refrigerant.

D. it contains refrigerant properties.

19. When testing for natural gas leaks, never use

A. a soap bubble solution.

B. an open flame.

C. electronic detectors.

D. a heavy soap solution containing chlorine.

20. To prevent danger caused by excessive pressures,

A. never overcharge an air conditioning system.

B. never undercharge an air conditioning system.

C. never allow an air conditioning system to operate above 200 psig.

D. never allow an air conditioning system to operate above 150 psig.

21. Before opening a refrigeration system to the atmosphere,

A. always purge air from the manifold gauge set.

B. connect a vacuum pump and evacuate the system to 500 microns.

C. remove the refrigerant until the system's pressure is 0 psig.

D. remove the refrigerant until the system's pressure is 30 psig.

22. To prevent the refrigerant from slugging the compressor,

A. never charge vapor into the high side while under a vacuum.

B. never charge liquid into the high side while under a vacuum.

C. never charge vapor into the low side.

D. never charge liquid into the low side.

23. If refrigerant vapor is vented into the atmosphere in a confined condition, it will

A. cause an explosive mixture.

B. be very flammable.

C. displace oxygen.

D. create phosgene gas.

24. Disposable refrigerant cylinders

A. are rarely color coded.

B. can be refilled with the proper equipment.

C. can be filled up to 80% of tank capacity.

D. should never be refilled.

25. A rechargeable refrigerant cylinder must

A. not be used as a storage tank.

B. be over 45 lb.

C. never be filled over 80% capacity.

D. be heated before it can be refilled with refrigerant.

26. When charging a rechargeable cylinder, always take into account the cylinder weight known as the

A. tare lb.

B. tank lb.

C. cylinder lb.

D. None of the above

27. Refrigerant cylinders should never be

A. heated with warm water.

B. heated with a torch.

C. stored at temperatures below 30°F.

D. refilled.

28. Any stored refrigerant cylinder must always be

A. properly labeled and placed in an upright position.

B. kept at temperatures below 40°F to prevent expansion.

C. no more than 30 lb.

D. at least 20% full at all times.

29. Refrigerant recovery cylinders are commonly color coded

A. as yellow.

B. as gray.

C. with a gray bottom and a yellow top.

D. with a yellow bottom and a gray top.

30. When heating a refrigerant cylinder using warm water, the temperature of the cylinder must never exceed

A. 60°F, or the rupture disc could open.

B. 90°F, or the rupture disc could open.

C. 125°F, or the rupture disc could open.

D. Cylinders must never be heated with warm water.

31. Refrigerant cylinders must be approved by

A. GAMA.

B. AGA.

C. DOT.

D. ACCA.

32. Which of the following are common types of pressure-relief devices?

A. fusible plug, rupture disc, and pressure-relief valve

B. fusible plug, vacuum break, and rupture disc

C. thermal disc, fusible disc, and rupture plug

D. fusible link, vacuum plug, and pressure-relief valve

33. Which of the following valves must never be capped?

A. the discharge service valve

B. the suction service valve

C. the king valve

D. the pressure-relief valve

34. Relief devices must

A. be installed in series with one another.

B. not be installed in parallel with one another.

C. not be installed in series with one another.

D. be inspected annually by DOT.

35. When pressurizing a low-pressure chiller, never exceed

A. 25 psig on the low side.

B. 25 psig on the high side.

C. 10 psig on the low side.

D. 10 psig on the high side.

36. When charging a liquid chiller,

A. refrigerant liquid is commonly added to the low side of the system.

B. refrigerant vapor is commonly added to the low side of the system.

C. refrigerant liquid is commonly added to the high side of the system.

D. refrigerant vapor is commonly added to the high side of the system.

37. A fill limit device used on a recovery system

A. shuts down the machine when a container reaches 50% full.

B. shuts down the machine when a container reaches 80% full.

C. shuts down the machine when a container reaches 90% full.

D. shuts down the machine when a container reaches 100% full.

38. According to EPA regulations, you must never

A. recover all refrigerants.

B. vent refrigerant to the atmosphere.

C. reuse refrigerant.

D. sell reclaimed refrigerant.

39. More caution should be given when handling class II refrigerants because

A. they are more expensive.

B. they are mostly used on industrial applications.

C. they are more toxic and flammable.

D. they are heavier.

40. Refrigerant cylinders should never be

A. subjected to extreme ambient temperatures.

B. labeled.

C. DOT or UL approved.

D. properly color coded.

41. When working with refrigerants, what is most likely to describe the flammability and the toxicity of a refrigerant?

A. the BOCA code

B. the EPA

C. the MSDS

D. None of the above

42. What precautions must be made when storing or transporting cylinders?

A. The cylinders must be kept under 80°F.

B. The valve and hood caps must always be secured.

C. The tanks must be inspected by the local mechanical inspector.

D. None of the above

43. When asbestos material is used as a pipe wrapping, the technician should

A. remove it immediately.

B. use eye protection, then remove the wrappings.

C. use eye, hand, and breathing apparatus protection, then remove wrappings.

D. leave it alone and allow qualified asbestos removal technicians to handle the removal.

THE FOLLOWING QUESTIONS MAY BE HELPFUL TO STUDY FOR THESE CERTIFICATION TESTS:
NATE—CR-SV, CR-IN, AC-SV, AC-IN, HP-SV, HP-IN, RC-SV, RC-IN, RL-SV, RL-IN.
ESCO INSTITUTE HVAC EXCELLENCE AIR CONDITIONING AND COMMERCIAL REFRIGERATION CERTIFICATION

REFRIGERATION THEORY

1. Temperature is

A. the intensity of molecular motion.

B. the measurement of heat in a liquid.

C. the measurement of heat in a gas.

D. the measurement of heat in an object.

2. What are the four common temperature scales?

A. Rankine, Fahrenheit, Celsius, and Centigrade

B. Centigrade, Fahrenheit, Celsius, and Kelvin

C. Fahrenheit, Rankine, Celsius, and Kelvin

D. Absolute, Kelvin, Centigrade, and Fahrenheit

3. Which of the following is the U.S. conventional standard scale?

A. Celsius

B. Fahrenheit

C. Kelvin

D. Rankine

4. Which of the following is the U.S. conventional absolute zero scale?

A. Centigrade

B. Fahrenheit

C. Kelvin

D. Rankine

5. Which temperature scale freezes water at 32° and boils water at 212°?

A. Centigrade

B. Fahrenheit

C. Kelvin

D. Absolute

6. To convert Fahrenheit to Rankine,

A. subtract 273°F.

B. add 273°F.

C. subtract 460°F.

D. add 460°F.

7. Which of the following is the SI unit standard scale?

A. Celsius

B. Fahrenheit

C. Kelvin

D. Rankine

8. Which of the following is the SI unit absolute zero scale?

A. Centigrade

B. Fahrenheit

C. Kelvin

D. Rankine

9. Which temperature scale freezes water at 0° and boils water at 100°?

A. Centigrade

B. Fahrenheit

C. Kelvin

D. Absolute

10. To convert Celsius to Kelvin,

A. subtract 273°Celsius.

B. add 273°Celsius.

C. subtract 460°Celsius.

D. add 460°Celsius.

11. To convert Celsius to Fahrenheit, which of the following is the correct formula?

A. $(1.8 \times C) + 32 = F$

B. $(1.8 - C) + 32 = F$

C. $(1.8 + C) - 32 = F$

D. $(1.8 \times C) - 32 = F$

12. To convert Fahrenheit to Celsius, which of the following is the correct formula?

A. $.56 \times (F - 32) = C$

B. $.56 \times (F + 32) = C$

C. $.56 - (F \times 32) = C$

D. $.56 + (F \times 32) = C$

13. What is the definition of specific heat?

A. the heat that can be measured by a thermometer

B. the amount of heat required to raise 1 lb of water 1°F

C. the amount of heat required to raise 1 lb of a substance 1°F

D. the heat used to cause a change of state

14. What is the definition of sensible heat?

A. the heat that can be measured by a thermometer

B. the amount of heat required to raise 1 lb of water 1°F

C. the amount of heat required to raise 1 lb of a substance 1°F

D. the heat used to cause a change of state

15. What is the definition of latent heat?

A. the heat that can be measured by a thermometer

B. the amount of heat required to raise 1 lb of water 1°F

C. the amount of heat required to raise 1 lb of a substance 1°F

D. the heat used that causes a change of state

16. What is the definition of a Btu?

A. the heat that can be measured by a thermometer

B. the amount of heat required to raise 1 lb of water 1°F

C. the amount of heat required to raise 1 lb of a substance 1°F

D. the heat used to cause a change of state

17. Heat energy is measured

A. by a thermometer.

B. by the temperature difference of a substance.

C. in Btu.

D. by the thermal mass of a substance.

18. To measure the amount of heat in a substance without the change of state, which formula is correct?

A. temperature × weight × volume

B. temperature × density × volume

C. specific heat × temperature difference × weight

D. specific heat × volume × temperature difference

19. If 2 lb of water are raised from 40 to 60°F, how many Btu are added to the water?

A. 20

B. 30

C. 40

D. 60

20. The amount of heat that it takes to change 1 lb of 212°F water to 1 lb of 212°F steam is

A. 144 Btu.

B. 200 Btu.

C. 746 Btu.

D. 970 Btu.

21. The amount of heat required to change 1 lb of 32°F ice to 1 lb of 32°F water is

A. 144 Btu.

B. 200 Btu.

C. 746 Btu.

D. 970 Btu.

22. The three states of matter are

A. solid, vapor, and gas.

B. solid, liquid, and gas.

C. solid, fluid, and liquid.

D. fluid, vapor, and solid.

23. Boyle's law states that

A. at a constant pressure, the volume of a gas varies directly with an absolute temperature.

B. the total pressure of a confined mixture of gases is the sum of the pressures of each of the gases in the mixture.

C. the volume of a gas varies indirectly with an absolute temperature.

D. a volume of gas varies inversely with the absolute pressure, if the temperature remains constant.

24. Charles' law states that

A. at a constant pressure, the volume of a gas varies directly with an absolute temperature.

B. the total pressure of a confined mixture of gases is the sum of the pressures of each of the gases in the mixture.

C. the volume of a gas varies indirectly with an absolute temperature.

D. a volume of gas varies inversely with the absolute pressure, if the temperature remains constant.

25. Dalton's law states that

A. at a constant pressure, the volume of a gas varies directly with an absolute temperature.

B. the total pressure of a confined mixture of gases is the sum of the pressures of each of the gases in the mixture.

C. the volume of a gas varies indirectly with an absolute temperature.

D. a volume of gas varies inversely with the absolute pressure, if the temperature remains constant.

26. The calculation for work is

A. time × distance.

B. mass × force.

C. force × distance.

D. force × rate.

27. One horsepower is equal to

A. 33,000 ft-lb of work in 1 min.

B. 12,000 ft-lb of work in 1 min.

C. 746 ft-lb of work in 1 min.

D. 200 ft-lb of work in 1 min.

28. One horsepower is equal to

A. 33,000 watts.

B. 746 watts.

C. 200 watts.

D. 3.41 watts.

29. One watt is equal to

A. 33,000 Btu.

B. 746 Btu.

C. 200 Btu.

D. 3.41 Btu.

30. Which method of heat transfer is known as conduction?

A. the transfer of heat from liquids to vapors

B. the transfer of heat from one molecule to another

C. the transfer of heat from one place to another by a fluid, commonly air or water

D. the transfer of heat from the heat source to a solid object through space

31. Which method of heat transfer is known as convection?

A. the transfer of heat from liquids to vapors

B. the transfer of heat from one molecule to another

C. the transfer of heat from one place to another by a fluid, commonly air or water

D. the transfer of heat from the heat source to a solid object through space

32. Which method of heat transfer is known as radiation?

A. the transfer of heat from liquids to vapors

B. the transfer of heat from one molecule to another

C. the transfer of heat from one place to another by a fluid, commonly air or water

D. the transfer of heat from the heat source to a solid object through space

33. One ton of refrigeration is

A. 33,000 Btu per hour.

B. 12,000 Btu per hour.

C. 746 Btu per hour.

D. 200 Btu per hour.

34. What type of heat is absorbed or extracted from a substance when the substance changes state?

A. specific heat

B. latent heat

C. sensible heat

D. superheat

35. When a substance changes from a liquid to a gas, this is known as

A. latent heat of condensation.

B. sensible heat of a vapor.

C. latent heat of sublimation.

D. latent heat of vaporization.

36. When a substance changes from a gas to a liquid, this is known as

A. latent heat of condensation.

B. sensible heat of a vapor.

C. latent heat of sublimation.

D. latent heat of vaporization.

37. When a substance changes from a solid to a gas, this is known as

A. latent heat of condensation.

B. sensible heat of a vapor.

C. latent heat of sublimation.

D. latent heat of vaporization.

38. In a refrigeration system, heat is absorbed by

A. the latent heat of condensation in the condenser.

B. the latent heat of vaporization in the evaporator.

C. the latent heat of vaporization in the condenser.

D. the latent heat of condensation in the evaporator.

39. In a refrigeration system, heat is rejected by

A. the latent heat of condensation in the condenser.

B. the latent heat of vaporization in the evaporator.

C. the latent heat of vaporization in the condenser.

D. the latent heat of condensation in the evaporator.

THE FOLLOWING QUESTIONS MAY BE HELPFUL TO STUDY FOR THESE CERTIFICATION TESTS:
NATE—CR-SV, CR-IN, AC-SV, AC-IN, HP-SV, HP-IN, RC-SV, RC-IN, RL-SV, RL-IN.
ESCO INSTITUTE HVAC EXCELLENCE AIR CONDITIONING AND COMMERCIAL REFRIGERATION CERTIFICATION

REFRIGERANTS

1. Which of the following refrigerants is considered an HCFC?

A. R-12

B. R-22

C. R-502

D. R-134A

2. What type of refrigerant is commonly used with alkylbenzene oil?

A. R-12

B. R-22

C. R-410A

D. R-134A

3. Refrigerant tanks must never be stored in an area where the temperature is above

A. 100 °F.

B. 115 °F.

C. 120 °F.

D. 125 °F.

4. A refrigerant that is replacing R-22 in residential and commercial air conditioning is

A. R-408A.

B. R-409A.

C. R-410A.

D. R-500.

5. The HFC refrigerant that is used as a replacement for R-12 is

A. R-409A.

B. R-134A.

C. R-22.

D. R-408A.

6. The blended refrigerant that is used as a replacement for R-12 is

A. R-414B.

B. R-134A.

C. R-22.

D. R-408A.

7. The blended refrigerant that is used as a replacement for R-502 is

A. R-409A.

B. R-134A.

C. R-22.

D. R-408A.

8. What type of refrigerant, when charged as a gas, will not cause fractionation?

A. a ternary blend

B. a zeotropic

C. a single component

D. a near-zeotropic

9. What type of refrigerant is considered a single-component refrigerant?

A. R-134A

B. R-502

C. R-409A

D. R-408A

10. Which of the following is a mixture of two or more refrigerants that have the same boiling point and act as one refrigerant?

A. an azeotropic

B. a ternary blend

C. a zeotropic

D. a near-azeotropic

11. Which of the following is a mixture of two or more refrigerants that do not have the same boiling points and that act close to an azeotropic blend?

A. an azeotropic
B. a ternary blend
C. a single component
D. a near-azeotropic

12. Which of the following is a mixture of two or more refrigerants that do not have the same boiling points and that do not act as one refrigerant?

A. an azeotropic
B. a zeotropic
C. a single component
D. a ternary blend

13. Which of the following is three refrigerants blended together?

A. an azeotropic
B. a zeotropic
C. a single component
D. a ternary blend

14. Zeotropic and near-azeotropic refrigerants must be charged into a system

A. as a gas.
B. as a liquid.
C. as a liquid or gas.
D. using the superheat method only.

15. What type of refrigerant is R-410A?

A. a single component
B. an azeotropic
C. a near-azeotropic
D. a zeotropic

16. What type of refrigerant is considered an azeotropic refrigerant?

A. R-12
B. R-22
C. R-502
D. R-409A

17. What type of refrigerant is considered a single-component refrigerant?

A. R-12
B. R-410A
C. R-502
D. R-409A

18. What type of refrigerant is considered a near-azeotropic refrigerant?

A. R-410A
B. R-408A
C. R-134A
D. R-503

19. What type of refrigerant is a zeotropic refrigerant?

A. R-22
B. R-502
C. R-409A
D. R-407A

20. Which of the following is not considered a ternary refrigerant?

A. R-407C
B. R-502
C. R-409A
D. None of the above

21. When charging a system using the superheat method, which value located on a pressure temperature chart would be used with R-409A?

A. the dewpoint line value
B. the bubble line value
C. the temperature glide value
D. the superheat value

22. When charging a system using the subcool method, which value located on a pressure temperature chart would be used with R-408A?

A. the dewpoint line value
B. the bubble line value
C. the temperature glide value
D. the superheat value

23. What is known as the difference between the vapor and liquid state of a near-azeotropic, a zeotropic, and a ternary blend refrigerant that occurs during evaporation and condensation at a constant pressure?

A. bubble line value
B. dewpoint value
C. temperature glide
D. ternary glide

24. An azeotropic refrigerant has a temperature glide of

A. 0°F.
B. above 0°F up to 10°F.
C. above 10°F up to 20°F.
D. over 20°F.

25. A near-azeotropic refrigerant has a temperature glide of

A. 0°F.
B. above 0°F up to 10°F.
C. above 10°F up to 20°F.
D. over 20°F.

26. A zeotropic refrigerant has a temperature glide of

A. 0°F.

B. above 0°F up to 10°F.

C. above 10°F.

D. None of the above

27. What does the term *fractionation* mean?

A. to separate a refrigerant mixture improperly by charging as a liquid

B. to separate a refrigerant mixture improperly by charging as a gas

C. to blend two types of refrigerants into a system for an increase of efficiency

D. to add small parts of refrigerant oil into the system as it is in operation

28. Using the pressure–temperature chart (TABLE 1–1), what pressure is R-502 at 26°F?

A. −6 psig

B. 26 psig

C. 50 psig

D. 60 psig

29. Using the pressure–temperature chart (TABLE 1–1), what is the temperature of R-22 at 60 psig?

A. 26°F

B. 34°F

C. 58°F

D. 62°F

TABLE 1–1 Pressure–Temperature Chart

Pressure in psig	Temperature in °F		
	R-12	**R-22**	**R-502**
25	25	1	−7
26	27	2	−6
27	28	4	−5
28	29	5	−3
29	31	6	−2
30	32	7	−1
31	33	8	0
32	34	9	1
33	35	10	2
34	37	11	3
35	38	12	4
36	39	13	5
37	40	14	6
38	41	15	7
39	42	16	8
40	43	17	9
42	45	19	11
44	47	21	13
46	49	23	15
48	51	24	16
50	53	26	18
52	55	28	20
54	57	29	21
56	58	31	23
58	60	32	24
60	62	34	26
62	64	35	27
64	65	37	29
66	67	38	30
68	68	40	32
70	70	41	33
72	72	42	34
74	73	44	36

THE FOLLOWING QUESTIONS MAY BE USED TO STUDY FOR THESE CERTIFICATION TESTS:
NATE—AC-SV, AC-IN, HP-SV, HP-IN, RC-SV, RC-IN, RL-SV, RL-IN.
ESCO INSTITUTE HVAC EXCELLENCE AIR CONDITIONING AND COMMERCIAL REFRIGERATION CERTIFICATION

LUBRICANTS

1. The properties of mineral oil are very common to the properties of

A. petroleum oil.

B. polyol ester oil.

C. alkylbenzene oil.

D. PAG oil.

2. Which refrigerant is not miscible with mineral oil?

A. R-12

B. R-11

C. R-409A

D. R-134A

3. Which lubricant is most commonly used with CFC refrigerants?

A. polyol ester

B. mineral

C. polyalkylene glycol

D. synthetic

4. Which lubricant is most commonly used with R-134A refrigerant?

A. polyol ester

B. mineral

C. alkylbenzene

D. petroleum-based oils

5. What are two compounds that are used to compose an alkylbenzene lubricant?

A. alcohol and benzene

B. glycol and alcohol

C. propylene and glycol

D. propylene and benzene

6. What types of refrigerants commonly use alkylbenzene lubricants?

A. CFCs and HFCs

B. HCFCs and HFCs

C. CFCs and HCFCs

D. HFCs only

7. What systems commonly use PAG lubricants?

A. automotive air conditioning

B. residential air conditioning systems

C. cascade refrigeration systems

D. chillers

8. What does PAG stand for?

A. polyurethane alkylene glycol

B. polymorphous alkylene glycol

C. polyalkylene glycol

D. None of the above

9. Hygroscopic oil

A. has a high attraction to moisture.

B. should be stored at ambient temperatures.

C. can be left open to the atmosphere.

D. has a low attraction to moisture.

10. Corrosion will start to form inside of an air conditioning system if the moisture content exceeds

A. 10 ppm.

B. 100 ppm.

C. 200 ppm.

D. 250 ppm.

11. What are two good characteristics of oil?

A. high floc point and low hygroscopic ability

B. low carbon-forming tendency and high hygroscopic ability

C. high floc point and high carbon-forming tendencies

D. low floc point and low hygroscopic ability

12. Which of the following refers to the temperature at which oil starts to produce a waxy substance?

A. floc point

B. pour point

C. sludge point

D. flow point

13. Which of the following refers to the ability of oil to flow at very low temperatures?

A. viscous point

B. flow point

C. pour point

D. floc point

14. A miscible fluid has

A. the ability to mix well with another fluid.

B. the ability to flow at low temperatures.

C. extremely thin qualities.

D. extremely dense qualities.

15. Why is it necessary to increase the dehydration time of a system using a synthetic oil compared to a system using a mineral oil?

A. because a synthetic oil has a higher evaporation point than a mineral oil

B. because a mineral oil has a higher evaporation point than a synthetic oil

C. because a synthetic oil has a higher hygroscopic capability

D. because a mineral oil has a higher hygroscopic capability

16. Which oil has the ability to absorb the most moisture?

A. a PAG oil

B. a polyol ester oil

C. a mineral oil

D. Both A and B

17. Polyol ester oil will tolerate a mixture of

A. 8% of mineral oil.

B. 3 to 5% of mineral oil.

C. 10 to 20% of mineral oil.

D. 40 to 50% of mineral oil.

18. High-viscosity oil

A. is used on low-temperature systems only.

B. is a thin oil.

C. is a thick oil.

D. is used on medium-temperature systems only.

19. What is the point at which the vapors of an oil will ignite but the flame will not persist?

A. burn point

B. flash point

C. floc point

D. pour point

THE FOLLOWING QUESTIONS MAY BE HELPFUL TO STUDY FOR THESE CERTIFICATION TESTS:
NATE—AC-SV, AC-IN, HP-SV, HP-IN, RC-SV, RC-IN, RL-SV, RL-IN.
ESCO INSTITUTE HVAC EXCELLENCE AIR CONDITIONING AND COMMERCIAL REFRIGERATION CERTIFICATION

COMPRESSORS

1. What is the purpose of the compressor?

A. It takes a low-pressure liquid and changes it into a high-pressure liquid.

B. It takes a high-pressure gas and changes it into a high-pressure liquid.

C. It takes a low-pressure gas and changes it into a high-pressure gas.

D. It takes a low-pressure liquid and changes it into a high-pressure gas.

2. A crankcase heater is used on a compressor to

A. thin the oil for increased lubrication.

B. evaporate liquid refrigerant out of the compressor oil.

C. warm the oil for increased lubrication of moving compressor components.

D. keep snow from building up on the compressor during winter months.

3. What are the two most common types of compressors used in residential and light commercial air conditioning systems?

A. reciprocating and rotary

B. rotary and scroll

C. centrifugal and reciprocating

D. reciprocating and scroll

4. Which of the following is considered a non-positive-displacement compressor?

A. reciprocating

B. rotary

C. screw

D. centrifugal

5. Which type of compressor is completely sealed and must be replaced if it has internal damage?

A. a semihermetic

B. a hermetic

C. an open type

D. a centrifugal type

6. Name two types of compressor lubrication systems.

A. forced and gravity-feed lubrication

B. forced and splash lubrication

C. gravity-feed and splash lubrication

D. plate and gravity-feed lubrication

7. How are compressors cooled in residential and light commercial air conditioning systems?

A. air motion and water

B. refrigerant and water

C. refrigerant and air

D. refrigerant and oil

8. What are the two types of compressor drives?

A. belt and pulley

B. direct and belt

C. direct and indirect

D. indirect and pulley

9. Which compressor is not internally driven by a motor?

A. the hermetic

B. the semihermetic

C. the open type

D. the closed type

10. What compressor can handle some liquid refrigerant floodback?

A. the reciprocating hermetic type

B. the reciprocating semihermetic type

C. the scroll type

D. the rotary type

11. How is the compressor ratio determined?

A. discharge psia/suction psia

B. suction psia/discharge psia

C. discharge psig/suction psig

D. suction psig/discharge psig

12. What is commonly used with a two-speed compressor?

A. a tachometer

B. a double riser

C. a two-stage thermostat

D. a dual pressure-relief valve

THE FOLLOWING QUESTIONS MAY BE HELPFUL TO STUDY FOR THESE CERTIFICATION TESTS:
NATE—AC-SV, AC-IN, HP-SV, HP-IN, RC-SV, RC-IN, RL-SV, RL-IN.
ESCO INSTITUTE HVAC EXCELLENCE AIR CONDITIONING AND COMMERCIAL REFRIGERATION CERTIFICATION

AIR-COOLED CONDENSERS

1. What is the purpose of the condenser?
   A. It takes a low-pressure liquid and changes it into a high-pressure liquid.
   B. It takes a high-pressure gas and changes it into a high-pressure liquid.
   C. It takes a low-pressure gas and changes it into a high-pressure gas.
   D. It takes a high-pressure liquid and changes it into a high-pressure gas.
2. Out of what materials are the tubes on a residential or light commercial air conditioning condenser fabricated?
   A. aluminum and steel
   B. iron and steel
   C. steel and copper
   D. aluminum or copper
3. Where in the system is the excess superheat added that must be removed before condensation in the condenser can take place?
   A. at the beginning of the evaporator and at the end of the receiver
   B. at the end of the receiver and at the beginning of the TXV
   C. at the beginning of the evaporator and in the compressor
   D. at the end of the evaporator and in the compressor
4. What section of the condenser coil is most likely to become dirty on a unit that has a side-air discharge with the fan pushing the air over the coil?
   A. the outside of the coil
   B. the top section of the coil
   C. the inside of the coil
   D. the bottom section of the coil
5. What is a common temperature difference from the air entering a condenser and the refrigerant condensing temperature?
   A. 5–10°F
   B. 10–15°F
   C. 20–30°F
   D. 35–45°F
6. A split-system condensing unit contains
   A. the compressor, condenser, and metering device.
   B. the condenser, compressor, and service valves.
   C. the condenser, service valves, and metering device.
   D. the condenser, evaporator, and service valves.
7. If air is trapped inside a condenser coil, what symptoms would be present?
   A. high discharge pressure, low amperage, and a decrease of efficiency
   B. low discharge pressure, high amperage, and excessive heat buildup
   C. a decrease in efficiency, a decrease of heat buildup, and high amperage
   D. an increase of heat buildup, high amperage, and high discharge pressure

THE FOLLOWING QUESTIONS MAY BE HELPFUL TO STUDY FOR THESE CERTIFICATION TESTS:
NATE—AC-SV, AC-IN, RC-SV, RC-IN, RL-SV, RL-IN.
ESCO INSTITUTE HVAC EXCELLENCE AIR CONDITIONING AND COMMERCIAL REFRIGERATION CERTIFICATION

WATER-COOLED CONDENSERS

1. What two factors affect the efficiency of a water-cooled condenser?
   A. the high-side pressure of the condenser and the inlet water temperature
   B. the high-side pressure of the condenser and the outlet water temperature
   C. the quantity of water and the temperature difference between the refrigerant and the water
   D. the quantity of water and the refrigerant condensing point compared to the water boiling point
2. The capacity of a water-cooled condenser increases
   A. as the temperature difference between the refrigerant and the water decreases.
   B. as the temperature difference between the refrigerant and the water increases.
   C. as the condenser outlet temperature increases.
   D. as the condenser outlet temperature decreases.
3. If the water quantity on a water-cooled condenser were increased, what effect would this have on the system?
   A. It decreases the power consumption of the compressor.
   B. It increases the power consumption of the compressor.
   C. It has no effect on the power consumption of the compressor.
   D. It increases the high-side pressure.

4. When using a water-cooled condenser, a decrease of water

A. decreases the power consumption of a compressor.

B. increases the fouling factor of the water.

C. decreases operating pressures.

D. increases the power consumption of the compressor.

5. The fouling factor of the water used in a water-cooled condenser

A. increases as water quantity increases.

B. is the measurement of heat transfer characteristics of the water.

C. decreases as water quantity decreases.

D. can be controlled by the water-regulating valve.

6. What must be done when the water of a water-cooled condenser has a high fouling factor?

A. The water-regulating valve must be turned counterclockwise.

B. The flow rate of the water must be decreased.

C. The flow rate of the water must be increased.

D. The system flow rate must not be adjusted.

7. The water of a water-cooled condenser

A. should never be discharged to storm drains or sewer drains.

B. should never be reused.

C. should always be reused.

D. should be discharged to storm drains and storm sewers if local codes permit.

8. What is used to control the flow of water on a water-cooled condenser?

A. a check valve

B. a pressure-reducing valve

C. a water-regulating valve

D. a float valve

9. Where on a water-cooled condenser should a water-regulating valve be located?

A. at the inlet of the condenser

B. at the outlet of the condenser

C. at the inlet or the outlet of the condenser

D. Water-regulating valves are not used on water-cooled condensers.

10. What two pressures control a water-regulating valve?

A. the outlet water pressure and the high-side pressure

B. the inlet water pressure and the high-side pressure

C. the high-side pressure and the crankcase pressure

D. the high-side pressure and the spring pressure

11. Which of the following is the opening force on a water-regulating valve?

A. the water outlet pressure

B. the water inlet pressure

C. the high-side pressure

D. the spring pressure

12. Which of the following is the closing force on a water-regulating valve?

A. the crankcase pressure

B. the water inlet pressure

C. the high-side pressure

D. the spring pressure

13. What is a common temperature difference between the high-side refrigerant temperature and the temperature of the leaving water in a water-cooled system?

A. 10°F

B. 20°F

C. 30°F

D. 35°F

14. What are the effects of mineral and scale deposits on the inside of a water-cooled condenser?

A. It causes the high-side pressure to decrease and a decrease of water flow.

B. It causes the high-side pressure to increase and a decrease of water flow.

C. It causes the leaving water temperature to decrease and an increase of water flow.

D. It causes an increase of water flow and high-side pressure.

15. Name three common water-cooled condensers.

A. a coiled-tube type, a vertical-tube type, and a tube-within-a-tube type

B. a shell-and-tube type, a vertical-tube type, and a horizontal-tube type

C. a tube within a tube type, a shell-and-coil type, and a shell-and-tube type

D. a shell-and-coil type, a horizontal-tube type, and a vertical-tube type

16. The tube-within-a-tube condenser uses

A. the outer tube for refrigerant and the inner tube for water.

B. the inner tube for refrigerant and the outer tube for water.

C. the inner tube for refrigerant on coil-type condensers and the inner tube for water on horizontal-tube types.

D. the inner tube for refrigerant on horizontal-tube-type condensers and the inner tube for water on coil-tube types.

17. The counterflow principle is

A. where the cooler water is in contact with the cooler refrigerant liquid and the warmer water is in contact with the hot gas.

B. where the warmer water is in contact with the cooler refrigerant liquid and the cooler water is in contact with the hot gas.

C. where the hot gas refrigerant meets the cool city water at its inlet.

D. where the subcooled refrigerant meets the leaving water at its outlet.

18. What are two common types of tube-within-a-tube condensers?

A. a sealed-coil type and a flanged-vertical-tube type

B. a sealed-tube type and an open-coil type

C. a flanged-horizontal-tube type and a sealed-coil type

D. a sealed-horizontal-tube type and flanged-vertical-tube type

19. What type of tube-within-a-tube condenser must always be chemically cleaned?

A. the sealed-coil type

B. the sealed-horizontal-tube type

C. the sealed-vertical-tube type

D. the open-coil type

20. What type of tube-within-a-tube condenser can be manually or chemically cleaned?

A. the flanged-coil type

B. the flanged-horizontal-tube type

C. the flanged-vertical-tube type

D. the sealed-coil type

21. The shell-and-coil type of water-cooled condenser

A. has the refrigerant in the coil, and water flows over the coil inside of the shell.

B. has subcooled refrigerant leaving the coil and superheated gas entering the coil.

C. has the refrigerant in the shell, and water flows through the coil located inside the shell.

D. can have refrigerant or water flowing inside of the coil.

22. The shell-and-coil condenser

A. must have the coil removed in order to be properly cleaned.

B. can be manually or chemically cleaned.

C. cannot be chemically cleaned—it must be manually cleaned.

D. cannot be manually cleaned—it must be chemically cleaned.

23. The shell-and-tube type of water-cooled condenser

A. has the refrigerant in the tubes, and water flows over the tubes inside of the shell.

B. has subcooled refrigerant leaving the tubes and superheated gas entering the tubes.

C. has the refrigerant in the shell, and water flows through the tubes located inside the shell.

D. can have refrigerant or water flowing inside of the tubes.

24. The shell-and-tube condenser

A. must be chemically cleaned.

B. can be manually or chemically cleaned.

C. cannot be manually cleaned.

D. is completely sealed with no flanges and must be chemically cleaned.

25. What is another name for the end cap of a shell-and-tube water-cooled condenser?

A. puddle plates

B. water plates

C. water box

D. water trap

26. The shell of a shell-and-tube or shell-and-coil condenser acts as a

A. liquid receiver.

B. water storage tank.

C. purge unit for noncondensables.

D. subcool tank.

THE FOLLOWING QUESTIONS MAY BE HELPFUL TO STUDY FOR THESE CERTIFICATION TESTS:
NATE—AC-SV, AC-IN, HP-SV, HP-IN, RC-SV, RC-IN, RL-SV, RL-IN.
ESCO INSTITUTE HVAC EXCELLENCE AIR CONDITIONING AND COMMERCIAL REFRIGERATION CERTIFICATION

INDOOR EVAPORATOR COILS

1. What is the purpose of the evaporator coil in a split-system air conditioner?

A. It removes heat from the system.

B. It absorbs heat into the system.

C. It decreases air velocity.

D. It is used to help increase the relative humidity.

2. An indoor evaporator coil must be installed

A. at least 1″ above the heat exchanger.

B. at least 2″ to 3″ above the heat exchanger.

C. at least 4″ to 5″ above the heat exchanger.

D. at least 6″ above the heat exchanger.

3. A direct-expansion evaporator coil

A. has the controlled medium in contact with the coil.

B. never uses a TXV-type metering device.

C. has the controlled medium in contact with a fluid that is cooled by the coil.

D. always uses a TXV-type metering device.

4. Air conditioning evaporator coils are commonly made of

A. stainless steel and aluminum.

B. copper and stainless steel.

C. galvanized steel and copper.

D. copper and aluminum.

5. What is the purpose of a distributor?

A. It feeds refrigerant to the compressor from the evaporator coil.

B. It feeds refrigerant to the evaporator coil.

C. It feeds refrigerant to the condenser coil.

D. It feeds refrigerant to a solenoid valve.

6. When an uncased evaporator coil is installed, metal plates are commonly installed around the coil. What is the purpose of these metal plates?

A. to increase the quantity of air in the system

B. to direct airflow around the coil

C. to decrease the flow of air through the coil

D. to force the air to flow through the coil

7. Why is it important to ensure that the evaporator coil on a split-system air conditioner is installed level?

A. to allow the proper flow of liquid refrigerant in the coil

B. to help the system properly return the oil back to the compressor

C. to allow for proper drainage

D. to direct the refrigerant gas back to the compressor at the proper rate

8. Split-system evaporator coils are commonly installed

A. on the positive-pressure side of the blower before the heat exchanger.

B. on the positive-pressure side of the blower after the heat exchanger.

C. on the negative-pressure side of the blower before the heat exchanger.

D. on the negative-pressure side of the blower after the heat exchanger.

9. What type of evaporator coil is commonly used in horizontal air-handling systems?

A. an "A" coil

B. an "N" coil

C. an "H" coil

D. a slant coil

10. What type of evaporator coil is commonly used in residential upflow systems?

A. an "A" coil

B. an "N" coil

C. an "H" coil

D. a slant coil

11. What type of evaporator coil is commonly used in residential upflow systems with limited plenum space?

A. an "A" coil

B. an "N" coil

C. an "H" coil

D. None of the above

12. Rooftop package unit evaporator coils are commonly installed

A. on the positive-pressure side of the blower before the heat exchanger.

B. on the positive-pressure side of the blower after the heat exchanger.

C. on the negative-pressure side of the blower before the heat exchanger.

D. on the negative-pressure side of the blower after the heat exchanger.

13. When is a "P" trap installed?

A. when an evaporator coil is located after a heat exchanger

B. when the evaporator is an "A" -type coil

C. when the evaporator is an "H" -type coil

D. when an evaporator coil is located on the negative side of the blower

14. What is the purpose of a "P" trap?

A. to prevent blockage of a condensate drain

B. to prevent condensation from entering the duct system

C. to aid in cleaning a condensate drain

D. to trap debris that enters a condensate drain

15. If an "A" coil has a $\frac{3}{4}''$ drain line leaving the drain pan, according to code, what would be the minimum size of the drain line?

A. $\frac{1}{2}''$

B. $\frac{3}{4}''$

C. $1''$

D. $1\frac{1}{4}''$

16. Which of the following is the minimum required slope for a drain line?

A. $\frac{1}{8}''$ per foot of horizontal run

B. $\frac{1}{4}''$ per foot of horizontal run

C. $\frac{1}{2}''$ per foot of horizontal run

D. $\frac{3}{4}''$ per foot

17. Cooling units larger than 6 tons must discharge to a

A. sanitary sewer drain, storm drain, or French drain.

B. gutter, roof drain, or on the roof.

C. Both A and B

D. None of the above

18. Cooling units 6 tons or smaller must discharge to a

A. sanitary sewer drain, storm drain, or French drain.

B. gutter, roof drain, or on the roof.

C. Both A and B

D. None of the above

19. Which of the following materials must not be used as a drain pipe?

A. galvanized steel

B. polybutylene, ABS, and CPVC

C. lead and aluminum pipe

D. copper and PVC

20. An auxiliary drain pan is required when cooling coils are located in areas where water damage can occur if the primary drain were to fail. What is the minimum required size of the auxiliary drain pan?

A. 20 deep and not less than 4″ larger than the unit on each side

B. $1\frac{1}{2}''$ deep and not less than 3″ larger than the unit on each side

C. 1″ deep and not less than 2″ larger than the unit on each side

D. None of the above

21. When using galvanized steel as an auxiliary drain pan, what is the minimum thickness of the steel?

A. 22 gauge

B. 24 gauge

C. 26 gauge

D. 30 gauge

22. When routing the secondary drain line, which of the following is allowed by code?

A. Route the line in a conspicuous place to serve as an alarm.

B. Route the line into the primary drain line.

C. Both A and B are allowed by code.

D. Neither A nor B is allowed by code.

23. Which of the following is an approved means as an alternative to the secondary drain line?

A. Use a normally opened float switch located in the secondary drain pan that is in series with the contactor coil of the condensing unit.

B. Use a normally closed float switch located in the secondary drain pan that is in series with the contactor coil of the condensing unit.

C. Neither A nor B is an approved method.

D. Both A and B are approved methods.

24. If a secondary drain pan is not possible to install on a system located in an attic, what other means is allowed by the code to prevent water damage?

A. Use a condensate pump system to remove the moisture.

B. There are no other means; a drain pan must be used.

C. A secondary drain line may be routed from the overflow drain to a conspicuous place.

D. None of the above

THE FOLLOWING QUESTIONS MAY BE HELPFUL TO STUDY FOR THESE CERTIFICATION TESTS
NATE—AC-SV, AC-IN, HP-SV, HP-IN, RC-SV, RC-IN, RL-SV, RL-IN.
ESCO INSTITUTE HVAC EXCELLENCE AIR CONDITIONING AND COMMERCIAL REFRIGERATION CERTIFICATION

METERING DEVICES

1. What condition is the refrigerant in before it enters the metering device?

A. a high-pressure, high-temperature saturated liquid

B. a high-pressure, low-temperature saturated liquid

C. a high-pressure, high-temperature subcooled liquid

D. a low-pressure, high-temperature subcooled liquid

2. What is the definition of flash gas?

A. the rapid condensation of a refrigerant gas

B. the rapid evaporation of a refrigerant liquid

C. the rapid sublimation of a refrigerant liquid

D. the rapid evaporation of a refrigerant gas

3. What component will help decrease flash gas?

A. a heat exchanger

B. an accumulator

C. a filter dryer

D. a liquid injecting valve

4. What are three common types of metering devices that are used in residential and light commercial air conditioning?

A. TXV, AXV, and low-side float

B. high-side float, low-side float, and capillary tube

C. low-side float, piston type, and AXV

D. capillary tube, piston type, and TXV

5. What controls the pressure drop of a capillary tube?

A. the type and length of the tubing

B. the outer diameter and the type of the tubing

C. the length and the inner diameter of the tubing

D. the inner diameter and the outer diameter of the tubing

6. The tool used to cut capillary tube would be

A. hand snips.

B. hacksaw.

C. tube cutter.

D. triangular file.

7. What are the operating pressures of a TXV?

A. spring pressure, TXV inlet pressure, and sensing-bulb pressure

B. TXV inlet pressure, TXV outlet pressure, and sensing-bulb pressure

C. spring pressure, evaporator pressure, and sensing-bulb pressure

D. spring pressure, TXV inlet pressure, and TXV outlet pressure

8. What is the purpose of the external equalizer on a TXV?
   A. to decrease the pressure drop across the TXV
   B. to increase the pressure drop across the TXV
   C. to drop the pressure across the evaporator
   D. to offset for the pressure drop across the evaporator

9. The external equalizer line should be located
   A. next to the TXV sensing bulb.
   B. 6″ to 8″ upstream from the TXV sensing bulb.
   C. 6″ to 8″ downstream from the TXV sensing bulb.
   D. None of the above

10. If a system using a TXV has a suction line of $\frac{3}{4}''$ diameter, where should the sensing bulb be placed?
    A. at the 4 o'clock and 8 o'clock position
    B. at the 12 o'clock position
    C. at the 6 o'clock position
    D. at either the 6 o'clock or the 12 o'clock position

11. If a system using a TXV has a suction line of $\frac{7}{8}''$ diameter, where should the sensing bulb be placed?
    A. at the 4 o'clock and 8 o'clock position
    B. at the 12 o'clock position
    C. at the 6 o'clock position
    D. at either the 6 o'clock or the 12 o'clock position

12. What is the formula for calculating superheat?
    A. suction line temperature + evaporator boiling point temperature
    B. liquid line temperature − evaporator boiling point temperature
    C. suction line temperature − evaporator boiling point temperature
    D. liquid line temperature + evaporator boiling point temperature

13. How is superheat increased on a TXV system?
    A. superheat cannot be increased
    B. by turning the TXV valve stem counterclockwise
    C. by turning the TXV valve stem clockwise
    D. by placing the sensing bulb farther downstream from the evaporator

THE FOLLOWING QUESTIONS MAY BE HELPFUL TO STUDY FOR THESE CERTIFICATION TESTS:

NATE—AC-SV, AC-IN, HP-SV, HP-IN, RC-SV, RC-IN, RL-SV, RL-IN.

ESCO INSTITUTE HVAC EXCELLENCE AIR CONDITIONING AND COMMERCIAL REFRIGERATION CERTIFICATION

THERMOSTATIC EXPANSION VALVES

1. The term *hunting* refers to which type of metering device?
   A. low-side float
   B. high-side float
   C. automatic expansion valve
   D. thermostatic expansion valve

2. MOP refers to which type of metering device?
   A. low-side float
   B. high-side float
   C. automatic expansion valve
   D. thermostatic expansion valve

3. In a valve having a MOP (maximum operating pressure), if the pressure in the evaporator increased higher than the MOP of the valve, the valve would then tend to
   A. open more.
   B. close until pressure is reduced.
   C. remain in equilibrium.
   D. have little effect.

4. Liquid charge, liquid cross-charge, gas charge, or gas cross-charge refer to
   A. types of charging methods.
   B. blend-type refrigerants.
   C. states of refrigerants in the evaporator.
   D. sensing bulb charges of a TXV.

5. A liquid charge-sensing bulb contains
   A. a mixture of liquid and gas.
   B. all gas.
   C. the same refrigerant as in the system.
   D. a volatile refrigerant gas mixture.

6. A liquid cross-charge-sensing bulb contains
   A. a liquid of the system's refrigerant.
   B. gas and liquid mixtures that are the same as in the system.
   C. a mixture of refrigerants that are different from that of the system.
   D. a gas mixture of blend-type refrigerants.

7. The type of sensing bulb that will prevent liquid floodback on system's startup is the
   A. liquid charge.
   B. liquid cross-charge.
   C. gas and gas cross-charge.
   D. gas charge.

8. The sensing-bulb type that has its own MOP is the

A. gas charge and gas cross-charge.

B. liquid charge.

C. liquid cross-charge.

D. liquid and liquid cross-charge.

9. To increase the superheat on a TXV one would

A. turn the adjusting stem counterclockwise.

B. turn the adjusting stem clockwise.

C. add refrigerant.

D. recover refrigerant.

10. To decrease the superheat on a TXV one would

A. turn the adjusting stem counterclockwise.

B. turn the adjusting stem clockwise.

C. add refrigerant.

D. recover refrigerant.

11. The superheat can be defined as a _______ that is _______ its saturation temperature for a given pressure.

A. gas; above

B. liquid; above

C. gas; below

D. liquid; below

12. When a TXV tends to overfeed or underfeed in an attempt to find a balance point and become stable, it is known to be

A. drifting.

B. hunting.

C. balancing.

D. stabilizing.

13. The main function of the thermostatic expansion valve is to _______ refrigerant flow and _______ a constant superheat.

A. control; modulate

B. regulate; modulate

C. regulate; maintain

D. modulate; regulate

14. When superheat increases at the evaporator outlet, the TXV will

A. close more.

B. stabilize.

C. balance the refrigerant.

D. open more.

15. The thermostatic expansion valve will decrease refrigerant flow when the superheat at the evaporator outlet

A. rises.

B. lowers.

C. balances.

D. Both A and C

16. When a thermostatic expansion valve is used, the system's refrigerant charge is generally not as critical as it would be with

A. an AXV.

B. an EEV.

C. a capillary tube.

D. a low-side float.

17. The sensing bulb is part of the TXV and is connected to the valve by a capillary tube to the

A. top of the valve's diaphragm.

B. bottom of the valve's diaphragm.

C. side of the valve's body.

D. suction line.

18. Pin, pin carrier, and pushrods are all components of the

A. AXV.

B. low-side float.

C. high-side float.

D. TXV.

19. The component of a TXV that actually adjusts superheat is called the

A. superheat spring.

B. push rods.

C. pin carrier.

D. diaphragm.

20. Which is *not* an operating pressure of a thermostatic expansion valve?

A. spring pressure

B. bulb pressure

C. condenser pressure

D. evaporator pressure

21. Which of the following pressures tends to close the TXV valve?

A. bulb pressure/spring pressure

B. evaporator pressure/spring pressure

C. evaporator pressure/bulb pressure

D. bulb pressure only

22. The pressure that tends to open the TXV valve is the

A. spring pressure.

B. evaporator pressure.

C. evaporator and spring pressure.

D. bulb pressure.

23. When the system is in normal operation the sensing-bulb pressure should equal the

A. spring pressure.

B. evaporator pressure.

C. spring and evaporator pressure.

D. temperature of the entire evaporator.

24. The function of the sensing bulb is to

A. sense the liquid refrigerant temperature.

B. sense the temperature of the refrigerant vapor as it leaves the evaporator coil.

C. sense the temperature of the entire evaporator.

D. sense and react to high subcooling temperatures.

25. Which pressure of the TXV will allow the valve pin to lift off the valve port?

A. bulb pressure

B. spring pressure

C. evaporator pressure

D. evaporator and spring pressure

26. When will the TXV valve pin move to close off the valve port?

A. when bulb pressure and temperature are increased

B. when bulb pressure increases

C. when evaporator pressure is reduced

D. when bulb temperature and pressure are reduced

27. When there is an increase in heat load on the evaporator, the refrigerant will tend to

A. evaporate at a slower rate.

B. evaporate at a constant rate.

C. evaporate at a faster rate.

D. flood the evaporator.

28. The term _______ refers to a method of bleeding off or allowing evaporator pressure to pass through a passageway within the valve body.

A. flooding

B. internally equalized

C. hunting

D. purging

29. The external equalizer line is connected to

A. a hot gas bypass line.

B. a liquid line.

C. a suction line.

D. the suction header.

30. Thermostatic expansion valves that are internally equalized should be used with

A. multicircuit evaporators.

B. double-row evaporators.

C. gravity-type evaporators.

D. single-circuit evaporators with no more than the equivalent of 2°F saturated temperature change.

31. The equalizer line should never be

A. capped off.

B. soldered to the suction line.

C. brazed to the suction line.

D. downstream of the sensing bulb.

32. The external equalizer uses

A. bulb pressure.

B. pressure at the beginning of the evaporator.

C. spring pressure.

D. evaporator pressure at the evaporator outlet.

33. The term *thermostatic charges* refers to

A. the temperatures of the suction line.

B. the substance in the sensing bulb that responds to suction line temperature.

C. the type of refrigerant in the system.

D. the oil and refrigerant mixture of a system.

34. Which is *not* classified as a thermostatic charge?

A. liquid charge

B. gas charge

C. liquid cross

D. lubricated charge

35. The liquid charge-sensing and gas charge-sensing bulb is commonly charged with

A. a blend-type refrigerant.

B. the same type of refrigerant that is in the system.

C. an azeotropic refrigerant.

D. a ternary mixture.

36. The _______ charge TXV will always have some liquid within the bulb, capillary, and diaphragm.

A. gas

B. gas cross-

C. liquid cross-

D. Both B and C

37. Which thermostatic charge element contains a mixture of refrigerant that will condense to a small quantity of liquid when the TXV is in its normal operating range?

A. gas cross-charge

B. gas charge

C. liquid cross

D. adsorption

38. Which thermostatic charge element has a noncondensable gas along with a special material within the sensing bulb?

A. gas charge

B. liquid charge

C. adsorption charge

D. liquid cross-charge

39. The term "MOP" refers to

A. more optimum pressure.

B. maximum operating pressure.

C. maximum optimum pressure.

D. maximum oscillating pressure.

40. Whenever a refrigerant distributor is used on an evaporator coil, a

A. TXV without an external equalizer should be used.

B. nonadjustable TXV should be used.

C. TXV with an external equalizer should be used.

D. de-superheating valve should be used.

41. The external equalizer line should be connected

A. downstream from the sensing bulb.

B. upstream from the sensing bulb.

C. after an EPR valve.

D. after an accumulator.

42. A de-superheating TXV is used in conjunction with

A. an electric defrost system.

B. a hot gas or discharge bypass valve.

C. gravity-type coils.

D. ammonia-type systems only.

43. The de-superheating TXV allows refrigerant to

A. cool the warm liquid refrigerant from the condenser.

B. allow refrigerant into the liquid line.

C. cool the hot discharge gas that enters the suction line.

D. subcool liquid refrigerant.

44. Specially designed thermostatic expansion valves must be used when the system's refrigerant is

A. R-717.

B. R-410a.

C. R-414b.

D. R-407.

45. Refrigeration systems will generally operate with _______ than air conditioning systems.

A. a higher superheat

B. a lower superheat

C. a higher subcooling

D. higher pressures

46. Thermostatic expansion valves can be installed

A. before the liquid line solenoid valve.

B. before the filter drier.

C. at least 12″ from the evaporator inlet.

D. in any position but as close to the evaporator as possible.

47. When a refrigerant distributor cannot be directly connected to the TXV outlet, the maximum distance between the valve outlet and the distributor should not exceed

A. 18″.

B. 24″.

C. 12″.

D. 6″.

48. The only component that should be installed between the TXV outlet and the evaporator is the

A. refrigerant distributor.

B. filter drier.

C. sight glass.

D. solenoid valve.

49. Static pressure will always be present in the liquid line when the TXV and the evaporator are located

A. below the receiver.

B. at the same level as the receiver.

C. above the receiver.

D. Static pressure will have no influence on performance.

50. The contact of the TXV sensing bulb on the suction line should be direct on the piping and connected with

A. a single bulb strap.

B. a solder connection.

C. a wrapping of cork tape.

D. two bulb straps.

51. The TXV sensing bulb needs to be placed at the 4 o'clock or the 8 o'clock position on the suction line when the line is

A. $\frac{5}{8}''$ OD.

B. $\frac{7}{8}''$ OD or larger.

C. $\frac{3}{8}''$ OD.

D. $\frac{3}{4}''$ OD.

52. The TXV sensing bulb should be installed on top of the suction line when the line is

A. 1″ OD.

B. $1\frac{1}{4}''$OD.

C. $1\frac{1}{2}''$OD.

D. smaller than $\frac{7}{8}''$OD.

53. When an application calls for brine tanks and water coolers to be used, the sensing bulb should be

A. above the liquid surface.

B. below the liquid surface.

C. connected to the outside of the tank.

D. partially exposed to the ambient air.

54. Which statement is true of a thermostatic expansion valve?

A. It controls subcooling.

B. It keeps the evaporator flooded.

C. Some valves are of the nonadjustable type.

D. It can function as a check valve.

55. What is the most accurate means of determining TXV performance?

A. superheat measurements

B. subcooling measurements

C. temperature measurements

D. pressure-drop measurements

56. In general, the correct superheat will be determined on the amount of temperature difference between the

A. suction line temperature and the liquid line temperature.

B. condensing temperature and the evaporator temperature.

C. refrigerant temperature and the temperature of the suction line.

D. suction gas temperature and the discharge gas temperature.

THE FOLLOWING QUESTIONS MAY BE HELPFUL TO STUDY FOR THESE CERTIFICATION TESTS:
NATE—AC-SV, AC-IN, HP-SV, HP-IN, RC-SV, RC-IN, RL-SV, RL-IN.
ESCO INSTITUTE HVAC EXCELLENCE AIR CONDITIONING AND COMMERCIAL REFRIGERATION CERTIFICATION

SERVICE VALVES

1. Where are service valves commonly located on a split air conditioning system?

A. the evaporator coil

B. the condensing unit

C. the indoor air-handling system

D. the discharge and suction line

2. How will back-seating a two-position service valve affect a system?

A. It will allow the system to operate normally and obtain gauge readings.

B. It will pump a system down.

C. It will block the gauge port.

D. None of the above

3. How will front-seating a two-position liquid line service valve affect a system?

A. It will allow the system to operate normally and obtain gauge readings.

B. It will pump a system down.

C. It will block the gauge port.

D. None of the above

4. How can the refrigerant in a split air conditioning system be stored in the condenser?

A. by back-seating the suction line service valve

B. by front-seating the discharge line service valve

C. by front-seating the liquid line service valve

D. by front-seating the suction line service valve

5. What is the purpose of a Schrader valve?

A. It blocks the refrigerant flow through a system.

B. It seals refrigerant in a system and connects a manifold gauge set.

C. Being located on a refrigerant tank, it aids in charging a refrigeration system.

D. It allows refrigerant to flow in one direction.

6. When using a three-position liquid line service valve, when is the back-seated position used?

A. when the manifold gauges are disconnected

B. when the manifold gauges are connected

C. when the system is to be pumped down

D. when checking the compressor valves or rings

7. When using a three-position suction line service valve, when is the midseated position used?

A. when the manifold gauges are disconnected

B. when the manifold gauges are connected

C. when the system is to be pumped down

D. when checking the compressor valves or rings

8. When using a three-position suction line service valve, when is the front-seated position used?

A. when the manifold gauges are disconnected

B. under normal operation

C. when the system is to be pumped down

D. when checking the compressor valves or rings

9. When using a three-position liquid line service valve, when is the front-seated position used?

A. when the manifold gauges are disconnected

B. under normal operation

C. when the system is to be pumped down

D. when checking the compressor valves or rings

10. Where is a king valve located?

A. at the outlet of the compressor

B. at the inlet of the compressor

C. at the outlet of the receiver

D. at the inlet of the receiver

11. Where is a queen valve located?

A. at the outlet of the compressor

B. at the inlet of the compressor

C. at the outlet of the receiver

D. at the inlet of the receiver

12. Where is a discharge service valve located?

A. at the outlet of the compressor

B. at the inlet of the compressor

C. at the outlet of the receiver

D. at the inlet of the receiver

13. Where is a suction service valve located?

A. at the outlet of the compressor

B. at the inlet of the compressor

C. at the outlet of the receiver

D. at the inlet of the receiver

THE FOLLOWING QUESTIONS MAY BE HELPFUL TO STUDY FOR THESE CERTIFICATION TESTS:
NATE—AC-SV, AC-IN, HP-SV, HP-IN, RC-SV, RC-IN, RL-SV, RL-IN.
ESCO INSTITUTE HVAC EXCELLENCE AIR CONDITIONING AND COMMERCIAL REFRIGERATION CERTIFICATION

LEAK DETECTION

1. What is an indication of a refrigerant leak in an air conditioning system?

A. an oil trace around copper fittings

B. excessively high discharge pressures

C. high discharge-line temperatures

D. low suction-line temperatures

2. What is commonly used to pressurize an air conditioning system for the purpose of leak detection?

A. air

B. nitrogen with a trace amount of R-22

C. nitrogen with a trace amount of R-12

D. nitrogen mixed with oxygen

3. When testing for a leak in an air conditioning system,

A. the system must be pressurized and electrically de-energized.

B. the system must be in full operation except for the condenser fan.

C. the system must be in full operation except for the indoor blower.

D. only the compressor must be energized.

4. If air is trapped in an air conditioning system, it has a tendency to stay in the

A. evaporator.

B. compressor.

C. condenser.

D. metering device.

5. When pressurizing an air conditioning system for the purpose of leak detection, the pressure in the system must never exceed

A. the high-side test pressure rating.

B. the low-side test pressure rating.

C. 500 psig.

D. 500 psia.

6. When pressurizing a system during a leak test, what must never be used?

A. carbon dioxide

B. nitrogen

C. R-22

D. oxygen

7. What must be used when pressure testing a system with nitrogen?

A. a vacuum pump and charging cylinder

B. a recovery system

C. a pressure regulator and an in-line pressure-relief valve

D. No accessories are needed to use nitrogen on a system.

8. How long must an air conditioning system operate after adding a refrigerant dye for leak detection?

A. 1 hr

B. 10 hr

C. 15 hr

D. 24 hr

9. Which leak detector is the most accurate and is used on systems with extremely small R-22 refrigerant leaks?

A. halide leak detector

B. soap bubbles

C. smoke test method

D. electronic leak detector

10. When using an electronic leak detector,

A. the probe must be held below the tubing being tested.

B. the probe must be held above the tubing being tested.

C. the probe must be held pointing in the *down* position for an accurate reading.

D. the probe must be held pointing in the *up* position for an accurate reading.

11. When is it inadvisable to use a halide leak detector?

A. on a system that uses halogenated fluorocarbons

B. on an inoperative system

C. near flammable materials

D. on a system that contains 10 lb of refrigerant or more

12. When using a halide leak detector, what color will the flame be when it is in the presence of a small refrigerant leak?

A. blue

B. green

C. red

D. purple

13. An ultrasonic leak detector senses a leak by

A. radar.

B. excessive refrigerant concentration.

C. sound waves.

D. a chemical reaction.

14. An ultrasonic leak detector senses a leak on a system that

A. is under a positive pressure only.

B. is under a negative pressure only.

C. is under a positive or a negative pressure.

D. uses CFC refrigerants only.

15. What is commonly added to a system when using an ultraviolet light leak detector?

A. a UV additive

B. nitrogen

C. a red dye additive

D. a fluorescent additive

THE FOLLOWING QUESTIONS MAY BE HELPFUL TO STUDY FOR THESE CERTIFICATION TESTS:

NATE—AC-SV, AC-IN, HP-SV, HP-IN, RC-SV, RC-IN, RL-SV, RL-IN.

ESCO INSTITUTE HVAC EXCELLENCE AIR CONDITIONING AND COMMERCIAL REFRIGERATION CERTIFICATION

RECOVERY EQUIPMENT

1. The term *recovery* means

A. to remove and store refrigerant from an appliance.

B. to remove and clean refrigerant from an appliance.

C. to remove, clean, and store refrigerant from an appliance.

D. to catch refrigerant in the air as it is being released.

2. What are the three types of recovery methods that are being used today?

A. charge migration, use of the appliance compressor, and use of a refrigerant recovery unit

B. use of appliance compressor, use of refrigerant recovery unit, and vacuum pump procedure

C. use of refrigerant recovery unit, charge migration, and liquid recovery

D. vapor recovery, liquid recovery, and use of appliance compressor

3. What are the three types of recovery machines that are being used today?

A. vapor/liquid recovery units, 10-cfm systems, and 15-cfm systems

B. 5-cfm systems, 10-cfm systems, and 15-cfm systems

C. vapor recovery units, liquid recovery units, and vapor/liquid recovery units

D. liquid recovery units, vapor recovery units, and 10-cfm systems

4. What are the three classifications of recovery equipment?

A. very-high-pressure, high-pressure, and medium-pressure equipment

B. high-pressure, medium-pressure, and low-pressure equipment

C. very-low-pressure, low-pressure, and high-pressure equipment

D. very-high-pressure, high-pressure, and low-pressure equipment

5. What type of equipment uses refrigerant that boils below −58°F at atmospheric pressure?

A. very high pressure

B. high pressure

C. medium pressure

D. low pressure

6. What type of equipment uses refrigerant that boils between −58 and 50°F at atmospheric pressure?

A. very high pressure

B. high pressure

C. medium pressure

D. low pressure

7. What type of equipment uses refrigerant that boils above 50°F at atmospheric pressure?

A. very high pressure

B. high pressure

C. medium pressure

D. low pressure

8. The term *recycle* means

A. to remove and store refrigerant from an appliance.

B. to remove and clean refrigerant from an appliance.

C. to remove and reuse refrigerant in another appliance.

D. to remove and take to a reclaim center.

9. After November 11, 1993, when recovering more than 200 lb of R-22,

A. the system must be taken down to 0 psig.

B. the system must be taken down to 4 psig.

C. the system must be taken down to 4″ Hg.

D. the system must be taken down to 10″ Hg.

10. After November 15, 1993, when recovering less than 200 lb of R-22,

A. the system must be taken down to 0 psig.

B. the system must be taken down to 4 psig.

C. the system must be taken down to 4″ Hg.

D. the system must be taken down to 10″ Hg.

11. Recovery cylinders must be approved by

A. the Environmental Protection Agency.

B. ASHRAE.

C. the Cylinder Inspectors of America.

D. the Department of Transportation.

12. A recovery cylinder must not be filled to more than
    A. 60% of its capacity.
    B. 70% of its capacity.
    C. 80% of its capacity.
    D. 90% of its capacity.

13. What component that is commonly used in a recycling machine is not found in a recovery machine?
    A. a filter
    B. an oil separator
    C. an expansion tank
    D. an automatic shutoff system

THE FOLLOWING QUESTIONS MAY BE HELPFUL TO STUDY FOR THESE CERTIFICATION TESTS:
NATE—CR-SV, CR-IN, AC-SV, AC-IN, HP-SV, HP-IN, RC-SV, RC-IN RL-SV, RL-IN.
ESCO INSTITUTE HVAC EXCELLENCE AIR CONDITIONING AND COMMERCIAL REFRIGERATION CERTIFICATION

SOLDERING AND BRAZING SAFETY

1. What safety device must always be present when using a torch?
   A. a striker
   B. a wet rag
   C. a fire extinguisher
   D. a respirator

2. When soldering or brazing, always
   A. adjust the acetylene to 40 psig.
   B. adjust the oxygen to 70 psig.
   C. use a lighter when igniting the torch.
   D. wear the proper safety glasses.

3. What is the purpose of a heat shield?
   A. It is used to absorb heat from a filter dryer when brazing.
   B. It is used to protect combustible material.
   C. It is used to add heat to a joint while brazing.
   D. It is used to help cool a joint after it has been soldered or brazed.

4. When using a torch to repair a system that is overhead,
   A. heat the male section of the joint only.
   B. heat the female section of the joint only.
   C. never stand directly under the torch.
   D. heat the center of the joint only.

5. Brazing materials should not contain
   A. lead.
   B. lead or tin.
   C. flux.
   D. cadmium.

6. When using torches while repairing copper tubing on a refrigeration system, always
   A. use flux.
   B. use silver solder.
   C. use brazing material.
   D. make sure the work area is well ventilated.

7. The pressure regulators on the oxygen/acetylene torch should
   A. always be tested for good working condition.
   B. be adjusted out for use.
   C. be adjusted in after each use.
   D. be disassembled and cleaned after each use.

8. Acetylene pressure should never exceed
   A. 5 psi.
   B. 10 psi.
   C. 15 psi.
   D. 20 psi.

9. Soldering or brazing should never be attempted when
   A. the system has been recovered.
   B. the system is under pressure.
   C. the system has been evacuated.
   D. the system has been retrofitted.

10. Oxygen/acetylene tanks and hoses should
    A. be shut off and have the hoses bled after each use.
    B. be correctly color coded.
    C. be correctly labeled and identified.
    D. All of the above

11. What is the proper source of ignition for an oxygen/acetylene torch?
    A. a match
    B. a lighter
    C. a striker
    D. another torch

12. A system must never be soldered or brazed
    A. when it is at 0 psig.
    B. when nitrogen is inside the system.
    C. when it is in a slight vacuum.
    D. when under pressure.

13. When soldering or brazing a sealed system, first recover the refrigerant, then

A. never open the system to the atmosphere.

B. open the system to the atmosphere.

C. there must be a minimum of 20 psig present.

D. draw a vacuum of 20″ Hg.

THE FOLLOWING QUESTIONS MAY BE HELPFUL TO STUDY FOR THESE CERTIFICATION TESTS:
NATE—CR-SV, CR-IN, AC-SV, AC-IN, HP-SV, HP-IN, RC-SV, RC-IN, RL-SV, RL-IN.
ESCO INSTITUTE HVAC EXCELLENCE AIR CONDITIONING AND COMMERCIAL REFRIGERATION CERTIFICATION

SOLDERING AND BRAZING

1. A copper joint must never be soldered or brazed

A. when nitrogen is present in the tubing.

B. when the system is opened to the atmosphere.

C. when the system is under pressure.

D. when air is present in the system.

2. At what temperature does Stay Brite #8 solder melt?

A. 360°F

B. 535°F

C. 635°F

D. 660°F

3. At what temperature does 95-5 solder melt?

A. 150°F

B. 250°F

C. 450°F

D. 650°F

4. Using flux while soldering will help

A. keep the joint that is being soldered clean.

B. retain heat on the joint that is being soldered.

C. stop oxides from forming.

D. absorb the solder into the joint.

5. What are the common melting points of brazing material and silver solder?

A. 500 to 800°F

B. 1,100 to 1,480°F

C. 2,000 to 2,500°F

D. 2,500 to 3,000°F

6. Leaks in coils made from aluminum are commonly repaired with

A. silver solder.

B. silver brazing material.

C. brass brazing material.

D. epoxy.

7. The term *capillary action* means to

A. spray water on a joint after it has been soldered.

B. expand a joint as it is being heated by a torch.

C. draw solder into a joint as it is being heated.

D. add flux to the male part of a joint that is to be soldered.

8. The gauges that are located on the oxyacetylene torch measure

A. the tank pressure (in psig) and the regulated pressure (in psig) to the torch tip.

B. the tank pressure (in psia) and the regulated pressure (in psia) to the torch tip.

C. the tank weight (in lb) and the regulated pressure (in psig) to the torch tip.

D. the volumetric pressure (in psig) and the regulated pressure (in psia) to the torch tip.

9. The operating pressure of an acetylene regulator ranges from

A. 0 to 15 psig.

B. 0 to 20 psig.

C. 0 to 40 psig.

D. 0 to 50 psig.

10. The operating pressure of an oxygen regulator ranges from

A. 0 to 20 psig.

B. 0 to 40 psig.

C. 0 to 50 psig.

D. 0 to 100 psig.

11. When adjusting pressure on a regulator,

A. turn the regulator clockwise to increase pressure for oxygen and counterclockwise to increase pressure for acetylene.

B. turn the regulator counterclockwise to increase pressure for oxygen and clockwise to increase pressure for acetylene.

C. turn the regulator clockwise to increase pressure for oxygen and clockwise to increase pressure for acetylene.

D. turn the regulator counterclockwise to increase pressure for oxygen and counterclockwise to increase pressure for acetylene.

12. What should be used to purge an air conditioning system's tubing while brazing?
 A. air
 B. oxygen
 C. nitrogen
 D. carbon dioxide

13. What is the purpose for purging an air conditioning system while brazing?
 A. to keep moisture out of the system
 B. to prevent oxidation
 C. to draw the heat from the brazed joint
 D. to aid the chemical reaction of the brazing compound

14. What could cause an explosion if it is used to purge an air conditioning system while brazing?
 A. oxygen
 B. air
 C. nitrogen
 D. carbon dioxide

15. The purpose for a reaming tool is to
 A. prepare the copper tubing before cutting.
 B. expand the copper tubing after cutting.
 C. smooth the outer edge of the tubing for a tighter-fitting joint.
 D. remove the burrs from the tubing after cutting.

16. What is used to expand the diameter of soft copper tubing?
 A. a swaging tool
 B. a flaring block
 C. a flaring tool
 D. an expansion joint

17. What would be the minimum bending radius on $\frac{1}{2}''$ soft copper tubing?
 A. $1''$
 B. $1\frac{1}{2}''$
 C. $2''$
 D. $2\frac{1}{2}''$

18. When using a spring-type tube bender,
 A. it must only be placed on the inside of the tubing.
 B. it must only be placed on the outside of the tubing.
 C. it can be placed on the inside or on the outside of the tubing.
 D. it must be oiled before each use.

19. What is the maximum bend that a lever bender can make?
 A. 45 degrees
 B. 90 degrees
 C. 160 degrees
 D. 180 degrees

20. Which tool holds the copper tubing in place when making a mechanical joint?
 A. the flaring tool
 B. the flaring block
 C. the swaging tool
 D. the reaming tool

21. When making a flare, what can be done to prevent the copper from splitting?
 A. cool the copper to 40°F
 B. heat the copper to 105°F
 C. tighten and loosen the flaring tool during the procedure
 D. quickly turn the flaring tool as the copper expands

22. Which material will create a stronger joint?
 A. brazing material
 B. 50:50 solder
 C. silver solder
 D. 45% silver

ESCO INSTITUTE-HVAC EXCELLENCE BRAZING AND SOLDERING CERTIFICATION EMPLOYMENT READY (ER)

1. The procedure of expanding the inside diameter of copper tubing is called
 A. enlarging.
 B. crimping.
 C. flaring.
 D. swaging.

2. The test pressure used to leak test the system's piping circuit should be
 A. more than the highest pressure indicated on the manufacturing plate.
 B. less than the lowest pressure indicated on the manufacturing plate.
 C. 10% less than the highest pressure indicated on the manufacturing plate.
 D. pressurized to full capacity controlled by the regulator.

3. The acetylene tank of an oxyacetylene kit
 A. is normally an "R" tank.
 B. is normally an "L" tank.
 C. is normally an "MC" tank.
 D. can use an "R" or "L" tank.

4. Common flames when brazing are
 A. carburizing, positive, and oxidizing.
 B. carburizing, neutral, and oxidizing.
 C. oxyacetylene, feather, and carburizing.
 D. neutral, acetylene, and oxidizing.

5. The air acetylene tank is a tank of

A. 40 cubic feet.

B. 18 cubic feet.

C. 20 cubic feet.

D. 30 cubic feet.

6. Soldering temperatures are normally done at

A. temperatures at 1200°F.

B. temperatures at 1050°F.

C. temperatures below 840°F.

D. temperatures not higher than 250°F.

7. The common types of copper tubing used in ACR are

A. L and K type.

B. M and K type.

C. thin wall and hard drawn type.

D. L wall and M type.

8. Once the chemical flux is heated it

A. will drip and must be wiped clean.

B. must have continuous heat applied.

C. must be quickly cooled.

D. must fill in the gaps of the joint.

9. The types of fuel used to solder are

A. acetylene and propane.

B. oxyacetylene and butane.

C. propane and butane.

D. propane and MAPP gas.

10. The most common copper tubing type used in HVAC/R is

A. soft and hard drawn.

B. untreated and hard drawn.

C. plumbing and universal.

D. "L" and rolled type.

11. ACR copper tubing is measured by

A. the outside diameter.

B. the inside diameter.

C. the length.

D. the nominal dimension.

12. To solder a joint, use the torch to

A. heat the fitting till it glows a cherry red color.

B. keep the flame at least 6 inches from the joint.

C. increase the flame intensity when applying the solder.

D. slowly heat at the base of the joint then evenly apply the solder.

13. What is used when soldering to prevent oxides from forming?

A. dry nitrogen

B. microfoam solution

C. chemical flux

D. increasing the torch flame

14. The gas that is mostly used when brazing is

A. butane.

B. acetylene.

C. propane.

D. methane.

15. The type of copper that has the thickest wall section is

A. Type L.

B. Type M.

C. Type K.

D. Types L or M.

16. The brazing procedure consists of

A. melting a filler material using a low heat.

B. melting a filler material with high propane fuel.

C. melting a filler material at a temperature of 640°F.

D. melting a filler material into the joint at a high temperature.

17. One method used to join copper tubing by using interchangeable jaws is

A. a swaging system.

B. a crimping system.

C. a flaring system.

D. a coupling system.

18. The temperatures used for soldering applications are

A. under 840°F.

B. over 840°F.

C. no more than 350°F.

D. at least 1100°F.

19. The nitrogen tank should always be equipped with a

A. tank gauge.

B. compound gauge.

C. pressure regulator.

D. relief valve.

20. An air-acetylene torch can provide temperatures around

A. 4500°F

B. 1650°F

C. 1500°F

D. 1900°F

21. The nitrogen tank has two gauges; the right-side gauge reads

A. tank pressure.

B. operating pressure.

C. liquid pressure.

D. outlet pressure.

22. Once a brazing procedure is completed, the next step should be

A. apply microfoam and check for leaks.

B. insert a dye leak solution and leak check.

C. pressurize with dry nitrogen and check for leaks.

D. pressurize using the system's refrigerant.

23. A microfoam solution, ultrasonic and electronic devices are used to

A. check for refrigerant leaks.

B. check for noncondensables.

C. check for moisture in the system.

D. check for contaminants.

24. The temperature at which ACR copper will melt is

A. 1900°F

B. 2200°F

C. 2500°F

D. 3000°F

25. In order to change the flame size on an air-acetylene torch

A. use a smaller regulator.

B. use longer hoses.

C. use shorter hoses.

D. change torch tips.

26. Which types of copper are used in A/C and refrigeration?

A. Type L

B. Types M and L

C. Type K

D. Types L and K

27. Air-acetylene torches that use a regulator should be

A. adjusted to 3–5 psig.

B. adjusted to 8–10 psig.

C. adjusted to 14–15 psig.

D. adjusted to 16–18 psig.

28. Once a system has passed a leak test, the next step is to

A. charge the system.

B. evacuate the system.

C. insert dry nitrogen.

D. purge the piping circuit.

29. When brazing with oxyacetylene a flame with 1:1 ratio oxygen to acetylene

A. is a neutral flame.

B. is a carburizing flame.

C. is an oxidizing flame.

D. is a feathering flame.

30. When joining copper tubing on mini-split systems, the

A. brazing procedure is preferred.

B. crimping procedure is preferred.

C. soldering procedure is preferred.

D. flaring procedure is preferred.

31. Which copper tubing type has the thinnest wall section?

A. Type M

B. Type L

C. Type K

D. Type G

32. Brazing temperatures on ACR copper tubing are

A. 450–550°F

B. 550–800°F

C. 1100–1600°F

D. 1650–1750°F

33. The procedure that can use compression, punch, and hydraulic refers to

A. flaring.

B. swaging.

C. coupling.

D. sweating.

34. Using oxyacetylene while brazing the flame that has more oxygen than acetylene in greater than 1:1 ratio is called

A. oxidizing flame.

B. carburizing flame.

C. neutral flame.

D. dry flame.

35. Nitrogen is used to purge copper lines to prevent

A. carbonization.

B. scale formation.

C. oxidation or scale formation.

D. rust formation.

36. For safety purposes, when using torches

A. have an abundance of water supply nearby.

B. have the nearest fire department number nearby.

C. have the appropriate fire extinguisher nearby.

D. have a co-worker nearby.

37. A good selection type of refrigerant lines for A/C, mini-splits, and heat pumps is

A. Type K.

B. Type L.

C. Type M.

D. Types L or M.

38. When cutting copper tubing in close areas

A. a small saw blade can be used.

B. macro tube cutters can be used.

C. nominal tube cutters can be used.

D. a small wheel tube cutter called an imp can be used.

39. After cutting copper tubing, the ends

A. should be blown out using compressed air.

B. should be deburred or reamed.

C. should be cleaned with a solution.

D. should be cleaned with sand cloth.

40. The procedure of flaring results in the end of the copper tubing forming a

A. bowl shape.

B. tunnel shape.

C. triangular shape.

D. funnel shape.

41. When brazing copper to copper, the regulators on the oxyacetylene tanks

A. should be set to produce a ratio of 1:1.5.

B. should be set to produce a ratio of 2:1.

C. should be set to produce a ratio of 2:1.5.

D. should be set to produce a ratio of 2:1.8.

42. The oxygen tank in a portable oxyacetylene kit is

A. an "MC" tank (10 cubic feet).

B. a "B" tank (40 cubic feet).

C. an "R" tank (20 cubic feet).

D. a "Q" tank (80 cubic feet).

43. The tank used for air-acetylene is

A. a "B" tank (40 cubic feet).

B. a "Q" tank (80 cubic feet).

C. an "FR" tank (18 cubic feet).

D. an "R" tank (20 cubic feet).

44. The flame that has more acetylene than oxygen greater than 1:1 ratio is

A. dry flame.

B. oxidizing flame.

C. carburizing flame.

D. feather flame.

45. The gauge on the left side of a nitrogen regulator reads

A. the outlet pressure.

B. the tank pressure.

C. the liquid pressure.

D. the vapor pressure.

46. When brazing or soldering, a safe practice would be to

A. keep the regulators open.

B. ensure there is proper ventilation available.

C. bleed tanks to zero psig.

D. purge all hoses.

47. Any defects or leaking joints on piping should be

A. reheated and leak tested.

B. cut out of the piping and replaced.

C. repaired with soldering material.

D. repaired with brazing material.

48. The flux chemical used when soldering or brazing should

A. be left on the fitting to secure the joint.

B. be applied to the inner portion of the fittings.

C. be applied on top of the fittings.

D. be cleaned from the copper after it has cooled.

49. When brazing or soldering, the amount of heat can be

A. adjusted by increasing or decreasing the flame distance.

B. adjusted by turning the regulator wide open.

C. adjusted by using a hotter fuel.

D. adjusted by using a dry flame.

50. Which step is the most critical when brazing?

A. using the correct flux

B. using a wire brush to clean the copper

C. adjusting to produce the correct flame

D. making sure the fitting or joint is very hot

THE FOLLOWING QUESTIONS MAY BE HELPFUL TO STUDY FOR THESE CERTIFICATION TESTS:
NATE—AC-SV, AC-IN, HP-SV, HP-IN, RC-SV, RC-IN, RL-SV, RL-IN.
ESCO INSTITUTE HVAC EXCELLENCE AIR CONDITIONING AND COMMERCIAL REFRIGERATION CERTIFICATION

MANIFOLD GAUGES

1. How is a temperature measurement obtained on a manifold gauge set?

A. by adding 14.7 to the pressure and comparing it to gauge temperature

B. by subtracting 14.7 from the pressure and comparing it to gauge temperature

C. by taking the temperature directly from the inner colored numbers

D. by taking the temperature directly from the outer black numbers

2. What would be the proper procedure for testing pressures on an air conditioning system?

A. Connect the blue hose to the high-side line, the red hose to the low-side line, and close the gauge valves.

B. Connect the blue hose to the low-side line, the red hose to the high-side line, and open the gauge valves.

C. Connect the yellow hose to the high-side line, the red hose to the low-side line, and open the gauge valves.

D. Connect the blue hose to the low-side line, the red hose to the high-side line, and close the gauge valves.

3. A manifold gauge set is used to check operating pressures in
   - **A.** psia.
   - **B.** psig.
   - **C.** inches WC.
   - **D.** psig and inches WC.
4. To convert pounds per square inch gauge (psig) to pounds per square inch absolute (psia),
   - **A.** add 14.7 inches of WC.
   - **B.** add 14.7 psig.
   - **C.** subtract 14.7 psia.
   - **D.** subtract 14.7 inches WC.
5. What is the purpose of the center hose on a manifold gauge set?
   - **A.** It is used to measure the pressure difference in a system.
   - **B.** It is used to connect to a heat pump system for reversing pressures.
   - **C.** It is used to charge a system.
   - **D.** It is used to measure the pressure on a suction line filter dryer.
6. On a manifold gauge set, the low-side pressure gauge commonly measures
   - **A.** 0 to 300 psig.
   - **B.** 0 to 250 psig.
   - **C.** 30″ Hg to 350 psig.
   - **D.** 30″ Hg to 500 psig.
7. On a manifold gauge set, the high-side pressure gauge commonly measures
   - **A.** 0 to 500 psig.
   - **B.** 0 to 550 psig.
   - **C.** 30″ Hg to 500 psig.
   - **D.** 30″ Hg to 700 psig.
8. What is the procedure for adjusting a manifold gauge set?
   - **A.** Connect the gauges to an R-12 refrigerant tank and adjust it to 70 psig.
   - **B.** After connecting the gauges to a system, adjust each gauge to 32°F.
   - **C.** Close the gauges from the atmosphere and adjust to zero.
   - **D.** Open the gauges to the atmosphere and adjust to zero.
9. What is commonly used with refrigerant gauges to prevent excessive refrigerant loss?
   - **A.** low-loss fittings
   - **B.** service port caps
   - **C.** refrigerant line purge device
   - **D.** pressure-reducing valve
10. What can cause excessively high discharge pressures on an air conditioning system?
    - **A.** insufficient airflow over the A-coil
    - **B.** noncondensables in the system
    - **C.** a restricted metering device
    - **D.** a kink in the suction line
11. What can cause a decrease in suction pressure on an air conditioning system?
    - **A.** a dirty condenser coil
    - **B.** too large of a TXV
    - **C.** a dirty evaporator coil
    - **D.** recirculated air over the condenser

THE FOLLOWING QUESTIONS MAY BE HELPFUL TO STUDY FOR THESE CERTIFICATION TESTS:
NATE—AC-SV, AC-IN, HP-SV, HP-IN, RC-SV, RC-IN, RL-SV, RL-IN.
ESCO INSTITUTE HVAC EXCELLENCE AIR CONDITIONING AND COMMERCIAL REFRIGERATION CERTIFICATION

SYSTEM EVACUATION

1. What is the purpose of evacuating an air conditioning system?
   - **A.** to remove air and moisture from the system
   - **B.** to depressurize the system for the expansion of refrigerant
   - **C.** to draw oil from the compressor during the dehydration process
   - **D.** to increase the rate of refrigerant flow through the expansion valve
2. When must a system be evacuated?
   - **A.** when the system has dirty evaporator and condenser coils
   - **B.** when the system has excessively high discharge pressures
   - **C.** when the system has been opened to the atmosphere
   - **D.** as a routine maintenance to ensure no moisture is present in the system
3. How often should the oil in a vacuum pump be replaced?
   - **A.** after every other evacuation
   - **B.** before and after every evacuation
   - **C.** after it has operated for 20 h according to manufacturer recommendation
   - **D.** after it has operated for 10 h according to manufacturer recommendation
4. Why must the oil of a vacuum pump be replaced?
   - **A.** The vacuum in the pump causes the oil to become thinner, which will cause excessive wear.
   - **B.** The refrigerant's pressure causes a breakdown of the vacuum pump oil.
   - **C.** The oil becomes full of moisture and acids, which will decrease the capability of the vacuum pump.
   - **D.** The high speed of the vacuum pump will cause a breakdown of vacuum pump oil.

5. What is commonly used to measure a deep vacuum?
   A. a mercury column located on the low-pressure manifold gauge
   B. a micrometer gauge
   C. a micron gauge
   D. a manometer
6. What will decrease the evacuation time of an air conditioning system?
   A. the removal of a Schrader valve core, smaller-diameter hoses, and higher ambient temperatures
   B. the removal of a Schrader valve core, larger-diameter hoses, and lower ambient temperatures
   C. the removal of a Schrader valve core, smaller-diameter hoses, and lower ambient temperatures
   D. the removal of a Schrader valve core, larger-diameter hoses, and higher ambient temperatures
7. When a system incorporates a three-way service valve,
   A. it must be in the mid position during evacuation.
   B. it must be in the back-seated position during evacuation.
   C. it must be in the front-seated position during evacuation.
   D. All of the above
8. During a system evacuation, always
   A. open the high-side gauge and close the low-side gauge.
   B. open the low-side gauge and close the high-side gauge.
   C. open both the high-side and the low-side gauge.
   D. close both the high-side and the low-side gauge.
9. A minimum evacuation level for an air conditioning system is
   A. 5,000 microns.
   B. 1,000 microns.
   C. 500 microns.
   D. 100 microns.
10. A triple evacuation is done by evacuating a system to
    A. 18″ Hg and charging the system to 50 psig three times.
    B. 28″ Hg and charging the system to 0 psig three times.
    C. 18″ Hg and charging the system to 0 psig three times.
    D. 28″ Hg and charging the system to 50 psig three times.
11. After evacuating a system, the manifold gauges are closed, then the vacuum pump is turned off. Using a micron gauge, the system is drawn down to 500 microns, the system is left off, and the pressure increases slightly then stops. What is the problem?
    A. The gauges are leaking.
    B. Moisture is in the system.
    C. The oil has expanded.
    D. A system leak is present.
12. When using a micron gauge, what is the proper level of vacuum that must be drawn on a system?
    A. 5,000 microns
    B. 500 microns
    C. 1,000 microns
    D. 3,000 microns

ESCO INSTITUTE-HVAC EXCELLENCE SYSTEM RECOVERY AND EVACUATION CERTIFICATION EMPLOYMENT READY (ER)

1. When using a new or empty recovery cylinder, it is good practice to
   A. evacuate the cylinder before using.
   B. flush the cylinder with nitrogen.
   C. place the cylinder on ice.
   D. warm the cylinder.
2. The EPA evacuation standard is
   A. 1200 microns.
   B. 1000 microns.
   C. 900 microns.
   D. 500 microns.
3. Reducing pressures below atmospheric pressure is referred to as
   A. evacuation.
   B. recycling.
   C. regenerating.
   D. reclamation.
4. When replacing a restricted liquid line drier, it is best to remove it from the system by
   A. using a torch to sweat it off.
   B. removing the drier by cutting it out with a tube cutter.
   C. using hand snips to cut the tubing.
   D. using a cutting torch.
5. The term hygroscopic refers to
   A. the condensing of refrigerant oil.
   B. refrigerant oil mixing good with vapor refrigerant.
   C. refrigerant oil's ability to absorb moisture.
   D. refrigerant oil vaporizing in the compressor.
6. Before the refrigeration unit is to be evacuated, the system's pressure must be reduced to
   A. 5 psig.
   B. 5–8 psig.
   C. 10 psig.
   D. 0 psig.
7. A technician removing a system's refrigerant to a DOT-approved container is called
   A. transferring.
   B. recovery.
   C. recycling.
   D. reclamation.

8. The two terms that describe a system's evacuation are
   - **A.** dehydration and degassing.
   - **B.** recycling and recovery.
   - **C.** reclamation and recovery.
   - **D.** dehydration and recovery.
9. Suction line driers and liquid line driers should be installed when
   - **A.** the system has experienced a compressor burnout and needs to be replaced.
   - **B.** a new TXV has been replaced.
   - **C.** the liquid receiver has been replaced.
   - **D.** the evaporator has been replaced.
10. The inner portion (core) of a filter drier is called
    - **A.** a type of desiccant.
    - **B.** a moisture remover.
    - **C.** a sponge absorber.
    - **D.** a contaminant collector.
11. The type of recovery that uses its own means to remove refrigerant from a sealed system is called
    - **A.** passive recovery.
    - **B.** mechanical recovery.
    - **C.** interdependent recovery.
    - **D.** self-contained recover.
12. The Department of Transportation mandates that a recovery tank should not be filled over
    - **A.** 75% of its capacity.
    - **B.** 85% of its capacity.
    - **C.** 80% of its capacity.
    - **D.** 65% of its capacity.
13. When moisture is dehydrated from a system, the process is referred to as
    - **A.** sublimation.
    - **B.** evaporation.
    - **C.** condensation.
    - **D.** recovery process.
14. When a recovery tank has been filled to its capacity, the tank
    - **A.** can be transferred into smaller size tanks.
    - **B.** can be transferred into a larger capacity storage tank.
    - **C.** should be recycled.
    - **D.** should be tagged for reclamation.
15. If more than one refrigerant type has been mixed in a recovery tank,
    - **A.** the mixture cannot be recycled or reclaimed and will need to be destroyed at the owner's expense.
    - **B.** the mixture can be recycled several times and reused.
    - **C.** the system should be flushed with nitrogen and then can be reused.
    - **D.** the mixture can be reused by its owner only.
16. The evacuation levels during a recovery procedure are regulated by the
    - **A.** Refrigerant Coalition.
    - **B.** Contractors Association.
    - **C.** Environmental Protection Agency.
    - **D.** state and local authorities.
17. To prevent vacuum pump oil from entering into the micron gauge, it should
    - **A.** be mounted horizontally.
    - **B.** be mounted near to the liquid line.
    - **C.** be connected on the manifold gauge.
    - **D.** be level with the vacuum pump.
18. Never allow an open filter drier to be exposed to the atmosphere because it
    - **A.** can form rust on its ends.
    - **B.** can freely be filled with moisture.
    - **C.** can be easily become corroded.
    - **D.** can cause acid build-up.
19. Recovery tanks must be certified and tested hydrostatically every
    - **A.** 3 years.
    - **B.** 5 years.
    - **C.** 8 years.
    - **D.** 10 years.
20. The type of liquid line filter driers that are used on heat pump applications
    - **A.** are one flow type.
    - **B.** must be installed in a vertical position flow.
    - **C.** must have double inside screens.
    - **D.** are the reverse flow drier type.
21. For a faster evacuation process, the technician
    - **A.** can keep the ballast open.
    - **B.** can close the low-side manifold gauge and open the high side.
    - **C.** can close the high-side manifold gauge and open the low side.
    - **D.** can remove the valve core pins before evacuating.
22. Cleaning a refrigerant to a new virgin purity level is called
    - **A.** recycling.
    - **B.** reclamation.
    - **C.** refrigerant flushing.
    - **D.** purifying.
23. To eliminate high pressures from rising in a recovery tank, the technician should
    - **A.** flush out the tank before use.
    - **B.** recover all refrigerant from the tank.
    - **C.** purge any air from the manifold hoses.
    - **D.** recycle refrigerant that is present in the tank.

24. When moisture in the system becomes mixed with the refrigerant and oil

A. the formation of acid can be found.

B. calcification of minerals can be found.

C. a clogged TXV will occur.

D. superheat will be reduced.

25. Moisture in an operating refrigeration system can cause

A. foaming of compressor oil.

B. mineral deposits.

C. a high superheat condition.

D. frosting at the receiver.

26. To read very low vacuum levels, the technician would use

A. the low-side gauge.

B. a digital micron gauge.

C. a recovery machine.

D. a standard pressure gauge.

27. A liquid line drier filters small particulates and

A. collects refrigerant oil.

B. collects moisture.

C. acts as a metering device.

D. acts as an accumulator.

28. A recovery is used when the system will need opening and

A. to add refrigerant oil.

B. to change out the evaporator fan motor.

C. to recharge the system with new refrigerant.

D. to release air from the condenser.

29. To prevent contaminants from entering into the recovery machine

A. a check valve could be installed.

B. a filter drier should be installed to the inlet of the recovery machine.

C. an oversized filter drier should be installed.

D. a low-loss fitting should be installed.

30. Nitrogen can be used between evacuation trials in a procedure known as

A. self-contained evacuation.

B. redundant evacuation.

C. recovered evacuation.

D. triple evacuation.

31. During the evacuation process and toward the last stages of a deep vacuum, the

A. technician should charge a small amount of refrigerant vapor into the system.

B. technician should charge a small amount of refrigerant liquid into the system.

C. technician should close off the ballast cap.

D. technician should close the low-side manifold gauge and keep the high side open.

32. When a vacuum is being performed, its efficiency is related to

A. the amperage rating of the pump.

B. the type of vacuum oil used.

C. the size of the pump and size of the exhaust port.

D. the diameter of the discharge line.

33. One method for a faster recovery is to lower the pressure inside the tank by

A. warming the recovery tank with heating blankets.

B. introducing some warm vapor into the recovery tank.

C. introducing warm vapor and liquid into the tank.

D. allowing the recovery tank to sit on ice.

34. When using a micron gauge, zero psig would be close to

A. 500 microns.

B. 760,000 microns.

C. 800,000 microns.

D. 960,000 microns.

35. The required recovery level of an air conditioning system containing less than 200 pounds of R-22 or R-410a is

A. 4 in Hg.

B. 10 in Hg.

C. 5 psig.

D. 0 psig.

36. Cleaning refrigerant by passing it through filter driers is called

A. recovery.

B. triple evacuation.

C. reclamation.

D. recycling.

37. Two common types of noncondensables are

A. nitrogen and air.

B. nitrogen and helium.

C. acetylene and moisture.

D. water and air.

38. Once refrigerant has been recovered, it can be reused in another system if it

A. is used on any similar equipment.

B. is owned by the same owner.

C. has met proper evacuation levels.

D. is resold.

39. When 3/8-inch manifold hoses are used for a recovery, they should be

A. 4 feet or less.

B. 8 feet or less.

C. 10 feet or less.

D. 6 feet or less.

40. Which stage tends to impede the time it takes to perform a proper evacuation?

A. recovering stage

B. dehydration stage

C. recycling stage

D. reclamation stage

41. To help minimize the loss of oil during a recovery, it is recommended that the technician

A. charge a small amount of vapor into the low side.

B. charge a small amount of nitrogen then purge.

C. recover in the liquid phase.

D. recover in the vapor phase.

42. The term noncondensable in a refrigeration system refers to

A. moisture in the system.

B. mixed refrigerants.

C. gases that will not condense from a vapor to liquid.

D. refrigerants that will not change from liquid to a vapor.

43. Refrigerants like HFO, HFC, CFC, and HCFC are regulated by the

A. Clean Air Act.

B. Refrigerant Act.

C. DOT.

D. Environmental Recovery Act.

44. Two common methods used to transfer refrigerants are

A. system-dependent and passive-dependent.

B. system-dependent (passive recovery) and self-contained (active recovery).

C. system-non-dependent and self-dependent.

D. system-pressurized and system-dependent.

45. Once a decay test has been performed, the evacuation pressure should not rise

A. over 250 microns.

B. more than 15,000 microns.

C. more than 20,000 microns.

D. more than 30,000 microns.

46. The transfer of refrigerant to a nonpressurized container using the help of a system's component is called

A. self-contained recovery.

B. self-dependent recovery.

C. passive recovery.

D. active recovery.

47. To help remove air and moisture from condensing with the vacuum pump oil, air is allowed to enter the pump through the

A. high-side hose.

B. gas ballast.

C. center manifold hose.

D. pump exhaust port.

48. When a refrigerant recovery is being performed on a large system

A. it is best to have two or more large storage tanks.

B. it is best to recover the refrigerant in a liquid state.

C. a vapor recovery is much faster than a liquid recovery.

D. recovery should start with vapor then followed by a liquid recovery.

49. The formula (Water Capacity × 0.8) × Specific Gravity + Tare Weight is used to calculate

A. refrigerant weight.

B. evacuation requirement per container.

C. maximum allowable recovery tank weight.

D. charging cylinder weight.

50. A refrigeration system operating with moisture and air will result in

A. low compressor amperage.

B. higher than normal operating pressures and temperatures.

C. lower than normal low-side pressures.

D. a frosted compressor crankcase.

THE FOLLOWING QUESTIONS MAY BE HELPFUL TO STUDY FOR THESE CERTIFICATION TESTS:
NATE—AC-SV, AC-IN, HP-SV, HP-IN, RC-SV, RC-IN, RL-SV, RL-IN.
ESCO INSTITUTE HVAC EXCELLENCE AIR CONDITIONING AND COMMERCIAL REFRIGERATION CERTIFICATION

REFRIGERANT CHARGING

1. When an electronic scale is being used to charge a system, a button must be pushed after turning the scale on. What is the button used for?

A. to set the scale to zero

B. to change the scale from pounds to ounces

C. to adjust for refrigerant pressure

D. to adjust for refrigerant type

2. What is a charging cylinder used for?

A. to charge a system by weight

B. to charge a system by quantity

C. to charge a system by sight

D. to charge a system by temperature

3. What component can aid a technician while charging an air conditioning system?
   A. a filter dryer
   B. a sight glass
   C. a TXV
   D. a receiver
4. When liquid charging a system,
   A. the system must be de-energized and have refrigerant added into the low side of the system through the low-side manifold gauge.
   B. the system must be in operation and have refrigerant added into the high side of the system through the high-side manifold gauge.
   C. the system must be de-energized and have refrigerant added into the high side of the system through the high-side manifold gauge.
   D. the system must be in operation and have refrigerant added into the low side of the system through the low-side manifold gauge.
5. When gas charging a system,
   A. the system must be de-energized and have refrigerant added into the low side of the system through the low-side manifold gauge.
   B. the system must be in operation and have refrigerant added into the high side of the system through the high-side manifold gauge.
   C. the system must be de-energized and have refrigerant added into the high side of the system through the high-side manifold gauge.
   D. the system must be in operation and have refrigerant added into the low side of the system through the low-side manifold gauge.
6. How will an overcharge affect a capillary tube system?
   A. It will have a decrease of superheat.
   B. It will have an increase of superheat.
   C. It will not affect the superheat.
   D. It will cause a starved evaporator coil.
7. What type of air conditioning system must be charged using the superheat method?
   A. a TXV system
   B. an AXV system
   C. a piston-type metering device
   D. a Pitot-tube system
8. When charging an air conditioning system using the superheat method, what factors need to be known?
   A. the indoor wet-bulb temperature and the outdoor dry-bulb temperature
   B. the indoor dry-bulb temperature and the outdoor wet-bulb temperature
   C. the indoor dry-bulb temperature and the outdoor relative humidity
   D. the indoor wet-bulb temperature and the outdoor relative humidity
9. To measure superheat,
   A. subtract the high-side condensing temperature from the suction line temperature.
   B. subtract the suction line temperature from the ambient temperature.
   C. subtract the low-side boiling point temperature from the suction line temperature.
   D. subtract the ambient temperature from the suction line temperature.
10. When charging a system using the subcool method, what type of metering device must be used on the system?
    A. a capillary tube
    B. a piston
    C. an AXV
    D. a TXV
11. To measure subcool,
    A. subtract the liquidline temperature from the high-side condensing temperature.
    B. subtract the high-side condensing temperature from the liquidline temperature.
    C. subtract the ambient temperature from the high-side condensing temperature.
    D. subtract the ambient temperature from the liquidline temperature.

ESCO INSTITUTE-HVAC EXCELLENCE BASIC REFRIGERATION AND CHARGING PROCEDURES CERTIFICATION EMPLOYMENT READY (ER)

1. The function of the condenser is to
   A. convert liquid to vapor.
   B. create superheat.
   C. change vapor to a liquid state.
   D. increase pressure.
2. The state of superheat refrigerant is
   A. liquid.
   B. vapor.
   C. liquid and vapor mix.
   D. subcooled liquid.
3. The function of the liquid receiver is to
   A. hold vapor refrigerant.
   B. prevent flash gas.
   C. store liquid refrigerant.
   D. prevent filter clogging.

4. The two components that contain both liquid and vapor are the
   A. compressor and condenser.
   B. accumulator and compressor.
   C. compressor and receiver.
   D. condenser and evaporator.
5. The component that prevents liquid refrigerant from entering the compressor is the
   A. equalizer tube.
   B. accumulator.
   C. liquid receiver.
   D. oil separator.
6. The state of refrigerant flowing through a filter drier is
   A. cooled vapor.
   B. superheated vapor.
   C. saturated vapor.
   D. liquid.
7. The range of compressor superheat temperature
   A. should be 20–30°F.
   B. should be 30–35°F.
   C. should never be over 40°F.
   D. The compressor should be cool to the touch.
8. A system that has been overcharged will result in
   A. lower suction pressure.
   B. lower head pressure.
   C. higher head and suction pressures.
   D. compressor amperage draw would be reduced.
9. The heat that changes liquid refrigerant to a vapor is called
   A. superheat.
   B. latent heat.
   C. subcooled.
   D. sensible.
10. The main components of a refrigeration system are
    A. metering device, condenser, evaporator, suction line, discharge line, receiver, and compressor.
    B. receiver, condenser, accumulator, condenser, and compressor.
    C. evaporator, condenser, compressor, and accumulator.
    D. compressor, discharge line, suction line, accumulator, condenser, and metering device.
11. A precise charging method for a capillary tube metering device would be
    A. weighing the correct amount into the system.
    B. charging by the subcooled method.
    C. charging with vapor only.
    D. charging by the ambient method.
12. The evaporator is where
    A. the refrigerant is collected from the refrigerated space.
    B. the refrigerant is subcooled.
    C. the vapor is converted to liquid.
    D. flash gas is created.
13. A capillary tube if partially restricted results in
    A. an overcharged system.
    B. a flooded evaporator.
    C. a starved evaporator.
    D. a system operating with very low suction pressure.
14. The method of checking refrigerant leaks on all high-pressure refrigerants would be using
    A. soap bubbles.
    B. open flame.
    C. litmus.
    D. halide torch.
15. The amount of cubic feet per minute across an air conditioning evaporator coil should be
    A. 550–600 cfm.
    B. 600–675 cfm.
    C. 400–450 cfm.
    D. 300–375 cfm.
16. A solenoid valve that is used on a pump-down system is located
    A. before the liquid receiver.
    B. before the metering device on the liquid line.
    C. after the evaporator.
    D. before the filter drier.
17. The majority of refrigerants exist in the
    A. vapor state.
    B. liquid state.
    C. A and B are correct.
    D. They are all in the liquid state.
18. The significance of subcooling is to
    A. prevent flash gas.
    B. lower suction pressure.
    C. increase head pressure.
    D. increase superheat.
19. The sightglass when located on the liquid line will
    A. bubble when the system is under-charged.
    B. show filled on the off cycle.
    C. be half filled during normal operation.
    D. not bubble until the system is shut down.
20. A sightglass would be found
    A. on the liquid line before the metering device.
    B. on the suction line before the accumulator.
    C. between the evaporator and the compressor.
    D. immediately after the condenser.
21. The two main components that separate the high-and low-side pressures are the
    A. evaporator and condenser.
    B. compressor and accumulator.
    C. compressor and metering device.
    D. liquid receiver and accumulator.

22. An air conditioning system uses a TXV; its superheat setting should be

A. around 20–30°F.

B. more than 30°F.

C. 5–10°F.

D. checked and possibly reset.

23. A common evaporator temperature range on an air conditioning application is

A. 40–50°F.

B. 50–60°F.

C. 25–35°F.

D. No more than 60°F.

24. In the operation of an automatic expansion valve, as evaporator pressure rises, the valve will

A. open.

B. close.

C. open halfway.

D. underfeed the evaporator.

25. When changing a high-pressure vapor to a high-pressure liquid, which pressure is measured on the manifold gauge?

A. Superheated pressure

B. Subcooled pressure

C. Sensible heat pressure

D. Condensing pressure

26. The term subcooling pertains to

A. increasing the temperature of vapor refrigerant.

B. cooling the liquid refrigerant below its saturated temperature.

C. the temperature reduction of the compressor.

D. increasing the saturated liquid temperature.

27. The refrigerant in a thermal bulb of a TXV

A. is cooled by an increase of condensing temperature.

B. is the same refrigerant type as is within the system.

C. can be changed when replacing the TXV.

D. can be changed to match any refrigerant type.

28. If not sure how to charge refrigerant when using a capillary tube or fixed piston metering device, you should

A. check with the manufacturer for the correct method to use.

B. charge by the subcooling method.

C. charge by the correct suction line temperatures.

D. charge by the ambient temperature method.

29. The operating pressures of a TXV are

A. condensing, bulb, and spring.

B. spring, bulb, and evaporator.

C. suction, discharge, and bulb.

D. compressor, spring, and evaporator.

30. The proper installation of the thermal bulb of the TXV should be

A. just before the evaporator.

B. about 6 inches after the receiver.

C. at the center of the evaporator.

D. at the outlet of the evaporator.

31. The pressure from the metering device to the compressor is

A. the suction or evaporating pressure.

B. high-side pressure.

C. condensing pressure.

D. subcooling pressure.

32. Reading a temperature pressure chart applies to when the refrigerant is

A. in a vapor state.

B. in a liquid state.

C. in a subcooled state.

D. in a liquid and vapor state.

33. A TXV metering device that uses an external equalizer tube is used on

A. all medium-temperature systems.

B. air conditioning systems less than 3 tons capacity.

C. high-temperature refrigeration systems.

D. large evaporators that have large pressure drops.

34. A system that uses a capillary tube for a metering device is referred to as being

A. a flooded system.

B. a critical charged system.

C. an easy to charge system.

D. charged only by the subcooling method.

35. One method used to charge capillary tube systems is

A. the superheat method using a charging table.

B. the subcooling method.

C. using suction line temperature.

D. the ambient method.

36. Refrigeration systems that use an accumulator will

A. prevent liquid refrigerant from entering the compressor.

B. act as a check valve.

C. hold subcooled liquid refrigerant.

D. allow high-temperature liquid to cool.

37. Suction line heat exchangers are used to

A. increase superheat temperature.

B. help increase liquid subcooling.

C. increase high-side temperatures.

D. help increase suction temperatures.

38. Which of the following are the conditions within a refrigeration system?

A. Superheated, saturated, and subcooled states of refrigerants

B. Subcooled and high-side pressure of the refrigerant

C. Superheated and low-side pressure of the refrigerant

D. Refrigerant temperature and pressure of the refrigerant

39. The term subcooling is best described as

A. increasing the suction line temperature.

B. the further cooling of 100% saturated liquid refrigerant.

C. the rise in high-side pressure.

D. reducing the suction temperature.

40. Following a system evacuation, which of the following is the proper charging procedure?

A. Starting the compressor, then liquid charge the system

B. Slowly charging liquid into the low side of the system

C. Charging with a small amount of vapor followed by liquid into the high side

D. Starting to charge with a small amount of vapor into the receiver

41. When charging an air conditioning system with a fixed piston metering device, which two conditions should be known?

A. indoor wet-bulb and subcooling temperatures.

B. superheat and dry-bulb temperatures.

C. indoor wet-bulb superheat and outdoor dry-bulb temperatures.

D. indoor wet-bulb and dry-bulb temperatures.

42. A loose TXV thermal bulb off the suction line will result in

A. lower than normal suction pressure and lower head pressure.

B. high suction pressure and higher than normal discharge pressure.

C. low suction and low head pressure.

D. low suction pressure and very high head pressure.

43. The state of refrigerant as it passes through the first rows of the condenser is

A. de-superheated.

B. a vapor until it starts to change state.

C. vapor and liquid.

D. dropping in temperature.

44. Superheat suppression is best described as

A. the process of injecting liquid refrigerant into the receiver.

B. the process of injecting liquid refrigerant into the suction line.

C. the process of injecting liquid refrigerant into the accumulator.

D. the process of injecting liquid refrigerant into the discharge line.

45. One great advantage of using a TXV metering device over a capillary tube is that

A. the TXV will never flood the evaporator.

B. the TXV can act as a check valve.

C. the TXV can adapt to load changes and control superheat more efficiently.

D. the TXV eliminates flash gas.

46. The function of liquid pressure amplification is

A. to help keep a higher head pressure.

B. to increase superheat.

C. to help eliminate flash gas and increase subcooling.

D. to lower suction line temperature.

47. The advantage of a balance port TXV is that

A. the TXV contains a needle valve that will cancel liquid pressure effects.

B. the TXV will help lower suction pressures.

C. the balance port TXV can be used on all applications.

D. the balance port TXV senses low liquid line pressures.

48. The AXV (automatic expansion valve) is best used

A. on systems where the load remains somewhat constant.

B. on systems with varying load conditions.

C. on large tonnage systems.

D. on systems using a semi-hermetic compressor.

49. Temperature differences before and after a filter drier would indicate

A. an overcharged system.

B. average operating conditions.

C. restriction at the filter.

D. an undercharged system.

50. A system using a TXV, a receiver, and a sight glass should be charged

A. under a low load condition.

B. by superheat.

C. under a high load condition.

D. under a cool temperature condition.

THE FOLLOWING QUESTIONS MAY BE HELPFUL TO STUDY FOR THESE CERTIFICATION TESTS:
NATE—CR-SV, CR-IN, AC-SV, AC-IN, HP-SV, HP-IN, RC-SV, RC-IN, RL-SV, RL-IN.
ESCO INSTITUTE HVAC EXCELLENCE EPA SECTION 608 CERTIFICATION

EPA SECTION 608 CERTIFICATION CORE SECTION

1. CFCs neither dissolve in water nor break down into compounds that dissolve in water, so they do not rain out of the atmosphere.

A. True

B. False

2. How many ozone molecules will one chlorine atom destroy?

A. 1

B. 100

C. 100,000

3. What does ODP stand for?

A. outdoor pollutants

B. ozone depletion potential

C. ozone distillation process

4. What is ODP?

A. the measurement of the density of the stratosphere

B. the measurement of the ozone layer

C. the measurement of CFC's and HCFC's ability to destroy ozone

5. Which refrigerant is the most harmful to the ozone layer?

A. R-12 CFC

B. R-22 HCFC

C. R-134A HFC

6. The Clean Air Act

A. prohibits venting of CFCs and HCFCs as of November 15, 1995.

B. prohibits venting of CFCs and HCFCs as of July 1, 1992.

C. prohibits venting of CFCs and HCFCs as of November 14, 1994.

7. The Clean Air Act

A. prohibits venting of HFCs as of November 15, 1995.

B. prohibits venting of HFCs as of July 1, 1992.

C. prohibits venting of HFCs as of November 14, 1994.

8. The Clean Air Act

A. requires a service technician to be EPA certified as of November 15, 1995.

B. requires a service technician to be EPA certified as of July 1, 1992.

C. requires a service technician to be EPA certified as of November 14, 1994.

9. When disposing of a refrigerant cylinder,

A. vent the refrigerant to the atmosphere and remove the top.

B. recover the refrigerant and render the cylinder useless.

C. evacuate the cylinder and use as an air tank.

10. When is it most likely that an appliance will need its refrigerant recovered?

A. before it can be discarded

B. after purging the system with dry nitrogen

C. after the system obtained a leak

11. What is known as *de minimis* releases?

A. It is a maximum amount of refrigerant that should not be released to the atmosphere when servicing equipment.

B. It is a maximum amount of refrigerant that can be released to the atmosphere when servicing equipment.

C. It is a minimum amount of refrigerant that must be recovered when servicing equipment.

12. What does the term *temperature glide* refer to?

A. the different refrigerant boiling and condensing points of a blended refrigerant

B. zeotropic refrigerant as it is charged into a system as a gas

C. the point at which the refrigerant temperature decreases as it exits an expansion valve

13. Service technicians who violate the Clean Air Act

A. may be fined.

B. may lose certification.

C. Both A and B

14. How much of a fine can be issued for venting refrigerants?

A. $5,000

B. $10,000

C. $44,539

15. An award of _______ is given to the person who supplies information about the venting of refrigerants.

A. $5,000

B. $10,000

C. $27,500

16. Fractionation is the separation of blended refrigerants caused by different refrigerant vapor pressures.

A. True

B. False

17. What is the pressure range of the high-side gauge that is located on a manifold set?

A. 30″ Hg to 300 psig

B. 0 psig to 300 psig

C. 0 psig to 800 psig

18. Which of the following mixtures is legal to add into a system for the purpose of leak testing?

A. a trace amount of R-12 and nitrogen

B. a trace amount of the refrigerant recommended in the system and nitrogen

C. a trace amount of R-22 and nitrogen

19. Hygroscopic oil is oil that

A. has a high attraction to moisture.

B. should be stored at ambient temperatures.

C. can be left open to the atmosphere.

20. What is the state of the refrigerant leaving the condenser?

A. low-pressure liquid

B. high-pressure liquid

C. low-pressure gas

21. What is the color code of refrigerant gauges?

A. high side is blue, low side is red

B. low side is blue, high side is red

C. low side is white, high side is red

D. high side is white, low side is red

22. Which of the following is considered recycling?

A. removing refrigerant from a system into an approved container

B. removing, filtering, and reusing refrigerant in a system

C. removing refrigerant from a system and transferring it to a center for a chemical analysis

23. Refrigerant cylinders should be

A. free of rust.

B. nondented and undamaged.

C. Both A and B

24. Which leak detector will help locate the general area of a system's leak?

A. electronic/ultrasonic

B. soap bubbles

C. halide torch

25. Which of the following makes it impossible to reclaim a refrigerant?

A. having air in the refrigerant

B. having acid in the refrigerant

C. having two different refrigerants in one container

26. Factors that affect the speed of evacuation are

A. size of equipment.

B. ambient temperature.

C. amount of moisture in the system.

D. All of the above

27. To prevent an explosion, never fill a refrigerant cylinder more than

A. 50% full.

B. 60% full.

C. 80% full.

D. 90% full.

28. Refrigerant can cause suffocation because

A. it is lighter than air and it displaces oxygen.

B. it is heavier than air and it displaces oxygen.

C. the chlorine causes a reaction to the oxygen molecule.

29. The Montreal Protocol is

A. a treaty regulating the production of CFCs.

B. a method of measuring refrigerant CFCs.

C. a method of measuring the CFCs in the stratosphere.

30. Which refrigerant is an HCFC?

A. R-12

B. R-22

C. R-134A

31. Which refrigerant is an HFC?

A. R-12

B. R-22

C. R-134A

32. A ternary blend is

A. a three-part blend.

B. a two-part blend.

C. an azeotropic mixture.

33. What type of oil does a ternary blend use?

A. polyol ester

B. mineral

C. alkylbenzene

34. Which of the following is considered a "system-dependent" (passive) recovery?

A. using the appliance compressor to remove refrigerant

B. using an EPA-approved recovery machine

C. None of the above

35. Which of the following is considered a "self-contained" (active) recovery?

A. using the appliance compressor to remove refrigerant

B. using an EPA-approved recovery machine

C. None of the above

36. A mechanical float device, an electronic shutoff device, and weight methods are devices used to

A. charge a system.

B. evacuate a system.

C. safely recover refrigerants.

37. A CFC and an HCFC have the same ozone depletion potential.

A. True

B. False

38. When pressure testing a system, never exceed

A. high-side test pressures.

B. low-side test pressures.

C. 100 psig.

39. Which agency has to approve refrigerant cylinders before they are transported?

A. AHRI

B. EPA

C. DOT

40. To prevent an explosion, never pressure test a system using

A. CO_2.

B. oxygen.

C. nitrogen.

41. A system is evacuated to 29.92″ Hg and the pump is turned off. The pressure increases and stops at 25″ Hg. What can be said about the system?

A. It has a leak.

B. The oil has vaporized.

C. It has moisture or refrigerant present in the system.

42. As of ________, the EPA requires technicians performing a sealed system repair to be certified.

A. November 15, 1993

B. July 1, 1992

C. November 14, 1994

43. When does the EPA allow for the venting of refrigerants?

A. When the EPA determines that venting the refrigerant has no effect on the environment such as HC refrigerants.

B. All refrigerants are never to be vented.

C. When HCFC and HFC refrigerants are vented.

44. A "self-contained" recovery device may only be used if the refrigeration equipment contains less than

A. 15 pounds of refrigerant.

B. 20 pounds of refrigerant.

C. 25 pounds of refrigerant.

45. Which oil can polyol ester oils be mixed with?

A. mineral oils

B. alkylbenzene oils

C. never mix polyol ester oils with any other oil.

46. What is temperature glide?

A. It is the temperature difference between the vapor and liquid state of a near-azeotropic, a zeotropic, and a ternary blend refrigerant that occurs at a constant pressure during evaporation and the condensation process.

B. A refrigerant that has a blend of bubble or dew points at a constant pressure.

C. Both A and B.

47. Which of the following refrigerants can be recovered in one cylinder?

A. HFC with HCFCs

B. CFCs with HFCs

C. Only recover one type of refrigerant into one cylinder at a time.

48. Always prevent trapping liquid refrigerant in the service hose between the closed service valve after

A. a liquid transfer between the recovery unit and the HVAC/R system.

B. a vapor transfer between the recovery unit and the HVAC/R system.

C. an evacuation from the entire HVAC/R system.

49. When recharging any flammable hydrocarbon or HFO refrigerant into a system, it is important to

A. always vent the remaining refrigerant to the atmosphere.

B. ensure that the refrigeration system is electrically grounded.

C. first add nitrogen to the system.

50. Which refrigerants are HFCs and require polyol ester oils?

A. R-134a and R-410A

B. R-404A and R-407C

C. All of the above

51. Which refrigerant is the baseline measurement for low global warming potential (GWP) refrigerants?

A. R-410A

B. R-744

C. R-600

52. All new refrigerants must be approved to which branch of the EPA?

A. CRACK

B. SNACK PACK

C. SNAP

53. Even though HFOs have no ozone depletion potential, they do have a very small effect on

A. GWP.

B. SNAP.

C. GPS.

54. HFOs are less flammable than hydrocarbon (HC) refrigerants category A3, but most are mildly flammable; they are classified under a new category, nontoxic slightly flammable, which is considered

A. A2L category.

B. B2L category.

C. B3 category.

55. Which HFO refrigerant is used on low-pressure chillers?

A. R-1234abc

B. R-1234zd

C. R-1234ze

56. Which HFO refrigerant is used on high-pressure chillers?

A. R-1234abc

B. R-1234zd

C. R-1234ze

57. What type of ODP and GWP do HC refrigerants have?

A. low ODP and low GWP

B. high ODP and low GWP

C. low ODP and high GWP

58. Which of the two are HC refrigerants?

A. R-744 and R-717

B. R-600 and R-290

C. R-290 and R-744

59. Why should propane cylinders for grilling never be used as a refrigerant in an R-290 system?

A. impurities in the propane can damage refrigeration equipment.

B. because they are more flammable.

C. because they are an A2L type of refrigerant.

60. What is the danger of working with a concentration of HFO and HC refrigerants in a confined space?

A. They are heavier than air.

B. If there is an ignition source, there could be an explosion.

C. They have a specific gravity that is less than a CFC refrigerant.

61. What is the indication of a flammable refrigerant on an HFO or HC refrigeration system?

A. a red color marking on all process tubes

B. a yellow caution marking on all process tubes

C. a flammable sticker connected to the compressor

62. An azeotrope blend is a combination of two or more refrigerants that

A. when blended, create one refrigerant that has its own characteristics, so it will not fractionize, such as zeotropes or near azeotropic mixtures (NARMs).

B. when blended, create one refrigerant that has two or more characteristics, so it will fractionize, such as zeotropes or near azeotropic mixtures (NARMs).

C. when blended, create a mixture that will fractionize.

63. When disposing of appliances with 5–50 pounds of refrigerant, which records does the EPA requires be kept?

A. type of refrigerant recovered and quantity of refrigerant

B. quantity recovered from disposed appliances in each calendar month, where it is sent for reclamation

C. all of the above

64. The acronym SNAP used with the new EPA program that identifies and approves new and substitute refrigerant stands for

A. Significant New Alternatives Policy.

B. Sustainable New Alternative Policy.

C. Substitutable New Alternatives Policy.

65. An HVAC/R servicing company must keep records for each disposed appliance with a charge between 5 and 50 pounds for

A. 1 year.

B. 2 years.

C. 3 years.

THE FOLLOWING QUESTIONS MAY BE HELPFUL TO STUDY FOR THESE CERTIFICATION TESTS:
NATE—CR-SV, CR-IN, AC-SV, AC-IN, HP-SV, HP-IN, RC-SV, RC-IN, RL-SV, RL-IN.
ESCO INSTITUTE HVAC EXCELLENCE EPA SECTION 608 CERTIFICATION

TYPE I CERTIFICATION

1. According to the EPA, a small appliance is

A. a system containing under 3 lb of refrigerant including PTACs.

B. a system containing 5 lb of refrigerant or less including PTACs.

C. a system containing 5 lb of refrigerant or less not including PTACs.

2. Under conditions of ARI 740-1993, recovery devices manufactured *before* November 15, 1993, must

A. be able to remove 80% of the refrigerant charge or achieve 4″ Hg vacuum.

B. be able to remove 90% of the refrigerant charge or achieve 4″ Hg vacuum, if the compressor is in operation.

C. be able to remove 80% of the refrigerant charge or achieve 4″ Hg vacuum, if the system's compressor is not operating.

3. Under conditions of ARI 740-1993, recovery devices manufactured *after* November 15, 1993, must

A. be able to remove 80% of the refrigerant charge or achieve 4″ Hg vacuum.

B. be able to remove 70% of the refrigerant charge or achieve 4″ Hg vacuum, if the system's compressor is in operation.

C. be able to remove 80% of the refrigerant charge or achieve 4″ Hg vacuum, if the system's compressor is not operating.

4. Technicians must certify their recovery equipment with the EPA after

A. November 15, 1993.

B. August 12, 1993.

C. November 14, 1994.

5. The EPA

A. does not require a leak in a small appliance to be repaired.

B. requires a leak in a small appliance to be repaired if it leaks within 6 months.

C. requires a leak in a small appliance to be repaired if it leaks within 1 year.

6. While recovering refrigerant, why is it necessary to install a piercing valve on both the high and low side of an appliance that has an inoperative compressor?

A. It will help drain the oil out of the system while recovering the refrigerant.

B. It will decrease the rate of refrigerant evaporation.

C. It will decrease the time of the refrigerant recovery.

7. When using a system-dependent (passive) recovery, the EPA requires

A. 80% of the refrigerant to be removed if the compressor is operating.

B. 80% of the refrigerant to be removed if the compressor is not operating.

C. 90% of the refrigerant to be removed if the compressor is not operating.

8. The pressure of an R-12 cylinder at 70°F ambient temperature will be

A. 100 psig.

B. 70 psig.

C. 50 psig.

9. Which of the following refrigerants must not be recovered with a current recovery system?

A. R-12 CFC

B. R-502 azeotrope

C. R-717 ammonia

10. Which of the following refrigerants is considered a replacement for R-12 refrigerant?

A. R-22

B. R-141A

C. R-134A

11. Which of the following is a direct replacement for R-12 when no oil change or system adjustments are required?

A. R-134A

B. R-141A

C. None of the above

12. When pressurizing a system with nitrogen, it is required to have

A. at least 50 lb of pressure in the system.

B. a pressure regulator and a relief valve.

C. the system in operation.

13. After installing a piercing valve onto a system,

A. leak-test the valve.

B. remove the valve after servicing the system.

C. remove the Schrader before capping the valve.

14. If refrigerants are exposed to open flames and hot surfaces,

A. sulfuric and phosphoric acids are produced.

B. ammonia and hydrogen are produced.

C. phosgene gas and hydrochloric hydrofluoric acids are produced.

15. What may cause a pungent odor in a CFC/HCFC system?

A. a burnt compressor

B. an overcharge

C. a defective filter dryer

16. Which of the following is required if a large amount of R-12 has been released?

A. safety glasses

B. a dust mask and safety glasses

C. a self-contained breathing apparatus

17. At what temperature is R-22 when its pressure is 121 psig?

A. 70°F

B. 90°F

C. 105°F

18. At what pressure is R-502 when its temperature is 80°F?

A. 40 psig

B. 161 psig

C. 205 psig

19. Relief valves must never be installed

A. in parallel.

B. in series.

C. on liquid receivers.

20. The color code for refrigerant recovery tanks is

A. yellow bottoms with gray tops.

B. yellow tops with gray bottoms.

C. yellow tops and bottoms with gray in the middle.

21. Relief valves must always be installed

A. in parallel.

B. in series.

C. on liquid receivers.

22. Which of the refrigerants must not be recovered with a current recovery system that may be vented?

A. hydrocarbon refrigerants

B. HFO refrigerants

C. HFC refrigerants

23. A "self-contained" recovery device may only be used if the refrigeration equipment contains less than

A. 15 pounds of refrigerant.

B. 20 pounds of refrigerant.

C. 25 pounds of refrigerant.

THE FOLLOWING QUESTIONS MAY BE HELPFUL TO STUDY FOR THESE CERTIFICATION TESTS:
NATE—CR-SV, CR-IN, AC-SV, AC-IN, HP-SV, HP-IN, RC-SV, RC-IN, RL-SV, RL-IN.
ESCO INSTITUTE HVAC EXCELLENCE EPA SECTION 608 CERTIFICATION

TYPE II CERTIFICATION

1. Type II technicians are certified to service
   - **A.** high-pressure equipment only.
   - **B.** high-pressure and very-high-pressure equipment, including small appliances and motor vehicles.
   - **C.** high-pressure and very-high-pressure equipment, excluding small appliances and motor vehicles.
2. After the installation of a refrigeration system, the technician should first
   - **A.** evacuate the system.
   - **B.** pressure test the system using R-22.
   - **C.** pressure test the system using nitrogen.
3. What maintenance task must always be done to recover/recycle equipment?
   - **A.** Pressure test the system after every system recovery.
   - **B.** Vacuum test the equipment after every system recovery.
   - **C.** Change the system's oil and filters on a daily basis.
4. What type of leak detection device is used to pinpoint a leak?
   - **A.** an electronic leak detector
   - **B.** a halide leak detector
   - **C.** soap bubbles
5. An open compressor that has sat for several months is most likely
   - **A.** not to operate because of seized bearings.
   - **B.** to leak from the shaft seal.
   - **C.** to need a hard start kit.
6. Two indications that a high-pressure system has a leak are
   - **A.** oil traces on flared fittings and high superheat.
   - **B.** low superheat and bubbles in the sight glass.
   - **C.** high head pressure and low superheat.
7. The EPA requires that all comfort cooling systems with more than 50 lb of refrigerant be repaired when the annual leak rate exceeds
   - **A.** 10%.
   - **B.** 20%.
   - **C.** 30%.
8. The EPA requires that all commercial refrigeration systems with more than 50 lb of refrigerant be repaired when the annual leak rate exceeds
   - **A.** 10%.
   - **B.** 20%.
   - **C.** 30%.
9. Why is it likely for a recovery and recycle machine to overheat when drawing a deep vacuum?
   - **A.** because a deep vacuum causes high head pressure
   - **B.** because the refrigerant is needed to cool the compressor
   - **C.** because the compressor oil is evaporated
10. What component is directly after the condenser on a system that uses a TXV-type metering device?
    - **A.** the TXV
    - **B.** the receiver
    - **C.** the filter dryer
11. Which refrigerant requires a separate set of gauges, a vacuum pump, and a recovery machine when servicing?
    - **A.** R-12
    - **B.** R-500
    - **C.** R-134A
12. What methods help to reduce recovery time of a system?
    - **A.** Pack ice around the recovery tank, heat the appliance, and recover liquid first.
    - **B.** Pack ice around appliance, heat the recovery tank, and recover gas first.
    - **C.** Pack ice around the recovery tank, heat the appliance, and recover gas first.
13. During a recovery, when should the refrigerant be removed from the condenser outlet?
    - **A.** when the condenser is below the receiver
    - **B.** when the condenser is above the receiver
    - **C.** when the condenser is below the compressor
14. When recovering refrigerant from a system that has a condenser on the roof and the evaporator on the first floor, the recovery should begin from
    - **A.** the suction line entering the evaporator.
    - **B.** the discharge line exiting the compressor.
    - **C.** the liquid line entering the evaporator.
15. After removing refrigerant from a system, the refrigerant may
    - **A.** be transported to another system on the premises.
    - **B.** be reused in the same system or on another system owned by the same person.
    - **C.** not be reused.
16. What must be done to an empty recovery cylinder before transferring refrigerant into the cylinder?
    - **A.** It must be evacuated.
    - **B.** It must be charged with nitrogen.
    - **C.** It must be purged.

17. What is commonly used to aid in minimizing the release of refrigerant to the atmosphere?

A. low-loss fittings and hand valves on the hoses

B. prepurged hoses and Schrader valves

C. vacuum pumps and purge valves

18. When using recovery and recycling equipment manufactured after November 15, 1993, technicians must evacuate an appliance containing more than 200 lb of HCFC-22 to what level before making a major repair?

A. 0 psig

B. 4″ Hg

C. 10″ Hg

19. When using recovery and recycling equipment manufactured before November 15, 1993, technicians must evacuate an appliance containing less than 200 lb of HCFC-22 to what level before making a major repair?

A. 0 psig

B. 4″ Hg

C. 10″ Hg

20. If a leak in an appliance makes the prescribed evacuation level unattainable, the appliance can be evacuated to

A. 0 psig.

B. 4″ Hg.

C. 10″ Hg.

21. Which of the following repairs is considered a major repair?

A. the replacement of an evaporator fan

B. the installation of a low-pressure control

C. the replacement of a condenser coil

22. When servicing an open system, which of the following should be replaced?

A. a filter dryer

B. a sight glass

C. a pressure control

23. In a refrigeration system, high-pressure liquid exits

A. the condenser and the metering device.

B. the receiver and the metering device.

C. the condenser and receiver.

24. When evacuating a system, it is required that the vacuum pump pull a minimum of

A. 500 microns.

B. 5,000 microns.

C. 1,000 microns.

25. A technician must never

A. front-seat the suction service valve while the compressor is in operation.

B. front-seat the king valve while the compressor is in operation.

C. operate a hermetic compressor under a deep vacuum.

26. No condensables (air) cause

A. low superheat.

B. high head pressure.

C. low suction pressure.

27. A technician must never

A. front-seat the discharge service valve while the compressor is in operation.

B. front-seat the king valve while the compressor is in operation.

C. back-seat the discharge service valve while the compressor is in operation.

28. Defined by ASHRAE Standard 15, a sensor and alarm are required for A–1 refrigerants (R-12, R-22, R-11)

A. to sense oxygen depletion.

B. to sense ozone depletion.

C. to sense CFCs.

29. A refrigeration system must always be protected by

A. a high-pressure control.

B. a safety head.

C. a pressure-relief device.

30. To prevent freezing when charging R-12 into an evacuated system, it must be in the vapor form to raise the pressure up to

A. 10″ Hg.

B. 0 psig.

C. 33 psig.

31. The sale of CFC and HCFC refrigerants has been restricted to certified technicians since

A. November 15, 1993.

B. July 1, 1992.

C. November 14, 1994.

32. Refrigerators built before 1950 often contain

A. R-12.

B. R-717.

C. SO_2.

33. The EPA requires that all industrial process refrigeration systems with more than 50 lb of refrigerant be repaired when the annual leak rate exceeds

A. 10%.

B. 20%.

C. 30%.

34. If an appliance with 50 or more pounds of a regulated refrigerant has exceeded the threshold leak rate, the equipment owner or operator has

A. 10 days to repair the appliance.

B. 20 days to repair the appliance.

C. 30 days to repair the appliance.

35. If an appliance with a charge of 200 pounds or more of a regulated refrigerant passes the initial verification leak test, a follow-up verification test must be conducted within

A. 10 days.

B. 20 days.

C. 30 days.

36. If after repairing a leak in an HCFC or HFC rack refrigeration system, the system fails the initial leak verification test, the technician has

A. 10 days to complete the repair or retrofit.

B. 20 days to complete the repair or retrofit.

C. 30 days to complete the repair or retrofit.

37. If a cold storage warehouse-operating appliance charged with HCFC or HFC refrigerants system leaks above the threshold rate, it will be allowed to operate without repair for

A. 10 months before it must be retrofitted or retired.

B. 12 months before it must be retrofitted or retired.

C. 20 months before it must be retrofitted or retired.

38. If the cold storage warehouse appliance system leak rate exceeds

A. 50% of its full charge within a year, a report must be submitted to the EPA describing the efforts to the leak repair.

B. 100% of its full charge within a year, a report must be submitted to the EPA describing the efforts to the leak repair.

C. 125% of its full charge within a year, a report must be submitted to the EPA describing the efforts to the leak repair.

39. What does the term mothballing a system refer to?

A. recovering all of the refrigerant in the system

B. isolating a leak within a refrigeration system

C. replacing a leaking refrigeration system

40. How many months can the repair deadline be extended to by temporarily shutting down the appliance and recovering the refrigerant from an isolated section or the entire system to atmospheric pressure?

A. 12 months

B. 15 months

C. 18 months

THE FOLLOWING QUESTIONS MAY BE HELPFUL TO STUDY FOR THESE CERTIFICATION TESTS:
NATE—CR-SV, CR-IN, AC-SV, AC-IN, HP-SV, HP-IN, RC-SV, RC-IN, RL-SV, RL-IN.
ESCO INSTITUTE HVAC EXCELLENCE EPA SECTION 608 CERTIFICATION

TYPE III CERTIFICATION

1. If EPA regulations change after a technician has been certified,

A. the technician will be grandfathered into the new changes.

B. this will only affect technicians that have not been certified.

C. it is the technician's responsibility to comply with future changes.

2. Why is it easier to have air enter gaskets and fittings on a low-pressure system than a high-pressure system?

A. because air is attracted to R-11 and R-123 more than R-12 and R-22

B. because a low-pressure system operates in a vacuum

C. because low-pressure systems often use centrifugal compressors

3. At what pressure does a rupture disc open on a low-pressure system?

A. at 10 psig

B. at 15 psig

C. at 20 psig

4. What is the easiest way to leak-test a low-pressure system?

A. connect gauges and add 15 psig of nitrogen

B. add warm water over the coil to raise the pressure to 10 psig

C. add R-22 and nitrogen to increase the pressure to 150 psig

5. Why is it important not to exceed 10 psig when leak-testing a low-pressure system?

A. because the relief valve on the receiver could blow

B. because the evaporator test pressure is 10 psig MAX

C. because the rupture disc, located on the evaporator, may fail

6. When using an electronic leak detector, where is a water box leak-tested after the water has been removed?

A. at the drain valve

B. at the rupture disc

C. at the inlet valve

7. What is used when leak-testing a water tube on a low-pressure chiller?

A. a hydrostatic tube kit

B. a hydrometer

C. a psychrometer

8. The EPA requires that all comfort cooling systems with more than 50 lb of refrigerant be repaired when the annual leak rate exceeds

A. 10%.

B. 20%.

C. 30%.

9. The EPA requires that all commercial refrigeration systems with more than 50 lb of refrigerant be repaired when the annual leak rate exceeds

A. 10%.

B. 20%.

C. 30%.

10. What pressure is a high-pressure cut-out set for on a recovery system that is used on low-pressure refrigeration systems?

A. 10 psig

B. 15 psig

C. 20 psig

11. When recovering R-11 or R-123 from a low-pressure system, always recover

A. liquid first.

B. vapor first.

C. vapor and liquid first.

12. What is commonly done to prevent a low-pressure system from freezing during evacuation?

A. Stop the water from running through the chiller tubes.

B. Run the water through the chiller tubes.

C. Remove the water from the chiller tubes.

13. What does the ASHRAE guideline 3-1990 state about a vacuum test?

A. If a 1-mmHg vacuum level rises to 2.5 mmHg, the system should be leak-tested.

B. If a 1-mmHg vacuum level rises to 2.5 mmHg, the system is not leaking.

C. If a 1-mmHg vacuum level rises to 2.5 mmHg, the chiller coil should be replaced.

14. How is refrigerant added to a low-pressure chiller system after an evacuation?

A. by the use of the king valve

B. by the use of the suction service valve

C. by the use of the evaporator charging valve

15. After a chiller system has had a deep system evacuation, what is commonly done to decrease the possibility of system freeze-up prior to a liquid charge?

A. Pressurize the system with liquid up to the 36°F saturation temperature.

B. Pressurize the system with vapor up to the 36°F saturation temperature.

C. Pressurize the system with liquid or vapor up to the 20°F saturation temperature.

16. When using recovery or recycling equipment manufactured or imported before November 15, 1993, the required level of evacuation for low-pressure equipment is

A. 25 mmHg absolute.

B. 25″ Hg.

C. 27″ Hg.

17. When using recovery or recycling equipment manufactured or imported on or after November 15, 1993, the required level of evacuation for low-pressure equipment is

A. 25 mmHg absolute.

B. 25″ Hg.

C. 27″ Hg.

18. What is used on low-pressure chillers to remove noncondensables from the system?

A. a vent system

B. an aerator

C. a purge unit

19. From where does the purge unit remove trapped air in a low-pressure system?

A. the end of the evaporator

B. the top of the receiver

C. the top of the condenser

20. The equipment room that contains a chiller using R-123 group B-1 refrigerant requires a(n)

A. oxygen deprivation sensor.

B. ozone depletion sensor.

C. refrigerant sensor.

21. It is important to recover the vapor refrigerant in a chiller after the liquid has been removed, since there is approximately ________ lb of vapor in a common 350-ton R-11 chiller.

A. 25

B. 50

C. 100

22. The EPA requires that all industrial process refrigeration systems with more than 50 lb of refrigerant be repaired when the annual leak rate exceeds

A. 10%.

B. 20%.

C. 30%.

23. If an appliance with 50 or more pounds of a regulated refrigerant has exceeded the threshold leak rate, the equipment owner or operator has

A. 10 days to repair the appliance.

B. 20 days to repair the appliance.

C. 30 days to repair the appliance.

24. If an appliance with a charge of 200 pounds or more of a regulated refrigerant passes the initial verification leak test, a follow-up verification test must be conducted within

A. 10 days.

B. 20 days.

C. 30 days.

25. If after repairing a leak in an HCFC or HFC rack refrigeration system, the system fails the initial leak verification test, the technician has

A. 10 days to complete the repair or retrofit.

B. 20 days to complete the repair or retrofit.

C. 30 days to complete the repair or retrofit.

26. If a cold storage warehouse-operating appliance charged with HCFC or HFC refrigerants leaks above, the threshold rate will be allowed to operate without repair for

A. 10 months before it must be retrofitted or retired.

B. 12 months before it must be retrofitted or retired.

C. 20 months before it must be retrofitted or retired.

27. If the cold storage warehouse appliance system leak rate exceeds

A. 50% of its full charge within a year, a report must be submitted to the EPA describing the efforts to the leak repair.

B. 100% of its full charge within a year, a report must be submitted to the EPA describing the efforts to the leak repair.

C. 125% of its full charge within a year, a report must be submitted to the EPA describing the efforts to the leak repair.

28. What does the term mothballing a system refer to?

A. recovering all of the refrigerant in the system

B. isolating a leak within a refrigeration system

C. replacing a leaking refrigeration system

29. How many months can the repair deadline be extended by temporarily shutting down the appliance and recovering the refrigerant from an isolated section or the entire system to atmospheric pressure?

A. 12 months

B. 15 months

C. 18 months

THE FOLLOWING QUESTIONS MAY BE HELPFUL TO STUDY FOR THESE CERTIFICATION TESTS:
NATE AC-SV, AC-IN, HP-SV, HP-IN, RC-SV, RC-IN, RL-SV, RL-IN.
ESCO INSTITUTE HVAC EXCELLENCE AIR CONDITIONING AND COMMERCIAL REFRIGERATION CERTIFICATION

REFRIGERANT PIPING

1. Connected between the compressor and the condenser is the

A. condensate line.

B. suction line.

C. inlet line.

D. discharge line.

2. Connected between the evaporator and the compressor is the

A. condensate line.

B. suction line.

C. inlet line.

D. discharge line.

3. Connected between the receiver and the metering device is the

A. condensate line.

B. suction line.

C. inlet line.

D. liquid line.

4. Connected between the metering device and the evaporator is the

A. discharge line.

B. suction line.

C. inlet line.

D. liquid line.

5. Connected between the condenser and the receiver is the

A. discharge line.

B. inlet line.

C. liquid line.

D. condensate line.

6. The line that is the hottest during operation is the

A. discharge line.

B. suction line.

C. liquid line.

D. condensate line.

7. The line that is the coldest during operation is the

A. discharge line.

B. suction line.

C. condensate line.

D. inlet line.

8. A horizontal discharge line should never have a velocity below

A. 300 fpm.

B. 500 fpm.

C. 800 fpm.

D. 1,000 fpm.

9. A vertical riser discharge line must be sized to have a minimum velocity of

A. 300 fpm.

B. 500 fpm.

C. 800 fpm.

D. 1,000 fpm.

10. Discharge line velocities must never exceed

A. 300 fpm.

B. 500 fpm.

C. 1,000 fpm.

D. 4,000 fpm.

11. To ensure proper oil return, a discharge line must be pitched ________ per 10′ of horizontal run away from the compressor down toward the condenser.

A. $\frac{1}{4}''$

B. $\frac{1}{2}''$

C. $\frac{3}{4}''$

D. 1″

12. A discharge line riser is required whenever the condenser is located

A. even with the compressor.

B. below the compressor.

C. above the compressor.

D. more than 10′ below the compressor.

13. A discharge line double riser is needed on refrigeration systems that use

A. rotary compressors.

B. single-stage compressors.

C. scroll compressors.

D. unloaded compressors.

14. A double riser used in the discharge line for full-load conditions can be used for velocities of

A. 500 fpm.

B. 800 fpm.

C. 900 fpm.

D. 1,000 fpm or more.

15. An oil trap is required at the base of a discharge line when the vertical rise is

A. 6′

B. 8′ or more.

C. 5′

D. less than 5′

16. A secondary oil trap must be placed in a discharge line vertical riser if it exceeds

A. 8′

B. 10′

C. 15′

D. 25′

17. A secondary oil trap must be placed in a discharge line riser approximately every

A. 8′ to 10′

B. 10′ to 12′

C. 15′ to 20′

D. 25′

18. The line that carries both refrigerant and oil mixture and is the least critical line is the

A. suction line.

B. discharge line.

C. condensate line.

D. liquid line.

19. Flash gas will form in the _______ line if it is undersized.

A. suction

B. discharge

C. liquid

D. condensate

20. Liquid-line friction loss is caused by

A. the resistance to flow from liquid line to components.

B. oversized filter driers.

C. high-efficiency condensers.

D. low head pressure.

21. Liquid-line static head losses cause a resistance to flow due to

A. the type of refrigerant.

B. the type of metering device.

C. the weight of the refrigerant that is flowing up a vertical riser.

D. too short of liquid-line length.

22. Which is not a result of flash gas in the liquid line?

A. reduced evaporator capacity

B. evaporator starvation

C. flooded coil

D. noisy metering device

23. A liquid-to-suction heat exchanger is used to

A. function as an accumulator.

B. reduce flash gas.

C. cool the suction line.

D. decrease subcooling.

24. A liquid-to-suction heat exchanger must be located

A. as close to the metering device as possible.

B. as close to the receiver as possible.

C. as far from the receiver as possible.

D. far from the system's accumulator.

25. When sizing refrigerant piping lines, the most critical would be

A. the discharge line.

B. the condensate line.

C. the liquid line.

D. the suction line.

26. To ensure proper oil flow, a horizontal suction line must never have a velocity below

A. 500 fpm.

B. 800 fpm.

C. 1,000 fpm.

D. 1,250 fpm.

27. Vertical suction lines must be sized to have a minimum refrigerant velocity of

A. 500 fpm.

B. 800 fpm.

C. 1,000 fpm.

D. 4,000 fpm.

28. A suction line must be pitched _______ per 10′ of horizontal run away from the evaporator down toward the compressor to ensure proper oil return.

A. $\frac{1}{4}''$

B. $\frac{1}{2}''$

C. $\frac{3}{4}''$

D. at least 1″

29. A suction line riser is required whenever the compressor is located

A. above the evaporator.

B. below the evaporator.

C. even with the evaporator.

D. It is not necessary if the suction line is slightly oversized.

30. A suction line double riser is needed for refrigeration systems that operate with _______ compressors.

A. rotary

B. scroll

C. single-stage

D. unloaders

31. When using a double riser in the suction line, the smaller line is used for unloaded conditions when the refrigerant flow is

A. less than 1,000 fpm.

B. 500 fpm.

C. more than 800 fpm.

D. more than 1,000 fpm.

32. A double riser in the suction line uses both the larger line and the smaller line for

A. partial-load conditions.

B. half-load conditions.

C. It depends on the type of metering device.

D. full-load conditions.

33. An oil trap may be used at the base of a suction line if the vertical rise is

A. less than 5′.

B. 6′.

C. more than 8′.

D. It is not needed if the line is slightly oversized.

34. A secondary oil trap must be placed in a suction line vertical riser if it exceeds

A. 8′.

B. 10′.

C. 15′ to 20′.

D. 25′.

35. A secondary oil trap must be placed in the suction line riser approximately every

A. 4′ to 8′.

B. 8′ to 10′.

C. 12′ to 15′.

D. 15′ to 20′.

36. An oversized suction line will cause a(n) _______ of oil return to the compressor due to the lack of refrigerant velocity.

A. increase

B. decrease

C. high temperature

D. satisfactory flow

37. An undersized suction line will cause the compression ratio to

A. decrease.

B. increase.

C. It has no effect if a TXV is used.

D. remain the same.

38. An oversized suction riser will cause

A. fast flow of oil return to the compressor.

B. more vapor and oil return to the compressor.

C. insufficient oil return to the compressor.

D. not enough superheat.

39. Suction lines must always be sized for saturation pressure drop equivalent to

A. 8°F.

B. 6°F.

C. less than 2°F.

D. 4°F.

THE FOLLOWING QUESTIONS MAY BE HELPFUL TO STUDY FOR THESE CERTIFICATION TESTS:
NATE—AC-SV, AC-IN, HP-SV, HP-IN, RC-SV, RC-IN, RL-SV, RL-IN.
ESCO INSTITUTE HVAC EXCELLENCE AIR CONDITIONING AND COMMERCIAL REFRIGERATION CERTIFICATION

REFRIGERATION ACCESSORIES

1. The component that is located on the discharge line and helps to minimize gas pulsations is called the

A. accumulator.

B. receiver.

C. heat exchanger.

D. hot gas muffler.

2. The _______ is installed in the liquid line and is responsible for removing moisture and foreign matter.

A. receiver

B. sight glass

C. solenoid

D. filter drier

3. The device that allows the technician to view the flow of refrigerant is called the
   A. sight glass.
   B. drier.
   C. solenoid.
   D. receiver.
4. Bubbles in the sight glass may indicate
   A. an overcharged system.
   B. refrigerant shortage.
   C. restriction.
   D. Both B and C
5. It is good practice to install a(n) ________ after a new compressor has been replaced, especially after a burnout.
   A. accumulator
   B. low-pressure control
   C. suction-line drier
   D. heat exchanger
6. Pressure valves, which are installed on systems, should be calibrated from 0 psi to no less than ________ above the normal operating suction and discharge pressures.
   A. 20–40 psi
   B. 40–60 psi
   C. 50–100 psi
   D. 100–120 psi
7. When pressure gauges are installed on refrigeration systems, they should be connected with
   A. solenoid valves.
   B. vibration absorbers.
   C. throttling valves.
   D. mufflers.
8. The type of valves that can divert the flow of refrigerant during service procedures are called
   A. manual shutoff valves.
   B. solenoid valves.
   C. check valves.
   D. king valves.
9. The type of valve that would use a packing gland to help ensure against leakage is called the
   A. manual shutoff valve.
   B. solenoid valve.
   C. check valve.
   D. king valve.
10. The type of valve that is located on the liquid line close to the condenser outlet is called the
    A. solenoid valve.
    B. king valve.
    C. queen valve.
    D. charging valve.
11. The king valve is located
    A. on the liquid receiver's inlet side.
    B. on the liquid receiver's outlet side.
    C. near the solenoid valve.
    D. near the check valve.
12. The component that functions as a storage tank for the system's refrigerant is called the
    A. accumulator.
    B. heat exchanger.
    C. receiver.
    D. separator.
13. The liquid receiver should be designed to hold liquid refrigerant, along with approximately ________% room for the expansion of vapors.
    A. 8–10
    B. 10–12
    C. 15–20
    D. 20–30
14. The type of valves that are considered to be safety valves are
    A. solenoid valves.
    B. receiver valves.
    C. relief valves.
    D. service valves.
15. The type of safety valve that will reseat itself once safe operating pressures are restored is called a
    A. solenoid valve.
    B. king valve.
    C. queen valve.
    D. pressure-relief valve.
16. The type of safety valve that contains a predetermined melting alloy is called the
    A. spring-loaded valve.
    B. EPR valve.
    C. CPR valve.
    D. fusible plug.
17. When a fusible plug opens, the system's refrigerant will be
    A. partially lost.
    B. one half lost.
    C. around three-fourths lost.
    D. entirely lost.
18. The check valve is designed to
    A. allow the flow of refrigerant in one direction only.
    B. allow the flow of refrigerant in two directions.
    C. lock refrigerant in the suction line.
    D. limit the amount of superheat.

19. Check valves are installed on multiple evaporator systems and are located on the

A. discharge line.

B. liquid line.

C. suction line of the warmest evaporator.

D. suction line of the coldest evaporator.

20. The valve that is located on the liquid line and is used to stop refrigerant flow to the evaporator is called the

A. CPR valve.

B. EPR valve.

C. check valve.

D. solenoid valve.

21. The valve that is controlled by a thermostat and can be used as part of a pump-down system is called the

A. CPR valve.

B. EPR valve.

C. check valve.

D. solenoid valve.

22. A(n) _______ is a device that is used to subcool liquid refrigerant.

A. accumulator

B. heat exchanger

C. receiver

D. check valve

23. The component that is installed on the suction line and helps to decrease flash gas is called the

A. accumulator.

B. heat exchanger.

C. receiver.

D. check valve.

24. The component that is installed on the suction line and is used to prevent liquid floodback to the compressor is called the

A. accumulator.

B. heat exchanger.

C. receiver.

D. check valve.

25. The refrigerant distributor is connected to a thermostatic expansion valve, and its purpose is to

A. flood the evaporator.

B. limit the refrigerant flow to the evaporator.

C. provide for proper refrigerant velocity to the evaporator.

D. provide little superheat.

26. The component that is located on the discharge line close to the compressor is called the

A. solenoid valve.

B. accumulator.

C. oil separator.

D. heat exchanger.

27. The component that helps to prevent the migration of oil and liquid refrigerant is called the

A. heat exchanger.

B. oil separator.

C. crankcase heater.

D. accumulator.

28. The valve that is installed on the suction line close to the evaporator and sometimes referred to as a "back pressure valve" is called the

A. check valve.

B. solenoid valve.

C. evaporator pressure regulator.

D. CPR valve.

29. The condenser pressure regulator is located

A. before the condenser.

B. after the condenser.

C. in series with the oil separator.

D. as close to the compressor as possible.

30. The condenser pressure regulator is used to

A. maintain adequate head pressure.

B. control suction superheat.

C. maintain low subcooling.

D. act as a hot gas bypass.

31. The oil failure safety control contains _______ bellows.

A. one

B. two

C. three

D. four

32. The function of the oil failure control is to

A. allow four more minutes of compressor run time.

B. allow five more minutes of compressor run time.

C. cycle the compressor off when oil pressure fails.

D. cycle the compressor on/off until a diagnosis is made.

33. The high-pressure control is a device that is responsive to high and unsafe head pressures; it is wired

A. in series with a solenoid valve.

B. in series with a compressor starter/contactor coil.

C. in series with the condenser fan motor.

D. in series with the crankcase heater.

34. The low-pressure control will stop the compressor when

A. suction pressure is stabilized.

B. suction pressure is 3 psi below the cut-in setting.

C. suction pressure falls too low.

D. enough superheat is established.

35. The high-pressure control protects against unsafe pressures. The contacts are

A. open on rise.

B. close on rise.

C. Low pressure is normally open.

D. High pressure is normally open.

THE FOLLOWING QUESTIONS MAY BE HELPFUL TO STUDY FOR THESE CERTIFICATION TESTS:
NATE—RC-SV, RC-IN, RL-SV, RL-IN.
ESCO INSTITUTE HVAC EXCELLENCE AIR CONDITIONING AND COMMERCIAL REFRIGERATION CERTIFICATION

MEDIUM- AND LOW-TEMPERATURE REFRIGERATION

1. Commercial case cabinets for low and medium temperatures are generally constructed of

A. plastic.

B. fiberglass.

C. porcelain and stainless steel.

D. plastic and stainless steel.

2. The type of insulation used in commercial cabinets is commonly

A. styrofoam.

B. blanket.

C. granule.

D. polystyrene or urethane.

3. To prevent condensation around the doors of case cabinets _______ are usually installed.

A. hot gas lines

B. mullion heaters

C. magnetic strips

D. felt fabric strips

4. Magnetic door gaskets are used to

A. prevent air leakage and infiltration.

B. prevent condensation.

C. prevent ice formation.

D. prevent fogging of windows.

5. The display case is designed to be either

A. chest or vertical with closed doors.

B. closed or open display type.

C. open or closed type with full viewing.

D. sliding or hinged door type.

6. The open display case uses a cold blanket of air that is

A. circulated beneath the merchandise.

B. circulated over the merchandise.

C. circulated on the sides of the cabinet.

D. blown through air inlets at the corners of the cabinet.

7. With open display cases, the evaporator is commonly installed at

A. the top of the case.

B. the front section of the case.

C. the sides of the case.

D. the bottom and back sections of the case.

8. A suitable location for an open display case would be

A. near an air duct.

B. close to an entryway.

C. near exhaust fans.

D. away from outdoor air movement.

9. Evaporators used on closed display cases are usually installed

A. on the back wall or top of the unit.

B. on the corner of the unit.

C. on the front of the unit.

D. on the bottom of the unit.

10. Frozen food display cases use evaporators that are of the forced-air type, and the compressors are located

A. with the case or in a remote area.

B. on rooftops only.

C. in a mechanical room only.

D. in areas that are remote to the outdoors.

11. Walk-in coolers or freezers use evaporators that are

A. gravity type.

B. bare-pipe type.

C. forced-air finned-and-tube type.

D. equipped with check valves.

12. Walk-in unit doors must have a(n) _______ on the inside of the cabinet that will allow anyone who might be trapped to reopen the door and get out.

A. alarm

B. light

C. safety latch

D. button switch

13. Some walk-in units use a fan control that will

A. allow the evaporator fans to continually operate.

B. turn evaporator fans off during defrost mode.

C. keep evaporator fans on during defrost.

D. turn evaporator fans off once the door is opened.

14. The condensate pan of a walk-in unit must be piped to the outside of the cabinet, and when temperatures are kept below freezing the drain line must be heated by

A. hot gas discharge.

B. electric heat.

C. heat exchanger.

D. compressor downtime.

15. The refrigerant piping of walk-in units is constructed of

A. steel and copper.

B. aluminum and copper.

C. hard-drawn copper and possibly with precharged lines.

D. annealed copper tubing.

16. The ice temperature at which ice machines operate while making ice is

A. −10°F.

B. 0°F.

C. 10°F

D. 20°F.

17. The harvest cycle of an ice machine is initiated by

A. a thermostat control.

B. a low-pressure control.

C. a timer clock.

D. an increase in temperature.

18. The ice machine that produces ice in the form of flakes uses

A. inverted cups.

B. an auger and gear motor.

C. linkage for water spray.

D. electric defrost.

19. Ice machines that produce ice in the form of cubes do not use

A. inverted cups.

B. spray linkage.

C. hot gas defrost.

D. auger and gear motor.

20. The harvest cycle most commonly used with ice machines is done by

A. timed shutdown.

B. electric heat.

C. hot gas.

D. hot water.

21. Ice machines occasionally will need cleaning and flushing out; this is accomplished with the use of

A. safe chemicals and clean water.

B. muriatic acid.

C. chlorine and water.

D. algicide treatment.

22. The ice machine is designed to operate where the location at ambient temperature is generally at

A. 40 to 115°F.

B. 50 to 125°F.

C. 25 to 100°F.

D. 50 to 145°F.

23. Compressors of vending machines are usually rated as

A. over 1 horsepower.

B. fractional horsepower.

C. hermetics with at least 1 horsepower.

D. semihermetics with potential relays.

24. Vending machines consist of different systems; one that is not part of a vending machine is

A. refrigeration.

B. refilling products.

C. dispensing.

D. coin return.

25. Generally, the electrical cord from a vending machine should plug into an electrical outlet rated at

A. 15 amps.

B. 20 amps.

C. 30 amps.

D. 35 amps.

26. Beverage coolers are designed to keep products at

A. slightly below freezing temperatures.

B. the same temperature as the evaporator.

C. above freezing temperatures.

D. consistently 45°F.

27. The evaporators of beverage coolers operate below freezing temperature; therefore, the system must include a(n)

A. check valve.

B. EPR valve.

C. defrost system.

D. CPR valve.

28. The metering device that commonly is used on beverage coolers is the

A. capillary tube.

B. automatic expansion valve.

C. thermostatic expansion valve.

D. TXV with an external equalizer.

29. Beverage coolers use forced-air evaporators that generally circulate the cold air

A. from the bottom of the evaporator below the products.

B. over the top of the products and back to the evaporator.

C. on sections of products at a time.

D. over the section first to be dispensed.

30. Beverage coolers that dispense ice into cups and prechill the entering water use

A. CPR valves.

B. automatic expansion valves.

C. two evaporators.

D. separate low-pressure controls.

31. Beverage coolers that refrigerate liquid and use a cylinder for carbonation locate the cylinder

A. at the top of the cabinet.

B. by the condensing unit.

C. behind the cabinet.

D. inside the vending machine.

32. Some water coolers use a large bottle that sits on top of the cooler that is then chilled. Another type of water cooler that dispenses chilled water through a bubbler is called the

A. pressure type.

B. temperature type.

C. bubbler type.

D. button type.

33. Water coolers use compressors that are rated at

A. fractional horsepower.

B. 1 hp.

C. 1–2 hp.

D. 1 hp semihermetic compressors.

34. The evaporators of water coolers are made of stainless steel, copper, or brass tanks and the refrigerant piping

A. enters into the tank.

B. is coiled on top of the tank.

C. is coiled at the bottom of the tank.

D. is coiled around the tank.

35. Ultra-low refrigeration is used to fast-freeze products for commercial use with temperatures around

A. −15°F.

B. −25°F.

C. −30°F.

D. −50°F.

36. With ultra-low-temperature refrigeration, when temperatures below 50°F are desired, generally

A. two-stage compression is used.

B. high-side floats are used.

C. low-side floats are used.

D. multiple evaporators are used.

37. Cascade systems may be used to maintain temperatures lower than (or as low as)

A. −20°F.

B. −30°F.

C. −160°F.

D. −40°F.

38. A popular solution used in transport refrigeration is

A. R-134a.

B. R-22.

C. R-502.

D. eutectic solution (brine).

THE FOLLOWING QUESTIONS MAY BE HELPFUL TO STUDY FOR THESE CERTIFICATION TESTS:
NATE—RC-SV, RC-IN, RL-SV, RL-IN.
ESCO INSTITUTE HVAC EXCELLENCE AIR CONDITIONING AND COMMERCIAL REFRIGERATION CERTIFICATION

DEFROST SYSTEMS

1. What type of evaporator coil is commonly used with a manual-shutdown type of defrost system?

A. spiral type

B. finned type

C. plate type

D. finned and plate type

2. When using a manual-shutdown defrost system, the evaporator coil usually operates

A. at 34 to 35°F.

B. below 32°F.

C. at 40 to 45°F.

D. at 50°F.

3. What control is commonly used to initiate a defrost on a pressure-operated defrost system?

A. a low-pressure control

B. a high-pressure control

C. a reverse-acting pressure control

D. a temperature pressure control

4. When adjusting a low-pressure switch that is used to control the frost buildup on an evaporator coil,

A. adjust the control to cut-in at a minimum of 33°F saturation temperature.

B. adjust the control to cut-in at a maximum of 33°F saturation temperature.

C. adjust the control to cut-in at 20°F saturation temperature.

D. adjust the control to cut-out at 32°F saturation temperature.

5. The typical storage temperature of a system that uses a pressure-operated defrost system is

A. 0°F.

B. 10°F.

C. 25°F.

D. above 35°F.

6. What is the method of defrost when using a pressure-operated defrost system?

A. an electric heater

B. a hot gas solenoid

C. warm water

D. evaporator fans

7. What is the length of the defrost period when using a pressure-operated defrost system?
   A. 20 min
   B. 30 min
   C. 40 min
   D. The length is determined by the cut-in and cut-out settings of the control.
8. When using a time shutdown defrost, what is commonly used to start and stop the defrost cycle?
   A. a temperature control
   B. trip pins
   C. a pressure control
   D. an electric heater
9. What type of switch contacts are located in the defrost timer and are used to control the condensing unit?
   A. a "normally closed" pressure type
   B. a "normally opened" pressure type
   C. a "normally closed" cam-operated type
   D. a "normally opened" cam-operated type
10. What should the length of time be when adjusting the stop and start times of a time shutdown defrost system?
   A. Always use the manufacturer's recommendation.
   B. Always set it for 20 min.
   C. Always set it for 30 min.
   D. Always set it for 40 min.
11. What are three common supplementary heat methods used to aid in the removal of frost on an evaporator coil?
   A. electric, radiant, and warm air convection
   B. hot gas, electric, and radiant
   C. warm water, electric, and warm air convection
   D. warm water, electric, and hot gas
12. When supplementary heat is used to aid frost removal during a defrost cycle, the
   A. products are commonly kept below 40°F.
   B. products are commonly kept above 32°F.
   C. products are commonly kept below 29°F.
   D. None of the above
13. On an electric defrost system using a 24-h defrost timer, what do the normally open contacts control?
   A. the defrost heater
   B. the evaporator fan
   C. the timer motor
   D. the condensing unit
14. On a hot gas defrost system using a 24-h defrost timer, what do the normally closed contacts control?
   A. the hot gas solenoid
   B. the evaporator fan
   C. the timer motor
   D. the condensing unit
15. When using a 24-h defrost timer that is used on systems with supplementary heat, the
   A. trip pins are used to start the defrost and an adjustable setting is used to end the defrost.
   B. trip pins are used to end the defrost and an adjustable setting is used to start the defrost.
   C. trip pins are used to start and end the defrost.
   D. adjustable settings are used to start and end the defrost without the use of trip pins.
16. What is the maximum amount of trip pins that can be installed in a 24-h defrost timer at one time?
   A. 12
   B. 24
   C. 36
   D. 48
17. On a 24-h timer, the length of the defrost cycle ranges from
   A. 20 to 40 min.
   B. 30 to 40 min.
   C. 10 to 90 min.
   D. 2 to 110 min.
18. During an electric, hot gas, or water defrost, what load must always be de-energized?
   A. the compressor
   B. the defrost timer
   C. the evaporator fan
   D. None of the above
19. On a hot gas defrost system, where does the defrost line connect?
   A. from the outlet of the condenser to the outlet of the evaporator
   B. from the inlet of the condenser to the outlet of the evaporator
   C. from the outlet of the condenser to the inlet of the evaporator
   D. from the inlet of the condenser to the inlet of the evaporator
20. What component is commonly added to a hot gas defrost system that is not on an electric or warm water defrost system?
   A. a reversing valve
   B. a drain valve
   C. a crankcase pressure regulator
   D. an accumulator
21. When using a time-initiated–temperature-terminated defrost system, what control is commonly used to de-energize the defrost circuit?
   A. a solenoid valve
   B. a switching relay
   C. a reverse-acting high-pressure control
   D. a termination thermostat

22. When a defrost system incorporates a termination thermostat, what temperature commonly controls its operation?

A. the evaporator coil temperature

B. the supply air temperature of the evaporator

C. the suction-line temperature

D. the return-air temperature of the evaporator

23. What are two common types of termination thermostats that are used with a time-initiated–temperature-terminated defrost system?

A. single-pole, double-throw and double-pole, single-throw types of controls

B. single-pole, single-throw and single-pole, double-throw types of controls

C. single-pole, single-throw and double-pole, single-throw types of controls

D. double-pole, single-throw and single-pole, double-throw types of controls

24. When a termination thermostat is used with a time-initiated–temperature-terminated defrost system, what is commonly used to override the timer motor?

A. a pressure control

B. a holding relay

C. a timer release solenoid

D. an override relay

25. When using a single-pole, double-throw termination thermostat, what loads are controlled by the switch contacts?

A. The "close on rise" contacts control the evaporator fan and the "open on rise" contacts control the compressor.

B. The "open on rise" contacts control the compressor and the "close on rise" contacts control the timer release solenoid.

C. The "open on rise" contacts control the defrost heater and the "close on rise" contacts control the defrost timer.

D. The "close on rise" contacts control the timer release solenoid and the "open on rise" contacts control the evaporator fan.

26. On a time-initiated–temperature-terminated defrost system, what would be the action of the system that is set for a 35-min defrost but the coil has defrosted after only 20 min?

A. The system is switched into the cooling mode after 20 min and the evaporator fan is delayed for a short period of time.

B. The system is switched into the cooling mode after 35 min, energizing the compressor, condenser fan, and evaporator fan.

C. The system is switched into the cooling mode after 20 min, energizing the compressor, condenser fan, and evaporator fan.

D. The system is switched into the cooling mode after 35 min and the evaporator fan is delayed for a short period of time.

THE FOLLOWING QUESTIONS MAY BE HELPFUL TO STUDY FOR THESE CERTIFICATION TESTS:

NATE—RC-SV, RC-IN, RL-SV, RL-IN.

ESCO INSTITUTE HVAC EXCELLENCE AIR CONDITIONING AND COMMERCIAL REFRIGERATION CERTIFICATION

COMMERCIAL REFRIGERATION ELECTRICAL DIAGRAMS

1. Using the temperature-controlled medium-temp cooler (Figure 1–1A, B), what would be the symptom of the system if the charge had leaked out of the TXV sensing bulb?

A. The system would shut down on low pressure.

B. The system would burn out the start capacitor.

C. The system would operate in a vacuum on the low side and overheat the compressor.

D. The system would have a low superheat and possible flood-back to the compressor could occur.

2. Using the temperature-controlled medium-temp cooler (Figure 1–1A, B), when is the start capacitor taken out of the circuit?

A. when the compressor motor reaches 75% of its operating speed

B. when the N.O. contacts of the current relay close

C. when the cooling thermostat is satisfied

D. when the overload protector closes its circuit

3. Using the temperature-controlled medium-temp cooler (Figure 1–1A, B), what type defrost system is being used?

A. an electric defrost

B. a timed shut-down defrost

C. an off-cycle defrost

D. None of the above

4. Using the temperature-controlled medium-temp cooler (Figure 1–1A, B), how can all of the refrigerant be stored in the receiver for service?

A. by front-seating the king valve

B. by front-seating the queen valve

C. by front-seating the suction service valve

D. by closing off the TXV

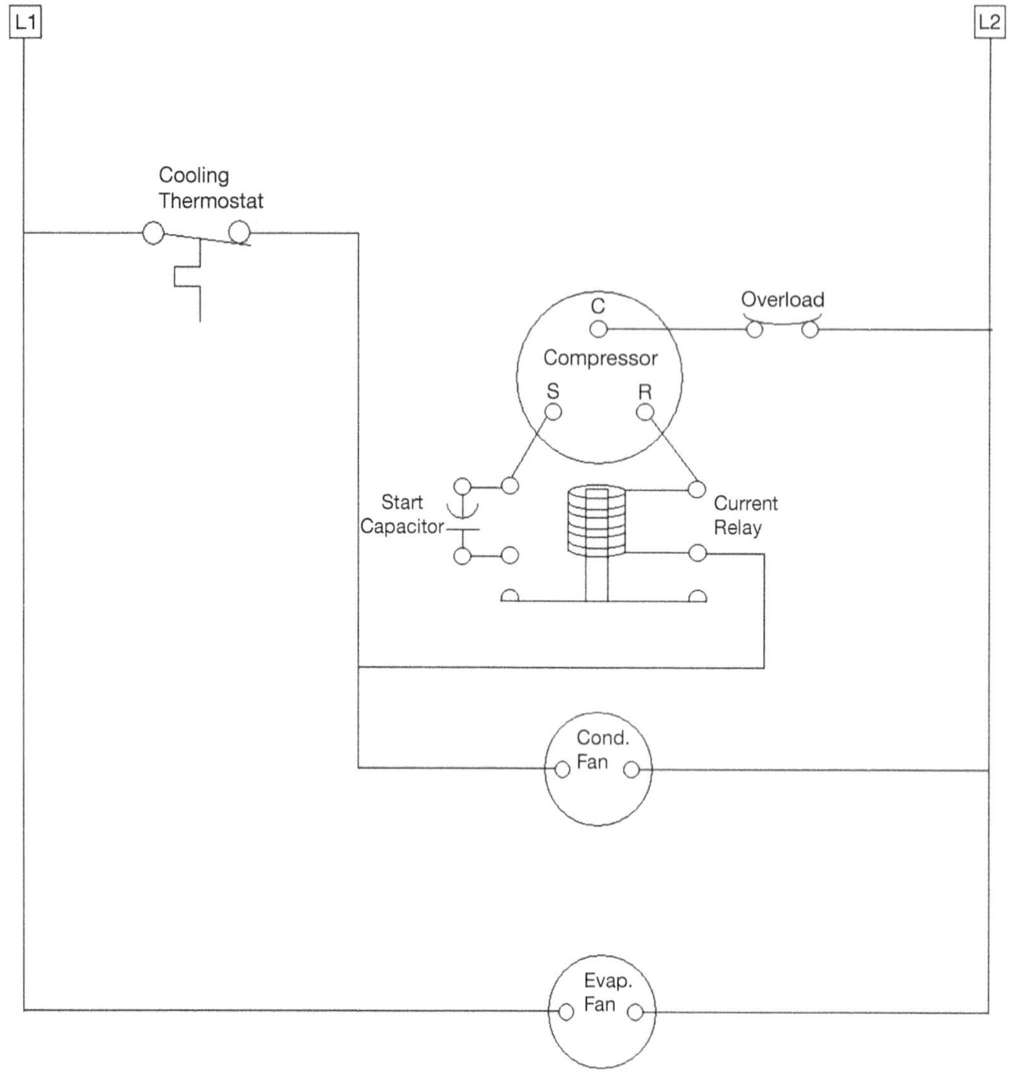

FIGURE 1–1A

Temperature-Controlled Medium-Temperature Cooler with a Capacitor Start Compressor Using a Current Relay

5. Using the temperature-controlled medium-temp cooler (Figure 1–1A, B), what would happen to the system if the cooling thermostat switch contacts were stuck in the open position?

   A. The compressor would shut down and there would be no cooling.

   B. The condenser fan would not operate.

   C. The cooler box temperature would drop below the 35°F box temperature.

   D. The evaporator fan would short cycle.

6. Using the temperature-controlled medium-temp cooler diagrams (Figure 1–1A, B), what type of compressor is being used in the circuit?

   A. a three-phase compressor

   B. a single-phase compressor

   C. a poly-phase compressor

   D. a dual-phase compressor

7. Using the temperature-controlled medium-temp cooler diagrams (Figure 1–1A, B), what are three mechanical system accessories?

   A. the TXV, sight glass, and filter dryer

   B. the compressor, TXV, and receiver

   C. the condenser, compressor, and evaporator

   D. the receiver, filter dryer, and sight glass

8. Using the temperature-controlled medium-temp cooler diagrams (Figure 1–1A, B), what operates the current relay?

   A. the start capacitor

   B. the LRA and FLA of the run winding

   C. the LRA of the start winding

   D. the overload protector

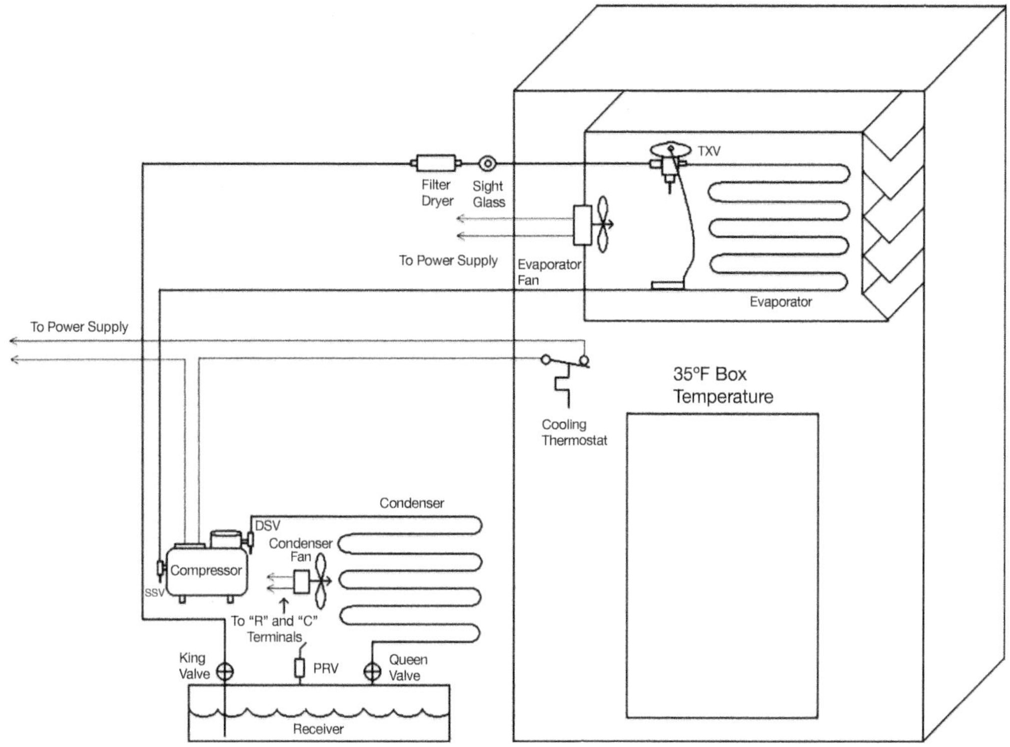

FIGURE 1–1B

Temperature-Controlled Medium-Temperature Cooler

9. Using the temperature-controlled medium-temp cooler diagrams (Figure 1–1A, B), which of the following valves should never be front-seated when the system is in operation?
   - **A.** the suction service valve
   - **B.** the discharge service valve
   - **C.** the queen valve
   - **D.** the king valve

10. Using the temperature-controlled medium-temp cooler diagrams (Figure 1–1A, B), if the system uses R-134a and the condenser split was 30 to 35°F what would the operating condensing temperature be if the ambient temperature were 75°F?
    - **A.** 75 to 80°F
    - **B.** 85 to 90°F
    - **C.** 95 to 100°F
    - **D.** 100 to 105°F

11. Using the pressure-controlled medium-temp cooler (Figure 1–2A, B), what would be the symptom of the system if the charge had leaked out of the TXV sensing bulb?
    - **A.** The system would shut down on low pressure.
    - **B.** The system would burn out the start capacitor.
    - **C.** The system would operate in a vacuum on the low side and overheat the compressor.
    - **D.** The system would have a low superheat and possible flood-back to the compressor could occur.

12. Using the pressure-controlled medium-temp cooler (Figure 1–2A, B), if the system used R-134a to achieve a cut-out point of 35°F with an evaporator split of 20°F what pressure would the cut-out point be on the low-pressure control? (This question requires a P/T chart.)
    - **A.** 30 psig
    - **B.** 20 psig
    - **C.** 18 psig
    - **D.** 15 psig

13. Using the pressure-controlled medium-temp cooler (Figure 1–2A, B), what type of defrost system is being used?
    - **A.** an electric defrost
    - **B.** a timed shut-down defrost
    - **C.** an off-cycle defrost
    - **D.** None of the above

14. Using the pressure-controlled medium-temp cooler (Figure 1–2A, B), the low-pressure control
    - **A.** always senses low-side pressures.
    - **B.** is used as a primary operating control in this system.
    - **C.** opens on a drop in low-side pressure only.
    - **D.** All of the above

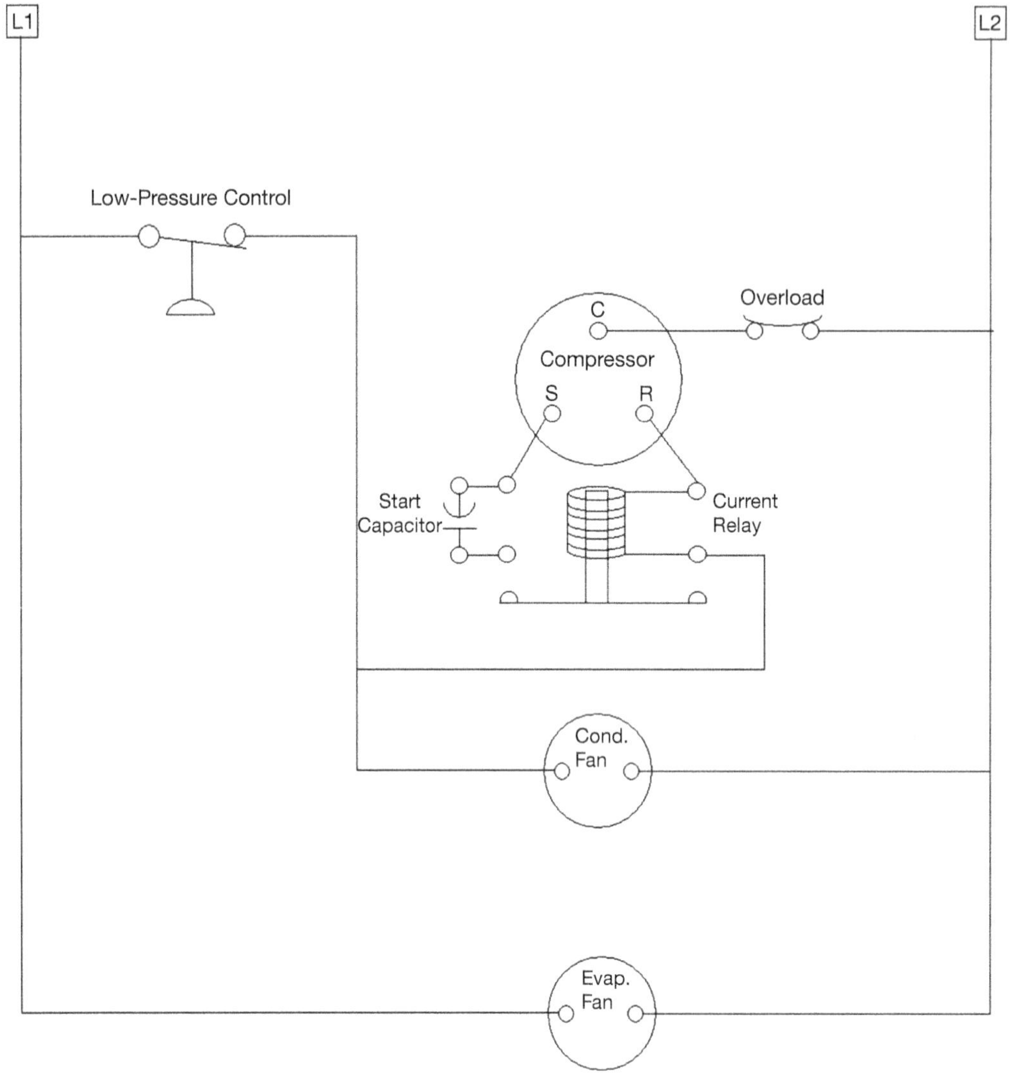

FIGURE 1–2A

Pressure-Controlled Medium-Temperature Cooler with a Capacitor Start Compressor Using a Current Relay

15. Using the pressure-controlled medium-temp cooler (Figure 1–2A, B), if the system used R-134a and a desired cut-in point of the low-pressure control were to correlate to a 38°F box temperature, to what cut-in pressure would the low-pressure switch be adjusted? (This question requires a P/T chart.)

A. 33 psig

B. 28 psig

C. 23 psig

D. 8 psig

16. Using the pressure-controlled medium-temp cooler (Figure 1–2A, B), if the system used R-134a and the cut-out point of the pressure control correlated to 35°F, what would the boiling point temperature of the evaporator be with a 20-evaporator split?

A. 55°F

B. 45°F

C. 25°F

D. 15°F

17. Using the pressure-controlled medium-temp cooler (Figure 1–2A, B), if the system used R-134a and the condensing pressure were 124 psig while the liquid line temperature was what 90°F, would be the system's subcool temperature?

A. 5°F

B. 10°F

C. 15°F

D. 20°F

18. Using the pressure-controlled medium-temp cooler (Figure 1–2A, B), what is the state of the refrigerant leaving the condenser?

A. low-pressure gas

B. high-pressure gas

C. low-pressure liquid

D. high-pressure liquid

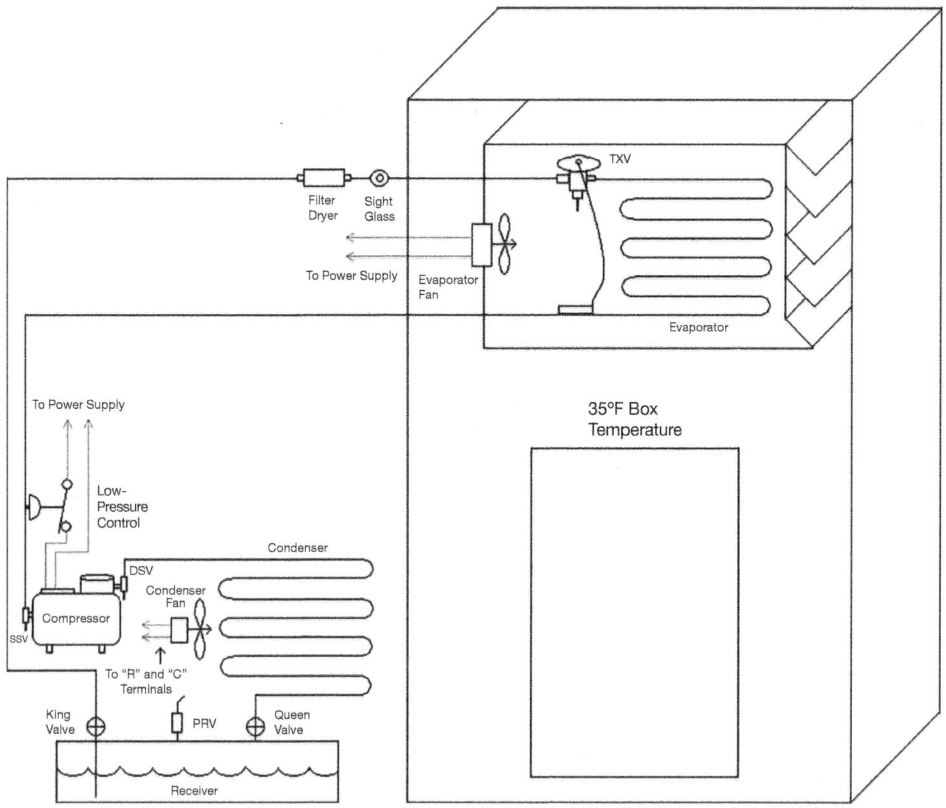

FIGURE 1–2B

Pressure-Controlled Medium-Temperature Cooler

19. Using the pressure-controlled medium-temp cooler (Figure 1–2A, B), if the sight glass had excessive bubbles, which of the following could cause this?

A. too much subcooling in the liquid line and a plugged filter dryer

B. not enough subcooling in the liquid line and a plugged filter dryer

C. a flooded condenser and excessive superheat

D. a flooded receiver and a TXV that lost its charge in the sensing bulb

20. Using the pressure-controlled medium-temp cooler (Figure 1–2A, B), if the sight glass had excessive bubbles, which of the following could cause this?

A. a TXV that lost its charge in the sensing bulb and too much subcooling in the liquid line

B. a receiver with a broken dip tube and a flooded condenser

C. low ambient condenser temperatures and a broken receiver dip tube

D. None of the above

21. Using the pressure-controlled medium-temp cooler (Figure 1–3A, B), which of the following would cause the compressor to short cycle on the low-pressure control?

A. an overcharge and a dirty condenser

B. an undercharge and a flooded evaporator coil

C. an undercharge and a differential adjustment that is set too low

D. a differential adjustment that is set too high and a faulty TXV

22. Using the pressure-controlled medium-temp cooler (Figure 1–3A, B), which of the following would cause the compressor to short cycle on the low-pressure control?

A. a flooded receiver and a TXV that lost its charge in the sensing bulb

B. low ambient temperatures and a partially restricted filter dryer

C. a differential that is set too low and a flooded condenser

D. low ambient condenser temperatures and a broken receiver dip tube

23. Using the pressure-controlled medium-temp cooler (Figure 1–3A, B), which two terminals on the potential relay, when energized, will cause the NC contacts to open?

A. terminals 1 and 2

B. terminals 2 and 5

C. terminals 1 and 5

D. Any of the above

24. Using the pressure-controlled medium-temp cooler (Figure 1–3A, B), which of the following will cause the compressor to start and then cycle on the overload protector?

A. contacts 1 and 2 stuck open

B. an open start capacitor

C. an open coil between terminals 2 and 5

D. All of the above

25. Using the pressure-controlled medium-temp cooler (Figure 1–3A, B), what symptoms would an inoperative evaporator fan cause in the system?

A. excessive frost on the evaporator coil

B. low suction pressure

C. insufficient cooling

D. All of the above

26. Using the pressure-controlled medium-temp cooler (Figure 1–3A, B), when is the start capacitor taken out of the circuit?

A. when the run winding produces the proper operating full-load amperage from the run winding

B. when the N.C. contacts of the potential relay close

C. when the pressure switch opens after the run cycle

D. when the start winding produces the correct amount of back EMF across terminals "2" and "5" on the potential relay

27. Using the pressure-controlled medium-temp cooler (Figure 1–3A, B), an ammeter is placed around the wire that is connected between terminal "1" and the start capacitor. What reading would be obtained if the relay and capacitor were operating properly on compressor motor startup?

A. The reading would be 0 amps.

B. The reading would be a continual starting amperage during motor operation.

C. The reading would be a measurable amperage, then it would drop to 0 amps.

D. The reading would be the running amperage of the motor.

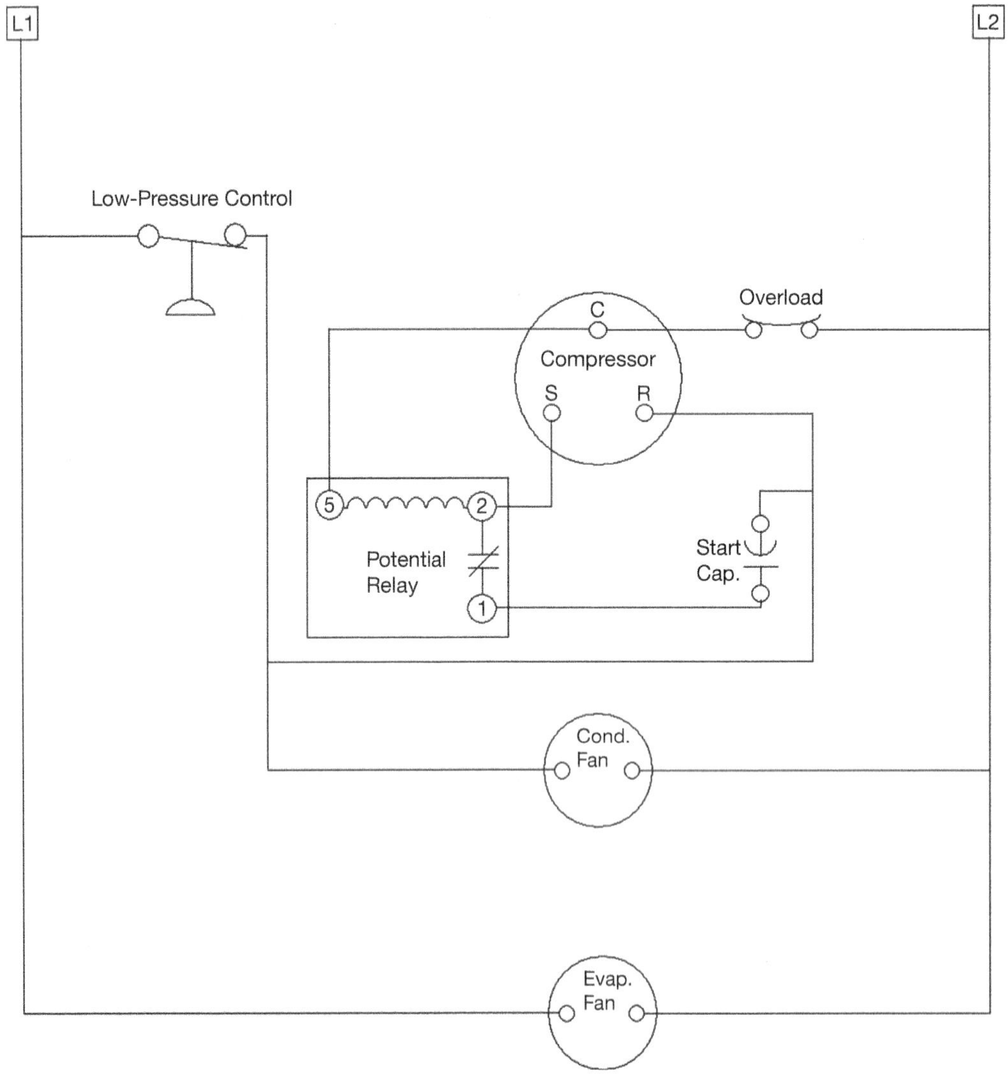

FIGURE 1–3A

Pressure-Controlled Medium-Temperature Cooler with a Capacitor Start Compressor Using a Potential Relay

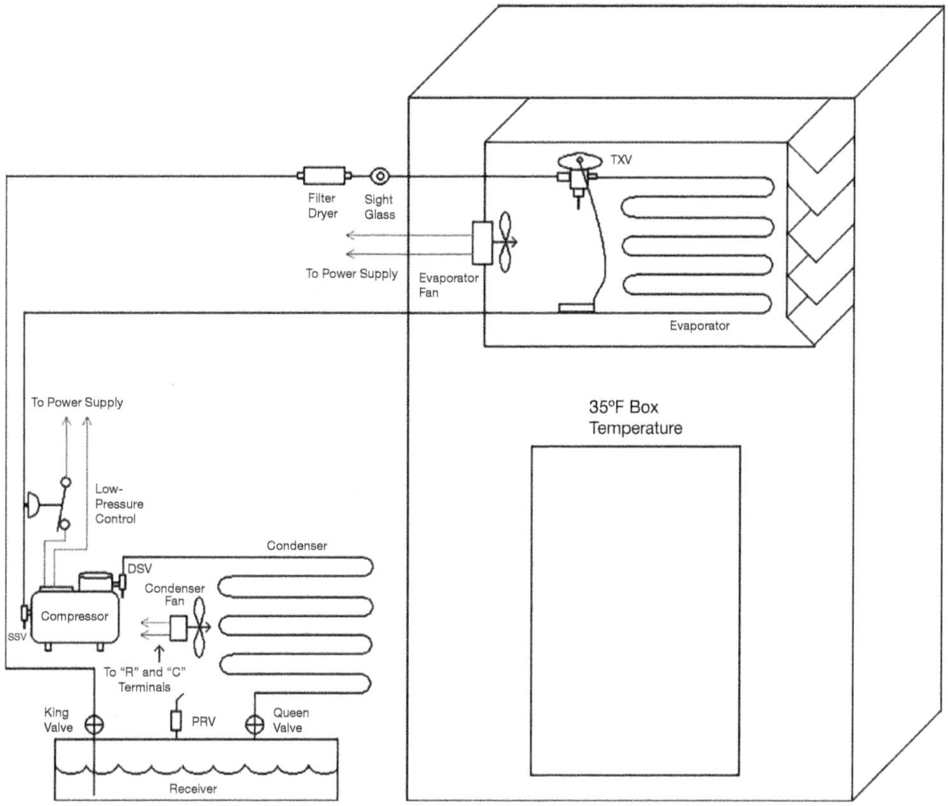

FIGURE 1–3B

Pressure-Controlled Medium-Temperature Cooler

28. Using the pressure-controlled medium-temp cooler (Figure 1–3A, B), if the system were using R-134a and the evaporator pressure were 22 psig and the sensing bulb temperature of the TXV were 30°F, what would have to be done to the TXV adjustment spring to properly adjust the evaporator superheat? (This question requires a P/T chart.)

A. Turn the TXV adjustment stem clockwise.

B. Turn the TXV adjustment stem counterclockwise.

C. Do not touch the adjustment spring.

D. Reposition the location of the TXV sensing bulb.

29. Using the pressure-controlled medium-temp cooler (Figure 1–3A, B), which of the following loads operates 24 hours a day?

A. the compressor

B. the evaporator fan

C. the condenser

D. the defrost timer

30. Using the pressure-controlled medium-temp cooler (Figure 1–3A, B), what could cause the compressor motor to hum without starting and then cause the overload to cycle to an open position?

A. an open relay coil between terminals "2" and "5" on the potential relay

B. the contacts of the potential relay terminals "1" and "2" stuck in the closed position

C. an open start capacitor

D. All of the above

31. Using the medium-temperature automatic pump-down cooler (Figure 1–4A, B), when does the compressor shut down?

A. when the thermostat opens and de-energizes the run winding

B. when the liquid-line solenoid is energized

C. when the low-pressure control opens

D. when N.C. switch contacts "1" and "2" on the potential relay open

32. Using the medium-temperature automatic pump-down cooler (Figure 1–4A, B), what is the purpose of the automatic pump-down system?

A. It prevents liquid floodback to the compressor during operation.

B. It aids in more precise control over cooler-box temperature.

C. It helps by increasing evaporator efficiency.

D. It prevents liquid migration during the off cycle.

33. Using the medium-temperature automatic pump-down cooler (Figure 1–4A, B), which of the following would cause the system to short cycle on the low-pressure control during the off cycle?

A. the thermostat cut-out point set too low

B. the pressure control cut-out set too low

C. a faulty start capacitor

D. a leaking liquid-line solenoid

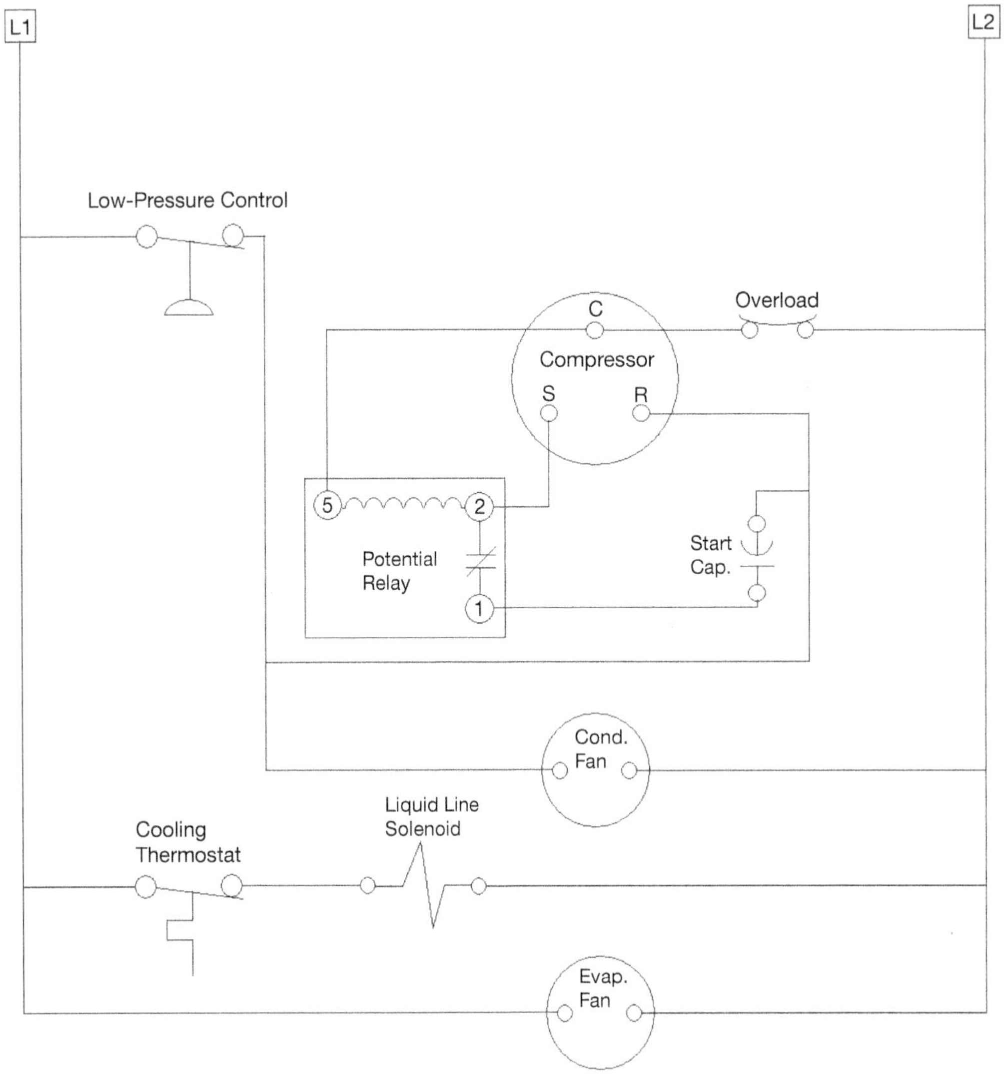

FIGURE 1–4A

Medium-Temperature Automatic Pump-Down System with a Capacitor Start Compressor Using a Potential Relay

34. Using the medium-temperature automatic pump-down cooler (Figure 1–4A, B), there is source voltage present at the liquid-line solenoid but the low-pressure switch is in the open position. What two problems could cause this condition?

A. a low system charge and a faulty cooling thermostat

B. a faulty liquid-line solenoid and a faulty potential relay coil

C. a liquid-line solenoid stuck in the open position and the TXV sensing bulb losing its charge

D. the TXV sensing bulb losing its charge and a low system charge

35. Using the medium-temperature automatic pump-down cooler (Figure 1–4A, B), the low-pressure control is set to cut-out at 2 psig and cut-in at 20 psig, the thermostat is set for 35°F, but the product in the cooler is freezing. What are some of the possible problems that could cause this condition?

A. The thermostat has failed in the open position or the low-pressure switch has failed closed.

B. The liquid-line solenoid has failed in the open position or the low-pressure switch has failed closed.

C. The thermostat has failed in the closed position or the liquid-line solenoid has failed closed.

D. None of the above

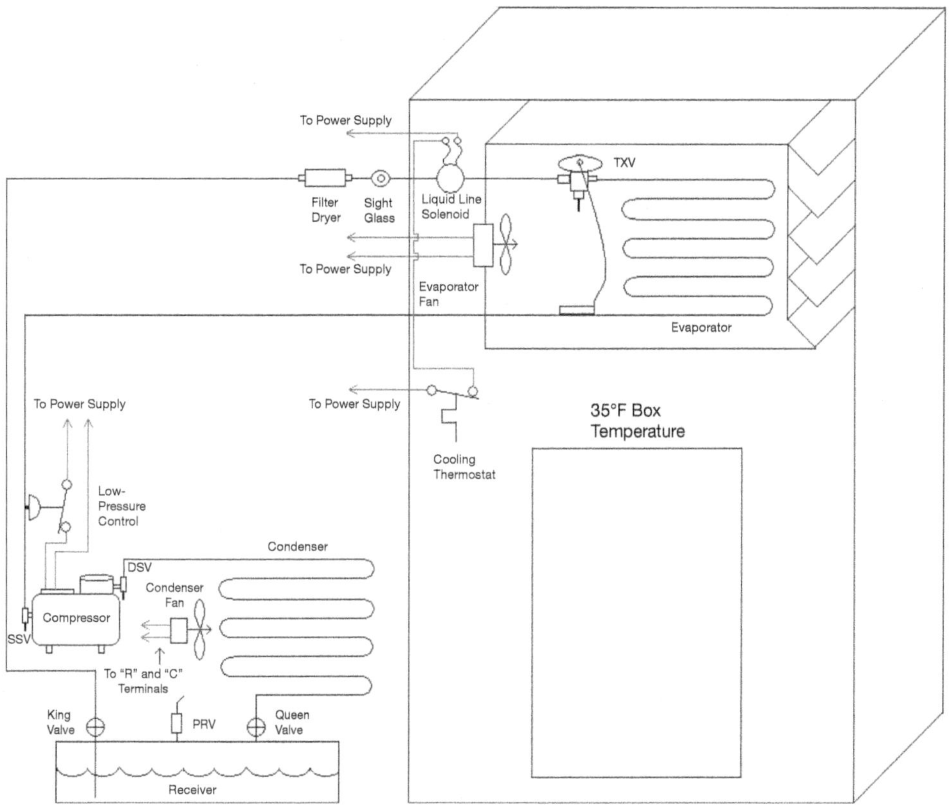

FIGURE 1–4B
Medium-Temperature Automatic Pump-Down System

36. Using the time shutdown defrost medium-temperature cooler (Figure 1–5A, B), which of the following could cause excessive frost buildup on the evaporator coil?

A. if the defrost timer contacts failed in the closed position

B. if the cooling thermostat failed in the closed position

C. if the current relay failed open

D. All of the above

37. Using the time shutdown defrost medium-temperature cooler (Figure 1–5A, B), which electrical loads operate 24 hours a day?

A. the evaporator fan motor and the condenser fan motor

B. the defrost timer motor and the evaporator fan motor

C. the compressor and the defrost timer motor

D. the condenser fan motor and the compressor

38. Using the time shutdown defrost medium-temperature cooler (Figure 1–5A, B), why would a defrost timer motor be added to this type of system?

A. to shut down the evaporator fan motor and prevent frost formation

B. to reduce downtime of the compressor

C. to prevent frost buildup on the evaporator coil, keeping the system operating efficiently

D. All of the above

39. Using the time shutdown defrost medium-temperature cooler (Figure 1–5A, B), what effect will excessive frost buildup on the evaporator coil have?

A. It causes lower operating evaporator pressures and temperatures.

B. It reduces heat transfer from the cooler into the evaporator.

C. It increases the operating time of the compressor.

D. All of the above

40. Using the time shutdown defrost medium-temperature cooler (Figure 1–5A, B), when testing the defrost timer using a voltmeter while the system is in the cooling mode, which of the following readings would be expected if the timer were operating correctly?

A. source voltage from "1" to "N," zero volts from "2" to "N," and source voltage from "1" to "2"

B. zero volts from "1" to "N," source voltage from "2" to "N," and zero volts from "1" to "2"

C. source voltage from "1" to "N" and "2" to "N" and zero volts from "1" to "2"

D. source voltage from "1" to "N" and "2" to "N" and source voltage from "1" to "2"

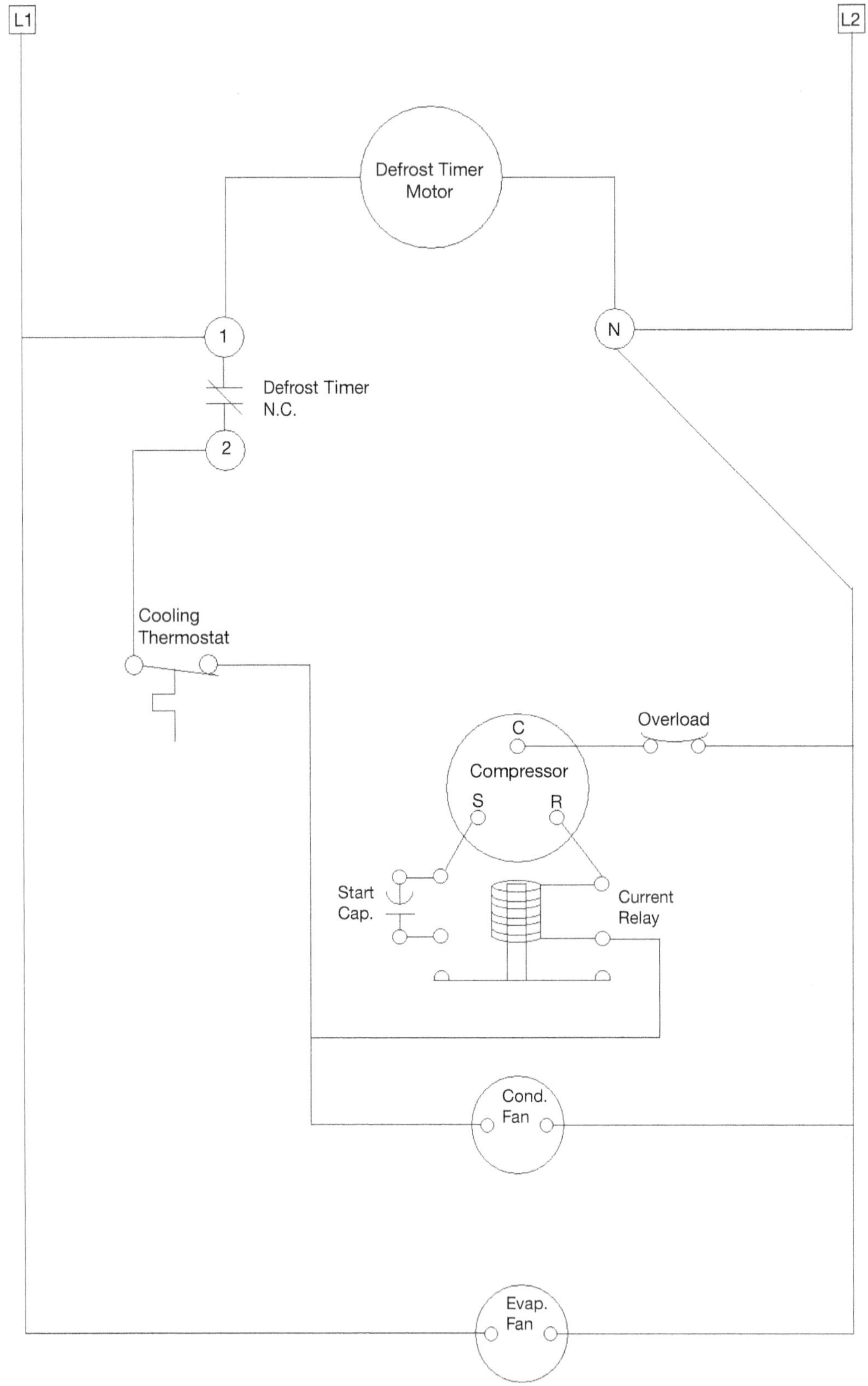

FIGURE 1–5A

Time Shutdown Defrost Medium-Temperature Cooler with a Capacitor Start Compressor Using a Current Relay

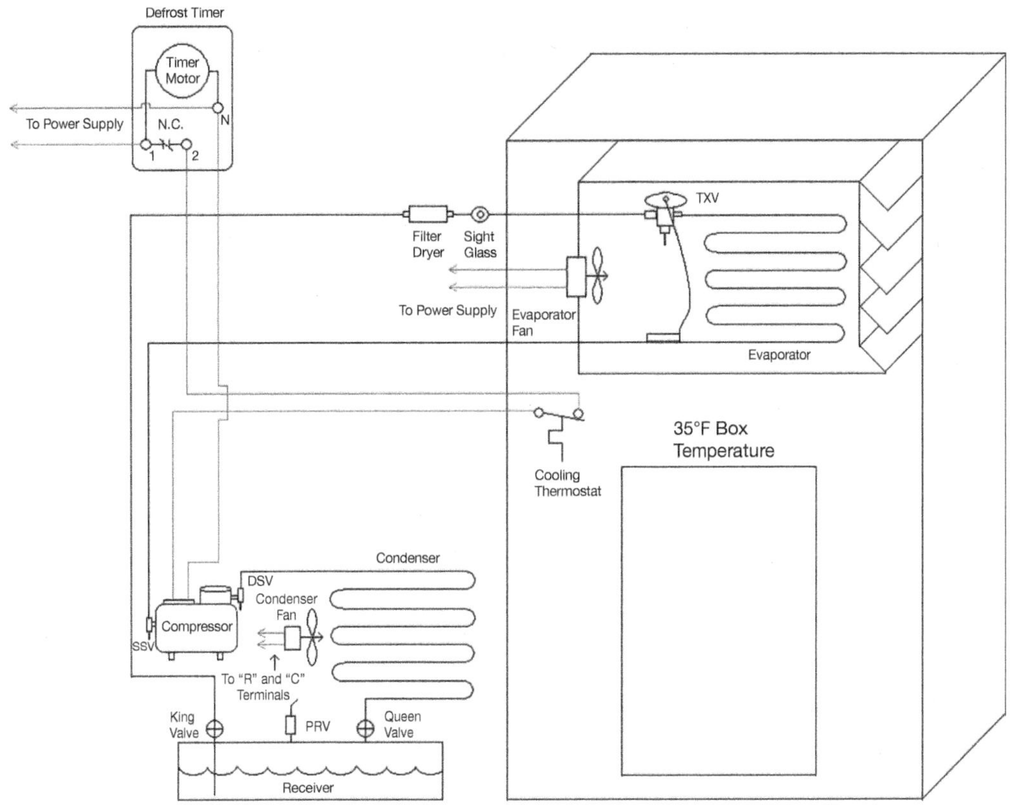

FIGURE 1–5B

Time Shutdown Defrost Medium-Temperature Cooler

41. Using the time shutdown defrost medium-temperature cooler (Figure 1–6A, B), the defrost timer is set in the cooling mode and the cooling thermostat is set for cooling but the compressor and condenser fan will not start. A voltage check from terminals "1" and "N" and "1" and "2" shows source voltage; which of the following is the problem with the system?

A. The thermostat has failed in the open position.

B. The low-pressure control has failed in the open position.

C. The liquid-line solenoid has failed closed.

D. The defrost timer contacts have failed open.

42. Using the time shutdown defrost medium-temperature cooler (Figure 1–6A, B), the defrost timer is set in the cooling mode and the cooling thermostat is set for cooling but the compressor and the condenser fan motor will not operate. Which of the following would be the most logical check when troubleshooting this system?

A. Measure voltage to the condenser fan.

B. Measure voltage to the potential relay.

C. Measure voltage to the liquid-line solenoid.

D. Measure voltage to the defrost timer motor.

43. Using the time shutdown defrost medium-temperature cooler (Figure 1–6A, B), what would be the symptoms of the refrigeration system if the N.C. contacts on the potential relay failed in the open position?

A. The compressor and condenser fan motor would operate for a short time and cycle off on the overload protector.

B. The compressor would not operate and the condenser fan motor would.

C. The compressor would not start—it would cycle on the overload protector—and the condenser fan motor would not operate.

D. The compressor would start, operate for a short time, and cycle on the overload protector and the condenser fan would operate.

44. Using the time shutdown defrost medium-temperature cooler (Figure 1–6A, B), what would be the symptoms of the refrigeration system if the N.C. contacts on the potential relay failed in the closed position?

A. The compressor and condenser fan motor would operate for a short time and cycle off on the overload protector.

B. The compressor would not operate and the condenser fan motor would.

C. The compressor would not start—it would cycle on the overload protector—and the condenser fan motor would not operate.

D. The compressor would start, operate for a short time, and cycle on the overload protector and the condenser fan would operate.

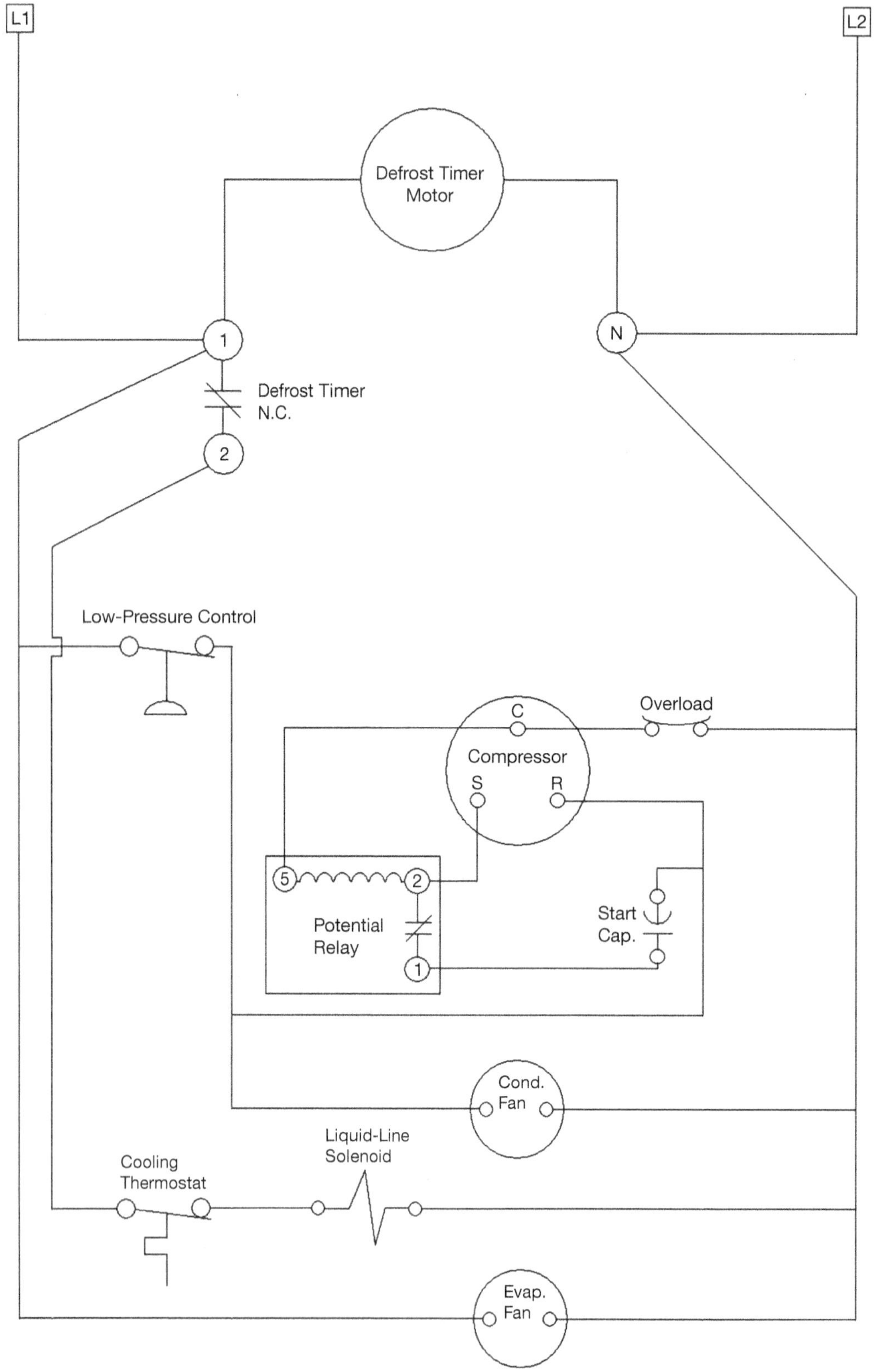

FIGURE 1–6A

Time Shutdown Defrost Medium-Temperature Cooler with Automatic Pump-Down Control

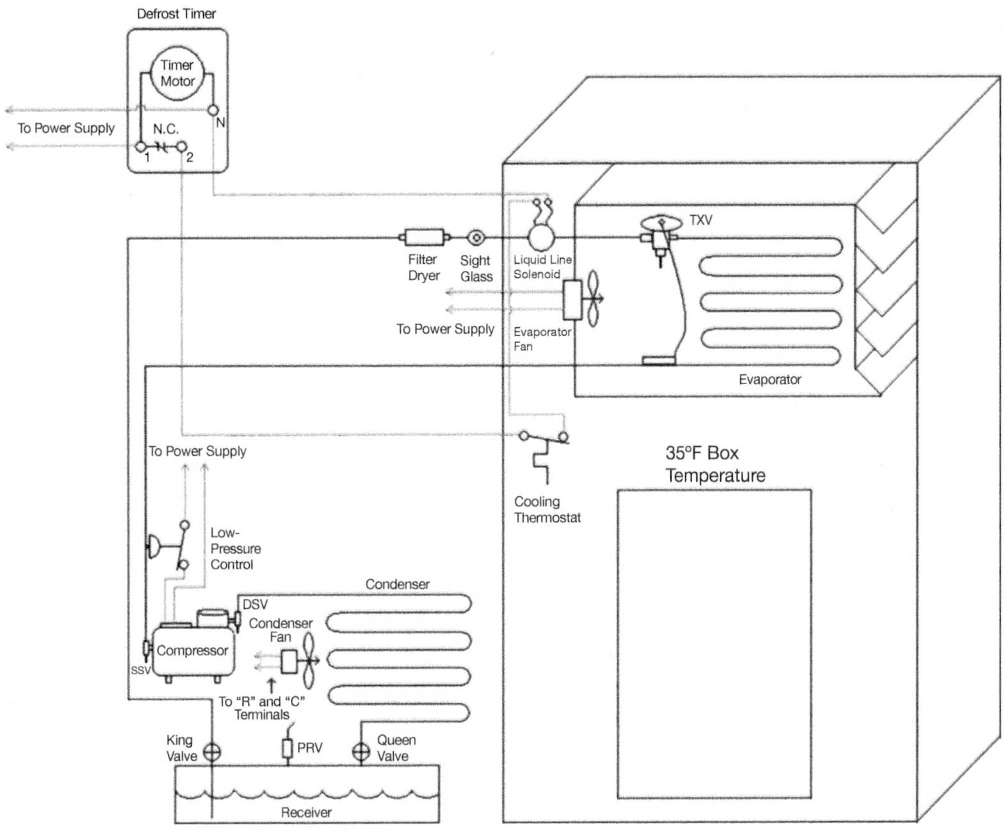

FIGURE 1–6B

Time Shutdown Defrost Medium-Temperature Cooler with Automatic Pump-Down Control

45. Using the time shutdown defrost medium temperature cooler (Figure 1–6A, B), which of the following would cause the low-pressure control to remain in the open position?

A. if the overload protector failed open and the cooling thermostat failed open

B. if the N.C. defrost contacts failed open and the liquid-line solenoid failed closed

C. if the refrigeration system lost its charge and the TXV sensing bulb lost its charge

D. Both B and C

46. Using the low-temperature refrigeration system with electric defrost (Figure 1–7A, B), there is excessive frost buildup on the evaporator coil, the timer is set to the defrost mode, and a voltage check at terminals N and 3 shows source voltage, but when a clamp-on ammeter is placed around the wire leading to the defrost heater, there is no amperage. Which of the following could be the cause of the problem?

A. The defrost heater is shorted.

B. The termination thermostat has failed open.

C. The N.O. defrost timer contacts have failed open.

D. The timer motor has burned open.

47. Using the low-temperature refrigeration system with electric defrost (Figure 1–7A, B), there is excessive frost buildup on the evaporator coil, the timer is set to the defrost mode, and a voltage check at terminals "N" and "3" shows source voltage, but when a clamp-on ammeter is placed around the wire leading to the defrost heater there is no amperage. Which of the following could be the cause of the problem?

A. The termination thermostat has failed closed.

B. The defrost heater has failed open.

C. The N.C. defrost contacts have failed open.

D. The defrost timer has failed in the cooling mode.

48. Using the low-temperature refrigeration system with electric defrost (Figure 1–7A, B), on the electric defrost timer, what do the normally opened contacts control?

A. the defrost heater

B. the evaporator fan

C. the timer motor

D. the condensing unit

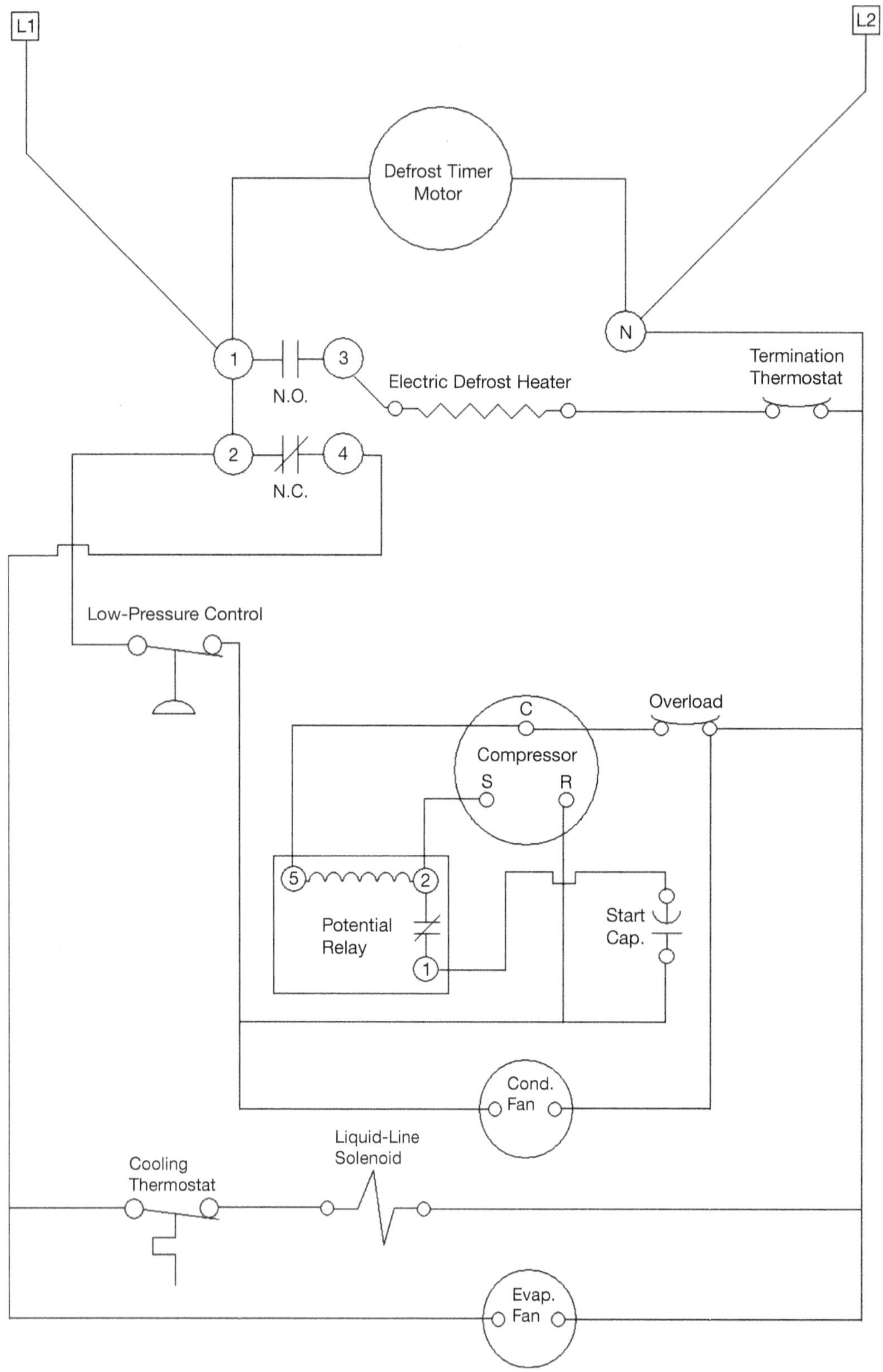

FIGURE 1–7A

Low-Temperature Refrigeration System with Electric Defrost and Automatic Pump-Down

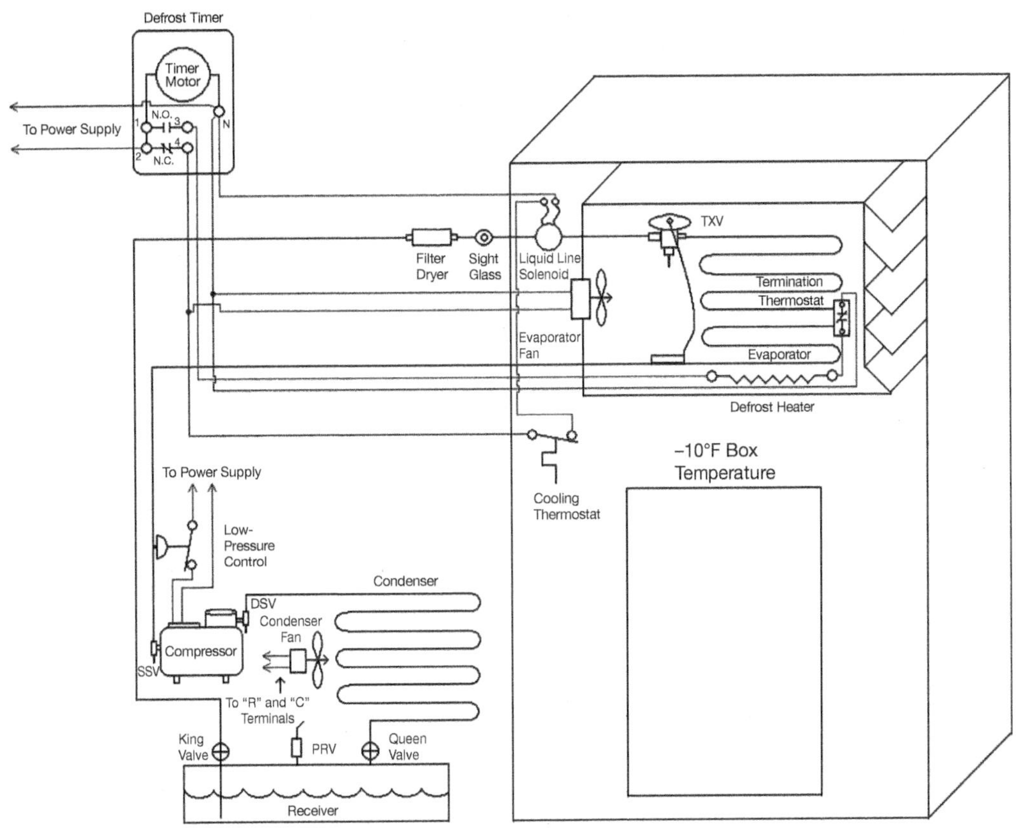

FIGURE 1–7B

Low-Temperature Refrigeration System with Electric Defrost and Automatic Pump-Down

49. Using the low-temperature refrigeration system with electric defrost (Figure 1–7A, B), what temperature controls the operation of the termination thermostat?

A. the evaporator coil temperature

B. the supply-air temperature of the evaporator

C. the suction-line temperature

D. the return-air temperature of the evaporator

50. Using the low-temperature refrigeration system with electric defrost (Figure 1–7A, B), there is excessive frost buildup on the evaporator coil, the timer is in the cooling mode, and when placed in the defrost mode and an ammeter is placed around the wire leading to the defrost heater, 10 amps are measured. A voltage check from terminals "N" and "3" shows source voltage. What is the most likely problem with the system?

A. The N.C. contacts have failed in the closed position.

B. The N.O. contacts have failed in the closed position.

C. The N.O. contacts have failed in the open position.

D. The defrost timer motor has failed open.

51. Using the low-temperature refrigeration system with hot gas defrost (Figure 1–8A, B), there is excessive frost buildup on the evaporator coil and the defrost timer is set to the defrost mode. Which of the following would be the next logical step in troubleshooting the system?

A. Measure voltage at terminals "1" and "N" at the defrost timer.

B. Measure voltage at terminals "2" and "N" at the defrost timer.

C. Measure voltage at terminals "3" and "N" at the defrost timer.

D. Measure voltage at terminals "4" and "N" at the defrost timer.

52. Using the low-temperature refrigeration system with hot gas defrost (Figure 1–8A, B), there is excessive frost buildup on the evaporator coil and the defrost timer is set to the defrost mode. Which of the following would be the next logical step in troubleshooting the system?

A. Measure voltage at the hot gas solenoid.

B. Measure amperage to the hot gas solenoid.

C. Determine if there is a temperature change on each hot gas solenoid bypass line.

D. All of the above

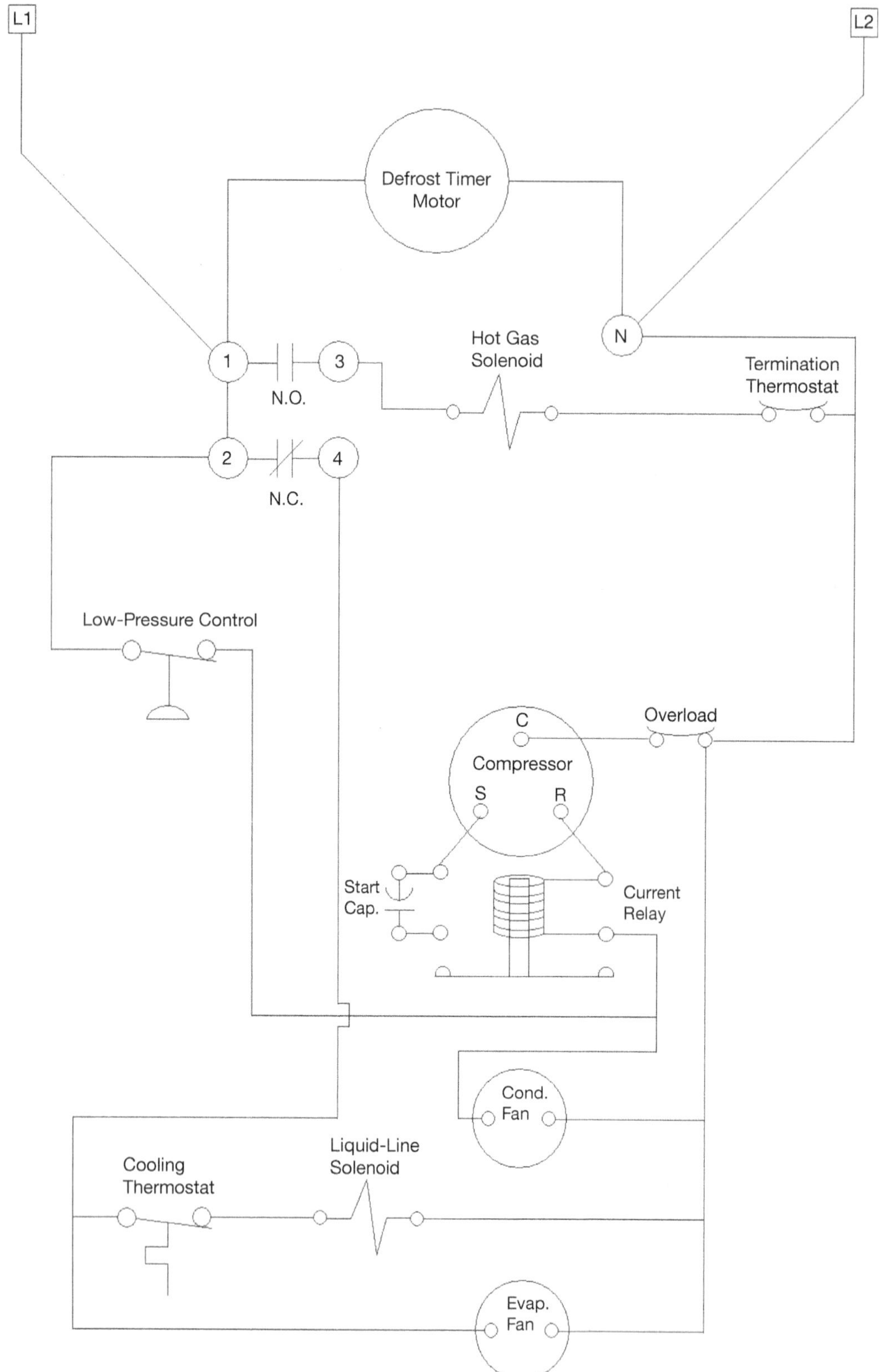

FIGURE 1–8A

Low-Temperature Refrigeration System with Hot Gas Defrost and Automatic Pump-Down

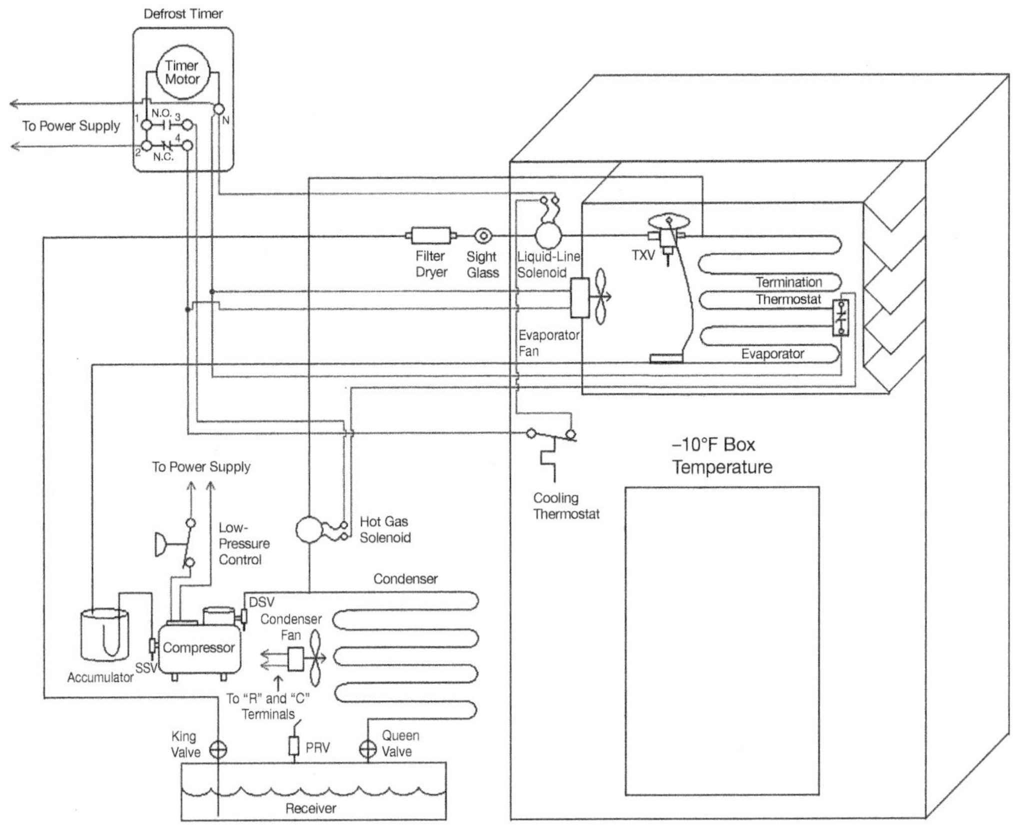

FIGURE 1–8B

Low-Temperature Refrigeration System with Hot Gas Defrost and Automatic Pump-Down

53. Using the low-temperature refrigeration system with hot gas defrost (Figure 1–8A, B), a customer complains that the freezer is not keeping the proper temperature. A visual inspection shows that the evaporator coil is clear and there is no frost buildup. The technician makes sure that the defrost timer is set in the cooling mode then connects the manifold gauge set and determines that the low-side pressure is higher than normal and the high pressure is lower than normal. What could be the problem with the system?

A. The internal compressor valves may be defective.

B. The hot gas solenoid valve may be stuck closed.

C. The hot gas solenoid valve may be leaking internally.

D. Both A and C

54. Using the low-temperature refrigeration system with hot gas defrost (Figure 1–8A, B), there is excessive ice buildup on the evaporator coil. Which of the following could be the cause of this problem?

A. The termination thermostat failed in the open position.

B. The N.O. defrost timer contacts failed in the open position.

C. The hot gas solenoid failed in the closed position.

D. A termination thermostat opened prematurely.

55. Using the low-temperature refrigeration system with hot gas defrost (Figure 1–8A, B), what is the purpose of the accumulator that is located on the suction line?

A. It prevents excessive hot gas from entering the compressor.

B. It prevents liquid floodback during and after a defrost.

C. It acts as a surge protection device during the hot gas defrost mode.

D. It is used as a superheat chamber for the refrigerant suction gases.

56. Using the low-temperature refrigeration system with electric defrost (Figure 1–9A, B), the refrigeration system uses R-404a refrigerant and the technician finds that the system is not cooling properly. The technician has found that the liquid-line solenoid is energized and the condenser fan is operating but the compressor and the evaporator fan are not operating. What are three possible problems that can cause these symptoms?

A. a faulty compressor motor winding, potential relay, or start capacitor

B. a faulty termination thermostat, defrost heater, or defrost timer

C. a low charge, defrost timer, or low-pressure switch

D. an overcharge, a faulty evaporator fan, or defrost timer

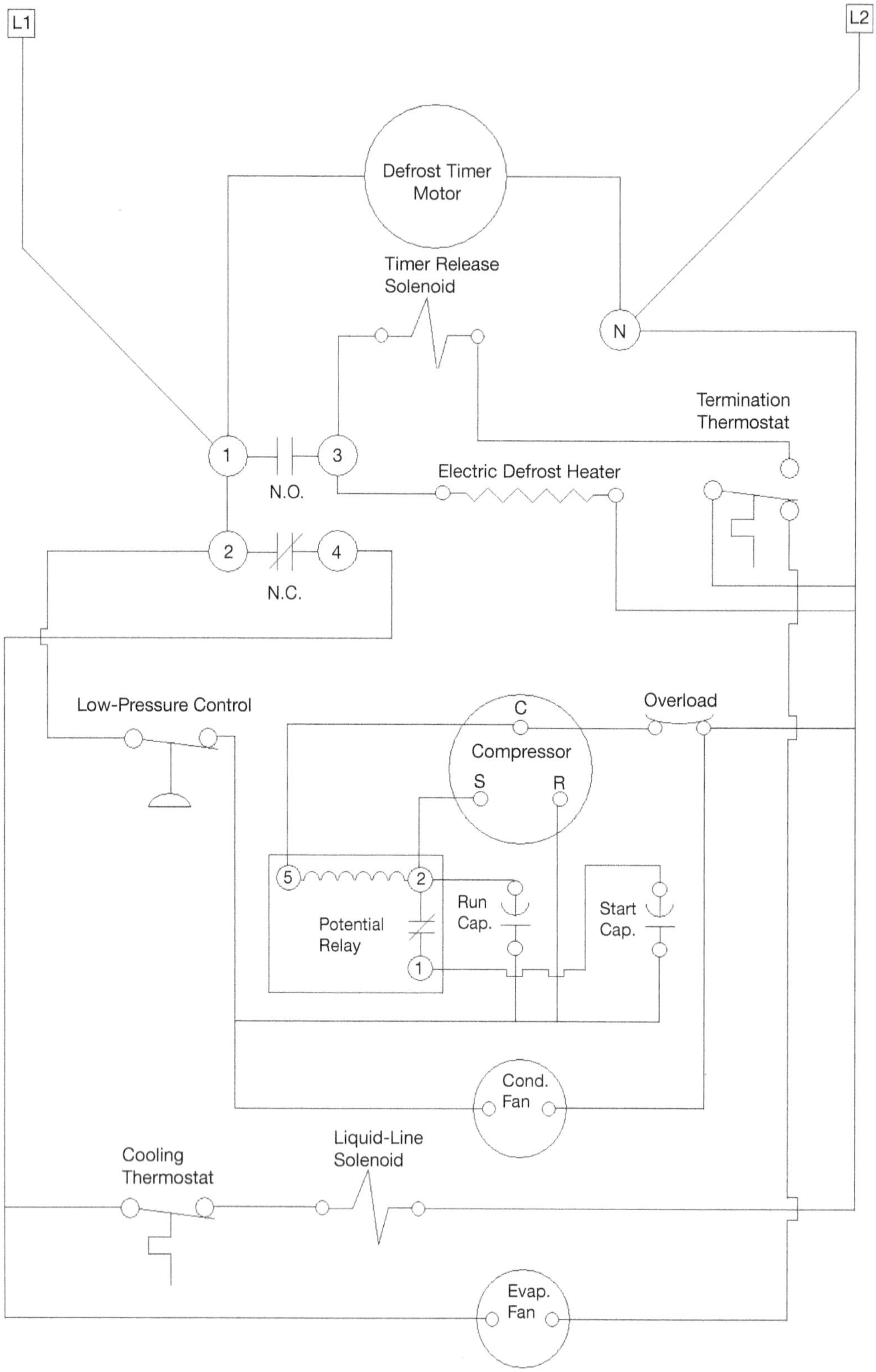

FIGURE 1–9A

Low-Temperature Refrigeration System with Electric Defrost Using a Termination Thermostat and a Timer Release Solenoid with an Automatic Pump-Down Control

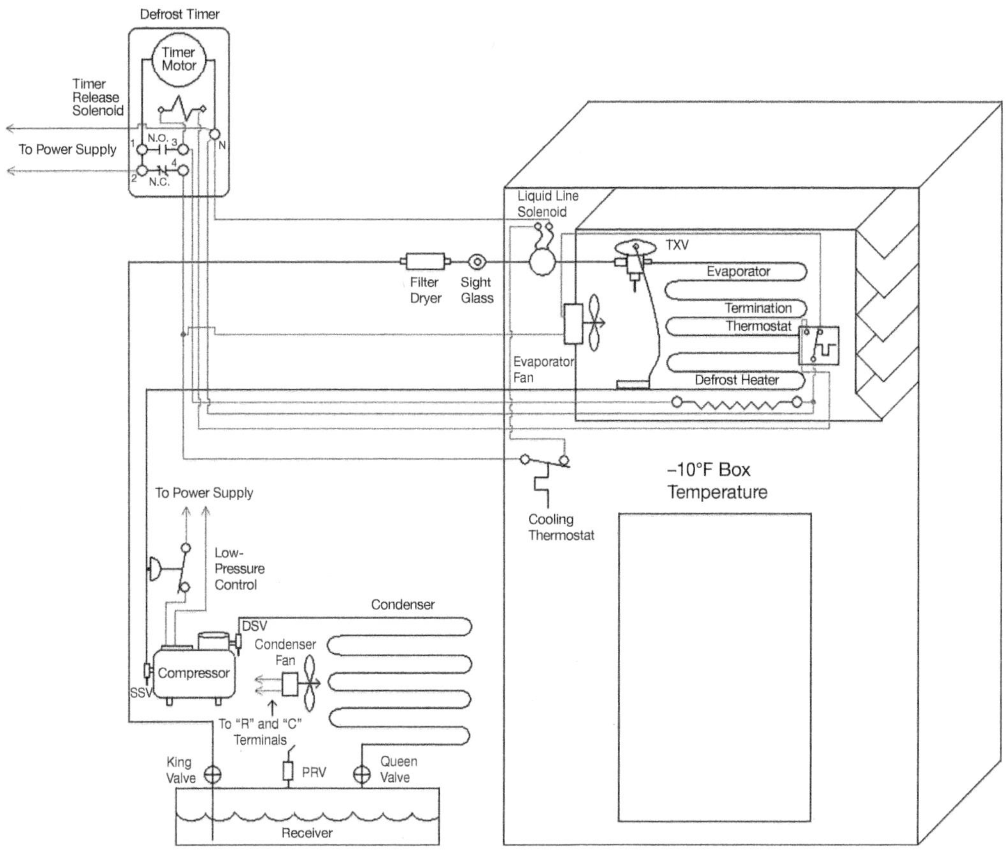

FIGURE 1–9B

Low-Temperature Refrigeration System with Electric Defrost Using a Termination Thermostat and a Timer Release Solenoid with an Automatic Pump-Down Control

57. Using the low-temperature refrigeration system with electric defrost (Figure 1–9A, B), the technician has found that there is excessive frost buildup on the evaporator coil, so the technician starts troubleshooting the defrost circuit. The technician first manually sets the timer to a defrost cycle. The system goes into a pump-down and the compressor and condenser fan shuts down on the low-pressure switch. What should be the next step that the technician should take when troubleshooting this system?

A. Measure voltage at terminals "1" and "N" at the defrost timer.

B. Measure voltage at the timer release solenoid.

C. Measure voltage at terminals "3" and "N" at the defrost timer.

D. Measure voltage at the termination thermostat.

58. Using the low-temperature refrigeration system with electric defrost (Figure 1–9A, B), the compressor, condenser fan, and evaporator fan will not operate. Voltage was checked at terminals "1" and "N" on the defrost timer and there is source voltage. Where would be the next logical test for voltage on the system?

A. Measure voltage at terminals "1" and "2" at the defrost timer.

B. Measure voltage at terminals "2" and "N" at the defrost timer.

C. Measure voltage at terminals "3" and "N" at the defrost timer.

D. Measure voltage at terminals "4" and "N" at the defrost timer.

59. Using the low-temperature refrigeration system with electric defrost (Figure 1–9A, B), what could be the problem if there was voltage to terminals "4" and "N" and the compressor, condenser fan, and evaporator fan were not operating?

A. a cooling thermostat that failed open or a liquid line solenoid that failed closed

B. a low-pressure control that failed open or a low system refrigerant charge

C. a TXV sensing bulb that lost its charge or the N.C. contacts on the defrost timer that failed open

D. All of the above

60. Using the low-temperature refrigeration system with electric defrost (Figure 1–9A, B), which of the following would cause excessive frost buildup on the evaporator coil?

A. The termination thermostat's "close on rise" contacts failed in the closed position.

B. The termination thermostat's "close on fall" contacts failed in the closed position.

C. The defrost timer N.O. contacts failed in the closed position.

D. The defrost timer N.C. contacts failed in the open position.

61. Using the low-temperature refrigeration system with electric defrost (Figure 1–9A, B), which of the following would cause excessive frost buildup on the evaporator coil?

A. The termination thermostat's "close on rise" contacts failed in the open position.

B. The termination thermostat's "close on fall" contacts failed in the closed position.

C. The defrost timer N.O. contacts failed in the open position.

D. The defrost timer N.C. contacts failed in the open position.

62. Using the low-temperature refrigeration system with electric defrost (Figure 1–9A, B), which of the following would cause excessive ice buildup on the evaporator coil?

A. the termination thermostat's "close on rise" contacts closing prematurely

B. a plugged drain line

C. a defrost drain pan heater that has failed open

D. All of the above

THE FOLLOWING QUESTIONS MAY BE HELPFUL TO STUDY FOR THESE CERTIFICATION TESTS:
NATE—RC-SV, RC-IN, RL-SV, RL-IN.
ESCO INSTITUTE HVAC EXCELLENCE AIR CONDITIONING AND COMMERCIAL REFRIGERATION CERTIFICATION

COMMERCIAL REFRIGERATION TROUBLESHOOTING

1. The efficiency of a compressor is affected if it becomes liquid slugged. Which of the following would also affect the compressor's efficiency?

A. defective discharge valves

B. leaking suction valves

C. high discharge pressure

D. All of the above

2. Which of the following is *not* a cause of why a compressor motor will not start?

A. open overload

B. defective start relay

C. overcharged unit

D. open motor winding

3. When testing compressor windings with an ohmmeter, zero between the common and start terminals indicates that

A. windings are good.

B. windings are shorted.

C. windings are opened.

D. windings are grounded.

4. The coil of a potential relay is located between

A. terminals 1 and 2.

B. terminals 2 and 5.

C. terminals 4 and 6.

D. terminals 2 and 6.

5. A run capacitor needs replacing; the microfarad rating is 35 mfd, and the voltage is 115. Which of the following would be acceptable to use as a replacement?

A. 50 mfd

B. 25 mfd

C. 45 mfd

D. 38 mfd

6. When using a run capacitor and wiring it in the system, the marked terminal on the capacitor terminal should be connected to the

A. run terminal of the motor.

B. start terminal of the motor.

C. common terminal of the motor.

D. Both B and C

7. The resistor on a start capacitor will bleed or discharge the capacitor

A. when the motor reaches full rpm.

B. when it is switched out of the circuit.

C. when the potential relay's contacts close.

D. after the motor has stopped.

8. Which of the following is *not* a symptom of a system being low on refrigerant?

A. frosted evaporator

B. bubbles in sight glass

C. frosted compressor

D. high superheat

9. Which of the following would *not* be a cause of high head pressure?

A. dirty condenser

B. system overcharge

C. system undercharge

D. defective condenser fan

10. Which of the following would *not* be a cause of a "starved evaporator" coil?

A. frosted coil

B. high superheat

C. low superheat

D. low system capacity

11. Which of the following would *not* be a cause of a "flooded" evaporator coil?

A. low superheat

B. high superheat

C. liquid slugging

D. overcharged system

12. A system having a restriction preventing refrigerant from flowing is indicated by

A. discharge pressures dropping.

B. suction pressures dropping.

C. compressor amperage dropping.

D. All of the above

13. A system's restriction is evident by

A. temperature difference at the restriction.

B. frosted liquid-line filter outlet.

C. suction pressure dropping.

D. All of the above

14. "Trip pins" on defrost timers are used for what purpose?

A. to initiate defrost

B. to terminate defrost

C. to initiate and terminate defrost cycle

D. to control length of harvest period

15. The two main refrigeration components needed to incorporate a pump-down system are

A. hot gas solenoid/check valve.

B. CPR valve/check valve.

C. temperature control/liquid-line solenoid.

D. EPR valve/check valve.

16. Which of the following will cause a compressor to hum but not start?

A. a blown fuse

B. open overload protector

C. defective start capacitor

D. suction-line restriction

17. Which of the following conditions will cause a compressor to start then cycle off on overload protector?

A. shorted run capacitor

B. open overload protector

C. defective thermostat

D. open run winding

18. Which of the following conditions will cause a refrigeration unit that uses a temperature control to run long and continuously?

A. a shortage of refrigerant

B. a shorted start winding

C. a low-voltage condition

D. a start relay that is stuck in the closed position

19. Which of the following conditions will cause the suction line to have excessive frost?

A. an undercharged system

B. bad compressor valves

C. plugged filter drier

D. insufficient evaporator airflow

20. Which of the following conditions will cause a refrigeration system to have excessive high-side pressure?

A. dirty condenser

B. air trapped in the condenser

C. bad condenser fan

D. All of the above

21. Which of the following would cause a refrigeration system to have excessively low discharge pressure?

A. too much refrigerant in the system

B. insufficient airflow over the condenser

C. too much water over a water-cooled condenser

D. a flooded evaporator

22. Which of the following would cause the high-side pressure to be too low and the low-side pressure to be too high?

A. an oversized filter drier

B. the float valve of an oil separator stuck open

C. an undersized filter drier

D. uneven airflow over the evaporator coil

23. Which of the following would cause a refrigeration system to have excessive subcooling?

A. refrigerant undercharge

B. refrigerant overcharge

C. a dirty condenser

D. Both A and C

24. Which of the following will cause a compressor to have a flooded start?

A. a defective crankcase heater

B. an oversized accumulator

C. moisture in the system

D. All of the above

25. Which of the following will cause high suction pressure?

A. a low refrigerant charge

B. excessive evaporator airflow

C. frost buildup on the evaporator coil

D. a plugged TXV

26. How will a faulty head pressure control affect the operation of a refrigeration system during low ambient conditions?

A. It will cause low high-side pressure.

B. It will cause low superheat.

C. It will cause high subcooling.

D. It will cause excessive amp draw.

27. Which of the following is a symptom of a flooded evaporator?

A. high superheat

B. frost buildup on the evaporator

C. low suction pressure

D. liquid floodback

28. Which of the following will cause a system to short cycle on the low-pressure control?

A. a leaking liquid-line solenoid

B. the oil separator float valve stuck in the open position

C. thermostat stuck in the open position

D. a faulty condenser fan

29. Which of the following would cause bubbles in a sight glass?

A. a restricted TXV

B. a faulty evaporator fan

C. a flooded condenser

D. a broken dip tube in the receiver

30. What symptom would a faulty crankcase pressure regulator (stuck open) have on a low-temperature system?

A. Suction pressure would be too low.

B. The system would cycle on overload protector.

C. The evaporator would frost up.

D. The amperage would drop.

31. When you reverse any two wire leads on a three-phase motor you would

A. have a power loss.

B. cause a shorted circuit.

C. change the direction of rotation.

D. blow a fuse.

32. In refrigerant piping, double risers are used on suction lines mainly to

A. mix well with the refrigerant vapor.

B. aid in the oil-return process.

C. reduce oil migration.

D. maintain the system's capacity.

33. The contacts of a potential relay are between terminals

A. 1 and 2.

B. 2 and 5.

C. 4 and 6.

D. 2 and 6.

34. Noncondensables (air) in a refrigeration system are generally accumulated in the

A. evaporator.

B. receiver.

C. condenser.

D. suction and discharge lines.

35. Air in a refrigeration system can be identified by

A. high suction pressure.

B. low superheat.

C. low subcooling.

D. high head pressure.

36. When checking for a restriction at a distributor feed with many circuits, an indication of a plugged circuit would be

A. a circuit that is sweating.

B. a frosted accumulator.

C. a circuit that is warmer than the rest.

D. extreme high head pressure.

37. Which of the following is true about dehydrating a refrigeration system?

A. Evacuate continuously for 20 min.

B. Double evacuate.

C. Change the filter drier after evacuating.

D. Generally, you cannot overevacuate.

38. When a refrigeration system's TXV is totally iced-up (snowballed over), the most probable cause would be

A. an overcharged system.

B. liquid-line restriction.

C. moisture in the system.

D. a dirty condenser.

39. Which of the following would not be the cause of a system having high suction pressure and low superheat?

A. oversized TXV

B. moisture, dirt, or wax plugging the TXV

C. poor thermal bulb contact

D. coil icing-up

40. Which of the following would not be a cause of a system having low suction pressure and low superheat?

A. dirty air filters

B. low refrigerant charge

C. coil icing-up

D. dirty evaporator coil

41. The defrost timer is equipped with two sets of contacts; one is normally closed (N.C.) and the other is normally open (N.O.). These contacts operate

A. N.C. for defrost, N.O. for cooling.

B. N.C. for defrost, N.O. for harvest operation.

C. N.C. for cooling, N.O. for defrost.

D. N.O. for evap, fans on, N.C. for defrost.

42. A thermostatic expansion valve would be best described as a component that maintains a constant

A. pressure.

B. superheat.

C. temperature.

D. subcooling.

43. A system requiring a cooling tower should include which of the following conditions?

A. water treatment

B. constant air movement

C. cleaning

D. All of the above

44. A pump-down system that is short cycling can be caused by

A. an overcharge.

B. leaking liquid-line solenoid.

C. bad discharge valves.

D. high superheat.

45. A pump-down system refers to a means of pumping the refrigerant to the

A. evaporator.

B. recovery tank.

C. condenser and receiver.

D. evaporator or receiver.

46. A system has three evaporators; one is maintained at 38°F, one at 28°F, and the other at 0°F. Which evaporator will need a check valve installed?

A. the evaporator at 38°F

B. the evaporator at 28°F

C. the evaporator at 0°F

D. All three should include a check valve.

47. Refrigerant oil will tend to

A. mix with vapor refrigerant.

B. mix with liquid refrigerant.

C. foam in the liquid line.

D. vaporize in the oil separator.

48. The refrigerant that will fractionize is

A. R-22.

B. R-717.

C. R-12.

D. R-409A.

49. The low-pressure control can be used to

A. act as a safety control.

B. act as a temperature control.

C. control subcooling.

D. Both A and B

50. A ternary blend refrigerant would be used with

A. a mineral oil lubricant.

B. alkylbenzene oil lubricant.

C. polyester oil lubricant.

D. wax-free oils only.

51. To control the capacity of large compressors, ________ should be used.

A. unloaders

B. multiple condenser fans

C. dampers

D. EPR valves

52. The oil safety switch is operated by

A. compressor discharge pressure.

B. suction pressure.

C. suction and discharge pressure.

D. oil pump discharge pressure and crankcase pressure.

53. The term *temperature glide* refers to

A. R-12.

B. R-22.

C. R-407A.

D. R-717.

54. The refrigerant line that is most critical when sizing is the

A. suction line.

B. discharge line.

C. liquid line.

D. condensate line.

55. A technician charging a system with R-407A wants to maintain 70 psig; what would be the approximate "bubble point" temperature?

A. 40°F

B. 30°F

C. 33°F

D. 35°F

56. The term *cryogenics* refers to very low temperatures. This range would start generally around

A. 0 to −40°F.

B. −40 to −60°F.

C. −250 to −450°F.

D. −100 to −150°F.

57. What operates a potential relay?

A. current flow on the run winding

B. back emf across the start winding

C. back emf across the run winding

D. the current on the start winding

58. What would be the effects of a loose sensing bulb on a TXV system?

A. a low superheat and possible floodback to the compressor

B. a high superheat and possible floodback to the compressor

C. low suction pressures and an overheated compressor

D. high discharge pressures and an overheated compressor

59. Which of the following controls would use a thermistor?

A. a current-sensing relay

B. a timed delay relay

C. an electronic temperature sensor

D. a potentiometer

60. When charging blend-type refrigerants, they should be

A. charged into the low side.

B. charged as a liquid.

C. charged as a vapor.

D. charged at the metering device in vapor form.

61. The purpose of an external equalizer is to

A. allow refrigerant to flow evenly.

B. compensate for pressure drop through the evaporator.

C. increase the system's capacity.

D. decrease flash gas.

62. The purpose of a head pressure control is to

A. flood the receiver during low ambient conditions.

B. flood the condenser during low ambient conditions.

C. flood the evaporator during low ambient conditions.

D. maintain a minimal flash gas.

63. What would be the main function of a heat exchanger?

A. to reduce flash gas and improve the system's efficiency

B. to produce flash gas at a constant

C. to decrease superheat

D. to help prevent liquid slugging

64. Which of the following refrigerants will leak at uneven rates due to different vapor pressures?

A. R-12

B. R-22

C. R-409A

D. R-717

65. An oil trap is used on a riser when the vertical lift is above

A. 6′.

B. 8′.

C. 10′.

D. 12′.

66. A second oil trap is required when the vertical riser height exceeds

A. 8′.

B. 10′.

C. 20′.

D. 25′.

67. Which of the following is not a cause of a hot liquid line?

A. high head pressure

B. refrigerant overcharge

C. air in the system

D. an undercharged system

68. Which of the following problems would not be a cause of a frosted liquid line?

A. restricted drier

B. solenoid valve leaking

C. receiver shutoff valve closed too far or clogged

D. restricted TXV

69. Which of the following problems would *not* be a cause of a compressor not starting nor even making a hum noise?

A. defective overload protector

B. incorrect wiring

C. charged with wrong refrigerant

D. compressor motor is defective

70. Which of the following problems would cause a run capacitor to burn out?

A. high voltage

B. overcharged system

C. undercharged system

D. defective overload protector

71. Which of the following problems would cause a start relay to burn out?

A. high or low voltage

B. incorrect rating capacitor

C. compressor short cycling

D. All of the above

72. Which of the following problems would *not* be a cause of a system having too low of head pressure?

A. leak in the system

B. restricted drier

C. refrigerant undercharged

D. oversized TXV

73. Which of the following would be used on a refrigeration system for the purpose of increasing subcooling and reducing flash gas?

A. an oil separator

B. an accumulator

C. liquid/suction heat exchanger

D. liquid-line solenoid

74. An undersized suction line will

A. increase the compression ratio.

B. decrease the compression ratio.

C. decrease oil return to the compressor.

D. cause an increase of high-side pressure.

75. Which of the following will cause ice buildup on an evaporator of a medium-temperature system?

A. an overcharged system

B. a faulty evaporator fan

C. a high saturation temperature

D. faulty compressor valves

76. The sensing bulb from a TXV has lost its charge; this will cause

A. the valve to remain open.

B. the valve to remain closed.

C. a low superheat condition.

D. an increase of flash gas.

77. Adjusting a TXV by turning the adjustment stem clockwise will

A. decrease superheat.

B. increase superheat.

C. increase flash gas.

D. decrease subcooling.

78. When a check valve is used on a two-temperature system, it should be located

A. in the liquid line.

B. at the outlet of the accumulator.

C. at the outlet of the coldest evaporator.

D. at the suction header.

79. A refrigeration system is experiencing the following symptoms: compressor short cycles, bubbles in the sight glass. Which of the following could be the cause?

A. undercharge

B. overcharge

C. excessive subcooling

D. None of the above

80. When the returning suction gas has too much superheat, which condition may occur?

A. liquid slugging

B. high amperage

C. high discharge-line temperatures

D. low suction-line temperatures

81. When testing a capacitor with an ohmmeter, the needle moves to zero then back to infinity and stays at that point. This indicates that

A. the capacitor is opened.

B. the capacitor is shorted.

C. the capacitor is satisfactory.

D. the capacitor microfarad rating is too high.

82. Turning the adjusting stem of a TXV counterclockwise will

A. increase superheat.

B. decrease superheat.

C. cause a high subcooling.

D. increase flash gas.

83. What terminals should have the lowest resistance reading of a compressor motor?

A. common to start

B. run to start

C. common to run

D. start to ground

84. In commercial applications, the low-pressure control is in series with

A. evaporator fans.

B. condenser fans.

C. compressor and condenser fans.

D. the hot gas solenoid.

85. A medium-temperature R-12 system compressor and condenser fan is short cycling but the evaporator fan continues to run. The possible causes may be

A. low charge.

B. low-pressure control not adjusted correctly.

C. partial restriction at the TXV.

D. All of the above

86. On a pump-down system the low-pressure and temperature controls operate which way?

A. The low-pressure control operates the liquid-line solenoid.

B. The temperature control operates the compressor cycle.

C. The low-pressure control operates the compressor cycle; temperature control operates the liquid-line solenoid.

D. Both controls act as safety devices.

87. When a pump-down system is used and the liquid line is energized

A. the refrigerant will stop flowing.

B. there will be a decrease the low side pressure.

C. the pressure will increase in the evaporator resulting in the closing of the low pressure switch, which will start the compressor.

D. there will be an increase of superheat opening the low pressure switch starting the compressor.

88. An R-409A pump-down system is not operating. There is correct voltage to the low-pressure control but it still will not start; both high- and low-side gauges read 8 psig. You bypassed the low-pressure control momentarily and the unit starts. Which of the following is most likely the problem?

A. overcharged system

B. off/on overload protector

C. low on refrigerant

D. too small of differential between cut-out and cut-in settings

89. An R-134A pump-down system's compressor is not operating correctly (shutting off on overload); the condenser fan is operating. Both high- and low-side pressures are at 75 psig. Which of the following would most likely *not* be the problem?

A. open motor windings

B. defective capacitors or start relay

C. low voltage

D. closed temperature control

90. Flash gas in the liquid line at the inlet of the TXV will

A. tend to flood the evaporator.

B. starve the evaporator.

C. increase subcooling.

D. lower superheat.

91. Which of the following would *not* be a symptom of flash gas in the liquid as it enters the TXV?

A. The system will lose some capacity.

B. The superheat will decrease.

C. The superheat will increase.

D. The TXV will operate inefficiently.

92. Which of the following is *not* a cause of liquid-line restriction?

A. filter drier is saturated with moisture

B. liquid-line solenoid is undersized

C. oversized filter drier

D. pipe joint fitting is partially filled with brazing alloy

93. When a system has excessive vertical lift, flash gas will form. For every foot of vertical lift using R-22 there is generally around _______ psi drop.

A. $\frac{1}{4}''$

B. $\frac{1}{2}''$

C. $\frac{3}{4}''$

D. 1″

94. Which of the following is *not* a cause of liquid-line pressure drop?

A. very long liquid line

B. too many fittings

C. liquid line is sized too large

D. liquid line is sized too small

95. An excessive amount of oil circulating in a system will

A. lubricate better.

B. reduce flash gas.

C. displace some refrigerant at the metering device.

D. increase subcooling.

96. Excessive oil in the evaporator will

A. increase the rate of evaporation.

B. decrease the rate of evaporation.

C. help the lubrication process.

D. mix thoroughly with the vapor.

97. Which of the following is *not* a cause of high discharge pressures?

A. refrigerant overcharge

B. refrigerant undercharge

C. dirty condenser

D. wrong airflow direction through the condenser

98. Which of the following is *not* a cause of evaporator coil frosting?

A. incorrect evaporator fan rotational direction

B. defective hot gas defrost solenoid valve

C. high case humidity

D. oversized evaporator

99. Which of the following would be a cause of fluctuating suction pressures?

A. restricted external equalizer line

B. insufficient charge

C. condenser fan cycling

D. oversized filter drier

100. Which of the following would cause high amperage draw?

A. defective capacitor

B. low-load conditions

C. oversized condenser

D. defective overload

101. Thermostatic expansion valves that are internally equalized generally are not used

A. above 2-ton capacity.

B. above $1\frac{1}{2}$-ton capacity.

C. above 1-ton capacity.

D. They can be used on any tonnage.

102. Which of the following is *not* a cause of oil pressure control trip-out?

A. faulty oil pump

B. kinked suction line

C. low refrigerant charge

D. high head pressure

103. A liquid-line solenoid valve will not open. Which one of the following is not a cause of this problem?

A. faulty thermostat contacts

B. valve plunger is restricted

C. incorrect wiring

D. undercharged system

104. A liquid-line solenoid valve will not close. Which one of the following is not a cause of this problem?

A. plunger is restricted

B. defective thermostat

C. valve outlet pressure is higher than inlet pressure

D. coil burnout

105. Which of the following would be a cause of a liquid-line solenoid coil burnout?

A. supply voltage to coil is below 85% of its rating

B. oversized solenoid valve

C. liquid line is undersized

D. liquid line contains too much flash gas

106. A thermostatic expansion valve is not feeding enough refrigerant. Which of the following is not a symptom of this problem?

A. Load temperature is too high.

B. Superheat is too high.

C. Low-side pressure is lower than normal and hot gas bypass is shut off.

D. Superheat is too low.

107. Gas in the liquid line can be caused by

A. pressure drops in the liquid line.

B. air or other noncondensables in the system.

C. long or undersized liquid lines.

D. All of the above

108. Which of the following contribute to insufficient pressure drop across a thermostatic expansion valve?

A. lower-than-normal condensing pressures

B. undersized distributor tubes

C. too much liquid-line pressure losses

D. All of the above

109. When the external equalizer line is capped or restricted, the thermostatic expansion valve will

A. remain closed.

B. remain open.

C. tend to hunt.

D. flood the evaporator.

110. When a compressor is inefficient and has low capacity, the suction pressure will

A. drop lower than normal.

B. operate higher than normal.

C. stabilize at normal pressures.

D. operate with low superheat conditions.

111. Which of the following is not a symptom of a thermostatic expansion valve that feeds too much refrigerant at startup only?

A. no superheat

B. low subcooling

C. liquid returning to the compressor

D. higher-than-normal suction pressures

112. Which of the following symptoms is not a reason for liquid slugging?

A. TXV hunting

B. overcharged system

C. iced-up coil

D. high superheat

113. Volumetric efficiency is mainly determined by

A. compressor horsepower.

B. type of refrigerant.

C. the system's tonnage.

D. the system's pressures.

114. The color code of a thermostatic expansion valve for R-500 would be

A. purple.

B. blue.

C. yellow.

D. orange.

115. Which of the following is not a symptom of a restricted liquid line?

A. low amperage draw

B. high amperage draw

C. high superheat

D. Both A and C

116. A compressor that is being starved of refrigerant vapor will result in

A. low compressor amp draw.

B. high compressor amp draw.

C. high suction pressure.

D. low superheat.

117. Low suction and discharge pressures with high superheats are symptoms of

A. an overcharge.

B. an undercharge.

C. high subcooling.

D. poor condenser airflow.

118. A refrigeration system operating with an overcharge will have

A. high subcooling.

B. a decrease in volumetric efficiency.

C. high amp draw.

D. All of the above

119. Which of the following is correct of a system that is overcharged and using a TXV?

A. The TXV will tend to overfeed.

B. The TXV will tend to underfeed.

C. The TXV will control superheat satisfactorily.

D. The evaporator pressure will be higher than normal.

120. Which of the following symptoms indicates an inefficient compressor?

A. a decrease in refrigerant flow

B. high suction pressure and low discharge pressures

C. high cabinet temperatures

D. All of the above

121. Which of the following is *not* a symptom of noncondensables in the system?

A. high subcooling

B. high discharge temperature

C. low subcooling

D. high discharge pressure

122. Which of the following is *not* a symptom of an undercharged system?

A. high superheat in the evaporator

B. low superheat in the evaporator

C. low condenser subcooling

D. high compressor superheat

123. Which is not a symptom of an overcharged system?

A. high condensing pressures

B. higher condenser temperature splits

C. low evaporator pressures

D. high evaporator pressures

124. Which of the following is not a symptom of a dirty condenser?

A. high condenser temperature split

B. high amp draw

C. low superheat

D. normal superheat

125. When a refrigeration compressor has been operating under high condensing pressures the result would be

A. lower-than-normal volumetric efficiency due to higher compression ratio.

B. higher-than-normal volumetric efficiency due to higher compression ratio.

C. high subcooling with low superheat.

D. low subcooling with low superheat.

126. A "no cooling" call on a refrigeration system is diagnosed and it is found that the suction pressure is remaining high and the head pressure is low. Technician A claims that the unit is undercharged with a possibility of a refrigerant leak. Technician B claims that, more than likely, the hermetic compressor has bad suction valves and should be replaced. Which technician is correct?

A. technician A

B. technician B

C. Neither technician is correct.

D. Both technicians are correct.

127. A "no cooling" call is diagnosed and it is found that a medium-temperature compressor is shutting off on internal overload. Technician A claims that the temperature rise across the condenser is 35°F and the condenser is dirty and should be cleaned. Technician B claims that the installation of the condensing unit is such that outlet air is hitting a barrier and as a result recirculating as inlet air—the unit should be reinstalled to another location. Which technician is correct?

A. technician A

B. technician B

C. Neither technician is correct.

D. Both technicians are correct.

128. A "no cooling" call is diagnosed and it is found that one evaporator on a multievaporator system is being starved of refrigerant. Technician A claims that the system needs charging because the superheat for that coil is high. Technician B claims that on checking the TXV, the strainer is clogged and needs cleaning. Which technician is most correct?

A. technician A

B. technician B

C. Neither technician is correct.

D. Both technicians are correct.

129. A "no cooling" call is diagnosed and it is found that the head pressure is very high with high compressor amperage draw and high suction pressures. Technician A claims that the reason for these symptoms is quite simple, the unit is overcharged. Technician B claims that the problem is a faulty TXV—the TXV is not feeding enough refrigerant. Which technician is correct?

A. technician A

B. technician B

C. Neither technician is correct.

D. Both technicians are correct.

130. A "no cooling" call is diagnosed and it is found that the suction pressures of a walk-in cooler are very low. Technician A claims the reasons for steady lower-than-normal suction pressures are: iced evaporator, restriction of evaporator air, and an undercharged system. Technician B claims that the reasons for low suction pressures are: dirty evaporator, low load conditions, and incorrect thermostat setting. Which technician is correct?

A. technician A

B. technician B

C. Neither technician is correct.

D. Both technicians are correct.

131. A "no cooling" call is diagnosed and it is found that on a low-temperature R-502 system the evaporator temperature cannot go lower than +4°F, and the desired evaporator temperature is −30°F. Technician A claims after troubleshooting the system that the problem is caused by excessive pressure drop in the suction line and possibly an oil-clogged evaporator. Technician B claims that the reason for the high evaporator temperature is that there is too much flash gas in the liquid line and this is preventing proper TXV operation. Which technician is correct?

A. technician A

B. technician B

C. Neither technician is correct.

D. Both technicians are correct.

132. A "no cooling" call is diagnosed and it is found that a medium-temperature unit is constantly running. The desired cabinet temperatures are 35 to 45°F. Technician A troubleshoots the system and finds that the evaporator fan motor is not operating and its fan blades are jammed against some ice buildup, and suggests replacing the fan motor, allowing the evaporator to thaw, and then rechecking system pressures. Technician B suggested allowing the evaporator to thaw, recovering refrigerant, and evacuating and then recharging the system. Which technician's suggestion would be more correct to follow?

A. technician A

B. technician B

C. Neither technician is correct.

D. Both technicians are correct.

133. A "no cooling" call is diagnosed and it is found that an R-502 walk-in freezer is cutting off on the high-pressure control. After troubleshooting the system, Technician A claims that the unit is overcharged and the condenser is dirty. Technician B claims that the reason for the compressor cutting off is because of the remote location of the condensing unit, in a small enclosed room with very little air infiltration or air to be circulated. Which technician is correct?

A. technician A

B. technician B

C. Neither technician is correct.

D. Both technicians are correct.

134. A "no cooling" call is diagnosed and it is found that the evaporator of a walk-in cooler is totally iced-up. After troubleshooting the system, Technician A claims that the system has definitely been overcharged and some needs to be recovered. Technician B claims that the superheat is too high and the TXV should be readjusted. Which technician is correct?

A. technician A

B. technician B

C. Neither technician is correct.

D. Both technicians are correct.

135. A "no cooling" call is diagnosed and it is found that a circuit breaker from a walk-in freezer has been tripped. The compressor is hot to touch and on resetting the breaker the compressor is reenergized but again trips the circuit breaker. Technician A further troubleshoots the system and concludes that the reason for the breaker tripping out is because of an overcharged system and that some needs to be recovered. Technician B concludes that the reason for tripping out is due to a short circuit at the run winding and suggests replacing the compressor. Which technician is correct?

A. technician A

B. technician B

C. Neither technician is correct.

D. Both technicians are correct.

136. A "no cooling" call is diagnosed and it is found that on a medium-temperature condensing unit the compressor is trying to start but continually shuts off on internal overload protector. The compressor uses a potential relay and start capacitor with a TXV as a metering device. Technician A troubleshoots and claims that noncondensables and/or an overcharged system is to blame for the condition. Technician B claims that the capacitor "ohms out" okay but the real fault is due to welded contacts stuck closed on the start relay. Which technician is correct?

A. technician A

B. technician B

C. Neither technician is correct.

D. Both technicians are correct.

137. A "no cooling" call is diagnosed and it is found that a medium-temperature system having dual evaporator fans is experiencing its compressor cutting off on low-pressure control. After troubleshooting, Technician A claims that the reason the compressor is cutting off on low-pressure control is because one of the evaporator fan motors is defective; therefore there is not enough load on the evaporator coil. Technician B claims the reason is due to an excessive evaporator load that is allowing the compressor to run too long. Which technician is correct?

A. technician A

B. technician B

C. Neither technician is correct.

D. Both technicians are correct.

138. A "no cooling" call is diagnosed on a refrigeration system that utilizes two hermetic compressors. The metering device is a TXV and the installation is piped as a pump-down system. The compressors will try to start but quickly shut down on low-pressure control. Technician A claims after troubleshooting the system that the fault is due to the excessive heat of compression and the many times the compressors have had to start and stop. Technician B claims after further diagnosis that the long TXV capillary has rubbed against metal and lost its charge, causing the TXV to remain closed. Which technician is more likely to be correct?

A. technician A

B. technician B

C. Neither technician is correct.

D. Both technicians are correct.

139. The proper method for checking the coil of a potential relay would be

A. voltage check.

B. amperage check.

C. resistance check.

D. bypassing the coil.

140. The type of device that contains a coil, main contacts, auxiliary contacts, and resistance heaters would best describe a

A. start relay.

B. starter.

C. contactor.

D. disconnect.

141. When two capacitors with the same mfd rating and voltage are connected in series, the

A. total capacitance doubles.

B. total capacitance is cut in half and voltage doubles.

C. total capacitance doubles but voltage is cut in half.

D. total capacitance and voltage are cut in half.

142. When two capacitors with the same mfd and voltage rating are connected in parallel, the

A. total capacitance and voltage double.

B. total capacitance doubles, and the voltage is cut by one half.

C. total capacitance doubles, and the voltage remains the same.

D. total capacitance is cut by one half, and the voltage is doubled.

143. A three-terminal external overload protector contains

A. a set of contacts and a heater.

B. two sets of contacts.

C. two heaters that must be checked with an ohmmeter.

D. a set of contacts with an extra terminal.

144. Internal overload protectors when open due to excessive temperatures can take as long as _______ to automatically reset.

A. 10 min

B. 15 min

C. 20 min

D. 45 min

145. A low-pressure control may be used to

A. control head pressure.

B. control box temperature.

C. control crankcase pressure.

D. stabilize superheat.

146. The type of system that is used to cool another part of a refrigeration system is called the

A. compound system.

B. multistaging system.

C. cascade system.

D. parallel system.

147. Open-drive-type compressors will tend to leak refrigerant from the

A. rotating shaft seal.

B. suction service valve.

C. discharge service valve.

D. oil separator.

148. The water-regulating valve is closed when

A. the compressor is not operating.

B. the compressor is running.

C. the cooling tower water is flowing.

D. low ambient conditions exist.

149. When long liquid lines are located on rooftops, they should be insulated to prevent

A. excessive flash gas.

B. high subcooling.

C. low superheat conditions.

D. oil foaming.

150. On a pump-down system a leaking liquid-line solenoid valve would cause

A. longer compressor run time.

B. compressor short cycling.

C. low superheat condition.

D. low cabinet temperatures.

151. A refrigeration system has been evacuated to 500 microns (approximately 29.92 in. Hg); the technician has waited for several minutes and observes that the pressure rises to 25 in. Hg and stops; this would indicate

A. a leak in the system.

B. that there is still liquid refrigerant in the oil and it should be reevacuated.

C. moisture in the system.

D. Both B and C

THE FOLLOWING QUESTIONS MAY BE HELPFUL TO STUDY FOR THESE CERTIFICATION TESTS:
NATE—AC-SV, AC-IN, HP-SV, HP-IN.
ESCO INSTITUTE HVAC EXCELLENCE AIR CONDITIONING AND COMMERCIAL REFRIGERATION CERTIFICATION

AIR CONDITIONING COMPONENTS

1. The part of an air conditioning system that changes the refrigerant from a low-pressure gas to a high-pressure gas is the

A. compressor.

B. condenser.

C. metering device.

D. evaporator.

2. The part of an air conditioning system that changes the refrigerant from a high-pressure gas to a high-pressure liquid is the

A. compressor.

B. condenser.

C. metering device.

D. evaporator.

3. The part of an air conditioning system that changes the refrigerant from a low-pressure liquid to a low-pressure gas is the

A. compressor.

B. condenser.

C. metering device.

D. evaporator.

4. The part of an air conditioning system that changes the refrigerant from a high-pressure liquid to a low-pressure liquid is the

A. compressor.

B. condenser.

C. metering device.

D. evaporator.

5. Which part of the system stores refrigerant until it is needed in the evaporator?

A. the accumulator

B. the surge drum

C. the receiver

D. the heat exchanger

6. Which part of the system prevents liquid from entering the compressor?

A. the moisture indicator

B. the receiver

C. the accumulator

D. the surge drum

7. A split-system air conditioning unit has

A. the evaporator, receiver, and compressor connected by a refrigeration line set to the condenser and metering device.

B. the metering device and compressor connected by a refrigeration line set to the condenser and evaporator.

C. the receiver, compressor, and metering device connected by a refrigeration line set to the condenser and evaporator.

D. the compressor and condenser connected by a refrigeration line set to the metering device and evaporator.

8. On the "System Component Diagram" (Figure 1–10), match the following components with their proper names.

1. ______
2. ______
3. ______
4. ______
5. ______
6. ______
7. ______
8. ______
9. ______
10. ______

Components:

A. accumulator

B. compressor

C. condenser

D. evaporator

E. filter dryer

F. heat exchanger

G. metering device

H. oil separator

I. receiver

J. sight glass

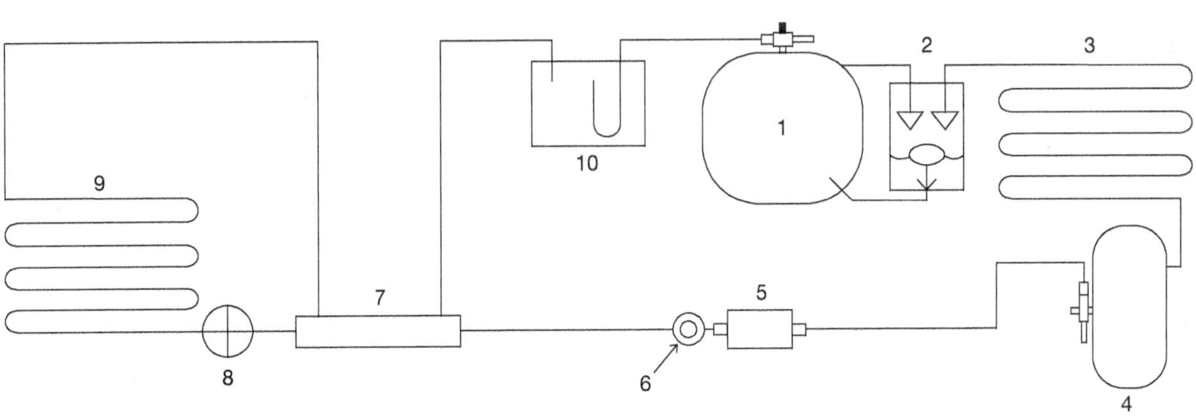

FIGURE 1–10
System Component Diagram

9. In a sealed system, as the pressure increases

A. the temperature increases.

B. the temperature decreases.

C. the temperature stays the same.

D. the refrigerant will always expand.

10. On the "Split-System Diagram" (Figure 1–11), match the following components with their proper names.

1. ______
2. ______
3. ______
4. ______
5. ______
6. ______
7. ______
8. ______
9. ______
10. ______

Components:

A. compressor

B. condenser

C. condenser fan

D. discharge line

E. evaporator

F. filter dryer

G. indoor blower

H. liquid line

I. metering device (restrictor)

J. suction line

11. In the "Split-System Diagram" (Figure 1–11) where does the refrigerant change from a high-pressure gas to a high-pressure liquid?

A. 2

B. 3

C. 8

D. 9

12. In the "Split-System Diagram" (Figure 1–11), where does the refrigerant change from a low-pressure liquid to a low-pressure gas?

A. 2

B. 3

C. 8

D. 9

13. In the "Split-System Diagram" (Figure 1–11), where does the refrigerant change from a low-pressure gas to a high-pressure gas?

A. 1

B. 3

C. 7

D. 8

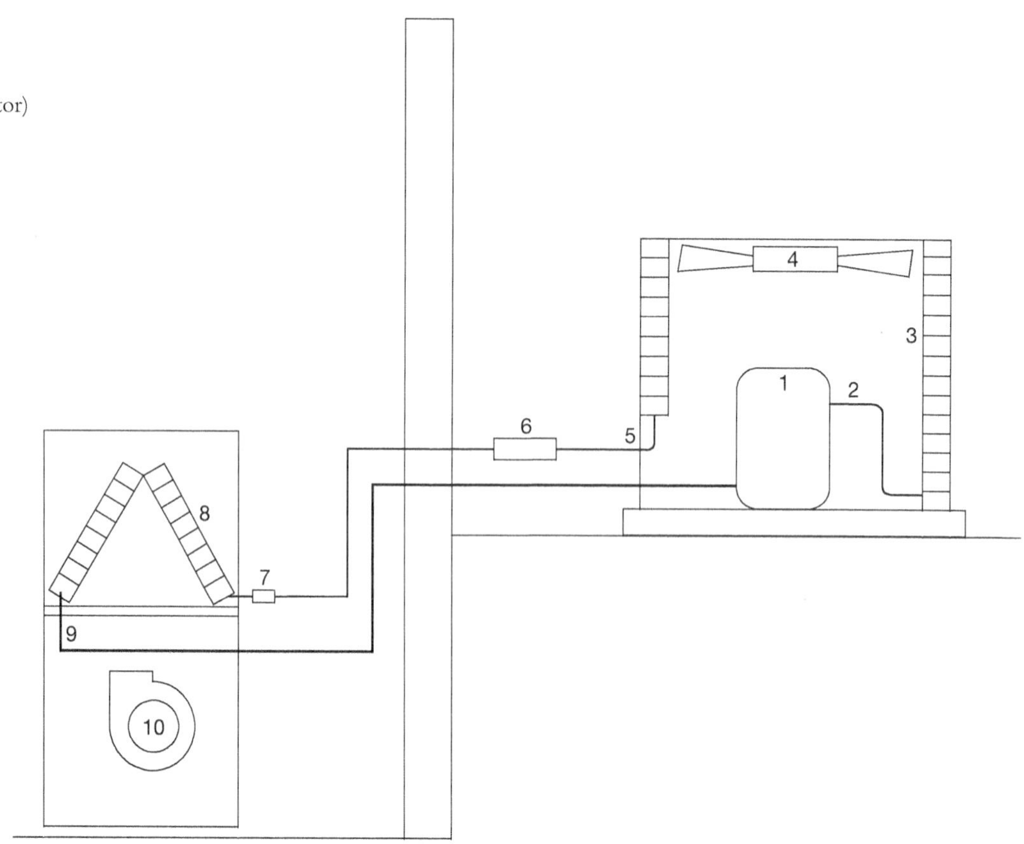

FIGURE 1–11

Split-System Diagram

14. In the "Split-System Diagram" (Figure 1–11), where does the refrigerant change from a high-pressure liquid to a low-pressure liquid?

A. 1

B. 3

C. 7

D. 8

15. Why is it recommended to add a gravel fill under the pad of a condensing unit?

A. This will help water to drain from inside the unit.

B. This can prevent vibration of the condensing unit.

C. This will prevent settling of the ground, which can cause refrigerant line damage.

D. This will allow the system to be free floating for installation flexibility.

16. Residential split-system outdoor condensing units are commonly wired with a

A. 208-volt/240-volt single-phase power supply.

B. 120-volt single-phase power supply.

C. 208-volt/240-volt three-phase power supply.

D. 208-volt/240-volt polyphase power supply.

17. Before energizing a newly installed system, what test should always be done?

A. Check operating pressures.

B. Check operating temperatures.

C. Check for tight connections at electrical terminals.

D. Check the resistance to the compressor terminals.

18. A room thermostat should be mounted

A. 4′ from the floor.

B. near a supply air register for accuracy.

C. on an outside wall.

D. near a return-air grill.

19. What is the secondary voltage on a step-down control transformer for a residential furnace?

A. 24 volts

B. 120 volts

C. 220 volts

D. 440 volts

20. What is the "R" terminal used for on the thermostat?

A. the power from the transformer

B. the blower relay coil

C. the cooling contactor coil

D. the heating control circuit

21. What is the "W" terminal used for on the thermostat?

A. the power from the transformer

B. the blower relay coil

C. the cooling contactor coil

D. the heating control circuit

22. What is the "Y" terminal used for on the thermostat?

A. the power from the transformer

B. the blower relay coil

C. the cooling contactor coil

D. the heating control circuit

23. What is the "G" terminal used for on the thermostat?

A. the power from the transformer

B. the blower relay coil

C. the cooling contactor coil

D. the heating control circuit

24. What is the purpose for insulating a suction line?

A. to prevent excessive subcooling of the refrigerant

B. to increase the refrigerant superheat entering the compressor

C. to decrease the system's vibration

D. to prevent condensation from forming

25. When installing a refrigeration line set through a masonry wall,

A. the lines must be covered by PVC to prevent contact with the masonry material.

B. the lines must use a flexible connector to protect against a foundation shift.

C. the lines must never be installed through a masonry wall.

D. the lines must be secured tightly against the top portion of the wall to prevent vibration.

26. To ensure proper oil return to the compressor, when installing a refrigeration line set

A. the suction line must be pitched away from the condensing unit.

B. the liquid line must be pitched away from the evaporator coil.

C. the suction line must be pitched toward the condensing unit.

D. the liquid line must be pitched toward the evaporator coil.

27. To ensure proper oil return to the compressor, a coiled refrigeration lineset

A. must be coiled horizontally and the vapor must flow from bottom to top.

B. must be coiled horizontally and the vapor must flow from top to bottom.

C. must never be coiled.

D. must be coiled vertically.

28. What must be done to the system when a condensing unit is precharged for 25′ of refrigerant line and uses 32′ of refrigerant line?

A. The operating high-side pressure must be reduced by 7 psig.

B. The operating high-side pressure must be increased by 7 psig.

C. Refrigerant must be added to the system.

D. Refrigerant must be removed from the system.

29. Quick-connect fittings are commonly used on

A. precharged refrigeration linesets.

B. soft copper tubing with the use of a flare nut.

C. hard-drawn copper tubing with the use of an O-ring.

D. None of the above

30. What type of fitting uses an O-ring to help seal its joint?

A. a compression fitting

B. a quick-connect fitting

C. a flare fitting

D. a sweat connection

31. When soldering a copper connection and solder is drawn into the joint, this is commonly called

A. conduction.

B. capillary action.

C. siphon action.

D. flow absorption.

32. When installing an evaporator coil, before soldering, the end of the evaporator lines must be opened to the atmosphere because

A. the coil is precharged with CO_2.

B. the coil is precharged with R-22.

C. the coil is precharged with nitrogen.

D. None of the above

33. What type of metering device is used on a 13 SEER split air conditioning system?

A. a capillary tube

B. a piston type

C. a TXV

D. an AXV

34. If ductwork is located in an attic or crawl space, it must be

A. supported every 18″

B. insulated.

C. supported with joist hangers.

D. a maximum of 15′ in length.

35. An auxiliary drain pan must

A. be a minimum of 3″ deep.

B. have a minimum perimeter of 5″ larger than the unit being served.

C. have a minimum perimeter of 3″ larger than the unit being served.

D. be made of aluminized steel.

36. A secondary drain line

A. is always necessary in an auxiliary drain pan.

B. must have no more than two 90° elbows.

C. must not be less than 1″ diameter.

D. must be located in a conspicuous place to serve as an alarm.

37. A normally closed float switch that is connected to an auxiliary drain pan should be wired to control

A. the blower motor.

B. the contactor coil.

C. the condensate pump.

D. the inducer motor.

THE FOLLOWING QUESTIONS MAY BE HELPFUL TO STUDY FOR THESE CERTIFICATION TESTS:
NATE—AC-SV, AC-IN, HP-SV, HP-IN.
ESCO INSTITUTE-HVAC EXCELLENCE AIR CONDITIONING AND COMMERCIAL REFRIGERATION CERTIFICATION

AIR CONDITIONING TROUBLESHOOTING

1. What are the symptoms of a TXV air conditioning system if it has lost its charge?

A. The superheat will decrease and the high-side pressure will increase.

B. The subcooling will increase and the low-side pressure will increase.

C. The superheat will increase and the low-side pressure will decrease.

D. The subcooling will increase and the low-side pressure will increase.

2. A customer who owns a split air conditioning system complains that his system has been running but his house doesn't seem to be cooling properly. The service technician arrives on the job and finds that the indoor blower is operating and the air filter is clean. The condenser fan on the outdoor unit is operating, but the technician notices that the liquid line is hot to the touch. Before connecting gauges to the system, the next step should be to

A. clean the condenser coil.

B. remove the air filter.

C. connect the TXV sensing bulb to the suction line.

D. make sure the condenser fan is turning counterclockwise.

3. What are symptoms of an air conditioning system that has a low charge?

A. The evaporator coil will be partially frosted and the discharge pressure will increase.

B. The suction line will sweat and the low-side pressure will decrease.

C. The low-side pressure will decrease and the evaporator coil will be partially frosted.

D. The discharge pressure will decrease and the suction pressure will increase.

4. A customer calls and says that her air conditioning system is not cooling properly. Upon arrival, a technician notices that the suction line and compressor have frost buildup.

   Technician A says that the problem is probably a low charge.

   Technician B says that it is possibly a dirty air filter.

   Which technician is more likely to be correct?

   A. technician A

   B. technician B

   C. Both technicians are correct.

   D. Both technicians are incorrect.

5. What are the operating pressures of an oil pressure safety switch?

   A. the oil pump pressure and the crankcase pressure

   B. the oil pump pressure and the discharge pressure

   C. the oil pump pressure only

   D. Both B and C

6. What is a common discharge temperature of an air-cooled air conditioning system that is used on a 90°F day?

   A. 230–250°F

   B. 110–120°F

   C. 90–100°F

   D. 80–90°F

7. What are the symptoms of an air conditioning system that has a bad compressor discharge valve?

   A. high suction pressure, high discharge pressure, and low amp draw

   B. high discharge pressure, low suction pressure, and high amp draw

   C. low amp draw, low suction pressure, and high discharge pressure

   D. low discharge pressure, high suction pressure, and low amp draw

8. What is the purpose of an EPR valve?

   A. It decreases the pressure to the compressor during an excessive load.

   B. It increases pressure in a high-temperature evaporator coil when it is used on a dual-evaporator system.

   C. It decreases pressure in a low-temperature evaporator coil when it is used on a dual-evaporator system.

   D. Both A and C

9. Which of the following is a possible cause for a compressor to overheat?

   A. a dirty air filter

   B. a plugged evaporator coil

   C. a low refrigerant charge

   D. Both A and C

10. What is a possible cause of a sweat or frost pattern downstream from a sight glass and filter dryer?

    A. an overcharged system

    B. a plugged metering device

    C. a leaking sight glass

    D. a plugged filter dryer

11. What causes the automatic expansion valve to close?

    A. a decrease in high-side pressure

    B. an increase in low-side pressure

    C. a decrease in low-side pressure

    D. an increase in high-side pressure

12. A customer calls with an insufficient cooling complaint on a 90°F day. Upon arrival it is found that the system, using R-22, is operating at 50 psig on the low side and 150 psig on the high side. Further investigation indicates that the evaporator coil is partially frosted on one circuit of the distributor.

    Technician A says that the system has a low charge and must be leak-tested.

    Technician B says that the capillary tube from the distributor is partially restricted.

    Which technician is correct?

    A. technician A

    B. technician B

    C. Both technicians are correct.

    D. Both technicians are incorrect.

13. What type of problems can be caused by an oversized air conditioning system?

    A. excessive cooling and dehumidification

    B. ice buildup at the condenser coil

    C. excessively high suction and discharge pressures

    D. short cycling and insufficient dehumidification

14. What types of symptoms will be present if the sensing bulb of a TXV is not properly connected?

    A. high suction pressure and low superheat

    B. low suction pressure and high superheat

    C. possible floodback and low discharge pressures

    D. Both B and C

15. An air conditioning unit has a suction line that is frosted up to the compressor. What faults could cause this to occur?

    A. a dirty evaporator coil

    B. a plugged air filter

    C. a bad indoor blower relay

    D. All of the above

16. If 24 volts are present to the contactor coil on an air conditioning system and the condenser fan or compressor is inoperative, which of the following could be at fault?

    A. a bad cooling thermostat

    B. a blown fuse

    C. a faulty control transformer

    D. All of the above

17. A customer calls with an insufficient cooling complaint. Upon arrival to the service call it is found that the system is using R-22 with the high-side pressure at 290 psig and the low-side pressure at 85 psig. The ambient temperature is 85°F and the indoor wet bulb (WB) is 60°F.

Technician A says that, more than likely, the condenser is dirty.

Technician B says that the system's pressures are common for this type of day.

Who is correct?

A. technician A

B. technician B

C. Both technicians are correct.

D. Both technicians are incorrect.

18. If a TXV sensing bulb had lost its charge, what effect would this have on the air conditioning system?

A. It would cause the high-side pressure to increase.

B. It would cause the low-side pressure to increase.

C. It would cause the low-side pressure to decrease.

D. Both A and C

19. A customer calls with a preventative maintenance service call on his R-22 water-cooled air conditioning system. The technician notices the high-side pressure is 195 psig and the leaving water temperature is 90°F. What is wrong with the system? (Use the pressure–temperature chart provided, Table 1–2.)

A. The condenser is dirty.

B. The water valve is closed too much.

C. The water valve is opened too much.

D. There is no problem.

20. What would be the approximate discharge temperature and pressure of an R-22 air-cooled condensing unit on an 80°F day? (Use the pressure–temperature chart provided, Table 1–2.)

A. 80–90°F at 145–170 psig

B. 100–110°F at 195–225 psig

C. 110–120°F at 225–260 psig

D. 120–130°F at 260–295 psig

21. A customer calls with an insufficient cooling complaint. Upon arrival to the service call it is found that the outdoor condensing unit is operating, but the indoor blower is not in operation.

Technician A says that it is possible that it is a bad fuse to the furnace.

Technician B says that the control transformer could also be at fault.

Who is correct?

A. technician A

B. technician B

C. Both technicians are correct.

D. Both technicians are incorrect.

TABLE 1–2

Pressure in psig	Temperature in °F	
	R-410A	**R-22**
100	31	58
105	34	62
110	36	64
115	38	66
120	40	68
180	63	94
185	65	96
190	66	98
195	68	100
200	70	102
205	71	104
210	73	107
215	74	108
220	76	109
225	77	110
230	78	112
240	81	114
250	84	118
260	86	120
270	89	122
280	91	126
290	94	128
300	96	132
310	98	134
320	100	136
330	103	138
340	105	142
350	107	144
360	109	146
365	110	147
370	111	148
375	112	149
380	113	150

22. A party-store owner tried to adjust the water valve on his water-cooled condensing unit, then he called for assistance from an HVAC/R service technician. It was found that when servicing the unit, the outlet water temperature was too low.

Technician A says that turning the adjustment stem on the water valve clockwise will close the valve, and this will cause an increase in high-side pressure.

Technician B says that turning the adjustment stem on the water valve counterclockwise will close the valve, and this will increase the high-side pressure.

Which technician is correct?

A. technician A

B. technician B

C. Both technicians are correct.

D. Both technicians are incorrect.

23. When connecting gauges to an R-410A air conditioning system on an 85°F day, the pressures read 365 psig on the high side and 105 psig on the low side. What is a possible problem with the system? Use the pressure–temperature chart provided earlier (Table 1–2) to answer this question.

A. dirty air filter

B. dirty condenser

C. insufficient compressor

D. There is no problem.

24. What are two common symptoms of an overcharged air conditioning system that uses a capillary-tube-type metering device?

A. an increase in discharge pressure and a decrease in suction pressure

B. an increase in suction pressure and an increase in discharge pressure

C. a decrease in discharge pressure and an increase in suction pressure

D. a decrease in suction pressure and a decrease in discharge pressure

25. Name two symptoms that are common to an overcharged air conditioning system and a system that has a dirty condenser.

A. The subcooling decreases and the amperage increases.

B. The suction pressure increases and the discharge pressure decreases.

C. The suction pressure decreases and the subcooling increases.

D. The discharge pressure increases and the amperage increases.

26. When an air conditioning system has an undercharge, what happens to the superheat and the subcooling of the system?

A. The superheat increases and the subcooling increases.

B. The superheat increases and the subcooling decreases.

C. The superheat decreases and the subcooling increases.

D. The superheat decreases and the subcooling decreases.

27. If a rooftop air conditioning system does not have any means of low ambient protection, what will be two symptoms of the system when operating under a low ambient condition?

A. high discharge pressure and low superheat

B. low discharge pressure and high suction pressure

C. high amperage and high superheat

D. high superheat and low discharge pressure

28. What are two common symptoms of a rooftop air conditioning system that operates with a reverse-acting high-pressure control stuck in the closed position?

A. condenser fans operating during a low ambient condition and a low suction pressure

B. high discharge pressure and high suction pressure

C. high discharge pressure and condenser fans de-energized

D. high suction pressure and low superheat

29. What are two problems that can cause an air conditioning system to have an excessively high superheat?

A. an oversized metering device and an overcharge

B. an uninsulated suction line and a restricted metering device

C. an undersized metering device and a disconnected TXV sensing bulb

D. a disconnected TXV sensing bulb and a restricted metering device

30. What are two common problems that can cause a sight glass in an air conditioning system to have bubbles?

A. an overcharge and a restricted metering device

B. air in the system and an undercharge

C. a dirty evaporator coil and a low ambient condition

D. an undercharge and low ambient conditions

31. When the CTH is in the closed position and the LPS is in the open position, what would the voltage be across terminal 1 at the ATH and T2? Use the "Single-Phase AC Diagram" (Figure 1–12) to find the answer.

A. 0 volts

B. 240 volts

C. 120 volts

D. Both A and B

32. When the CTH is in the closed position and the LPS is in the closed position, what would be the voltage across terminals 1 and 2 of FR1? Use the "Single-Phase AC Diagram" (Figure 1–12) to find the answer.

A. 0 volts

B. 240 volts

C. 120 volts

D. None of the above

33. On a call for cooling, the COMP is in operation but the CF is not operating. A 240-volt reading is obtained across terminal 2 on the ATH and T2. What two malfunctions could be present? Use the "Single-Phase AC Diagram" (Figure 1–12) to find the answer.

A. a defective ATH and/or RC

B. a defective RC and/or CF

C. a defective CF and/or ATH

D. a defective C2 contact and/or ATH

34. On a call for cooling, the CF is operating but the COMP is not. A clamp-on ammeter is placed between "T1" and the "R" terminal on the compressor and a reading of 0 amps is obtained. It is then placed between terminal 1 on the PR and the SC and 0 amps is obtained. What is most likely to be the problem? Use the "Single-Phase AC Diagram" (Figure 1–12) to find the answer.

A. a defective SC

B. an open OL

C. a defective run winding

D. a defective C1 contact

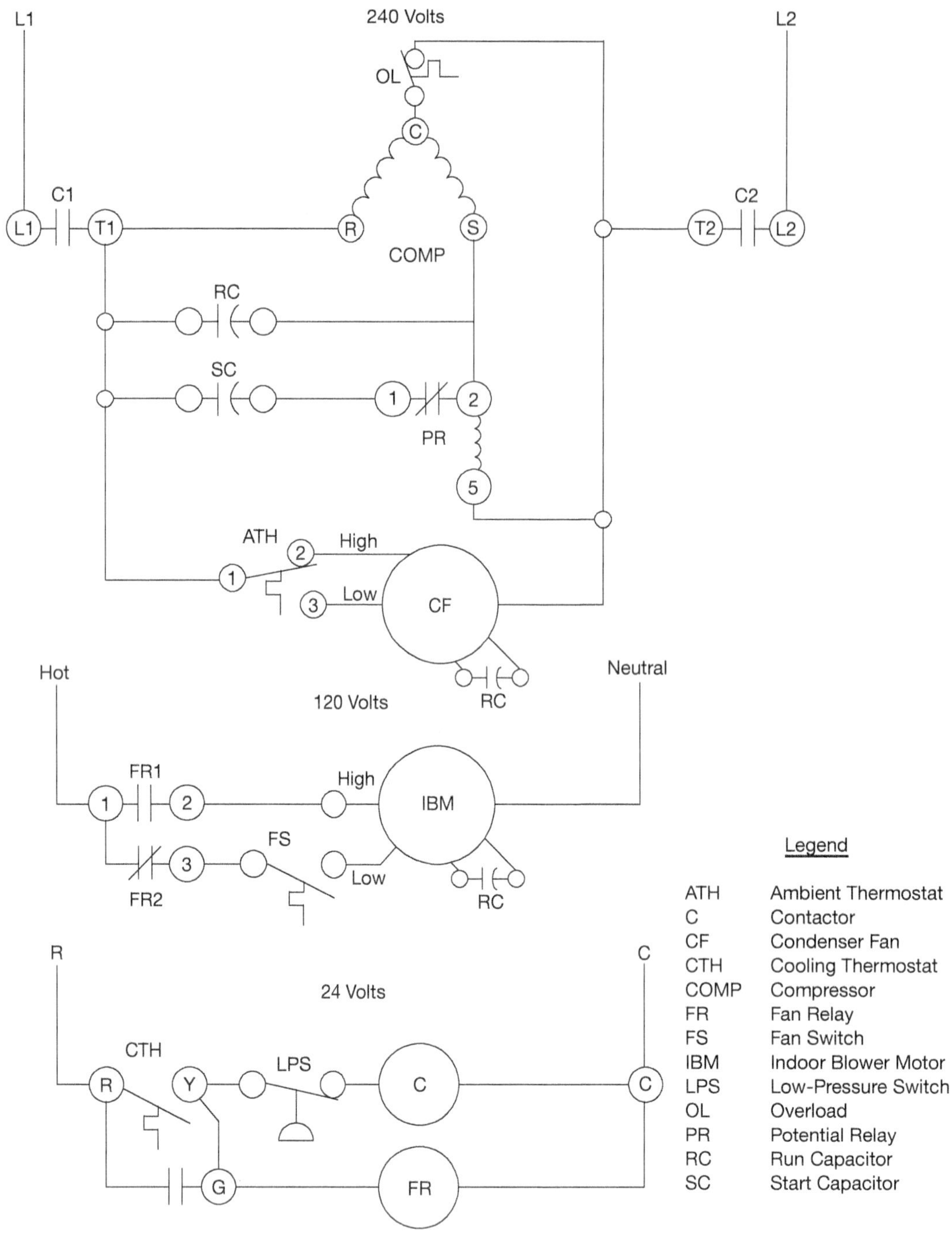

FIGURE 1–12
Single-Phase AC Diagram

35. A technician arrives on the job and the customer complains that there is no cooling on his rooftop unit. Troubleshooting the system, the technician finds that the compressor is faulty. The technician notices that there is another faulty compressor laying next to the unit. A further check shows that the crankcase heater is electrically burnt open. What type of damage is most likely to have been done to the compressor?

A. broken rings

B. broken valves

C. seized bearings

D. All of the above

36. After a compressor motor has been severely burnt, what accessory must be added as an aid to rid the air conditioning system of acid?

A. an acid-core liquid-line drier

B. an acid-core receiver

C. an acid-core suction line drier

D. All of the above

37. When troubleshooting a split-system air conditioner, it is found that the compressor motor is humming but not starting.

Technician A says that it could be an open internal overload protector.

Technician B says that it could be a blown fuse.

Which technician is correct?

A. technician A

B. technician B

C. Both technicians are correct.

D. Both technicians are incorrect.

38. A technician receives an air conditioning service call stating that the system is operating but isn't cooling properly. When the technician arrives, the technician notices that the compressor has recently been changed and it is an R-22 system. The technician checks the indoor air filter and the condenser coil. Both are clean and both coils seem to have adequate airflow. When the technician connects gauges, it is noticed that the suction pressure is 95 psig, the high-side pressure is 285 psig, and the ambient temperature is 80°F. What is most likely to be the problem with the system?

A. The system was improperly serviced and it has air trapped in the condenser.

B. The unit has been overcharged.

C. The unit has an ice blockage at the metering device caused by improper service procedures.

D. Either A or B

39. An air conditioning system that uses a fan center has an inoperable indoor blower. When the thermostat is set for cooling and dialed below room temperature, a test with the voltmeter shows that there is 24 volts across terminals "R" and "G" at the terminal block. What is the problem with the system?

A. a bad blower

B. a bad relay

C. a bad thermostat

D. Any of the above

40. When testing an indoor blower control circuit and the cooling circuit on a fan center, what terminals should read 24 volts if the air conditioning system is in operation?

A. ("R" and "Y") and ("R" and "G")

B. ("Y" and "G") and ("R" and "C")

C. ("R" and "W") and ("R" and "G")

D. ("Y" and "C") and ("G" and "C")

41. Which of the following problems can cause a decrease in a system's pressures and possible icing of the evaporator coil?

A. a missing air filter

B. improper rotation of the indoor blower

C. blower too small

D. All of the above

42. A split-system air conditioner that uses a piston-type metering device is to be charged with R-22.

Technician A says that the outdoor dry-bulb temperature is needed to determine the proper superheat.

Technician B says that the indoor wet-bulb temperature is needed to determine the proper superheat.

Which one is correct?

A. technician A

B. technician B

C. Both technicians are correct.

D. Both technicians are incorrect.

43. A 220-volt 30-amp motor circuit keeps tripping the breaker after the motor is started. The starting amperage is 60 amps. When measuring the amperage of the circuit, this is drawn for a brief moment. After startup, the amperage drops to 22 amps and then the breaker trips. What is the problem with the circuit?

A. a shorted run winding

B. an open run winding

C. a grounded motor winding

D. a defective circuit breaker

44. A rooftop package system keeps tripping out on the high-pressure switch. What are three possible causes for this problem?

A. condenser air being recirculated, plugged metering device, and overcharge

B. excessively bent condenser fins, air in the system, and undercharge

C. dirty condenser, incorrect rotation of the condenser fan, and bad compressor valves

D. dirty condenser, bent condenser fins, and air recirculating over the condenser

45. What are two common symptoms of refrigerant leaks in an air conditioning system?

A. oil traces around flared fittings and low superheat

B. partially frosted evaporator and high superheat

C. low discharge pressure and high suction pressure

D. low suction pressure and a high compressor amp draw

46. A system that is being serviced has been evacuated down to 500 microns. When the gauges are closed and the vacuum pump is turned off, the pressure starts to increase and stops at 1,000 microns.

Technician A says that the system has a leak and needs to be repaired.

Technician B says that the system has moisture present and it needs to be evacuated longer.

Which technician is correct?

A. technician A

B. technician B

C. Both technicians are correct.

D. Both technicians are incorrect.

47. What are common methods that are used for leak-testing a system?

A. a vacuum test

B. oil traces around fittings

C. soap bubble solution

D. All of the above

48. Using the "Three-Phase AC Diagram" (Figure 1–13), determine when the crankcase heater operates.

A. when the thermostat is in the closed position

B. when the thermostat is in the open position

C. as long as there is power to the 240-volt circuit

D. Both B and C

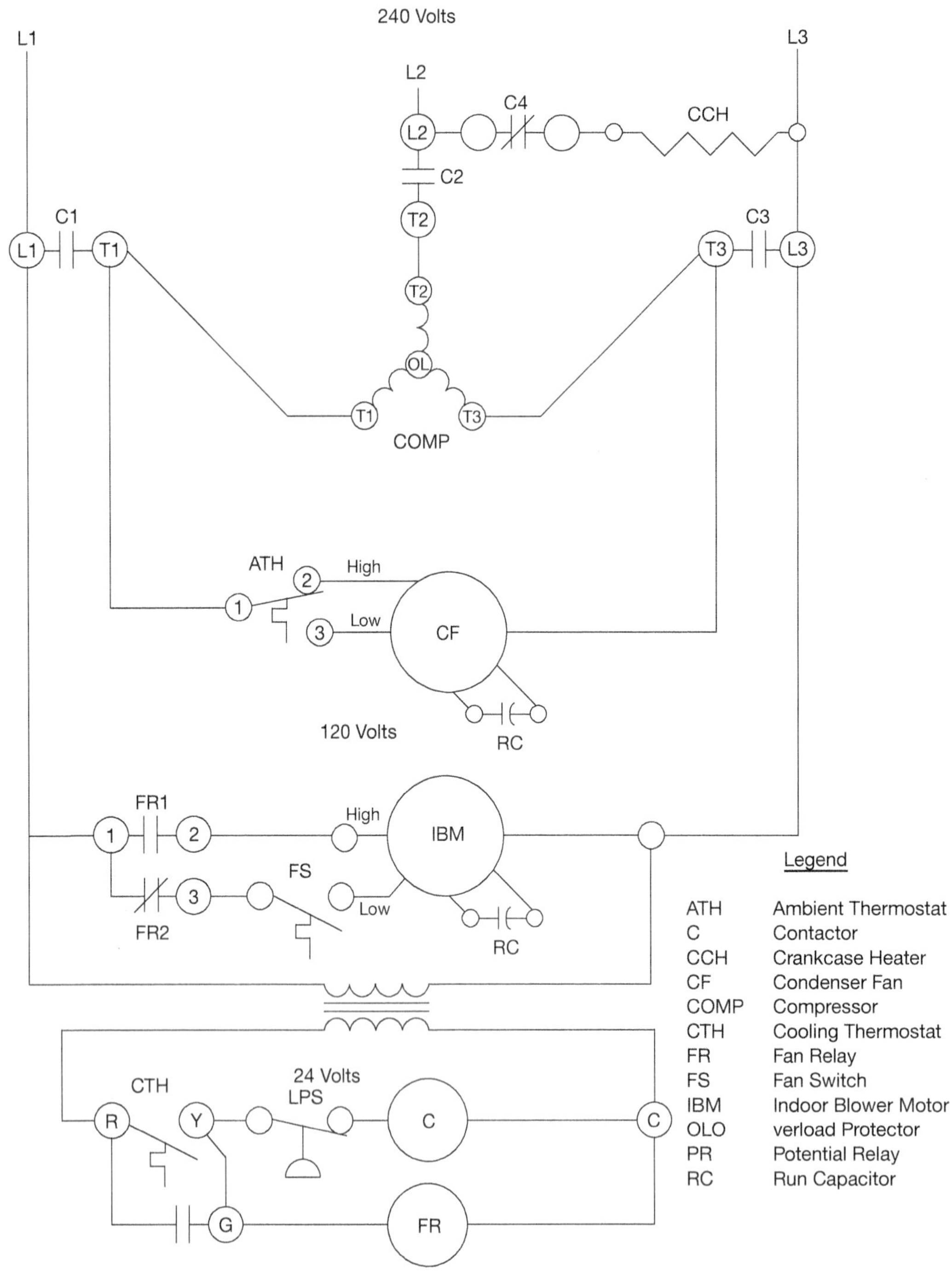

FIGURE 1–13

Three-Phase AC Diagram

49. If L2 lost its source of power due to a blown fuse, what 240 loads could still operate? Use the "Three-Phase AC Diagram" (Figure 1–13) to find the answer.

A. COMP

B. CCH

C. IBM

D. None of the above

50. What would be the resistance readings across T1, T2, and T3 on the COMP if OL were in the open position? Use the "Three-Phase AC Diagram" (Figure 1–13) to answer this question.

A. T1–T2 infinite resistance, T1–T3 zero resistance, and T2–T3 zero resistance

B. T1–T2 zero resistance, T1–T3 zero resistance, and T2–T3 zero resistance

C. T1–T2 infinite resistance, T2–T3 infinite resistance, and T2–T3 measurable resistance

D. T1–T2 infinite resistance, T2–T3 infinite resistance, and T2–T3 infinite resistance

51. On an 85°F outdoor ambient day, the air conditioning system's indoor coil and suction line have frost buildup. A voltage check from "Y" to "C" and "G" to "C" shows 24 volts. What are the possible problems with the system? Use the "Three-Phase AC Diagram" (Figure 1–13) to answer this question.

A. a defective C, FR, and/or ATH

B. a defective RC, CF, and/or ATH

C. a defective FR, IBM, or RC

D. a defective FR, C, or LPS

52. On a 90°F outdoor ambient day, the system's condenser fan is inoperable. The discharge pressure is at 350 psig and the compressor overheats, causing it to cycle on the internal overload protector. A voltage check from "Y" to "C" and "G" to "C" shows 24 volts. What are the possible problems with the system? Use the "Three-Phase AC Diagram" (Figure 1–13) to answer this question.

A. a defective C, FR, and/or ATH

B. a defective RC, CF, and/or ATH

C. a defective FR, IBM, or RC

D. a defective C, CF, or ATH

53. What is the purpose of the ATH? Use the "Three-Phase AC Diagram" (Figure 1–13) to answer this question.

A. It de-energizes the CF if the outdoor ambient air temperature is too high.

B. It de-energizes the CF if the outdoor ambient air temperature is too low.

C. It decreases the speed of the CF if the outdoor ambient air temperature is too high.

D. It decreases the speed of the CF if the outdoor ambient air temperature is too low.

54. What is the purpose of a CPR valve?

A. It is used to increase the pressure in the evaporator coil of a two-temperature system.

B. It is used to control refrigerant pressure in the compressor crankcase.

C. It keeps a constant evaporator pressure on a single-pipe system.

D. It floods the receiver during low ambient temperatures.

55. What is the common cause of a compressor burnout?

A. improper system evacuation during installation

B. short cycling of the system

C. excessively dirty condenser

D. All of the above

56. Which of the following is a common cause of an excessively low superheat on a TXV system?

A. a sensing bulb improperly connected to the suction line

B. a low system charge

C. the superheat spring adjusted too far in the clockwise position

D. Both A and C

57. What are three methods for controlling a system's high-side pressure during low ambient conditions?

A. cycling condenser fans, EPR valves, and CPR valves

B. flooding the condenser, cycling condenser fans, and dampering condenser air

C. pump-down system, flooding the condenser, and oversized receivers

D. reverse-acting pressure switch, cycling evaporator fans, and dampering condenser air

58. What is a common temperature difference between the condensing temperature on an air-cooled condensing unit and the outdoor ambient temperature?

A. 5 to 10°F

B. 10 to 15°F

C. 20 to 30°F

D. 40 to 50°F

59. What controls are used on air conditioning or refrigeration systems that incorporate an automatic pump-down system?

A. A high-pressure switch cycles the liquid-line solenoid, and a thermostat controls the contactor coil.

B. A low-pressure switch cycles the condenser fans, and a thermostat controls the compressor.

C. A thermostat cycles the condenser fans, and a reverse-acting pressure switch controls the compressor.

D. A thermostat cycles the liquid-line solenoid, and a low-pressure switch controls the compressor.

60. A customer calls and complains that his air conditioning system is not operating properly. It is noticed that the compressor is not operating. A check on the run capacitor shows that there is $33\mu f$ and the rating on the capacitor is $40\mu f$.

Technician A says that the capacitor is good because it is within its normal 20% range.

Technician B says that the capacitor is bad because it is not within its 10% range.

Which technician is correct?

A. technician A

B. technician B

C. Both technicians are correct.

D. Both technicians are incorrect.

61. A system that is operating with a clear sight glass shows

A. the sight glass is full of liquid.

B. the sight glass is full of gas.

C. the system is always running efficiently.

D. Both A and B

62. Technician A says "When charging a system using the superheat method, adding refrigerant to the system will increase the superheat."

Technician B says "When charging a system using the superheat method, adding refrigerant will decrease the superheat."

Which technician is correct?

A. technician A

B. technician B

C. Both technicians are correct.

D. Both technicians are incorrect.

63. What is the purpose of the TXV?

A. It keeps a constant superheat in the evaporator.

B. It keeps a constant pressure in the evaporator.

C. It keeps a constant pressure drop through the evaporator.

D. Both A and C

64. A liquid-charged TXV bulb will

A. always have some liquid present in the bulb and diaphragm.

B. have 100% liquid present in the bulb and diaphragm at all times.

C. have liquid present only until the valve is fully open.

D. Both A and B

65. When using a vapor-charged TXV bulb, it is necessary to have

A. the valve body at a lower temperature than the sensing bulb.

B. the sensing bulb at a lower temperature than the valve body.

C. an external equalizer on the TXV if the evaporator has a 1- to 2-psig pressure drop.

D. an internal equalizer on the TXV if the evaporator has an over 5-psig pressure drop.

66. A customer calls and complains that her high-efficiency air conditioning system is operating, but is not cooling properly. Further investigation shows that the system is using a TXV type of metering device. A subcooling check shows that the system appears to have the proper charge. A test shows that the superheat is 30°F when it is supposed to be 15°F. What could be the problem?

A. The TXV valve spring adjustment could be turned counterclockwise too far.

B. The sensing bulb could be loose.

C. The external equalizer could be plugged.

D. The system could have improper evaporator airflow.

67. A system that uses a thermal electric expansion valve

A. must use a liquid-line solenoid to pump the system down during the off cycle.

B. can use the thermal electric expansion valve instead of a liquid-line solenoid as part of the pump-down system.

C. must not use a pump-down system during the off cycle.

D. closes the valve as voltage increases and opens the valve as voltage decreases.

68. What are possible problems that can cause the discharge-line temperature on an air conditioning system to be cool to the touch?

A. The system could be undercharged.

B. The compressor could have bad valves.

C. The system could have a restricted metering device.

D. All of the above

69. A customer calls to complain that his central air conditioning system is not cooling properly. The technician arrives on the job and notices that the indoor blower has adequate airflow and the outdoor condenser fan is operating, but the compressor is not operating. After pulling out the electrical disconnect, the technician measures resistance across the compressor motor windings. The readings are as follows: run to common, infinite ohms; start to common, infinite ohms; and run to start, 10 ohms. What could have caused this condition?

A. an open start winding

B. an open run winding

C. an open internal overload protector

D. Both A and B

70. A commercial customer calls complaining that her rooftop air conditioning system is not operating. The technician arrives on the job and sets the thermostat to cooling, then notices that the indoor fan is not operating. The technician then sets the thermostat to the fan on and it still does not operate. What two terminals must be tested for 24 volts to ensure that the problem is not the thermostat?

A. "R" to "C"

B. "Y" to "C"

C. "R" to "G"

D. "G" to "C"

71. A commercial customer calls on a 90°F day, complaining that his rooftop unit is running but not operating properly. The technician arrives on the job and the customer shows the technician an invoice from the spring season. The invoice states that 2 lb of R-22 were added to the system. After connecting the appropriate gauges, the pressure reads 88 psig on the low side and 300 psig on the high side. What is the most likely system problem?

A. The system has been overcharged.

B. The system has air trapped in the condenser.

C. The system has a small leak.

D. Both A and B

72. Which of the following could be the cause for a system's discharge and suction pressure being low?

A. a restricted metering device

B. bad compressor valves

C. an inoperative compressor

D. Both A and C

73. What problem or problems can cause low discharge and suction pressure?

A. an oversized TXV

B. an undercharged system

C. insufficient airflow over the evaporator coil

D. Both B and C

74. What problem can cause an increase of discharge pressure and an increase of suction pressure?

A. bad compressor valves

B. an overcharged system

C. an inoperative compressor

D. an oversized compressor

75. What problem or problems can cause the discharge pressure to be low and the suction pressure to be high?

A. an oversized metering device

B. bad compressor valve

C. too low of an outdoor ambient temperature

D. Both B and C

76. Which of the following can cause excessive subcooling?

A. an undercharged system

B. an overcharged system

C. bad compressor valves

D. Both A and C

77. Which of the following can cause low superheat in a TXV system?

A. an overcharge

B. an undercharge

C. a TXV that is turned clockwise too far

D. an uninsulated TXV sensing bulb

78. What three things can prevent an air conditioning system from releasing heat at the proper rate?

A. a dirty evaporator coil, an improperly adjusted metering device, and a dirty condenser

B. a dirty condenser, recirculation of condenser air, and a defective condenser fan

C. a dirty evaporator coil, a defective evaporator fan, and an iced-up evaporator coil

D. All of the above

79. What three things can prevent an air conditioning system from absorbing heat at the proper rate?

A. a dirty evaporator coil, an improperly adjusted metering device, and a dirty condenser

B. a dirty condenser, recirculation of condenser air, and a defective condenser fan

C. a dirty evaporator coil, a defective evaporator fan, and an iced-up evaporator coil

D. All of the above

80. What component is most often the cause of a start capacitor failure?

A. a start relay

B. a run capacitor

C. a faulty start winding

D. a seized compressor

81. A compressor motor starts and operates for a short period of time and cycles on the external overload protector. What are two possible problems that could cause this?

A. an open start winding and low voltage

B. an open start capacitor and a faulty start relay

C. a shorted run capacitor and an open start capacitor

D. low voltage and a faulty start relay

82. Technician A says that to manually pump down an air conditioning system with a two-position service valve, the liquid line must be front-seated and the high- and low-side pressure will decrease to 0 psig when the pump-down is complete.

Technician B says that to manually pump down an air conditioning system with a two-position service valve, the liquid line must be front-seated and only the low-side pressure will decrease to 0 psig when the pump-down is complete and the high side will drop slightly.

Which technician is correct?

A. technician A is correct

B. technician B is correct

C. Both technicians are correct.

D. Both technicians are incorrect.

83. Technician A says that if you connect 208 volts to the primary side of a 240-volt/24-volt transformer, the transformer will burn out.

Technician B says that if you connect 208 volts to the primary side of a 240-volt/24-volt transformer, the control circuit voltage will drop.

Which technician is correct?

A. technician A

B. technician B

C. Both technicians are correct.

D. Both technicians are incorrect.

84. When a high-efficiency split air conditioning system has a plugged metering device, what symptoms might occur?

A. a high amp draw, an increase in high-side pressure, and a decrease in low-side pressure

B. a low amp draw, a decrease in high-side pressure, and an increase in low-side pressure

C. a high amp draw, a decrease in high-side pressure, and a decrease in low-side pressure

D. a low amp draw, a decrease in high-side pressure, and a decrease in low-side pressure

85. Why is it unsafe to operate an air conditioning system with a low charge?

A. It can seize the compressor because of excessive heat.

B. It can seize the compressor because of lack of lubrication.

C. It can seize the compressor because of excessive system pressures.

D. Both A and B

86. The technician observes a pressure drop of 25 degrees across an air conditioning evaporator coil at start-up. This drop would indicate

A. lack of airflow.

B. low refrigerant charge.

C. overcharged system.

D. excessive airflow.

87. Air filters should be replaced

A. whenever necessary.

B. every three months.

C. every six months.

D. once monthly.

88. On an R-22 air conditioning system, the suction-line filter drier should be changed when the pressure drop dictates

A. $\frac{1}{2}$ psig.

B. 3 psig.

C. 5 psig.

D. 7 psig.

89. A compressor having tight bearings would cause the current value to

A. rise more than normal.

B. decrease more than normal.

C. not be affected.

D. rise then decrease to normal.

90. The technician observes that the compressor is not operating. The compressor windings read 5 ohms start to run, run to common is opened, and start to common is opened. The technician concludes that

A. the internal overload is opened.

B. the compressor is grounded.

C. the contactor is defective.

D. the run winding is opened.

91. The temperature range of the suction line from an air conditioning system should be

A. 30 to 42°F.

B. 50 to 65°F.

C. 70 to 80°F.

D. above 80°F

92. Superheat is found in an air conditioning system in the

A. condenser.

B. liquid receiver.

C. evaporator.

D. evaporator and compressor.

93. A hard-start kit for an air conditioning system

A. increases resistance in the run winding.

B. increases resistance in the start winding.

C. increases starting torque.

D. decreases starting torque.

94. An air conditioning system using a fixed-orifice metering device and with noncondensables in the system will cause

A. the system's pressures to balance during shut-down.

B. the system to operate with low head pressure.

C. the system to operate with low head and high suction pressure.

D. the system to operate with high head and high suction pressure.

95. The internal overload of an air conditioning compressor senses winding temperature and

A. current in the compressor circuit.

B. current in the control circuit.

C. voltage in the control circuit.

D. voltage in the compressor circuit.

96. A restricted air filter in an air conditioning system can be identified by

A. high evaporator temperature.

B. high evaporator pressure.

C. low superheat.

D. high superheat.

97. The best description of a relay would be

A. it is usually larger than a contactor.

B. it contains a solenoid coil and one or more sets of contacts.

C. it contains a solenoid coil and one set of contacts.

D. it is energized by primary voltage.

98. Insufficient airflow can be caused by

A. an increase of air leakage from ductwork.

B. increased cooling capacity.

C. a defective compressor.

D. refrigerant restrictions.

99. A hard-start kit consists of

A. a run and start capacitor.

B. a start capacitor.

C. a start relay.

D. a start capacitor and a solid-state relay.

100. Which of the following will cause excessive discharge pressures on an air conditioning system?

A. condenser air being recirculated over the condenser

B. air trapped in the condenser due to poor service procedures

C. an overcharged air conditioning system

D. All of the above

THE FOLLOWING QUESTIONS MAY BE HELPFUL TO STUDY FOR THESE CERTIFICATION TESTS:

NATE—AC-SV, AC-IN, HP-SV, HP-IN

ESCO INSTITUTE HVAC EXCELLENCE AIR CONDITIONING AND HEAT PUMP

HEAT PUMPS

1. Using "Heat Pump Diagram" (Figure 1–14), match the following components to their proper names.

1. ____

2. ____

3. ____

4. ____

5. ____

6. ____

7. ____

8. ____

Components:

A. accumulator

B. check valve

C. compressor

D. de-ice coil

E. indoor coil

F. metering device

G. outdoor coil

H. reversing valve

2. What is the purpose of the reversing valve on an air-to-air heat pump?

A. It reverses the direction of the compressor.

B. It changes the flow of the refrigerant, making the outdoor coil the condenser during the winter months.

C. It changes the flow of the refrigerant, making the indoor coil the condenser during the winter months.

D. It changes the flow of the refrigerant, making the indoor coil the condenser during the summer months.

3. On a reversing-valve solenoid, what is connected to the top tube?

A. the compressor discharge line

B. the compressor suction line

C. the indoor coil

D. the outdoor coil

4. On a reversing-valve solenoid, what is connected to the bottom center tube?

A. the compressor discharge line

B. the compressor suction line

C. the indoor coil

D. the outdoor coil

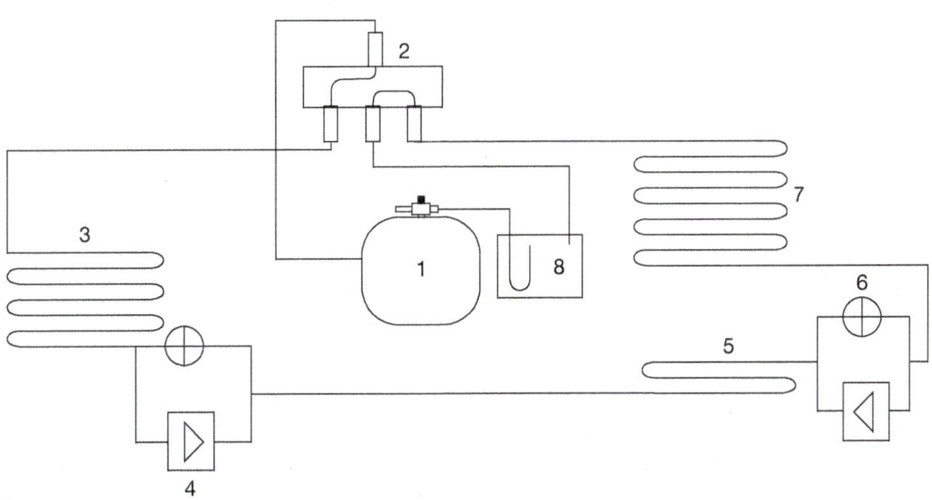

FIGURE 1–14

Heat Pump Diagram

5. What pressure difference must be between the low side and the high side of the heat pump system for the reversing valve to operate properly?
   A. 25 psig
   B. 50 psig
   C. 75 psig
   D. 100 psig
6. Where are the check valves located on an air-to-air heat pump?
   A. on the outdoor coil
   B. on the indoor coil
   C. on the outdoor coil and indoor coil
   D. on the suction line
7. How are the check valves connected in a heat pump system?
   A. on the inlet of the accumulator
   B. in parallel with both metering devices
   C. in series with one metering device and parallel with the other
   D. Both B and C
8. A ball-type check valve must be installed
   A. horizontally, with the seat on the outlet of the valve.
   B. vertically, with the seat on the bottom of the valve.
   C. horizontally, with the seat on the inlet of the valve.
   D. vertically, with the seat on the top of the valve.
9. When installing a ball-type check valve, what must be done to prevent warping?
   A. A heat sink must be used when brazing it into the line.
   B. The ball must be removed when brazing it into the line.
   C. A brass coupling must be used if connecting it to the liquid line.
   D. The valve must be cooled quickly after brazing it into the liquid line.
10. Where is the accumulator located on a heat pump system?
    A. between the bottom center tube of the reversing valve and the compressor
    B. between the top tube of the reversing valve and the compressor
    C. between the bottom right tube of the reversing valve and the compressor
    D. between the bottom left tube of the reversing valve and the compressor
11. When would you most likely find liquid in the vapor line on a heat pump?
    A. during the cooling mode
    B. during the heating mode
    C. during the beginning of a defrost
    D. during the heating mode after a defrost
12. What component must always be de-energized during the defrost cycle on a heat pump?
    A. the reversing valve solenoid
    B. the outdoor fan
    C. the indoor fan
    D. the defrost relay coil
13. Why are the auxiliary strip heaters energized during the defrost cycle?
    A. to add an extra heat load to the indoor coil
    B. to increase the pressure of the system for maximum defrost efficiency
    C. to increase supply-air temperature for comfort
    D. The auxiliary strip heaters are not energized during the defrost cycle.
14. What position must the reversing valve be in during the defrost cycle?
    A. the cooling mode
    B. the heating mode
    C. in the open position
    D. in the closed position
15. When a heat pump switches into the defrost cycle, it must
    A. remain in defrost until the temperature rise across the indoor coil reaches 50°F.
    B. remain in defrost until the indoor coil has reached a predetermined temperature.
    C. remain in defrost until the outdoor coil has reached a predetermined temperature.
    D. remain in defrost until the indoor thermostat is satisfied.
16. A defrost termination thermostat is commonly located
    A. on the vapor line, and opens at 55°F.
    B. on the liquid line, and closes at 55°F.
    C. on the indoor coil, and closes at 55°F.
    D. on the outdoor coil, and opens at 55°F.
17. What type of system uses a temperature and a pressure sensor to control the defrost cycle?
    A. a time-initiation–temperature-termination defrost system
    B. a temperature-initiation–temperature-termination defrost system
    C. a pressure-initiation–temperature-termination defrost system
    D. a pressure-time-initiation/temperature-termination defrost system
18. A pressure switch, used on the defrost cycle of a heat pump, commonly closes at a preset pressure difference of
    A. 1″ to 2″ WC.
    B. .5″ to .65″ WC.
    C. .02″ to .05″ WC.
    D. .2″ to 2.45″ WC.
19. Temperature split refers to
    A. the temperature difference of the air across the outdoor coil.
    B. the temperature difference between the indoor and the outdoor of the home.
    C. the temperature difference of the air across the indoor coil.
    D. the temperature difference between the air entering a coil and the temperature of the coil.

20. The outdoor ambient thermostat is commonly used to

A. energize the first stage of heating in an air-to-air heat pump.

B. de-energize the compressor during the second stage of heating in an air-to-air heat pump.

C. energize the auxiliary strip heaters during the first stage of heating.

D. energize the auxiliary strip heaters during the second stage of heating.

21. A hold-back thermostat is used to

A. hold back compressor operation during the defrost cycle.

B. keep the system energized during a defrost cycle, when the room thermostat is satisfied.

C. prevent the operation of the auxiliary strip heaters during the defrost cycle.

D. energize the auxiliary strip heaters during the heating cycle.

22. A discharge temperature thermostat is used

A. on air-to-air heat pump systems that use electric auxiliary strip heaters.

B. on air-to-air heat pump systems that use fossil fuel.

C. on air-to-air heat pump systems that use electric auxiliary strip heaters or fossil fuel.

D. on air-to-air heat pump systems to control the defrost sequence.

23. When sizing a heat pump, it must initially be sized for

A. the heating load.

B. the cooling load.

C. the heating and cooling load.

D. the total capacity.

24. When sizing for the auxiliary heat for a heat pump system, the auxiliary heat must

A. be capable of handling the entire heat load of the structure minus the heat pump capacity.

B. be capable of handling the entire heat load of the structure.

C. be electric strip heaters only.

D. be fossil fuel only.

25. Auxiliary strip heaters are

A. used to defrost the outdoor coil during low ambient conditions.

B. located at the indoor unit upstream from the indoor coil.

C. located at the indoor unit downstream from the indoor coil.

D. wired in parallel with the compressor.

26. What controls the auxiliary strip heaters in a heat pump system?

A. the first stage of a two-stage thermostat

B. the second stage of a two-stage thermostat

C. the air pressure switch

D. the auxiliary switch on the contactor

27. What is the purpose of the de-ice coil on an air-to-air heat pump?

A. It prevents the formation of ice on the indoor coil during the cooling cycle.

B. It prevents the formation of ice on the indoor coil during the heating cycle.

C. It prevents the formation of ice on the upper section of the outdoor coil.

D. It prevents the formation of ice on the lower section of the outdoor coil.

28. The coefficient of performance (COP) is

A. a heat pump's rating in the cooling cycle, comparing the heat input by the electrical input.

B. a heat pump's rating in the heating cycle, comparing the heat input by the electrical output.

C. a heat pump's rating in the cooling cycle, comparing the heat output by the electrical output.

D. a heat pump's rating in the heating cycle, comparing the heat output by the electrical input.

29. When installing a heat pump system, the unit should always be located as close to the indoor unit as possible because

A. during the cooling cycle a longer liquid line will cause the system to operate less efficiently.

B. during the cooling cycle a longer vapor line will cause the system to operate less efficiently.

C. during the heating cycle a longer liquid line will cause the system to operate less efficiently.

D. during the heating cycle a longer vapor line will cause the system to operate less efficiently.

30. When installing an outdoor coil, it must

A. always be placed under the overhang of the building.

B. never be placed under the overhang of the building.

C. have at least 1″ of clearance around all sides.

D. have a maximum distance from the exterior wall of 6″.

31. What are common supply-air static pressures for a heat pump duct system?

A. 2″ WC.

B. 1″ WC.

C. .5″ WC.

D. .15″ WC.

32. What is a common temperature rise over the indoor coil of a heat pump during the heating cycle?

A. 70°F

B. 50°F

C. 35°F

D. 20°F

33. When installing an indoor unit, a clearance that is commonly required for service accessibility is

A. 20″.

B. 16″.

C. 10″.

D. 6″.

34. How much of an air space is required between the supply-air duct of a heat pump and combustible material?

A. 6″

B. 4″

C. 2″

D. 1″

35. What is required on a downflow-type indoor unit used with an air-to-air heat pump?

A. a smaller return-air duct system

B. a noncombustible floor base

C. an air pressure switch located in the return-air duct

D. a 70°F temperature rise

36. What is required on a horizontal-flow-type indoor unit used with an air-to-air heat pump?

A. a filter rack downstream from the auxiliary heaters

B. a backward curved blower

C. an auxiliary drain pan

D. a de-ice coil

37. How many strands and what size of wire are commonly used for the 24-volt control circuit on an air-to-air heat pump?

A. 4 strand 18 gauge

B. 6 strand 18 gauge

C. 6 strand 22 gauge

D. 8 strand 18 gauge

38. On a common heat pump thermostat, what connects to the "Y1" terminal?

A. the fan relay coil

B. the auxiliary heaters

C. the reversing valve

D. the compressor contactor

39. On a common heat pump thermostat, what connects to the "W1" terminal?

A. the fan relay coil

B. the auxiliary heaters

C. the reversing valve

D. the compressor contactor

40. On a common heat pump thermostat, what connects to the "W2" terminal?

A. the thermostat trouble light

B. the auxiliary heaters

C. the reversing valve

D. the compressor contactor

41. On a common heat pump thermostat, what connects to the "X" terminal?

A. the thermostat trouble light

B. the auxiliary heaters

C. the reversing valve

D. the emergency heat light

42. On a common heat pump thermostat, what connects to the "E" terminal?

A. the thermostat trouble light

B. the auxiliary heaters

C. the reversing valve

D. the emergency heat light

THE FOLLOWING QUESTIONS MAY BE HELPFUL TO STUDY FOR THESE CERTIFICATION TESTS:
NATE—AC-SV, AC-IN, HP-SV, HP-IN.
ESCO INSTITUTE HVAC EXCELLENCE AIR CONDITIONING AND HEAT PUMP

HEAT PUMP TROUBLESHOOTING

1. On an air-to-air heat pump system, the suction pressure and discharge pressure are low during the cooling months. What are two possible problems?

A. a dirty air filter and a dirty condenser

B. a low charge and a dirty condenser

C. a restricted indoor metering device and a dirty air filter

D. a restricted outdoor metering device and a restricted indoor check valve

2. On an air-to-air heat pump system, the discharge pressure and suction pressure are high during the heating months. What are two possible problems?

A. a restricted indoor metering device and a restricted outdoor check valve

B. a dirty outdoor coil and a restricted outdoor metering device

C. a defective defrost system and a dirty outdoor coil

D. a dirty indoor filter and noncondensables in the system

3. A service call is made on an air-to-air heat pump system during the winter. When servicing the system, it is found that the outdoor coil has frost buildup and the system is operating on emergency heat.

Technician A says that the system might have a defrost problem.

Technician B says that the reversing valve might be leaking.

Which technician is correct?

A. technician A

B. technician B

C. Both technicians are correct.

D. Both technicians are incorrect.

4. A service call is made on an air-to-air heat pump system during the winter months. The technician arrives on the job and notices that the discharge pressure is low and the suction pressure is high. What two problems could cause this symptom?

A. a restricted outdoor metering device and a defective defrost system

B. an inefficient compressor and a dirty indoor air filter

C. an excessive heat load and a dirty condenser

D. a leaking reversing valve and bad compressor valves

5. How can a system be tested to determine if the reversing valve solenoid is leaking internally?

A. A temperature measurement can be taken at the inlet and the outlet of the valve in the heating and cooling mode.

B. The valve can only be changed because it is impossible to test the valve for internal leaks.

C. A temperature-drop reading across the indoor coil during the cooling season will determine if the reversing valve is leaking.

D. The reversing valve often has a high-pitched whistle if it is leaking internally.

6. What component must always be de-energized during the defrost cycle on a heat pump?

A. the reversing-valve solenoid

B. the outdoor fan

C. the indoor fan

D. the defrost relay coil

7. When an air-to-air heat pump is in the cooling mode, what can cause a decrease in high-side and low-side pressure?

A. a restricted indoor metering device

B. a defective indoor coil check valve

C. bad compressor valves

D. a defective outdoor metering device

8. Using the "Heat Pump Diagram Cooling Mode" (Figure 1–15), what is connected to the top tube of the reversing valve?

A. the compressor discharge line

B. the compressor suction line

C. the indoor coil

D. the outdoor coil

9. Using the "Heat Pump Diagram Cooling Mode" (Figure 1–15), what is connected to the bottom-left tube of the reversing valve?

A. the compressor discharge line

B. the compressor suction line

C. the indoor coil

D. the outdoor coil

10. Using the "Heat Pump Diagram Heating Mode" (Figure 1–16), what is connected to the bottom-left tube of the reversing valve?

A. the compressor discharge line

B. the compressor suction line

C. the indoor coil

D. the outdoor coil

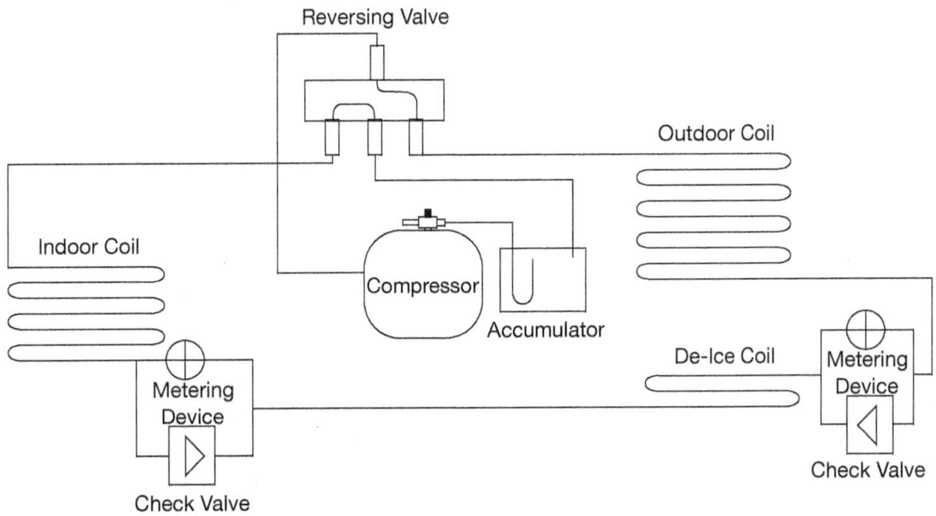

FIGURE 1–15

Heat Pump Diagram Cooling Mode

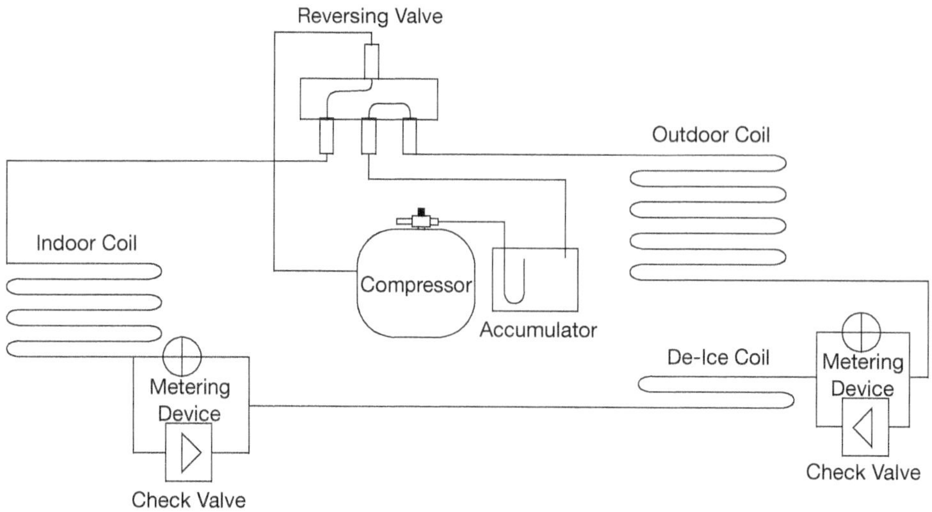

FIGURE 1–16
Heat Pump Diagram Heating Mode

11. What is connected to the bottom center tube of the reversing valve? Use the "Heat Pump Diagram Heating Mode" (Figure 1–16) to answer this question.

A. the compressor discharge line

B. the compressor suction line

C. the indoor coil

D. the outdoor coil

12. Technician A says that the check valves in a heat pump system are located in series with both metering devices.

Technician B says that the check valves in a heat pump system are located in parallel with both metering devices.

Which technician is correct?

A. technician A

B. technician B

C. Both technicians are correct.

D. Both technicians are incorrect.

13. A service technician notices that a check valve needs to be replaced in a heat pump system. Because it is a ball-type check valve, it is critical for the technician to install the valve

A. horizontally, with the seat on the outlet of the valve.

B. vertically, with the seat on the bottom of the valve.

C. horizontally, with the seat on the inlet of the valve.

D. vertically, with the seat on the top of the valve.

14. After troubleshooting a heat pump system, the technician concludes that the check valve is defective. He connects a recovery system and removes the refrigerant from the unit. He then proceeds to unseat the old check valve and remove it from the system. He cleans the area where the new valve will be installed, places it into position, and brazes it into the system while not protecting the valve. After a thorough evacuation, he starts the system and notices that the new valve is defective. What could have been the problem with the new valve?

A. A heat sink was not used and the valve warped.

B. The valve may have been defective from the factory.

C. The problem was misdiagnosed.

D. All of the above

15. Technician A says that the auxiliary heaters found on an air-to-air heat pump are used to increase the system's pressure during a defrost cycle.

Technician B says that the auxiliary heaters found on an air-to-air heat pump are used to increase supply-air temperature for comfort during a defrost cycle.

Which technician is correct?

A. technician A

B. technician B

C. Both technicians are correct.

D. Both technicians are incorrect.

16. When a heat pump switches into the defrost cycle, it must

A. remain in defrost until the temperature rise across the indoor coil reaches 50°F.

B. remain in defrost until the indoor coil has reached a predetermined temperature.

C. remain in defrost until the outdoor coil has reached a predetermined temperature.

D. remain in defrost until the indoor thermostat is satisfied.

17. A defrost termination thermostat is commonly located

A. on the vapor line, and opens at 55°F.

B. on the liquid line, and closes at 55°F.

C. on the indoor coil, and closes at 55°F.

D. on the outdoor coil, and opens at 55°F.

18. Temperature split refers to

A. the temperature difference of the air across the outdoor coil.

B. the temperature difference between the inside and the outside of the home.

C. the temperature difference of the air across the indoor coil.

D. the temperature difference between the air entering a coil and the temperature of the coil.

19. Technician A says that the outdoor ambient thermostat commonly de-energizes the compressor during the second stage of heating in an air-to-air heat pump.

Technician B says that the outdoor ambient thermostat used with an air-to-air heat pump commonly energizes the auxiliary strip heaters during the first stage of heating.

Which technician is correct?

A. technician A

B. technician B

C. Both technicians are correct.

D. Both technicians are incorrect.

20. If there is malfunction in a hold-back thermostat that is used with an air-to-air heat pump, how would this affect the system?

A. The system would not hold back the compressor operation during the defrost cycle. This could cause excessive power consumption.

B. The system would not stay energized during defrost if the room thermostat is satisfied. This could cause freezing of the outdoor coil.

C. The system would not de-energize the auxiliary strip heaters during the defrost cycle. This could cause excessive heat buildup.

D. The system would not energize the auxiliary strip heaters during the heating cycle. This could cause insufficient heating.

21. A technician arrives on a service call for an air-to-air heat pump during the heating season. She notices that the system operates normally until the defrost circuit is energized. The system stays in defrost for approximately 15 min when there is no ice or frost buildup on the coil. What could be the problem with the system?

A. The defrost timer is defective.

B. The defrost termination switch is stuck closed.

C. The defrost termination switch is stuck open.

D. The reversing valve solenoid is stuck open.

22. A technician receives a service call for an air-to-air heat pump. The customer is complaining about insufficient heat during low ambient conditions. Upon arrival to the job, the technician notices that the mechanical heat pump system is operating properly but the auxiliary heat is not operating. What circuit should be checked next?

A. the first-stage heating circuit

B. the second-stage heating circuit

C. the air-pressure control circuit

D. the auxiliary contactor circuit

23. A technician arrives at a home on a service call for an air-to-air heat pump. After measuring the air temperature over the indoor coil and finding a temperature rise of 35°F, the technician determines that

A. the indoor air filter might be dirty.

B. the indoor blower is running too fast.

C. the outdoor coil might be dirty.

D. the temperature rise is sufficient.

24. With ODT closed between terminals 2 and 3 and the thermostat in the heating mode with W2 in the closed position, what voltage would be expected across terminal 2 on LIM and terminal 1 on FU in the H3 circuit? Use the "Heat Pump Electrical Diagram" (Figure 1–17) to answer this question.

A. 0 volts

B. 24 volts

C. 120 volts

D. 240 volts

25. What should be the resistance across terminals 1 and 2 on TM1 during a normal heating operation? Use the "Heat Pump Electrical Diagram" (Figure 1–17) to answer this question.

A. 0 ohms

B. 50 ohms

C. 1,000 ohms

D. infinite ohms

26. What voltage should be across terminal 1 on LIM and terminal 1 on FU in the H1 circuit during the first stage of heating? Use the "Heat Pump Electrical Diagram" (Figure 1–17) to answer this question.

A. 0 volts

B. 24 volts

C. 120 volts

D. 240 volts

27. If FU was in the open position in the H2 circuit when the first-stage heating was energized, what voltage would be across terminals 1 and 2? Use the "Heat Pump Electrical Diagram" (Figure 1–17) to answer this question.

A. 0 volts

B. 24 volts

C. 120 volts

D. 240 volts

28. What will happen to the system if the HP switch opens? Use the "Heat Pump Electrical Diagram" (Figure 1–17) to answer this question.

A. The system will remain de-energized until HP closes and then the system will energize again.

B. The system will short cycle on the HP control.

C. The system will be locked out until the power is disconnected and then reconnected.

D. Both A and B

29. What are symptoms of an air-to-air heat pump system that has a low charge when in the cooling mode?

A. The indoor coil will be partially frosted and the discharge pressure will increase.

B. The suction line will sweat and the low-side pressure will decrease.

C. The low-side pressure will decrease and the indoor coil will be partially frosted.

D. The discharge pressure will decrease and the suction pressure will increase.

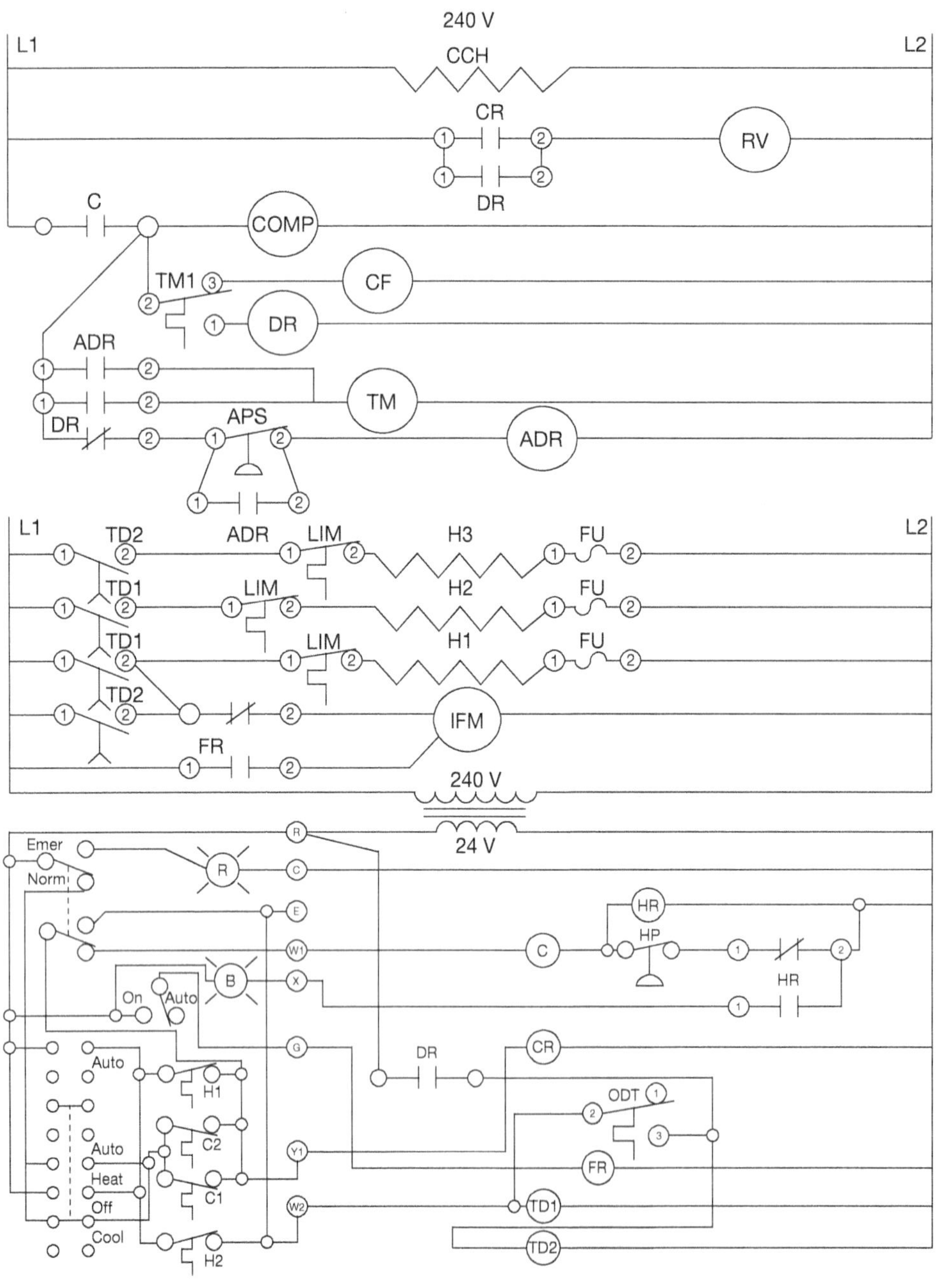

Legend

ADR	Auxiliary Defrost Relay
APS	Air Pressure Switch
C	Contactor
CCH	Crankcase Heater
CF	Condenser Fan
COMP	Compressor
CR	Control Relay
CT1	Cooling Thermostat 1st Stage
CT2	Cooling Thermostat 2nd Stage
DR	Defrost Relay
FR	Fan Relay
FU	Fusible Link
H1	Auxiliary Strip Heater 1
H2	Auxiliary Strip Heater 2
H3	Auxiliary Strip Heater 3
HP	High-Pressure Switch
HR	Holding Relay
HT1	Heating Thermostat 1st Stage
HT2	Heating Thermostat 2nd Stage
IFM	Indoor Fan Motor
LIM	Limit Switch
ODT	Outdoor Thermostat
RV	Reversing Valve
TD1	Timed Delay Relay 1
TD2	Timed Delay Relay 2
TM	Timer Motor

FIGURE 1–17
Heat Pump Electrical Diagram

30. A customer calls and says that her air-to-air heat pump system is not cooling properly. Upon arrival, a technician notices that the suction line and compressor have frost buildup.

Technician A says that the problem is probably a low charge.

Technician B says that it is possibly a dirty air filter.

Which technician is more likely to be correct?

A. technician A

B. technician B

C. Both technicians are correct.

D. Both technicians are incorrect.

31. Name two symptoms that are common to an overcharged air-to-air heat pump system with a system that has a dirty condenser.

A. The subcooling decreases and the amperage increases.

B. The superheat increases and the discharge pressure increases.

C. The suction pressure increases and the subcooling increases.

D. The discharge pressure increases and the amperage increases.

32. When an air-to-air heat pump system has an undercharge, what happens to the superheat and the subcooling of the system?

A. The superheat increases and the subcooling increases.

B. The superheat increases and the subcooling decreases.

C. The superheat decreases and the subcooling increases.

D. The superheat decreases and the subcooling decreases.

33. A heat pump operating in the heating mode with a dirty filter would result in

A. very low side pressure.

B. iced-up indoor coil.

C. excessive head pressure.

D. low head pressure.

34. What is the best method to recharge the heat pump system during the heating season?

A. weight in the exact charge

B. the subcooling method

C. the superheat method

D. the ambient method

35. After a heat pump installation, the system will heat but not cool. The probable cause could be

A. the outdoor metering device is not metering.

B. the indoor check valve is open.

C. the indoor check valve is closed.

D. the outdoor check valve is not closing.

36. The technician diagnoses that in the cooling mode the compressor runs but the outdoor fan is not operating. After checking the fan motor and capacitor to be good, the technician concludes that

A. the compressor contactor is defective.

B. the reversing valve is defective.

C. the defrost control is defective.

D. the defective relay is defective.

37. A heat pump reversing valve is sticking in the midway position; the manifold gauges would indicate

A. higher than normal high-side pressure and higher than normal low-side pressure.

B. lower than normal high-side pressure and higher than normal low-side pressure.

C. lower than normal high-side pressure and lower than normal low-side pressure.

D. Both high and low side pressures will be higher than normal.

38. An electric furnace has four 5-kw heating elements rated at 240 volts. The voltage at which the furnace has to operate is 208 volts. What will be the effect on the Btu/hr output of the furnace?

A. It will reduce the Btu/hr output of the furnace.

B. It will increase the Btu/hr output of the furnace.

C. If the elements are wired in series there will be no reduction of Btu/hr output.

D. There will be an increase in Btu/hr output if the elements are wired in series.

39. With a heat pump operating in the cooling mode, the check valve at the evaporator coil is

A. wide open.

B. modulating.

C. in mid-position.

D. closed.

40. An electric furnace using a single-stage thermostat is recording a low temperature rise. To correct this problem the technician should

A. reduce the volume of air through the furnace.

B. increase the volume of air through the furnace.

C. change the air filter.

D. install a two-stage thermostat.

41. The difference of the ON heating cycles between fossil fuel furnaces and heat pumps during the heat mode is that the cycles are

A. There is no difference; they are equal.

B. shorter.

C. longer.

D. very short.

42. A technician has just charged a heat pump system. She finds that the compressor does not want to start but instead hums then opens on internal overload. The compressor is a CSCR type. The more common problem would be a(n)

A. overcharged system.

B. undercharged system.

C. defective compressor.

D. defective potential relay.

43. The thermostat second-stage heating contacts of an all-electric heat pump system serve to

A. shut down the compressor and condenser fan motor.

B. turn on the defrost heaters.

C. shut down the compressor.

D. turn on the auxiliary heaters and the compressor.

44. A technician is reading a high suction pressure at the compressor during the heating mode of heat pump system. The likely cause would be

A. a leaking reversing valve.

B. a leaking indoor check valve.

C. a loose bulb at TXV location.

D. low refrigerant charge.

45. Performing temperature measurements at the four-pipe connection of the reversing valve may indicate

A. reversing valve leakage.

B. an overcharged system.

C. an undercharged system.

D. indoor check valve leaking.

46. A heat pump application that is using an oversized piston orifice metering device will result in

A. high head pressure and low suction pressure in the cooling mode.

B. high head pressure and high suction pressure in the cooling mode.

C. low head pressure and low suction pressure in the heating mode.

D. low head pressure and high suction pressure in the cooling cycle.

47. When speaking to a customer of a heat pump installation, you should also inform him that to conserve electricity he should

A. replace air filters often.

B. install a night set-back thermostat.

C. close off dampers to rooms not used.

D. install ceiling fans.

48. In order to add cooling efficiency, a fan delay may be used to

A. turn on the indoor fan motor after the compressor is off.

B. start the outdoor fan motor after the compressor turns on.

C. start the indoor motor after compressor start-up.

D. turn on the condenser fan motor and the indoor fan motor before the compressor.

49. A 12 × 8 return-air duct has been installed to a 3-ton heat pump system. What will be the most likely result while operating in the heating mode?

A. operate satisfactorily

B. overheat the premises

C. underheat the premises

D. shutdown on high-pressure control

SECTION 2

Gas and Oil Forced-Air and Hot Water Heating Systems

THE FOLLOWING QUESTIONS MAY BE HELPFUL IN STUDYING FOR THESE CERTIFICATION TESTS:
NATE—GS-SV, GS-IN, HO-SV, HO-IN, HG-SV, HG-IN.
ESCO INSTITUTE HVAC EXCELLENCE GAS HEAT, OIL HEAT, HYDRONIC HEAT SERVICE

FUNDAMENTALS OF GAS COMBUSTION

1. What fuel is heavier than air?

A. natural gas

B. LP gas

C. ethane

D. None of the above

2. What is the heating value of natural gas?

A. 800 Btu/cu ft

B. 1,000 Btu/cu ft

C. 1,500 Btu/cu ft

D. 2,500 Btu/cu ft

3. What is the heating value of LP gas?

A. 800 Btu/cu ft

B. 1,000 Btu/cu ft

C. 1,500 Btu/cu ft

D. 2,500 Btu/cu ft

4. What are the three requirements needed for combustion?

A. carbon, oxygen, and ignition

B. air, carbon, and fuel

C. fuel, carbon, and ignition

D. oxygen, fuel, and ignition

5. Complete combustion is accomplished when

A. one part methane is combined with two parts oxygen.

B. two parts methane are combined with one part oxygen.

C. one part carbon is combined with two parts methane.

D. two parts carbon are combined with one part methane.

6. What are two products of complete combustion?

A. CO and H_2O

B. CO_2 and H_2O

C. CH_4 and O_2

D. CH_4 and H_2O

7. Air is approximately

A. 21% oxygen and 79% nitrogen.

B. 51% oxygen and 49% nitrogen.

C. 70% oxygen and 21% nitrogen.

D. 49% oxygen and 51% nitrogen.

8. The proper air and fuel mixture for complete combustion using natural gas is

A. 1 part fuel and 1 part air.

B. 1 part fuel and 4 parts air.

C. 1 part fuel and 8 parts air.

D. 1 part fuel and 10 parts air.

9. To ensure complete combustion,

A. an additional 2 parts of excess air is added to the fuel–air mixture.

B. an additional 5 parts of excess air is added to the fuel–air mixture.

C. an additional 10 parts of excess air is added to the fuel–air mixture.

D. an additional 15 parts of excess air is added to the fuel–air mixture.

10. What can cause carbon monoxide during the combustion process?
    A. incomplete combustion and flame impingement on the cold surface of a heat exchanger
    B. not enough fuel mixture and leaks in the vent pipe
    C. too much primary air and incomplete combustion
    D. too much primary and secondary air

11. Natural gas is also known as
    A. butane.
    B. propane.
    C. methane.
    D. None of the above

12. What are three methods of heat transfer?
    A. evaporation, radiation, and convection
    B. convection, air motion, and conduction
    C. conduction, convection, and radiation
    D. condensation, convection, and conduction

13. What is the maximum allowable CO in flue gases of a furnace?
    A. 200 ppm
    B. 400 ppm
    C. 600 ppm
    D. 800 ppm

14. What is the maximum allowable concentration of CO for continuous 24-hr exposure, according to ASHRAE?
    A. 9 ppm
    B. 2 ppm
    C. 50 ppm
    D. 80 ppm

15. Where should CO measurements be taken while testing a gas-fired furnace?
    A. in the vent pipe only
    B. only around the burner
    C. at the draft diverter on a 90% efficient furnace
    D. in the vent pipe leaving the furnace and around the burner area and draft diverter

16. A common percentage of CO_2 in the flue gases of a natural gas furnace ranges from
    A. 4 to 6%.
    B. 7 to 9%.
    C. 11 to 12%.
    D. 14 to 15%.

THE FOLLOWING QUESTIONS MAY BE HELPFUL TO STUDY FOR THESE CERTIFICATION TESTS:
NATE—GS-SV, GS-IN, HG-SV, HG-IN.
ESCO INSTITUTE HVAC EXCELLENCE GAS HEAT SERVICE

GAS FURNACE CONTROLS AND COMPONENTS

1. What is the purpose of a pilot?
    A. It ignites the main burners.
    B. It proves the existence of a flame.
    C. It sends a signal to the gas valve.
    D. It opens a pilot valve.

2. What type of pilot sensor is used on a standing pilot furnace?
    A. a mercury flame sensor
    B. a flame rod
    C. a hot surface igniter
    D. a thermocouple

3. What is the purpose of a thermocouple?
    A. It energizes the system using 750 millivolts.
    B. It ignites the main burner.
    C. It proves the existence of a pilot flame.
    D. It ignites the pilot.

4. How far must the tip of a thermocouple be placed in the pilot flame?
    A. $\frac{1}{8}''$ to $\frac{1}{4}''$
    B. $\frac{3}{8}''$ to $\frac{1}{2}''$
    C. $\frac{3}{4}''$ to $1''$
    D. $1\frac{1}{2}''$

5. How many millivolts does a thermocouple produce?
    A. 750
    B. 550
    C. 30
    D. 15

6. How many millivolts does a power-pile produce?
    A. 750
    B. 350
    C. 150
    D. 30

7. What three components make up a spark ignition system?
    A. a cable, an electrode, and a sensor
    B. an ignition transformer, a cable, and an electrode
    C. an electrode, an ignition transformer, and a sensor
    D. a cable, a sensor, and an ignition transformer

8. In a furnace that uses an intermittent pilot,

A. the pilot is continually lit.

B. the pilot is lit with each cycle of the furnace.

C. it is considered a standing pilot system.

D. it often uses a thermocouple.

9. An intermittent pilot system uses

A. a spark igniter to light the pilot.

B. a hot surface igniter to light the pilot.

C. a match to light the pilot.

D. a spark or hot surface igniter to light the pilot.

10. What is another name for a hot surface igniter?

A. a cadmium igniter

B. a silicon carbide igniter

C. a tungsten igniter

D. Both A and B

11. Which of the following would be the quietest ignition system?

A. an indirect spark ignition

B. a direct spark ignition

C. a hot surface igniter

D. an intermittent spark igniter

12. The voltage of a hot surface igniter can be

A. 120 volts.

B. 50 volts.

C. 24 volts.

D. Both A and C

13. What type of pilot is commonly used on a direct ignition system?

A. a nonaerated pilot only

B. an aerated pilot only

C. an aerated or nonaerated pilot

D. Pilots are not used on direct ignition systems.

14. What is the setting of the spark gap on a spark ignition system?

A. $\frac{1}{8}''$

B. $\frac{1}{4}''$

C. $\frac{3}{8}''$

D. $\frac{1}{2}''$

15. How far must the tip of a flame igniter/sensor be placed from the main burner or pilot flame?

A. $\frac{1}{8}''$ to $\frac{1}{4}''$

B. $\frac{3}{8}''$ to $\frac{1}{2}''$

C. $\frac{3}{4}''$ to 1″

D. $1\frac{1}{2}''$

16. The voltage created by an ignition control module is commonly in the range of

A. 2,000 to 3,000 volts.

B. 7,000 to 15,000 volts.

C. 20,000 to 50,000 volts.

D. None of the above

17. What are four common furnace designs?

A. downflow, lowboy, highboy, and multipositional

B. upflow, lowboy, downflow, and multipositional

C. horizontal, lowboy, highboy, and multipositional

D. upflow, downflow, horizontal, and multipositional

18. Which of the following are not located on a combination gas valve?

A. a pilot solenoid and a main valve solenoid

B. a pressure regulator and a pilot solenoid

C. a main shutoff and a pilot adjustment

D. a spark igniter and a sensor wire

19. What is the purpose of the pressure regulator that is located on a combination gas valve?

A. It takes an incoming gas pressure of 3.5″ WC and increases it to 7″ WC.

B. It takes an incoming gas pressure of 7″ WC and reduces it to 3.5″ WC.

C. It increases the pressure when the screw is turned clockwise.

D. Both B and C

20. What component is located in the gas manifold that is used to distribute gas to each burner?

A. a distributor

B. an equalizer

C. an orifice

D. None of the above

21. What is the purpose of primary air?

A. It is air needed for proper combustion. It mixes with the main burner fuel at the point of ignition.

B. It is air needed for proper combustion. It mixes with the main burner fuel before the point of ignition.

C. It is air needed for proper combustion. It mixes with the main burner fuel after the point of ignition.

D. Both A and B

22. What is secondary air?

A. It is air that mixes with the main burner fuel at the point of ignition. This ensures proper combustion.

B. It is air that mixes with the main burner fuel before the point of ignition. This ensures proper combustion.

C. It is air that mixes with the main burner fuel after the point of ignition. This ensures proper combustion.

D. Both A and B

23. In what part of the burner does air and fuel mix before combustion?

A. the crossover section

B. the venturi

C. the orifice

D. the outlet

24. If gas burners are not installed properly, what problem could this cause?

A. erratic limit cycles

B. improper heat exchanger contraction

C. delayed ignition

D. decreased blower operation

25. Which of the following is *not* considered a type of gas burner design?

A. blast tube

B. slotted

C. ribbon

D. inshot

26. Which part of a gas furnace allows the air within the furnace location to blend with the products of combustion?

A. the draft diverter

B. the vent connector

C. the barometric damper

D. the flue pipe

27. The purpose of a fan control is to

A. start the blower.

B. stop the blower.

C. start and stop the blower.

D. start the inducer motor.

28. What type of fan control operates the blower by temperature only?

A. a bimetal fan control

B. a timed delay fan control

C. a solid-state fan control

D. Both A and B

29. What type of fan control uses an electric heater as a means of operating the blower motor?

A. a bimetal fan control

B. a timed delay fan control

C. a solid-state fan control

D. Both A and B

30. A pressure switch on a noncondensing furnace is used to

A. prove the operation of the indoor blower.

B. prove the operation of the induced-draft blower.

C. sense the gas pressure to the gas valve.

D. sense the gas pressure to the pilot valve.

31. A limit control is designed to

A. cycle the system on and off depending on space temperature.

B. operate the blower when the thermostat calls for heat.

C. de-energize the gas valve if the furnace overheats.

D. limit the operation of the induced draft motor.

32. Which control is commonly found on downflow and multipositional furnaces but is not normally used on upflow furnaces?

A. an air pressure control

B. a primary limit control

C. a pilot safety control

D. an auxiliary limit control

33. What control is commonly used on high-efficiency, noncondensing furnaces to protect against high stack temperatures?

A. a limit control

B. a flame roll-out control

C. a vent safety control

D. an auxiliary limit control

34. What control is commonly used on high-efficiency, noncondensing furnaces to protect against a plugged heat exchanger?

A. a limit control

B. a flame roll-out control

C. a vent safety control

D. an auxiliary limit control

35. The part of the system that contains the products of combustion primarily for a thermal transfer is the

A. combustion blower.

B. heat exchanger.

C. burner.

D. accumulator.

36. Two common methods used to seal a sectional heat exchanger are

A. mechanical crimp and iron banding.

B. welded seam and Pittsburgh seam.

C. iron banding and Pittsburgh seam.

D. welded seam and mechanical crimp.

37. A bimetal flame sensor

A. proves the existence of a pilot flame.

B. proves the existence of the main gas flame.

C. uses a normally open set of contacts to control the pilot valve.

D. uses a normally closed set of contacts to control the main valve.

38. A flame rod sensor

A. proves the presence of a pilot flame only.

B. proves the presence of a main burner flame only.

C. proves the presence of a pilot or main burner flame.

D. proves the presence of the spark from an ignition system.

39. A normally open float switch should be wired to control

A. the blower motor.

B. the contactor coil.

C. the condensate pump.

D. the inducer motor.

40. An induced draft blower used in a noncondensing gas furnace

A. produces a positive pressure in the vent system when properly installed.

B. produces a positive pressure in the heat exchanger.

C. pulls the products of combustion from the heat exchanger and pushes them into the venting system.

D. pushes the products of combustion into the heat exchanger then into the venting system.

41. In what type of furnace is a secondary heat exchanger used?

A. a high-efficiency noncondensing type

B. a high-efficiency condensing type

C. a standing pilot system

D. a mid-efficiency noncondensing type

42. What is the purpose of the secondary heat exchanger?

A. to increase the temperature of the products of combustion for added efficiency

B. to increase the quantity of the products of combustion for added efficiency

C. to increase the velocity of the products of combustion for added efficiency

D. to condense the products of combustion for added efficiency

43. Which control is used to disconnect all electrical power supply to the furnace, which will prevent the products of combustion from mixing with supply air?

A. the limit control

B. the auxiliary limit control

C. the pressure switch

D. the interlock switch

44. What control is used to start the operation of the induced draft blower of a high-efficiency noncondensing furnace?

A. the pressure switch

B. the auxiliary switch

C. the thermostat

D. the end switch

45. What control is used to energize the ignition control on the startup of a high-efficiency noncondensing furnace?

A. the end switch

B. the auxiliary limit control

C. the pressure switch

D. the pilot safety switch

46. A technician suspects a crack in the heat exchanger of a high-efficiency condensing furnace; the technician blocks off the intake and exhaust pipe and connects a manometer to the small hose leading to the pressure switch. The technician then energizes the blower on high speed at the thermostat (On Fan). What reading will be obtained if there is a crack in the heat exchanger?

A. a negative pressure reading

B. a positive pressure reading

C. a reading of 0 in wc

D. Any of the above

THE FOLLOWING QUESTIONS MAY BE HELPFUL TO STUDY FOR THESE CERTIFICATION TESTS:

NATE—GS-SV, GS-IN, HG-SV, HG-IN, HO-SV, HO-IN.

ESCO INSTITUTE HVAC EXCELLENCE GAS HEAT SERVICE

COMBUSTION AIR

1. Dilution air is

A. air that mixes with the fuel for combustion.

B. air that is brought into a return air duct system.

C. air that mixes with combustion gases in a natural draft furnace.

D. air that mixes with combustion gases in an induced draft noncondensing furnace.

2. Infiltration is

A. air that is never allowed to mix with the combustion air.

B. air that is exhausted out of a gas-fired appliance area.

C. air that is brought into tight-constructed homes through a combustion air system.

D. air that is brought into loose-constructed homes around windows and doors.

3. How is infiltration measured?

A. by cubic foot per minute

B. by foot per minute

C. by air changes per hour

D. by calculating the square footage of the home by a factor of .5

4. How much combustion air is required for every 1 cubic foot of natural gas used on a gas-fired appliance?

A. 5 cu ft

B. 10 cu ft

C. 15 cu ft

D. 20 cu ft

5. How much dilution air is required for every 1 cubic foot of gas used on a natural draft appliance?

A. 6 cu ft

B. 8 cu ft

C. 10 cu ft

D. 14 cu ft

6. Which statement is true?
   A. A fan-assisted draft furnace only requires 5 cu ft of combustion air for every 1 cu ft of natural gas.
   B. A fan-assisted draft furnace does not use dilution air.
   C. A fan-assisted draft furnace does not require combustion air.
   D. A fan-assisted draft furnace only requires 5 cu ft of combustion air for every 2 cu ft of natural gas.
7. What is a common cause of heat exchanger and vent system corrosion on an induced draft noncondensing furnace?
   A. not enough dilution air for the vent system
   B. too much outdoor air brought in for combustion
   C. household chemicals
   D. not enough outdoor air brought in for combustion
8. An unconfined space is
   A. a space whose volume is greater than 50 cu ft/1,000 Btu/h of total input of all gas-fired appliances.
   B. a space whose volume is less than 50 cu ft/1,000 Btu/h of total input of all gas-fired appliances.
   C. a space whose volume is greater than 50 cu ft/4,000 Btu/h of total input of all gas-fired appliances.
   D. a space whose volume is less than 50 cu ft/4,000 Btu/h of total input of all gas-fired appliances.
9. Combustion air
   A. can be taken from an unconfined space if the construction of the home is tight.
   B. can be taken from an unconfined space if the construction of the home is loose.
   C. can be taken from an unconfined space only if a mechanical ventilation system is installed.
   D. can never be taken from an unconfined space.
10. When introducing outdoor combustion air into an unconfined space,
   A. one 100 sq in. opening is generally required.
   B. two equal-sized openings are required.
   C. never place the opening near the ceiling.
   D. never place the opening near the floor.
11. When introducing outdoor combustion air into an unconfined space,
   A. each opening must have a total free area of at least 1 sq in. per 1,000 Btu/h input and not less than 100 sq in.
   B. each opening must have a total free area of at least 1 sq in. per 2,000 Btu/h input and not less than 100 sq in.
   C. each opening must have a total free area of at least 1 sq in. per 4,000 Btu/h input and not less than 100 sq in.
   D. each opening must have a total free area of at least 1 sq in. per 5,000 Btu/h input and not less than 100 sq in.
12. When horizontal ducts are used to introduce outdoor air into an unconfined space,
   A. the openings must be 1 sq in. for every 1,000 Btu/h input and not less than 100 sq in.
   B. the openings must be 1 sq in. for every 2,000 Btu/h input and not less than 100 sq in.
   C. the openings must be 1 sq in. for every 4,000 Btu/h input and not less than 100 sq in.
   D. the openings must be 1 sq in. for every 5,000 Btu/h input and not less than 100 sq in.
13. What is the total free area of a 10″ × 15″ combustion air opening used with a grill that has a 70% free area?
   A. 70 sq in.
   B. 105 sq in.
   C. 120 sq in.
   D. 150 sq in.
14. A confined space is
   A. a space whose volume is greater than 50 cu ft/1,000 Btu/h of total input of all gas-fired appliances.
   B. a space whose volume is less than 50 cu ft/1,000 Btu/h of total input of all gas-fired appliances.
   C. a space whose volume is greater than 50 cu ft/4,000 Btu/h of total input of all gas-fired appliances.
   D. a space whose volume is less than 50 cu ft/4,000 Btu/h of total input of all gas-fired appliances.
15. When introducing combustion air into a confined space,
   A. two equal openings must be provided, within 12″ of the floor.
   B. two equal openings must be provided, within 12″ of the ceiling.
   C. at least one opening must be provided near the ceiling or the floor.
   D. two equal openings must be provided, one at or within 12″ of the ceiling and the other at or within 12″ from the floor.
16. When introducing outdoor combustion air into a confined space,
   A. each opening must have a total free area of at least 1 sq in. per 1,000 Btu/h input and not less than 100 sq in.
   B. each opening must have a total free area of at least 1 sq in. per 2,000 Btu/h input and not less than 100 sq in.
   C. each opening must have a total free area of at least 1 sq in. per 4,000 Btu/h input and not less than 100 sq in.
   D. each opening must have a total free area of at least 1 sq in. per 5,000 Btu/h input and not less than 100 sq in.

17. When horizontal ducts are used to introduce outdoor air into a confined space,

A. the openings must be 1 sq in. for every 1,000 Btu/h input and not less than 100 sq in.

B. the openings must be 1 sq in. for every 2,000 Btu/h input and not less than 100 sq in.

C. the openings must be 1 sq in. for every 4,000 Btu/h input and not less than 100 sq in.

D. the openings must be 1 sq in. for every 5,000 Btu/h input and not less than 100 sq in.

18. The minimum dimension of a horizontal duct that is used to introduce combustion air into a confined space must not be less than

A. 1″.

B. 2″.

C. 3″.

D. 5″.

19. When a vertical duct is used to introduce outdoor combustion air into a confined space, the opening must be

A. 1 sq in. for every 1,000 Btu/h total input and not less than 100 sq in.

B. 1 sq in. for every 2,000 Btu/h total input and not less than 100 sq in.

C. 1 sq in. for every 4,000 Btu/h total input and not less than 100 sq in.

D. 1 sq in. for every 5,000 Btu/h total input and not less than 100 sq in.

20. When a vertical duct is used to introduce combustion air into a confined space, it must terminate

A. 1′ above the floor.

B. 4′ above the floor.

C. 2′ below the ceiling.

D. 1′ below the ceiling.

21. What should never be allowed to mix in a confined space when installing a furnace?

A. the dilution with the combustion air

B. the supply with the return air

C. the combustion with the supply and return air

D. the dilution and the exhaust air

22. Direct-vent high-efficiency condensing furnaces

A. must be vented according to the manufacturer's recommendation only.

B. must be vented by using AGA vent tables only.

C. must use AGA vent tables and the manufacturer's recommendation.

D. must never use polyvinyl chloride pipe for combustion air.

23. Most direct-vent high-efficiency condensing furnaces

A. only allow for vertical venting.

B. only allow for horizontal venting.

C. allow for vertical and horizontal venting.

D. must be vented into a chimney whenever possible.

24. When installing combustion air for a direct-vent high-efficiency condensing furnace, the combustion air and vent pipes must be supported at least every

A. 5′.

B. 10′.

C. 15′.

D. 20′.

25. The combustion air and vent pipes used on a direct-vent high-efficiency furnace must be located

A. at two different areas to prevent the system from operating under the same air-pressure zones.

B. at two different areas to prevent the system from operating under different air-pressure zones.

C. near the same area to prevent the system from operating under the same air-pressure zones.

D. near the same area to prevent the system from operating under different air-pressure zones.

THE FOLLOWING QUESTIONS MAY BE HELPFUL TO STUDY FOR THESE CERTIFICATION TESTS:
NATE—GS-SV, GS-IN, HG-SV, HG-IN, OL-SV, OL-IN, HO-SV, HO-IN.
ESCO INSTITUTE HVAC EXCELLENCE GAS HEAT SERVICE A. CO_2, H, N, O_2

VENTS AND CHIMNEYS

1. What gases are commonly found in products of combustion?

A. CO_2, H, N, and O_2

B. Cl, O_2, H, CO

C. H_2O, CO and H

D. None of the above

2. What two problems can cause carbon monoxide?

A. too long of an "on" cycle and soot

B. too much primary air and flame impingement

C. flame impingement and incomplete combustion

D. incomplete combustion and water vapor

3. What is meant by the term *buoyancy*?

A. It is the increase of flue gas temperatures during a furnace startup.

B. It is the sooting of a heat exchanger.

C. It is the lifting of the flame off the burners.

D. It is the moving of gases through a vent system due to high temperatures.

4. Buoyancy

A. increases as combustion gas temperature increases.

B. increases as combustion gas temperature decreases.

C. is not affected by combustion gas temperatures.

D. is controlled by the primary air adjustment.

5. For a natural draft venting system to operate, there must be a difference in ____________ between the ambient air and the flue gas.

A. volume

B. density

C. velocity

D. viscosity

6. What can cause a vent to have a minimized static pressure, reducing its effectiveness?

A. increase of flue gas velocity

B. heat loss through the vent

C. the use of a type "B" vent

D. having too short of a lateral run

7. A category 1 venting system

A. operates in a positive pressure, less than 140°F above flue gas dewpoint temperature, and requires a fan-powered airtight vent to remove the flue gases.

B. operates in a negative pressure, at least 140°F above flue gas dewpoint temperature, and depends on a natural draft to remove the flue gases.

C. operates in a positive pressure, below 140°F, and requires a fan-powered airtight vent to remove the flue gases.

D. operates in a negative pressure, less than 140°F above flue gas dewpoint temperature, and depends on a natural draft to remove the flue gases.

8. A category 2 venting system

A. operates in a positive pressure, less than 140°F above flue gas dewpoint temperature, and requires a fan-powered airtight vent to remove the flue gases.

B. operates in a negative pressure, at least 140°F above flue gas dewpoint temperature, and depends on a natural draft to remove the flue gases.

C. operates in a positive pressure, below 140°F, and requires a fan-powered airtight vent to remove the flue gases.

D. operates in a negative pressure, less than 140°F above flue gas dewpoint temperature, and depends on a natural draft to remove the flue gases.

9. A category 4 venting syste

A. operates in a positive pressure, less than 140°F above flue gas dewpoint temperature, and requires a fan-powered airtight vent to remove the flue gases.

B. operates in a negative pressure, at least 140°F above flue gas dewpoint temperature, and depends on a natural draft to remove the flue gases.

C. operates in a positive pressure, below 140°F, and requires a fan-powered airtight vent to remove the flue gases.

D. operates in a negative pressure, less than 140°F above flue gas dewpoint temperature, and depends on a natural draft to remove the flue gases.

10. What is the most common venting system used with a direct-vent high-efficiency condensing furnace?

A. category 1

B. category 2

C. category 3

D. category 4

11. What is the most common venting system used with an induced draft noncondensing furnace?

A. category 1

B. category 2

C. category 3

D. category 4

12. What type of material is most commonly used for venting a category 4 system?

A. type "B" vent only

B. standard single-wall vent pipe only

C. standard single-wall or type "B" vent

D. schedule 40 PVC

13. What does the term *wet-time* refer to?

A. It is the middle of a furnace cycle on an induced draft non-condensing furnace.

B. It is the end of a furnace cycle where a cool-down process could cause vent condensation.

C. It is the beginning cycle of a furnace where condensate could form in a vent system.

D. Both B and C

14. *Wet-time* usually lasts no more than

A. 30 s.

B. 2 min.

C. 5 min.

D. 10 min.

15. When dilution air is brought into the draft diverter on a natural draft furnace, what effect does this have on the products of combustion?

A. It decreases the dewpoint temperature.

B. It increases the dewpoint temperature.

C. It increases the velocity of the gases leaving the vent.

D. There is no effect on flue gases.

16. What is the main purpose of a type "B" vent pipe?

A. It aids in helping to keep flue gases from reaching their dewpoint temperature.

B. It decreases heat transfer from the flue gases.

C. It is used when a vent system passes through floors and ceilings.

D. All of the above

17. The outer wall of a type "B" vent is constructed of

A. galvanized steel.

B. aluminum.

C. galvanized metal.

D. aluminized steel.

18. The inner wall of a type "B" vent is constructed of

A. galvanized steel.

B. aluminum.

C. galvanized metal.

D. aluminized steel.

19. The added air space located in the middle of the inner and outer wall of a type "B" vent

A. decreases heat loss by 25% compared to a standard single-wall vent pipe.

B. decreases heat loss by 50% compared to a standard single-wall vent pipe.

C. allows for the formation of condensation to occur.

D. decreases heat loss by 75% compared to a standard single-wall vent pipe.

20. Common clearances between type "B" vents and combustible materials are

A. 5″ minimum.

B. $\frac{1}{2}$″ to 1″.

C. 4″ to 6″.

D. 1″ to 3″.

21. What extra precautions are taken when a type "B" vent passes through a ceiling or floor?

A. A smoke detector must be placed near the vent system.

B. The vent system must always be placed outside of the building whenever possible.

C. A firestop must be added around the vent.

D. A type "B" vent must never pass through a ceiling or floor.

22. Common clearances between standard single-wall vents and combustible materials are

A. 3″.

B. 4″.

C. 5″.

D. 6″

23. Vent connectors must have a minimum pitch of

A. $\frac{1}{2}$ inch per 1 foot of horizontal run.

B. 1 inch per 1 foot of horizontal run.

C. $2\frac{1}{2}$ inches per 10 feet of horizontal run.

D. 1 inch per 10 feet of horizontal run.

24. A 6″ single-wall vent connector must be at least

A. 30-gauge galvanized metal.

B. 24-gauge galvanized metal.

C. 22-gauge galvanized metal.

D. 30-gauge aluminized steel.

25. A 10″ single-wall vent connector must be at least

A. 30-gauge galvanized metal.

B. 26-gauge galvanized metal.

C. 22-gauge galvanized metal.

D. 30-gauge aluminized steel.

26. The maximum length of a 6″ single-wall vent connector that is used on a category 1 appliance is

A. 6′.

B. $7\frac{1}{2}$″.

C. 9′.

D. 12′.

27. All vent connectors entering a chimney must

A. never enter more than $7\frac{1}{2}$″ into the chimney.

B. be cemented at the chimney with an approved cement.

C. be constructed of a 30-gauge galvanized metal (minimum thickness).

D. always be made of type "B" venting.

28. One 3″ 90-degree elbow has the same resistance as

A. 2′ to 3″ pipe.

B. 3′ to 3″ pipe.

C. 4′ to 3″ pipe.

D. 5′ to 3″ pipe.

29. What will happen to a system that has too small of a vertical vent when using an induced draft motor?

A. An excessive draft will be in the vent pipe.

B. Condensation will form inside of the vent pipe.

C. Pressurization will form inside the vent pipe.

D. This will prevent downdrafts.

30. What can cause condensation to form in a venting system?

A. too low of a temperature rise across the heat exchanger

B. too high of a temperature rise across the heat exchanger

C. too large of a vent system

D. Both A and C

E. Both B and C

31. What can cause pressurization in a venting system?

A. too long of a lateral run

B. too short of a lateral run

C. blower speed set too high

D. Both A and C

E. Both B and C

32. Which is *not* acceptable when installing a main vent system?

A. a straight vertical pipe with no elbows

B. a vertical pipe with two 45-degree elbows

C. a vertical pipe with two 90-degree elbows

D. All of the above

33. What is the minimum height of a main vent (Figure 2–1) from the nearest connected appliance?

A. 3′

B. 5′

C. 8′

D. 10′

34. The main vent (Figure 2–1) must be

A. 2′ higher than any vertical wall within 15′.

B. 3′ higher than any vertical wall within 10′.

C. 2′ higher than any vertical wall within 10′.

D. 3′ higher than any vertical wall within 5′

35. What is the minimum height of a main vent that is used on a $\frac{7}{12}$ pitched roof? Use the "roof pitch" chart (Figure 2–1) to help determine your answer.

A. 1′3′

B. 1′6′

C. 2′

D. 2′6′

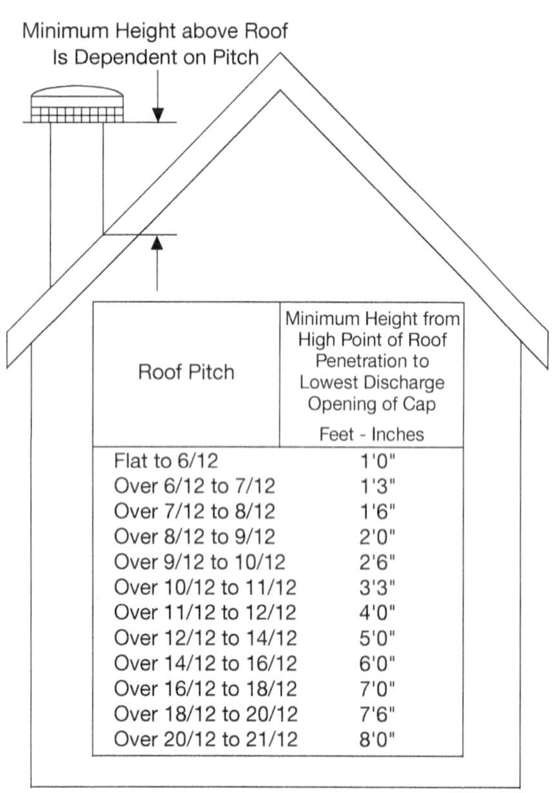

Roof Pitch	Minimum Height from High Point of Roof Penetration to Lowest Discharge Opening of Cap Feet - Inches
Flat to 6/12	1'0"
Over 6/12 to 7/12	1'3"
Over 7/12 to 8/12	1'6"
Over 8/12 to 9/12	2'0"
Over 9/12 to 10/12	2'6"
Over 10/12 to 11/12	3'3"
Over 11/12 to 12/12	4'0"
Over 12/12 to 14/12	5'0"
Over 14/12 to 16/12	6'0"
Over 16/12 to 18/12	7'0"
Over 18/12 to 20/12	7'6"
Over 20/12 to 21/12	8'0"

FIGURE 2–1
Roof Pitch Chart

36. When venting a direct-vent high-efficiency condensing furnace horizontally, the vent pipe should

A. be pitched toward the furnace.

B. be pitched away from the furnace.

C. be insulated.

D. never be schedule 40 PVC.

37. When venting a direct-vent high-efficiency condensing furnace horizontally, the vent pipe should

A. be pitched a minimum of $\frac{1}{2}''$ per foot of length.

B. be pitched a minimum of $\frac{1}{4}''$ per foot of length.

C. be pitched a minimum of 1″ per foot of length.

D. be completely level.

38. When venting a direct-vent high-efficiency condensing furnace horizontally, the vent pipe should

A. have an airtight seal and be supported every 5′.

B. have an airtight seal and be supported every 10′.

C. have an airtight seal and be supported every 15′.

D. have an airtight seal and be supported every 20′.

39. What is the minimum height of a masonry chimney?

A. 1′ above the peak

B. 2′ above the peak

C. 3′ above the peak

D. 4′ above the peak

40. If a masonry chimney is to be used for a gas appliance,

A. it must have a clay tile liner.

B. it must always have an aluminum liner.

C. it must be located in the center of the home away from an outside wall.

D. only one appliance is permitted per chimney.

41. Which statement is *false*?

A. The maximum lateral run of a 5″ vent connector is $7\frac{1}{2}'$.

B. An induced draft noncondensing furnace must have a chimney liner when used in a masonry chimney.

C. Chimneys with clay tile liners can be used to vent gas appliances.

D. The space surrounding a chimney liner can be used to vent an existing appliance.

42. When common venting a 100,000-Btu/h furnace with a 35,000-Btu/h water heater,

A. the furnace vent must be the shortest of the two and located above the water heater vent.

B. the furnace vent must be the longest of the two and located above the water heater vent.

C. the water heater vent must be the shortest of the two and located above the furnace vent.

D. the water heater vent must be the longest of the two and located above the furnace vent.

TABLE 2–1 Capacity of Type "B" Double-Wall Vents with Single-Wall Metal Connectors Serving a Single Category 1 Appliance

Height	Lateral	3″			4″			5″			6″		
H (ft)	L (ft)	FAN MIN	FAN MAX	NAT MAX	FAN MIN	FAN MAX	NAT MAX	FAN MIN	FAN MAX	NAT MAX	FAN MIN	FAN MAX	NAT MAX
6	0	38	77	45	59	151	85	85	249	140	126	373	204
	2	39	51	36	60	96	66	85	156	104	123	231	156
	4	NR	NR	33	74	92	63	102	152	102	146	225	152
	6	NR	NR	31	83	89	60	114	147	99	163	220	148
8	0	37	83	50	58	164	93	83	273	154	123	412	234
	2	39	56	39	59	108	75	83	176	119	121	261	179
	5	NR	NR	37	77	102	69	107	168	114	151	252	171
	8	NR	NR	33	90	95	64	122	161	107	175	243	163
10	0	37	87	53	57	174	99	82	293	165	120	444	254
	2	39	61	41	59	117	80	82	193	128	119	287	194
	5	52	56	39	76	111	76	105	185	122	148	277	186
	10	NR	NR	34	97	100	68	132	171	112	188	261	171
15	0	36	93	57	56	190	111	80	325	186	116	499	283
	2	38	69	47	57	136	93	80	225	149	115	337	224
	5	51	63	44	75	128	86	102	216	140	144	326	217
	10	NR	NR	39	95	116	79	128	201	131	182	308	203
	15	NR	NR	NR	NR	NR	72	158	186	124	220	290	192

43. Using Table 2–1, determine what the first column under "Height" and "Lateral" represents.

A. the height of the main vent and the height of the vent connector

B. the length of the chimney and the height of the vent connector

C. the height of the chimney and the length of the vent connector

D. the height of the main vent and the length of the vent connector

44. Using Table 2–1, determine what the numbers under "FAN MIN," "FAN MAX," and "NAT MAX" represent.

A. appliance input rating of thousands of Btu/h

B. appliance output rating of hundreds of Btu/h

C. appliance input rating of hundreds of Btu/h

D. appliance output rating of thousands of Btu/h

45. When venting a 150,000-Btu/h induced draft noncondensing furnace that has a 10′ wall lateral connector that is connected to a 15′ vent, what size should the main vent be? Use Table 2–1 to find your answer.

A. 4″

B. 5″

C. 6″

D. This cannot be accomplished according to the vent tables.

46. Using Table 2–2, determine what the first column under "Height" and "Rise" represents.

A. the height of the main vent and the height of the vent connector

B. the length of the chimney and the height of the vent connector

C. the height of the chimney and the length of the vent connector

D. the height of the main vent and the length of the vent connector

47. When common venting two or more appliances, the vent connector capacity tables that are used assume

A. that the length of the vent connector is no more than 3 feet.

B. that the maximum length of the vent connector is no more than 112 feet per inch of connector diameter.

C. that the maximum length of the vent connector is no more than 2 feet per inch of connector diameter.

D. that the minimum length of the vent connector is no more than 1 foot.

48. According to Table 2–2, what is the minimum vent diameter that can be used on a 150,000-Btu/h induced draft noncondensing furnace using a 3′ connector rise and a main vent that is 20′ high?

A. 4″

B. 5″

C. 6″

D. This cannot be accomplished according to the vent tables.

49. Using Table 2–2, determine the minimum vent diameter that can be used on a 40,000-Btu/h natural draft water heater using a 2′ connector rise if the main vent is 20′ high.

A. 4″

B. 5″

C. 6″

D. This cannot be accomplished according to the vent tables.

50. According to Table 2–2 and Table 2–3, what is the minimum diameter of a 20′ main vent when common venting a 150,000-Btu/h induced draft noncondensing furnace using a 3′ connector rise with a 40,000-Btu/h natural draft water heater using a 2′ connector rise?

A. 4″

B. 5″

C. 6″

D. This cannot be accomplished according to the vent tables.

TABLE 2–2 Capacity of Type "B" Double-Wall Vents with Type "B" Double-Wall Connectors Serving Two or More Category 1 Appliances common Vent Diameter in Inches

Connector		Vent Connector Diameter in Inches											
Height	Rise	3″			4″			5″			6″		
H (ft)	L (ft)	FAN MIN	FAN MAX	NAT MAX	FAN MIN	FAN MAX	NAT MAX	FAN MIN	FAN MAX	NAT MAX	FAN MIN	FAN MAX	NAT MAX
8	1	22	40	27	35	72	48	49	114	76	64	176	109
	2	23	44	32	36	80	57	51	128	90	66	195	129
	3	24	47	36	37	87	64	53	139	101	67	210	145
10	1	22	43	28	34	78	50	49	123	78	65	189	113
	2	23	47	33	36	86	59	51	136	93	67	206	134
	3	24	50	37	37	92	67	52	146	104	69	220	150
15	1	21	50	30	33	89	53	47	142	83	64	220	120
	2	22	53	35	35	96	63	49	153	99	66	235	142
	3	24	55	40	36	102	71	51	163	111	68	248	160
20	1	21	54	31	33	99	56	46	157	87	62	246	125
	2	22	57	37	34	105	66	48	167	104	64	259	149
	3	23	60	42	35	110	74	50	176	116	66	271	168

TABLE 2–3 Common Vent Diameter in Inches

	4″			5″			6″		
Vent Height (ft)	FAN +FAN	FAN +NAT	NAT +NAT	FAN +FAN	FAN +NAT	NAT +NAT	FAN +FAN	FAN +NAT	NAT +NAT
8	101	90	73	155	129	114	224	178	163
10	110	97	79	169	141	124	243	194	178
15	125	112	91	195	164	144	283	228	206
20	136	123	102	215	183	160	314	255	229

THE FOLLOWING QUESTIONS MAY BE HELPFUL TO STUDY FOR THESE CERTIFICATION TESTS:

NATE—GS-SV, GS-IN, HG-SV, HG-IN.

ESCO INSTITUTE HVAC EXCELLENCE GAS HEAT SERVICE

GAS PIPING

1. A drip leg should be installed

A. on the vertical gas line.

B. on a horizontal gas line.

C. before the main gas shutoff.

D. at the gas meter.

2. A manual gas cock valve shutoff should be installed at which part of the furnace?

A. internal

B. external

C. at the main gas line

D. at the gas meter

3. A common pressure of _______ WC is needed from the meter to the furnace when using natural gas.

A. $3\frac{1}{2}''$

B. 9″

C. 7″

D. 11″

4. A common pressure of _______ WC is needed from the gas regulator to the burner orifices for natural gas.

A. $3\frac{1}{2}''$

B. 9″

C. 7″

D. 11″

5. Liquefied petroleum gas (LPG) must maintain a pressure of _______ WC from the regulator at the tank when it is being used.

A. $3\frac{1}{2}''$

B. 9″

C. 7″

D. 1″

6. _______ refers to the relative weight of a gas as compared to the weight of an equal amount of air.

A. Specific gravity

B. Relative humidity

C. Specific heat

D. Density

7. The specific gravity of natural gas is approximately

A. 0.25.

B. 0.65.

C. 0.75.

D. 2.01.

8. To install or alter fuel gas piping,

A. a license is always required to obtain a permit.

B. a permit is required by most municipalities.

C. a permit is not required by most municipalities.

D. None of the above are true.

9. The gas supply line to a mobile home is commonly

A. $\frac{3}{4}''$

B. $\frac{1}{2}''$

C. 1″

D. $1\frac{1}{4}''$

10. The gas piping to supply mobile homes should be installed

A. at least 10″ below grade.

B. at least 12″ below grade.

C. at least 18″ below grade.

D. at least 16″ below grade.

11. When installing gas pipe near structures such as porches, steps, breezeways, etc., gas pipe installations should be kept at least

A. 4″ above grade or structure.

B. 6″ above grade or structure.

C. 8″ above grade or structure.

D. 16″ above grade or structure.

12. All gas piping must be

A. vacuum tested.

B. tested with CO_2.

C. tested with air pressure.

D. tested with oxygen.

13. Gas piping is commonly constructed of

A. wrought iron.

B. steel.

C. black pipe.

D. All of the above

14. Gas piping that is not protected can be

A. at least 10″ below grade.

B. embedded in 4″ of cement.

C. embedded in 6″ of cement.

D. installed above grade in recesses or channels that are approved.

15. Gas shutoff valves should be installed

A. 6′ above floor level.

B. 8′ above floor level.

C. on a horizontal pipe.

D. within 3′ of the appliance.

16. Leaks in gas piping should be located with

A. nitrogen.

B. oxygen.

C. a halide torch.

D. a soapy solution.

17. All manual gas cock valves need to be

A. constructed of brass.

B. approved for gas fuel only.

C. constructed of aluminum casting.

D. approved by GAMA.

18. Never use an open flame

A. while using a soap solution.

B. to locate gas leaks.

C. around an electronic leak detector.

D. around a combustible leak detector.

19. Plastic gas piping should be

A. at least 18″ below grade.

B. at least 10″ below grade.

C. at least 12″ below grade.

D. at least 14″ below grade.

20. Supports for hanging gas piping should be constructed of

A. plastic straps.

B. thin wire.

C. copper hooks.

D. metal straps or hooks at approved intervals.

21. Underground gas piping that is ferrous and metallic should be at least

A. 6″ below grade.

B. 8″ below grade.

C. 10″ below grade.

D. 12″ below grade.

22. Underground piping should be supported by

A. compacted soil or sand.

B. concrete.

C. top soil.

D. gravel and soil.

23. On the "Gas Piping Diagram," Figure 2–2, and in Table 2–4, what is the total amount of gas volume used in the system?

A. 250 sq ft

B. 250,000 Btu/h

C. 275 sq ft

D. 275,000 Btu/h

24. On the "Gas Piping Diagram," Figure 2–2, and Table 2–4, what length of gas pipe will be used to size the system?

A. The sections vary depending on where the appliance is placed in the system.

B. 40′

C. 50′

D. 60′

25. When sizing pipe section "C" to "F" in the system, what length will be used? (See Figure 2–2 and Table 2–4.)

A. 10′

B. 20′

C. 50′

D. 60′

26. Using the "Gas Piping Diagram," Figure 2–2, and Table 2–4, determine the correct size for the main gas pipe from points "A" to "B."

A. $\frac{3}{4}''$

B. 1″

C. $1\frac{1}{4}''$

D. $\frac{1}{4}''$

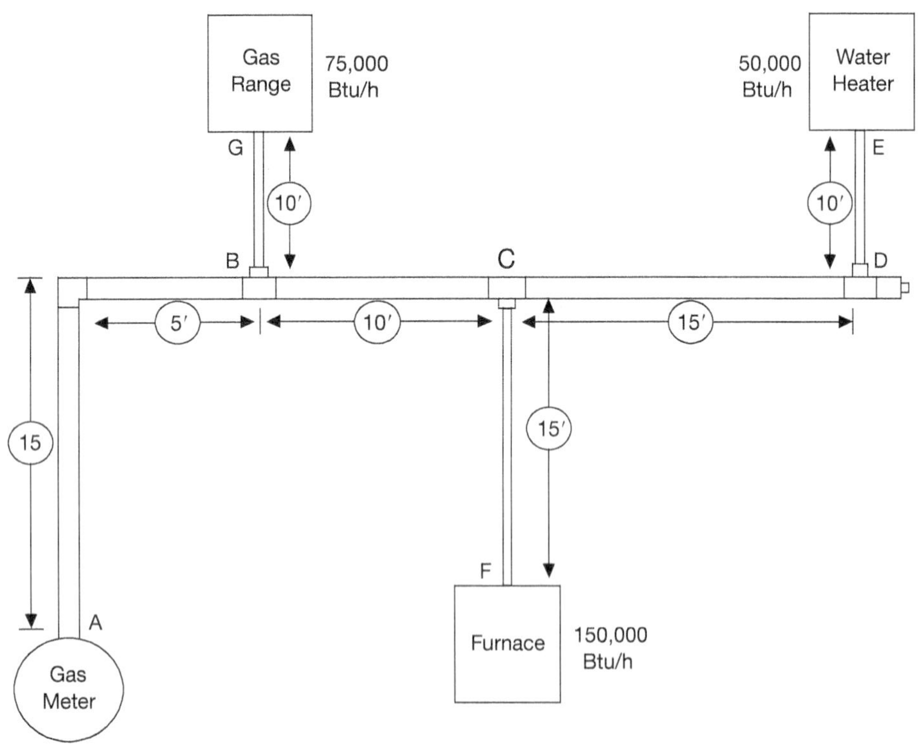

FIGURE 2–2

Gas Piping Diagram

TABLE 2–4 Pipe Equivalent Chart

	LENGTH OF PIPE IN FEET					
Nominal Iron Pipe Size in Inches	10′	20′	30′	40′	50′	60′
$\frac{1}{4}''$	32	22	18	15	14	12
$\frac{3}{8}''$	72	49	40	34	30	27
$\frac{1}{2}''$	132	92	73	63	56	50
$\frac{3}{4}''$	278	190	152	130	115	105
1″	520	350	285	245	215	195
$1\frac{1}{4}''$	1,050	730	590	500	440	400
$1\frac{1}{2}''$	1,600	1,100	890	760	670	610

Maximum Capacity of Pipe in Cubic Feet per Hour

27. Using the "Gas Piping Diagram," Figure 2–2, and Table 2–4, determine the correct size for the main gas pipe from points "B" to "C."

A. $\frac{1}{2}''$

B. $\frac{3}{4}''$

C. 1″

D. $1\frac{1}{4}''$

28. Using Figure 2–2 and Table 2–4, determine the correct size for the main gas pipe from points "C" to "D."

A. $\frac{1}{2}''$

B. $\frac{3}{4}''$

C. 1″

D. $1\frac{1}{4}''$

29. Using Figure 2–2 and Table 2–4, determine the correct size for the gas pipe from points "D" to "E."

A. $\frac{1}{2}''$

B. $\frac{3}{4}''$

C. 1″

D. $1\frac{1}{4}''$

30. Using Figure 2–2 and Table 2–4, determine the correct size for the gas pipe from points "B" to "G."

A. $\frac{1}{2}''$

B. $\frac{3}{4}''$

C. 1″

D. $1\frac{1}{4}''$

31. Using Figure 2–2 and Table 2–4, determine the correct size for the gas pipe from points "C" to "F."

A. $\frac{1}{2}''$

B. $\frac{3}{4}''$

C. 1″

D. $1\frac{1}{4}''$

32. Black iron pipe is used in areas where corrosion will not affect it, and the element that gives the pipe its color is

A. copper.

B. carbon.

C. silver.

D. The heat-treating process gives the pipe its color.

33. When measuring black iron pipe, it is measured by its

A. outside diameter.

B. amount of threads per inch.

C. outside diameter and length.

D. nominal size.

34. A description of nominal size would be:

A. outside diameter and length.

B. amount of threads per inch and diameter.

C. an approximation of inside diameter for pipe up to 12″ in length.

D. an estimation of pipe weight per linear feet.

35. Pipe thread types are taper and straight and are classified by

A. SMACNA.

B. ACCA.

C. BOCA Code.

D. NPT.

36. Pipe thread compound is used on pipe threads for the purpose of

A. acting as a lubricant.

B. protecting the threads from corrosion.

C. acting as a lubricant and providing a seal.

D. protecting against contaminants within the gas.

THE FOLLOWING QUESTIONS MAY BE HELPFUL TO STUDY FOR THESE CERTIFICATION TESTS:
NATE—GS-SV, GS-IN.
ESCO INSTITUTE HVAC EXCELLENCE GAS HEAT SERVICE

GAS FURNACE TROUBLESHOOTING

1. On a standing pilot furnace, the thermocouple must be replaced when

A. it is not positioned properly.

B. the millivolt reading is 28 or higher.

C. the millivolt reading is 22 or lower.

D. the tip of the thermocouple is in the pilot flame $\frac{3}{8}''$.

2. Natural gas pressure at the main line is recorded at 7″ WC, but at the gas manifold it reads 1.5″ WC. What is most likely to be the cause of this?

A. a defective main line regulator

B. too large of a gas valve

C. plugged burner orifices

D. a defective pressure regulator at the gas valve

3. A technician servicing a gas-fired furnace notices four out of five ribbon burners are firing simultaneously. What could be the problem with this system?

A. too much gas pressure

B. too little gas pressure

C. burners not aligned properly

D. a defective gas valve

4. Technician A says that a combination gas valve needs to be replaced if there are 24 volts present to the valve and it will not open.

Technician B says that if a combination gas valve has a new thermocouple and the pilot will not stay lit, the gas valve needs to be replaced. Which technician is correct?

A. technician A

B. technician B

C. Both technicians are correct.

D. Both technicians are incorrect.

5. What should be the first step that a technician should take if gas pressure at the manifold is too high?

A. Adjust the pressure regulator clockwise.

B. Adjust the pressure regulator counterclockwise.

C. Replace the pressure regulator.

D. Replace the combination gas valve.

6. If the flame is lifting off a burner port, what should a technician do to correct this problem?

A. Adjust primary air by closing off the air shutter.

B. Adjust primary air by opening the air shutter.

C. Replace the burner.

D. Decrease main gas pressure.

7. A step-down transformer is reading 0 ohms at its secondary winding. What is a possible cause for this problem?

A. an open primary winding

B. an open secondary winding

C. a shorted primary winding

D. a shorted secondary winding

8. When a fan center relay needs to be tested for 24 volts, what two terminals are used when measuring for transformer output voltage?

A. terminals "R" and "W"

B. terminals "R" and "G"

C. terminals "R" and "Y"

D. terminals "R" and "C"

9. On a call for heat, what two terminals should be closed on the thermostat?

A. terminals "R" and "W"

B. terminals "R" and "G"

C. terminals "R" and "Y"

D. terminals "R" and "C"

10. When a thermostat subbase fan selector switch is moved to fan "on," what terminals will close in the thermostat?

A. terminals "R" and "W"

B. terminals "R" and "G"

C. terminals "R" and "Y"

D. terminals "R" and "C"

11. A new thermostat was recently replaced. The technician observes that the burner cycle is not long enough. The most likely cause would be

A. a defective thermostat.

B. low gas pressure.

C. that the anticipator needs adjusting.

D. a defective gas valve.

12. A customer has replaced a mercury bulb thermostat but complains that the burners will not shut off and the house is getting too hot. The most likely cause would be

A. a defective limit control.

B. a defective gas valve.

C. that the thermostat is not properly level.

D. that the anticipator is set too low.

13. A bimetal fan control will not turn the indoor blower motor off. As a result, cool air passes through the registers between furnace cycles. What is a possible cause of this problem?

A. The heating circuit is opening on limit.

B. The fan control contacts are stuck open.

C. The fan control contacts are stuck closed.

D. There is a defective gas valve.

14. Residents are complaining of experiencing symptoms of headaches, watery eyes, and nausea. Technician A says that the furnace heat exchanger possibly is cracked.

Technician B says that the furnace may only need cleaning and that the heat exchanger is probably plugged. Which technician is correct?

A. technician A

B. technician B

C. Both technicians are correct.

D. Both technicians are incorrect.

15. A belt drive motor has recently been replaced but the furnace is shutting off on limit control. The most likely cause is that

A. the motor is running in the wrong direction.

B. the motor is defective.

C. the motor is too large.

D. the motor mount is not correct.

16. Upon testing a motor run winding for continuity, the technician records 3 ohms. The winding resistance should read 8 ohms. The technician concludes that

A. the winding is open.

B. the winding is shorted.

C. the winding is grounded.

D. the motor is within satisfactory operating conditions.

17. A run capacitor is tested with an analog meter. The needle moves toward zero then deflects back to infinity. What is the condition of the capacitor?

A. The capacitor is open.

B. The capacitor is good.

C. The capacitor should not be used.

D. The capacitor needs to be of a higher mfd rating.

18. The temperature rise of a particular furnace states that it should be between 35 and 65°F. The furnace is operating at 20°F temperature rise with the proper gas pressure at the manifold. What could be done to solve this problem?

A. Add more outside air to the basement.

B. Increase the indoor blower speed.

C. Decrease the indoor blower speed.

D. Replace the air filter with a more efficient type.

19. A belt drive motor has been replaced, but quickly gets hot and stops just short of burning out. Which of the following could cause this problem?

A. a fan belt that is too tightly adjusted

B. a fan belt that has only 1″ of slack between pulleys

C. the pulley on the motor is too small

D. None of the above

20. A combination fan/limit control has been replaced. The indoor blower motor runs too long and the resident complains of cooler airflow. The technician should

A. replace the control again.

B. adjust the limit setting.

C. lengthen the fan-off setting.

D. check and readjust the fan on/off differential.

21. A service call is made to a customer who complains that when his furnace ignites, the flame rolls out in front of the heat exchanger.

Technician A says that it's due to a poor draft, possibly a blocked flue or heat exchanger.

Technician B says that it has to do with the limit control or the fan switch control circuit. Which technician is correct?

A. technician A

B. technician B

C. Both technicians are correct.

D. Both technicians are incorrect.

22. A customer complains of headaches, sleepiness, and nausea.

Technician A finds a crack in the heat exchanger and says that the furnace should be shut down.

Technician B says that the crack isn't too serious, and that the furnace can operate temporarily but it must be cleaned. Which technician is correct?

A. technician A

B. technician B

C. Both technicians are correct.

D. Both technicians are incorrect.

23. A technician is servicing a standing pilot gas-fired furnace. The complaint is that the burners are not lighting. Where should the technician start troubleshooting the system?

A. at the transformer secondary and at the combination gas valve

B. at the transformer primary only

C. at the primary limit control

D. at the fan/limit control

24. On a spark ignition system, the spark for ignition continues with no ignition of gas at the burners.

Technician A concludes that the thermostat is defective.

Technician B concludes that it is possibly the control module. Which technician is correct?

A. technician A

B. technician B

C. Both technicians are correct.

D. Both technicians are incorrect.

25. Which of the following is the correct test for a faulty cable on a direct spark ignition system?

A. a voltage check

B. a continuity check

C. an amperage check

D. None of the above

26. A customer complains that the burners light on her system, but then immediately shut down. The problem is most likely to be

A. in the control module.

B. in the control transformer.

C. in the limit control.

D. in the gas valve.

27. Which of the following would not be a proper procedure to perform on a direct spark ignition system where the spark continues, the burners light, and the flames are immediately extinguished?

A. check that the sensor is positioned in the flame correctly

B. check for proper flame rectification current

C. check for a proper grounding terminal

D. check that there is proper gas flow to the pilot

28. A hot surface ignition system does not ignite the main burner. After measuring voltage to the igniter, the technician should

A. perform a resistance test on the igniter.

B. replace the flame sensor.

C. replace the main gas valve.

D. replace the control module.

29. The burners of a gas furnace that is equipped with a 115-volt hot surface igniter light for approximately 7 s and then go out. What is a problem that could cause this symptom?

A. a faulty igniter

B. a faulty transformer secondary winding

C. burners shutting off on high-limit control

D. a faulty sensor

30. The burners on a furnace having a hot surface ignition system light up, but quickly shut off.

Technician A states that after testing the gas valve, it has been determined that it is shorted.

Technician B visually checked the flame sensor, cleaned it, and performed a continuity test. The technician faults a dirty sensor.

Which technician is most likely to be correct?

A. technician A

B. technician B

C. Both technicians are correct.

D. Both technicians are incorrect.

31. The thermostat on a mid-efficiency gas furnace with a hot surface ignition system is placed on a call for heat. The burners do not light, but the igniter does glow. Which of the following would be the next step to take when troubleshooting the system?

A. Test the flame sensor.

B. Test the gas valve and gas manifold pressure.

C. Test the primary limit control and the fan control.

D. Test the thermostat.

32. The draft pressure switch on a mid-efficiency furnace is not allowing the ignition system to energize. Which of the following is a possible problem?

A. a dirty filter

B. a defective indoor blower

C. a defective induced draft motor

D. a defective flame roll-out sensor

33. A high-efficiency gas furnace will not stay in operation. The unit runs for approximately 5 min then is shut down by the main furnace controller with an "open pressure switch" flash code. Which of the following is not a proper test to perform in this situation?

A. Check the induced draft motor.

B. Check the furnace filter.

C. Check for a restricted flue pipe.

D. Both A and C

34. A service call is made on a mid-efficiency furnace. On a call for heat, the induced draft motor will not operate. What is the proper troubleshooting procedure that should be taken?

A. Check for control voltage at the "G" and "C" terminals at the circuit board.

B. Check for control voltage across terminals "W" and "C" at the circuit board.

C. Check for control voltage at the induced draft relay.

D. Both B and C

35. A high-efficiency furnace that uses a furnace controller will not energize the blower motor after a preselected 45-s "time on" setting. Which of the following would be the most correct troubleshooting procedure?

A. Measure the resistance across the furnace controller low-speed terminal and common.

B. Measure the resistance across the furnace controller high-speed terminal and common.

C. Measure voltage to the motor and test the run capacitor.

D. Measure amperage to the blower motor and test the run capacitor.

36. When servicing a mid-efficiency furnace, the thermostat has been turned to the "off" position, but the burners remain lit, and the induced draft motor continues to operate. Which of the following would be the most correct troubleshooting procedure?

A. Determine if the thermostat is open and check voltage at the gas valve.

B. Check for voltage at the indoor blower motor.

C. Replace the control module.

D. Check the ignition system.

37. On a high-efficiency furnace that uses a furnace controller, the indoor blower motor will not de-energize after its preselected "Off Delay" setting. Which of the following is a logical repair for the system?

A. Replace the run capacitor.

B. Replace the blower motor.

C. Replace the control module.

D. Change the speed settings on the control module.

38. The burners on a furnace cycle for approximately 3 to 4 min. During a call for heat, the system does not provide adequate heat for the home.

Technician A has determined that a new thermostat has been installed and the heat anticipator was incorrectly adjusted.

Technician B cleans and positions the flame sensor and claims that this was the fault.

Which technician is correct?

A. technician A

B. technician B

C. Both technicians are correct.

D. Both technicians are incorrect.

39. A gas furnace has been recently converted to LPG gas. On a call for heat, the furnace is not firing correctly and is starting to soot up. What would be the most likely cause for this problem?

A. low LPG pressure

B. too long of a burner cycle

C. wrong orifice size installed

D. All of the above

40. When inspecting a gas furnace, it is noticed that the burner flames are very small in height. What action should be taken by the technician?

A. Replace the gas valve.

B. Replace the orifices.

C. Check gas pressures and adjust primary air shutters if needed.

D. Contact the utility company to replace the gas meter.

41. A customer is experiencing loud, delayed ignitions from his furnace.

Technician A says that it's possible that the gas valve is not opening or closing properly.

Technician B claims that it's possible that after the furnace had been cleaned, the burners were not properly aligned.

Which technician is correct?

A. technician A

B. technician B

C. Both technicians are correct.

D. Both technicians are incorrect.

42. A customer called in complaining that his indoor blower would not stop running. What is the most possible cause of this problem?

A. a defective blower motor run capacitor

B. contacts stuck open on the fan control

C. contacts stuck closed on the fan control

D. Both A and C

43. After diagnosing a no-heat call, it has been determined that the induced draft motor is not operating.

Technician A claims that it is the fault of an induced draft pressure switch.

Technician B states that the indoor blower relay module is defective.

Which technician is more likely to be correct?

A. technician A

B. technician B

C. Both technicians are correct.

D. Both technicians are incorrect.

44. The burners on a gas fired furnace are continuously lit, and setting the thermostat to the "off" position does not make any difference. What is a possible problem for this fault?

A. a defective gas valve

B. a defective fan control

C. a limit control stuck in the closed position

D. a shorted secondary winding on the transformer

45. A customer complains that she has no heat and that the furnace does nothing at all. She says that it is a mid-efficiency furnace with an induced draft motor and a spark ignition system. Which of the following would be a likely place to start troubleshooting?

A. Check the fan door switch.

B. Check for a blown fuse.

C. Check for a defective furnace switch.

D. All of the above

46. On a hot surface ignition system, the igniter does not glow. What would be the most likely cause of this fault?

A. a bad flame sensor

B. a bad gas valve

C. a bad closed pressure switch

D. a bad igniter

47. On a spark ignition system, the spark is occurring but its pilot will not light. Which of the following would you check?

A. Test terminals PV and MV/PV for 24 volts.

B. Test for continuity of PV coil.

C. Test for gas to the pilot line.

D. All of the above

48. The thermocouple of a furnace has just been replaced, but the pilot light continues to go out. What test should a technician perform next?

A. Ensure that the pilot flame is engulfing the tip of thermocouple by $\frac{3}{8}''$.

B. Test continuity of the main valve.

C. Test for voltage to the main gas valve.

D. Check for an open limit control.

49. An existing thermocouple has been tested with a millivolt meter and has registered at 13 millivolts. Which of the following would be the correct action to take?

A. Make sure that the thermostat is calling for heat.

B. Replace the thermocouple.

C. Replace the pilot gas valve.

D. There is no problem with the thermocouple, and it should be left alone.

50. What is an acceptable voltage of a thermocouple?

A. 30 millivolts

B. 20 millivolts

C. 10 millivolts

D. 5 millivolts

51. An indoor blower motor will run, but it cycles on an internal overload protector. Using a digital meter, technician A states that the meter reads zero when placing it across the motor's run capacitor.

Technician B states that the meter reads "OL" when placing it across the run capacitor. Which technician would have the correct diagnosis?

A. technician A

B. technician B

C. Both technicians are correct.

D. Both technicians are incorrect.

52. A furnace has been operating well during mild winters. During the very cold days, the customer complains that the 3-year-old furnace does not provide continuous heat to satisfy the demand. The technician does a Btu input measurement by clocking the gas meter and finds the system within specifications. The temperature rise is also within system limits. Which of the following would be the most correct diagnosis?

A. The furnace was undersized in heating capacity.

B. The gas pressure needs to be increased.

C. The ductwork needs redesigning.

D. None of the above

53. During furnace operations, the thermostat is quickly satisfied. The new furnace heats the entire house very fast, then shuts off on thermostat demand. Which would be the most correct statement of this situation?

A. The thermostat needs replacing.

B. The blower speed should be increased.

C. The furnace heating capacity is oversized.

D. The duct system is too large.

54. Two gas furnaces were installed. Both furnaces were 150 MBtu/h (150,000 Btu/h). Both furnaces may operate at the same time. When one furnace is running, all operations are satisfactory. When both furnaces are operating at the same time, the burner flames of both furnaces decrease in size. Which procedure most likely would be correct?

A. Replace the heating thermostat.

B. Increase the gas pressure regulator of the first furnace.

C. Replace both gas valves.

D. Determine if the gas piping layout is properly sized.

55. On a natural-gas-to-LPG furnace conversion, technician A states that only the burner orifices need changing.

Technician B states that the orifice's gas valve regulator spring and pilot orifice should be changed.

Which technician is correct?

A. technician A

B. technician B

C. Both technicians are correct.

D. Both technicians are incorrect.

56. A draft test has been performed and it is indicating that there is an improper draft for the gas furnace that is being used. Which of the following would be the correct procedure to perform?

A. Increase the gas pressure.

B. Increase the induced draft motor speed.

C. Check for chimney obstructions.

D. Replace the heat exchanger.

57. On an LPG furnace installation, the burner flames are low and the new furnace is sooting up. Which of the following would be the most correct procedure to solve this problem?

A. Replace the gas valve.

B. Measure gas pressure and install a low-gas-pressure switch.

C. Call the LPG fuel provider for service.

D. Clean the heat exchanger and operate the furnace.

58. A crack has been detected on a heat exchanger. The furnace is only 5 years old. Which of the following would be the most correct procedure to perform?

A. Replace the entire furnace.

B. Replace the heat exchanger under manufacturer's warranty.

C. Replace the complete ductwork and furnace.

D. None of the above

59. On a belt-drive indoor blower motor, a fan belt needs replacing.

Technician A says that if you adjust the belt to ride higher on the pulley, you will increase the blower speed.

Technician B says that if you adjust the belt to ride lower into the pulley, you will increase the blower speed.

Which technician is correct?

A. Technician A

B. Technician B

C. Both technicians are correct.

D. Both technicians are incorrect.

60. A technician replaces a fan belt and a fan motor. The new motor is operating with too high of an amp draw. What is most likely the cause of this fault?

A. The belt was adjusted too tightly.

B. The belt was adjusted too loosely.

C. The motor horsepower is too low.

D. Both A and C

61. On a no-heat call, a fan motor is not operating.

Technician A measures voltage at the fan relay's normally open contacts.

Technician B measures voltage at the bimetal fan control.

Which technician performed the correct procedure?

A. Technician A

B. Technician B

C. Both technicians are correct.

D. Both technicians are incorrect.

62. The amperage of an indoor blower motor must be determined during the heating operation of a furnace. Where should this reading be taken?

A. after the N.O. contacts at the fan relay

B. before the N.O. contacts at the fan relay

C. at the high speed on the motor

D. at the red or white wire on the motor

63. Where should a resistance reading be taken when measuring the low speed of an indoor blower motor?

A. at the red wire and green ground wire on the motor

B. at the red wire and the white wire on the motor

C. at the black wire and the white wire on the motor

D. at the yellow wire and the green ground wire on the motor

64. A frequently used fireplace satisfies a thermostat located in the same room as the fireplace. This shuts down the call for heat while the greater portion of the home remains cold. What should the technician do to solve this problem?

A. Compensate by turning the thermostat higher.

B. Damper down the fireplace control.

C. Relocate the thermostat.

D. Increase gas pressure at the furnace.

65. What will an extremely dirty air filter do to furnace operation?

A. cause the furnace to shut off on the limit control

B. cause the indoor blower motor to overheat

C. cause an increase of furnace temperature rise

D. All of the above

66. Where can a limit control be wired in a gas-fired furnace?

A. in the high-voltage circuit

B. in the low-voltage circuit

C. in the high- or low-voltage circuit

D. None of the above

67. How are limit controls and flame roll-out sensors wired into their circuits?

A. in series with each other

B. in parallel with each other

C. in series parallel with each other

D. None of the above

68. A limit control should never be

A. wired in the high-voltage circuit.

B. wired in the low-voltage circuit.

C. bypassed permanently.

D. None of the above

69. Most limit controls that are wired on the high-voltage side of a circuit are in series with the transformer. Where would a voltage check be made to see if there is proper source voltage to the transformer?

A. after the limit control and at the common of transformer

B. before the limit control and at the common of transformer

C. at both leads of the transformer

D. Both A and C

70. Adjusting a heat anticipator to a higher setting will

A. provide a longer burner cycle.

B. provide a shorter burner cycle.

C. cause the indoor fan motor to short cycle.

D. prematurely shut off the gas valve.

71. A service call is made to a customer who complains that the furnace circuit breaker will not stay in the "on" position.

Technician A says that the problem can be in the blower motor or the primary side of the transformer.

Technician B says that the problem is most likely in the heating controls circuit.

Which technician is most likely correct in the diagnosis?

A. Technician A

B. Technician B

C. Both technicians are correct.

D. Both technicians are incorrect.

72. An induced draft motor is not operating when the thermostat is calling for heat. What would be the most correct test to make?

A. an amperage check at the motor

B. a resistance check of the motor

C. a voltage check on circuit board at induced draft motor terminals

D. a voltage check at the primary of the transformer

73. Which of the following residential furnace components would be checked for 24-volt operation?

A. inducer motor

B. blower motor

C. pressure switch

D. primary of transformer

74. Technician A says, "To ensure that a heating thermostat is in the closed position, a voltage check at 'R' and 'W' should read 24 volts."

Technician B says, "To ensure that a heating thermostat is in the closed position, a voltage check at 'C' and 'W' should read 24 volts."

Which technician is correct?

A. Technician A

B. Technician B

C. Both technicians are correct.

D. Both technicians are incorrect.

75. Which of the following is *not* a thermostat problem?

A. improper anticipator setting

B. subbase not level

C. furnace switch turned off

D. loose connection at subbase

76. Which of the following can cause a limit control to open?

A. a dirty filter

B. too high of supply-air static pressure

C. not enough return air

D. All of the above

77. A direct short-to-ground of a primary wire on the transformer would most likely cause

A. a circuit breaker to trip.

B. a transformer to become defective.

C. a gas valve to become defective.

D. a thermostat to become defective.

78. To correctly set a heat anticipator, the control amperage reading should be

A. performed at the gas valve.

B. performed at the secondary of the transformer.

C. performed at the thermostat subbase.

D. All of the above

79. When a natural gas furnace has flames that are more yellow than blue, what corrective action should a technician take?

A. Replace the gas valve.

B. Increase the gas pressure.

C. Open the primary air ports.

D. Close the primary air ports.

80. What are the normally open set of contacts used for on a fan center relay?

A. high-speed operation of the blower motor

B. low-speed operation of blower motor

C. humidifier circuit switching

D. isolation contacts for the heating circuit

81. A customer complaint is that the furnace is not operating. After inspection of the system, the technician notices a small hairline crack in the ceramic of the sensing probe. What action must the technician take to repair this problem?

A. Repair the probe by sealing the crack with silicone.

B. Replace the sensing probe.

C. No action; sensing probes can handle small cracks.

D. None of the above

82. How should a limit safety check be performed on a gas furnace?

A. Temporarily jump out the limit control for at least 10 min, then remove the jumper.

B. Stop the blower operation for at least 10 min or until the limit control opens.

C. Block off cold-air return for 5 min, or until the limit control closes.

D. Jump out the thermostat for 15 min until the furnace cycles on the limit control.

83. How is an induced draft pressure switch tested for proper operation?

A. Jump out the pressure switch after setting the thermostat to the heat position and test the control with a manometer.

B. Decrease the indoor motor speed and test the control with a manometer.

C. De-energize the induced draft motor and test the switch with a manometer.

D. Energize the heating system while testing the pressure switch with a manometer.

84. Which of the following are input signals to a high-efficiency furnace control board?

A. the thermostat, the flame roll-out sensor, and the igniter

B. the pressure switch, the pilot valve, and the limit control

C. the flame sensor circuit, the main gas valve, and the thermostat

D. the thermostat, the flame sensor circuit, and the limit control

85. Which of the following are output signals from a high-efficiency furnace control board?

A. the thermostat, the main gas valve, and the igniter

B. the main gas valve, the pilot valve, and the limit control

C. the flame sensor circuit, the main gas valve, and the igniter

D. the igniter, the pilot gas valve, and the main gas valve

86. Where do high-efficiency condensing furnaces get their combustion air?

A. from outside of the home only

B. from inside of the home only

C. from inside or outside of the home

D. None of the above

87. Which of the following is not a cause of insufficient low gas input?

A. a plugged orifice

B. a faulty regulator

C. a faulty heat exchanger

D. a bad gas valve

88. A customer complains that there is not enough supply air coming through the registers.

Technician A checked the furnace temperature rise, inspected the air filter, and adjusted a loose fan belt.

Technician B inspected the air filter, observed the blower motor operation, checked for duct obstructions, and made sure that the balancing dampers were in the correct position.

Which technician performed the proper procedure?

A. technician A

B. technician B

C. Both technicians are correct.

D. Both technicians are incorrect.

89. A technician was called because not enough heat was being generated at the furnace. The technician checked the heat exchanger, flue pipe, and burners. The technician also tested the inlet and outlet gas pressure to the furnace. The measurement was 4″ WC to the main gas valve and 2″ out of the gas valve. Which of the following could be the problem with the system?

A. a faulty gas valve

B. a faulty gas meter regulator

C. a faulty heat exchanger

D. a faulty gas control circuit

90. Which of the following would not be a cause of a faulty thermostat on a standing pilot furnace?

A. The gas valve will not de-energize.

B. The gas valve will not energize.

C. The furnace is short cycling.

D. The pilot will not stay lit.

91. A new high-efficiency furnace has been installed, and the furnace seems to come on too often. Which of the following would most likely be the problem?

A. a defective fan control

B. an improperly adjusted heat anticipator

C. a defective blower motor

D. too much airflow

92. Which of the following would *not* cause the burners of a gas furnace to shut down?

A. excessive airflow

B. a defective gas valve

C. a faulty thermostat

D. a defective induced draft motor

93. Which of the following is *not* a cause for the burners of a gas furnace to short cycle?

A. a dirty air filter

B. poor thermostat location

C. heat anticipator set too low

D. fan control contacts stuck closed

94. A customer complains that her house is too warm. The technician arrives on the job and notices that the gas valve is stuck open. The technician taps the side of the valve and the gas flow stops. What action should the technician take?

A. Replace the gas valve.

B. Lubricate the sticking valve.

C. Turn the valve on and off several times to free the problem.

D. Any of the above

95. A pilot light has been established, but it is too small and soon goes out. Which of the following procedures would be correct to remedy this condition?

A. Replace the pilot valve.

B. Repair the pilot valve coil.

C. Clean the pilot orifice.

D. Replace the entire gas valve.

96. What is the sequence of operation of a mid-efficiency furnace using a hot surface indirect ignition system?

A. The thermostat calls for heat, the ignition system lights the pilot, the inducer motor is energized, the pressure switch closes, the main gas valve is energized, and the blower motor is energized.

B. The thermostat closes, the inducer motor is energized, the main burner is lit from the igniter, and the blower motor is energized.

C. The thermostat closes, the inducer motor is energized, the pilot is lit from the igniter, the pilot is proved, the main gas valve is energized, and the blower motor is energized.

D. The thermostat opens, the inducer motor is energized, the main gas valve is energized, and the blower motor is energized.

97. On an intermittent spark ignition system, the pilot valve is energized, but there is no spark from the igniter. Which of the following could be the problem?

A. a faulty ignition cable

B. a faulty control module

C. a bad ground

D. All of the above

98. On a direct-spark ignition system, the igniter is sparking, but the main burners will not light. Which of the following is an *incorrect* test to perform?

A. Check for 24 volts across the gas terminals.

B. Check for correct position of spark gap.

C. Check for a cracked insulation on the igniter ceramic.

D. Check from terminals "W" to "C" to make sure that the thermostat is closed.

99. On a direct hot surface ignition system, the igniter is not glowing, but the main gas valve is energized during a trial for ignition. Which of the following is an *incorrect* test to perform?

A. Measure voltage to the igniter.

B. Measure resistance on the igniter.

C. Measure resistance of the limit control.

D. Measure input voltage to the control module.

100. A furnace that uses a direct hot surface ignition system will not ignite the burners.

Technician A measures 50 ohms on the hot surface igniter and says that the igniter is defective.

Technician B also measures 50 ohms on the igniter and says that the igniter is not defective; the problem is probably a bad control module.

Which technician is correct?

A. technician A

B. technician B

C. Both technicians are correct.

D. Both technicians are incorrect.

101. What terminals should measure 24 volts if the thermostat is calling for heat? Use the "Bimetal Flame Sensor Pilot Ignition System Diagram" (Figure 2–3) to answer this question.

A. LIM 1 and LIM 2

B. "R" and "W"

C. GAS 1 and GAS 2

D. "GC" and "C"

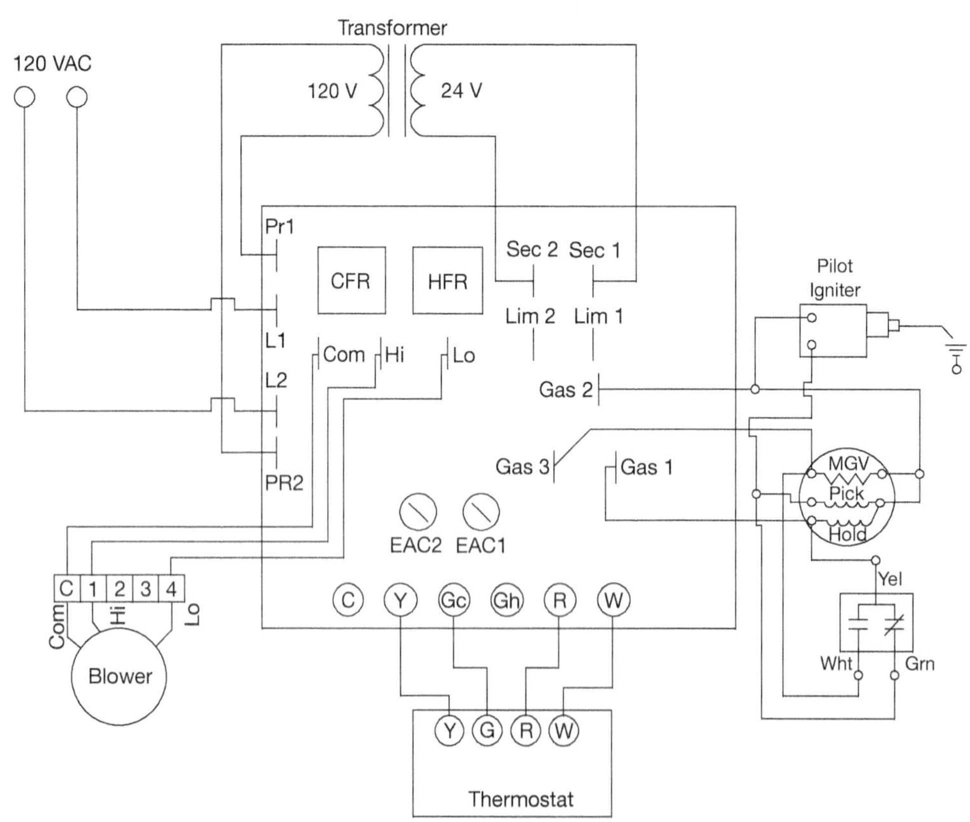

FIGURE 2–3

Bimetal Flame Sensor Pilot Ignition System Diagram

102. What terminals should measure 24 volts if the pilot has been proven? Use the "Bimetal Flame Sensor Pilot Ignition System Diagram" (Figure 2–3) to answer this question.

A. "Wht" at the sensor and GAS 2

B. "Yel" at "Wht" at the sensor

C. GAS 3 and GAS 1

D. "R" and "W"

103. Using the "Bimetal Flame Sensor Pilot Ignition System Diagram" (Figure 2–3), what two terminals energize the fan motor during the heating cycle?

A. terminals EAC 1 and EAC 2

B. terminals 4 and "C"

C. terminals 1 and "C"

D. terminals "R" and "GC"

104. Using the "Bimetal Flame Sensor Pilot Ignition System Diagram" (Figure 2–3), what terminals should always measure 24 volts if power is connected to the system?

A. terminals GAS 1, GAS 2, SEC 1, and SEC 2

B. terminals SEC 1, SEC 2, "R," and "W"

C. terminals SEC 1, SEC 2, "R," and "C"

D. All of the above

105. Where should a measurement of 24 volts be made when the heating thermostat is in the closed position? Use the "Indirect Spark Ignition System Diagram" (Figure 2–4) to answer this question.

A. at the "TH" terminal and terminal before "PS"

B. at the "W" terminal and any of the "HL" terminals

C. at the "W" terminal and "SL" terminal

D. at the "TH" terminal and "TR" terminal

106. The indoor blower motor will not start on a call for heat. The technician suspects a faulty set of FR contacts. Where should a voltage reading occur if the contacts are in good condition? Use the "Indirect Spark Ignition System Diagram" (Figure 2–4) to answer this question.

A. across the FR coil and from the load side of the normally open FR contacts to the common side of the motor

B. across the FR coil and across the normally open contacts

C. across the FR coil and across the FS contacts

D. across the FR coil and from the load side of the normally open contacts to the line side of the motor

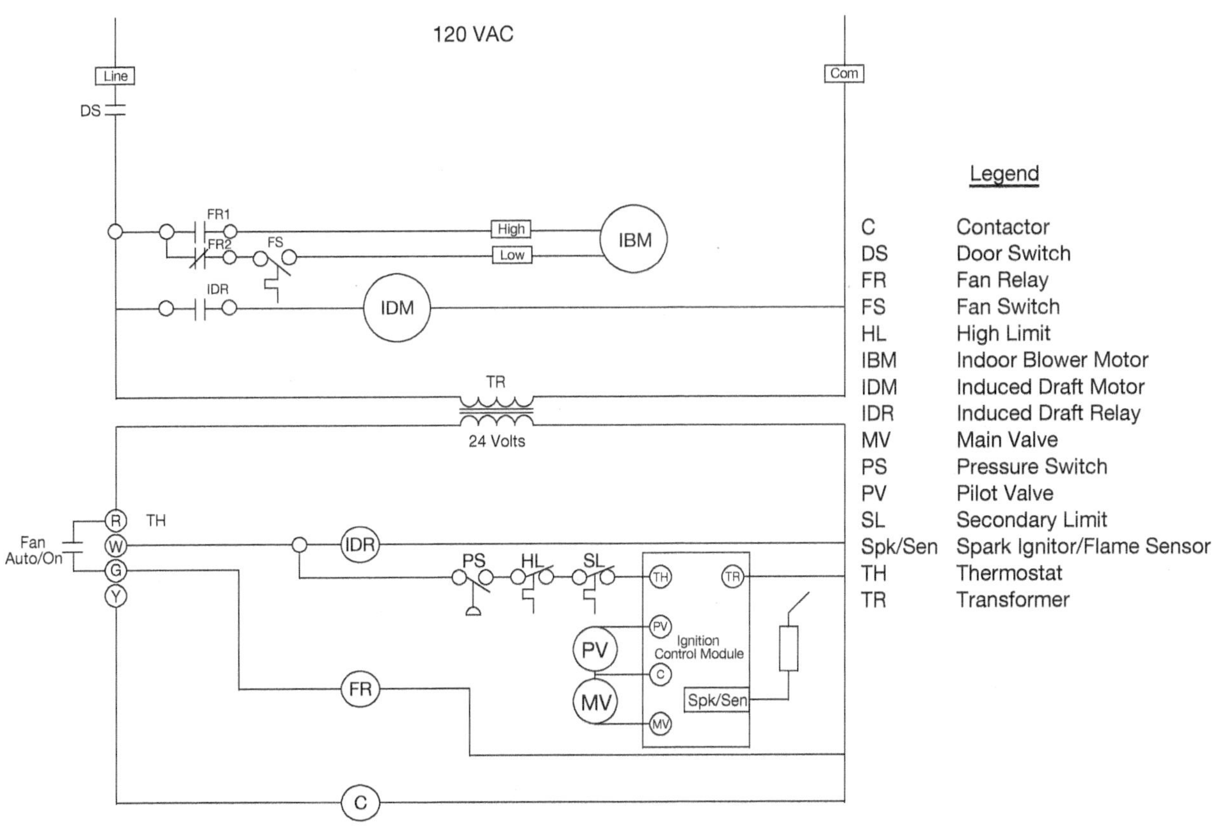

FIGURE 2–4
Indirect Spark Ignition System Diagram

107. What are the correct terminals to measure voltage to the main gas valve? Use the "Indirect Spark Ignition System Diagram" (Figure 2–4) to answer this question.

A. "TH" and "TR"

B. "PV" and "C"

C. "TH" and "C"

D. "MV" and "C"

108. A technician observes a furnace with an intermittent pilot retrofit ignition system and finds that although a pilot light exists, the main burners will not ignite; he concludes that

A. the gas valve pilot solenoid is defective.

B. the thermostat is not set high enough.

C. the thermostat is located too high on the wall.

D. the ignition system is not proving a pilot flame.

109. When a measurement is taken to record temperature rise, the supply air temperature should be taken

A. 10 inches above the evaporator coil.

B. above and out of direct line from the heat exchanger.

C. in the supply branch duct.

D. at the first supply run off the main trunk.

110. One of the reasons that a pressure switch fails to close to continue the ignition system is

A. manifold gas pressure is too high.

B. the blower motor is not operating adequately.

C. restriction in the flue vent.

D. manifold gas pressure is too low.

111. When there is not sufficient air available for the combustion process, one element of incomplete combustion that can become fatal is

A. soot.

B. aldehydes.

C. carbon monoxide.

D. carbon dioxide.

112. A technician observes a gas furnace in operation in which the inducer motor is operating but no ignition is occurring. The component that is *not* at fault is the

A. transformer.

B. flame sensor.

C. limit switch.

D. pressure switch.

113. To check a gas furnace with flame rectification, the technician would check

A. AC milliamps.

B. DC microamps.

C. AC microamps.

D. DC milliamps.

114. The type of voltage that a thermocouple produces is

A. small DC voltages.

B. large DC voltages.

C. large AC voltages.

D. small AC voltages.

115. When installing thermostat wire to a gas furnace using a millivolt thermopile generator, care should be taken to prevent extreme lengths of wire because

A. kinks may form in the wire.

B. of the extra expense of the wire.

C. the thermostat will not work properly.

D. of the possibility of voltage drops.

116. To help reduce noise transmitting from a furnace installed in an attic, the installers could

A. screw the furnace to the attic floor.

B. install many flex connectors.

C. suspend the furnace from the roof trusses.

D. not allow the blower motor to operate on high speed.

117. The reason for the International or National Fuel Gas Code is to

A. satisfy the local building codes.

B. identify all safety controls.

C. identify model and serial numbers.

D. abide by OSHA regulations.

118. When installing an electromechanical heating and cooling thermostat, which of the following is true?

A. A programmable thermostat is not interchangeable with an electromechanical thermostat.

B. A programmable thermostat is interchangeable with an electromechanical thermostat.

C. To operate properly, a programmable thermostat must be completely level.

D. All programmable thermostats require a common terminal.

119. When installing a programmable thermostat that is used on a heating and cooling system that uses only one control circuit transformer,

A. a common wire must always be supplied to the thermostat.

B. the jumper between the RC and RH terminals must be removed.

C. the jumper between the RC and RH terminals must not be removed.

D. there will only be an RC terminal and not an RH in the thermostat.

120. Which of the following will cause excessive temperature rise on a forced-air heating system?

A. insufficient gas flow

B. excessive airflow

C. an oversized duct system

D. a dirty air filter

ESCO INSTITUTE-HVAC EXCELLENCE CARBON MONOXIDE SAFETY CERTIFICATION EMPLOYMENT READY (ER)

1. The abbreviation for carbon monoxide is

A. CA

B. CO

C. CM

D. CN

2. What is a standard therapy for CO poisoning?

A. administer supplemental O_2

B. provide ventilator support

C. monitor the heart rate

D. all of the above

3. What can low concentrations of CO cause?

A. death within 10 minutes

B. no consequences

C. it can compound pre-existing health conditions

D. all of the above

4. If CO is inhaled into the lungs, what does it bond to in the blood?

A. O_2 (Oxygen) in the blood

B. CO_2 in the blood

C. H_2O in the body

D. Nitrogen

5. When inhaling CO, what will it produce in the bloodstream?

A. CO_2

B. CCH

C. Carboxyhemoglobin (COHb)

D. CH_4

6. What are symptoms of CO poisoning?

A. headache

B. sinus head pressure

C. constant stuffiness

D. all of the above

7. How is carbon monoxide CO measured?

A. PPM

B. PPG

C. PPD

D. PPO

8. To produce complete combustion in a heating system a burner needs to mix:

A. fuel and air-oxygen

B. fuel and nitrogen

C. fuel and CO_2

D. all of the above

9. What does parts per million refer to when discussing carbon monoxide within the air of a home?

A. the number of CO molecules within million molecules of air

B. the number of CO molecules within 5 million molecules of air

C. the number of CO molecules within 10 million molecules of air

D. none of the above

10. What are the required components must be present for complete combustion?

A. fuel

B. air/oxygen

C. ignition

D. all of the above

11. What are the three "Ts" of combustion?

A. time, temperance, temperature

B. time, temperature, turbulence

C. temperance, turbulence, tempo

D. turbulence, tempo, time

12. What human organ do medical studies show damage to when low levels of CO are present?

A. nerves

B. liver

C. endothelium

D. spinal cord

13. Where is COHb percentage measured?

A. liver

B. muscle

C. brain

D. blood

14. What are two ways that COHb can be measured in a human?

A. Take a blood sample.

B. Use a breath analyzation module.

C. Take a urine sample.

D. Both A and B

15. What will a healthy person exposed to 200 ppm of CO experience?

A. headaches

B. dizziness

C. nausea

D. all of the above

16. What is the concentration of CO after 2–3 hours of exposure that will be life-threatening?

A. 100 ppm

B. 200 ppm

C. 300 ppm

D. 400 ppm

17. Which of the following can cause various symptoms and health effects of CO poisoning within an individual?

A. age

B. state of health

C. weight

D. all of the above

18. There are many causes of CO poisoning within a building. What is the number one cause?

A. a plugged vent pipe on a high efficient furnace

B. a cracked heat exchanger on a gas-fired furnace

C. back drafting of a gas-fired appliance or an appliance that is not drafting at all

D. positive pressure within the building

19. What is the least common cause of CO poisoning in homes today?

A. CO in an attached garage

B. a cracked heat exchanger on a gas-fired furnace

C. back drafting of a gas-fired appliance or an appliance that is not drafting at all

D. none of the above

20. What type of appliance always recommends the installation of a CO detector?

A. an electric heating system

B. a solar heating panel

C. a HC (hydrocarbon) appliance is installed

D. a detached garage is over 300 feet away

21. CO alarms are meant to meet standard safety principles; however, it is possible that they may not have satisfactory protection of:

A. children

B. older people

C. people with deteriorating health issues

D. all of the above

22. What is the typical level in ppm of CO_2 in outdoor air?

A. 350–450

B. 500–600

C. 600–1000

D. 800–1000

23. What is the level in ppm of CO_2 that will cause complaints of stuffiness and odors with indoor air?

A. 350–450

B. 500–600

C. 600–1000

D. 800–1000

24. What should a homeowner be instructed to do as soon as their CO alarm goes off?

A. vacate the premises immediately.

B. call the fire department.

C. call the police department.

D. none of the above

25. What should CO test equipment be calibrated to before use?

A. 400 ppm

B. always follow the manufacturer's recommendations.

C. 600 ppm

D. between 700 and 1000 ppm

26. What can be the source of a false CO reading?

A. a high concentration of CO outside a building

B. a cracked heat exchanger

C. excessive CO_2 within a building

D. back drafting of an appliance

27. What must a technician do when they answer a CO alarm where no one is outside the building?

A. Carefully enter the building using self-contained breathing apparatus (SCBA) and have the CO test instruments in operation.

B. Go into the building and open all windows.

C. Call the fire department.

D. None of the above

28. What are the situations that make CO a hazard to humans?

A. 21% of O_2 in the air with nitrogen

B. 350 ppm of CO_2 in the air and 78% nitrogen

C. 21% of O_2 in the air with 50 ppm CO

D. a strong enough concentration of CO that can harm and that it can get into breathable air

29. What are the two ways that carbon monoxide is produced during the combustion process?

A. excess air during the combustion process and flame impingement

B. flame impingement and insufficient primary air

C. excess primary air and insufficient excess air

D. direct contact with the primary flame on a heat exchanger and too much O_2 for combustion

30. A flammable gas must always be leak detected by

A. a halide leak detector.

B. an open flame.

C. using soap bubbles.

D. all of the above

31. What will measuring the pressures of the home as well as HVAC equipment ensure with all gas-fired equipment?

A. Exhaust

B. Proper Venting

C. Safe operation

D. All of the above

32. What is the air pressure at sea level?

A. 14.7 psi

B. 30 millimeter mercury

C. 24 inches mercury

D. All of the above

33. What type of pressure is on the supply side of a blower?

A. positive

B. negative

C. neutral

D. any of the above

34. When testing a pressure within a building, what are technicians comparing?

A. inside pressure to internal pressure

B. building pressure to the outside pressure

C. external pressure to vent pressure

D. back pressure to internal pressure

35. What kind of pressure would a windward side of a structure experience?

A. atmospheric

B. neutral pressure

C. negative pressure

D. positive pressure

36. What will excessive tightness of a building cause when comparing indoor and outdoor pressure?

A. Increase pressure difference

B. Decrease pressure difference

C. Has no effect on pressure difference

D. None of the above

37. What is the definition of pressure?

A. $P = F \times A$

B. $P = A/F$

C. $P = F/A$

D. All the above

THE FOLLOWING QUESTIONS MAY BE HELPFUL IN STUDYING FOR THESE CERTIFICATION TESTS:

NATE—OL-SV. OL-IN, HO-SV, HO-IN.

ESCO INSTITUTE HVAC EXCELLENCE OIL HEAT SERVICE

OIL FURNACES

1. What is the heating value of a number 2 heating oil?

A. 80,000 Btu/gallon

B. 100,000 Btu/gallon

C. 140,000 Btu/gallon

D. 180,000 Btu/gallon

2. What is the common operating pressure of a high-pressure oil burner?

A. 25 psig

B. 50 psig

C. 75 psig

D. 100 psig

3. What is the secondary voltage of the ignition transformer used on a high-pressure oil burner?

A. 5,000 volts

B. 10,000 volts

C. 15,000 volts

D. 20,000 volts

4. What is the purpose of an electrode in a high-pressure gun burner?

A. It is used in conjunction with the blower to create a static electrical charge for ignition.

B. It is used to generate the voltage to the ignition transformer.

C. It is used to carry the spark to the point of ignition.

D. Electrodes are only used in low-pressure gun burners.

5. What is the measurement of the spark gap on an electrode assembly?

A. $\frac{1}{8}''$

B. $\frac{5}{16}''$

C. $\frac{1}{4}''$

D. $\frac{3}{8}''$

6. The capacity of a high-pressure oil burner nozzle is rated in

A. gph (gallons per hour).

B. gpm (gallons per minute).

C. Btu/h (British thermal units per hour).

D. ppg (pounds per gallon).

7. What are the three different spray patterns that are used on a nozzle?

A. solid, semisolid, hollow

B. solid, semihollow, hollow

C. uniform, semisolid, hollow

D. nonuniform, semihollow, hollow

8. The spray angle of an oil nozzle ranges from

A. 60 degrees to 80 degrees.

B. 45 degrees to 80 degrees.

C. 60 degrees to 90 degrees.

D. 30 degrees to 90 degrees.

9. What type of motor is used on a residential high-pressure oil burner?

A. shaded pole

B. permanent split capacitor

C. split phase

D. capacitor start–capacitor run

10. What is the speed of the burner motor on a high-pressure gun burner?

A. 1,725 rpm

B. 3,450 rpm

C. 1,725 or 3,450 rpm

D. 1,050 rpm

11. What is a common capacity of an indoor oil tank?

A. 200 gallons

B. 275 gallons

C. 325 gallons

D. 375 gallons

12. What is the size of a fill pipe on an oil storage tank?

A. $1\frac{3}{4}''$ diameter

B. $2''$ diameter

C. $2\frac{1}{2}''$ diameter

D. $3''$ diameter

13. What is the size of an air vent on an oil storage tank?

A. $\frac{3}{4}''$ diameter

B. $1''$ diameter

C. $1\frac{1}{4}''$ diameter

D. $1\frac{1}{2}''$ diameter

14. What safety feature is used on an inside oil storage tank that will protect against a house fire?

A. a fuse-type globe valve

B. a rupture disc

C. a pressure-relief valve

D. a fusible plug

15. What effect, if any, would a decrease of primary air have on combustion?

A. It would decrease efficiency.

B. It would decrease CO_2 percentage.

C. It would increase CO_2 percentage.

D. There would be no effect on CO_2 percentage.

16. What effect, if any, would an increase of primary air have on combustion?

A. It would increase efficiency.

B. It would decrease CO_2 percentage.

C. It would increase CO_2 percentage.

D. There would be no effect on CO_2 percentage.

17. If an oil has a high viscosity,

A. it mixes easily with water.

B. it does not mix easily with water.

C. it is a thicker oil under low temperatures.

D. it is a thinner oil under low temperatures.

18. What type of a system uses a single-stage fuel oil pump?

A. an aboveground one-pipe system

B. an underground one-pipe system

C. an aboveground two-pipe system

D. an underground two-pipe system

19. What type of a system uses a two-stage fuel oil pump?

A. an aboveground one-pipe system

B. an underground one-pipe system

C. an aboveground two-pipe system

D. an underground two-pipe system

20. When a cad-cell is located in the light, its resistance is

A. above 100,000 ohms of resistance.

B. below 1,500 ohms of resistance.

C. above 1,600 ohms of resistance.

D. zero resistance.

21. When a cad-cell is located in the dark, its resistance is

A. above 100,000 ohms of resistance.

B. below 1,600 ohms of resistance.

C. below 100,000 ohms of resistance.

D. zero resistance.

22. The orange wire on a cad-cell relay is used to

A. feed line voltage to the relay.

B. feed the common line to the relay.

C. connect the cad-cell to the circuit.

D. feed line voltage to the burner motor.

23. The number 3 terminal on the stack mount relay is used to

A. feed line voltage to the relay.

B. feed the common line to the relay.

C. connect the cad-cell to the circuit.

D. connect the burner motor circuit to the relay.

24. What is the purpose of the barometric damper used on an oil-fired furnace?

A. It measures the pressure created during operation.

B. It controls the draft through the system.

C. It controls the quantity of primary air during the combustion process.

D. It restricts the flow of flue gases through the vent.

25. What is the most common size oil line used on a residential oil-fired system?

A. $\frac{1}{4}''$ OD

B. $\frac{3}{8}''$ OD

C. $\frac{1}{2}''$ OD

D. $\frac{5}{8}''$ OD

26. What type of lubrication should be used in the burner motor of a high-pressure gun burner?

A. 10-weight detergent-type oil

B. 30-weight nondetergent-type oil

C. 10-weight nondetergent-type oil

D. 30-weight detergent-type oil

27. What is the purpose of a delayed oil valve?

A. It delays the ignition of a gun burner.

B. It delays the oil pressure to the nozzle.

C. It shuts down the oil flow if there is a house fire.

D. It delays the oil pressure to the oil pump.

28. How is a delayed oil valve wired in a heating circuit?

A. It is connected across the black and white wire on the cad-cell relay.

B. It is connected across terminals 2 and 3 on a stack mount relay.

C. It is connected across terminals 1 and 2 on a stack mount relay.

D. It is connected across the limit control.

29. What are three products produced in the combustion process?

A. NH_3, H_2O, and CO

B. O_3, H_2O, and CO_2

C. H_2O, CO_2, and N

D. NH_3, CO_2, and N

30. A high CO_2 reading in an oil-fired furnace indicates

A. a hot fire.

B. a cool fire.

C. too much excess air.

D. a high O_2 reading also.

31. A common CO_2 measurement for an oil-fired furnace is from

A. 8.5 to 9.5%.

B. 10 to 12%.

C. 13 to 15%.

D. 15 to 20%.

32. A common CO_2 measurement for a gas-fired furnace is from

A. 8.5 to 9.5%.

B. 10 to 12%.

C. 13 to 15%.

D. 15 to 20%.

33. A common fire draft reading for an oil furnace would be

A. −1″ to −2″ WC.

B. −.05″ to −.1″ WC.

C. −.03″ to −.06″ WC.

D. −.01″ to −.02″ WC.

34. A common breech draft for an oil furnace is

A. −1″ to −2″ WC.

B. −.05″ to −.1″ WC.

C. −.03″ to −.06″ WC.

D. −.01″ to −.02″ WC.

35. If the draft through a furnace is too high, what adjustment should be made?

A. Increase the primary air adjustment.

B. Adjust the barometric damper to more of an open position.

C. Adjust the barometric damper to more of a closed position.

D. Decrease the burner motor adjustment.

36. A common smoke reading for an oil-fired furnace is

A. #0.

B. #1 to #2.

C. #3 to #4.

D. #10.

37. What does a high smoke reading indicate on an oil-fired furnace?

A. too much of an over-the-fire draft

B. not enough of an over-the-fire draft

C. too much primary air

D. not enough primary air

38. What does a low smoke reading indicate on an oil-fired furnace?

A. too much of an over-the-fire draft

B. not enough of an over-the-fire draft

C. too much primary air

D. not enough primary air

39. A stack thermometer commonly has a range from

A. 0 to 500°F.

B. 300 to 500°F.

C. 200 to 1,000°F.

D. 500 to 1,000°F.

40. Common stack temperatures for an oil-fired furnace are

A. 600 to 800°F.

B. 800 to 1,000°F.

C. 200 to 450°F.

D. 500 to 550°F.

41. If combustion air must be supplied to an oil furnace, there must not be

A. less than 1 sq in. per 5,000 Btu/h.

B. less than 2 sq in. per 5,000 Btu/h.

C. less than 1 sq in. per 10,000 Btu/h.

D. less than 2 sq in. per 10,000 Btu/h.

THE FOLLOWING QUESTIONS MAY BE HELPFUL TO STUDY FOR THESE CERTIFICATION TESTS:
NATE—OL-SV, OL-IN, HO-SV, HO-IN.
ESCO INSTITUTE HVAC EXCELLENCE OIL HEAT SERVICE

OIL FURNACE TROUBLESHOOTING

1. What is the Btu input rating of an oil-fired furnace that uses a nozzle rated at .85 gpm?

A. 16,470 Btu/h

B. 93,333 Btu/h

C. 119,000 Btu/h

D. 210,000 Btu/h

2. What are two problems that can cause an oil-fired gun burner to operate below 100 psig?

A. insufficient oil and improperly adjusted electrodes

B. a plugged nozzle and a bad oil pump

C. a plugged in-line filter and a bad ignition transformer

D. a bad oil pump and a plugged in-line filter

3. Technician A says that the input voltage of an ignition transformer used on a gun-type oil burner is 208 volts, and the output voltage is 15,000 volts.

Technician B says that the input voltage of an ignition transformer used on a gun-type oil burner is 120 volts, and the output is 10,000 volts.

Which technician is correct?

A. technician A

B. technician B

C. Both technicians are correct.

D. Both technicians are incorrect.

4. Which of the following can cause a weak spark in a high-pressure gun burner?

A. a cracked electrode

B. a bad ignition transformer

C. a bad thermocouple

D. Both A and B

5. When adjusting the electrode assembly, what is the proper measurement from the center of the nozzle to the tip of the electrode?

A. $\frac{1}{8}''$ to $\frac{1}{4}''$

B. $\frac{5}{16}''$ to $\frac{1}{4}''$

C. $\frac{1}{2}''$ to $\frac{9}{16}''$

D. None of the above

6. The capacity of a high-pressure oil burner nozzle is rated in

A. gph (gallons per hour).

B. gpm (gallons per minute).

C. Btu/h (British thermal units per hour).

D. ppg (pounds per gallon).

7. What type of spray pattern does a residential oil-fired furnace commonly use?

A. solid

B. semihollow

C. uniform

D. hollow

8. What would the spray angle be on a residential oil-fired furnace?

A. 30 degrees

B. 45 degrees

C. 60 degrees

D. 80 degrees

9. Technician A says that the motor that is commonly used on a gun-type oil burner is a capacitor start motor.

Technician B says that the motor that is commonly used on a gun-type oil burner is a split-phase motor.

Which technician is correct?

A. technician A

B. technician B

C. Both technicians are correct.

D. Both technicians are incorrect.

10. When replacing a burner motor that is used on a high-pressure gun burner, it is critical to have the proper speed. Burner motors have two common speeds that are used in the field.

What are they?

A. 650 and 1,300 rpm

B. 750 and 1,400 rpm

C. 1,725 and 3,450 rpm

D. 2,025 and 2,525 rpm

11. Technician A says that the size of a fill pipe on an oil storage tank is 2″ in diameter. Technician B says that the size of a fill pipe on an oil storage tank is $2\frac{1}{2}''$ in diameter. Which technician is correct?

A. technician A

B. technician B

C. Both technicians are correct.

D. Both technicians are incorrect.

12. Technician A says that the size of an air vent on an oil storage tank is 2″ in diameter. Technician B says that the size of an air vent on an oil storage tank is $1\frac{1}{2}''$ in diameter. Which technician is correct?

A. technician A

B. technician B

C. Both technicians are correct.

D. Both technicians are incorrect.

13. A technician performs a combustion efficiency test on an oil-fired furnace. He notices that the CO_2 reading is high. What could cause this problem?

A. not enough fuel in the fuel–air mixture

B. too much air in the fuel–air mixture

C. not enough air in the fuel–air mixture

D. Both A and B

14. A technician is servicing an oil-fired furnace and she notices that the single-stage oil pump has to be replaced. What type of system is the technician servicing?

A. an aboveground one-pipe system

B. an underground one-pipe system

C. an aboveground two-pipe system

D. an underground two-pipe system

15. A technician is servicing an oil-fired furnace that uses a cad-cell primary control. He notices that the control locks out after approximately 45 s. Which of the following can cause this to happen?

A. The cad-cell is measuring above 100,000 ohms of resistance in the dark.

B. The cad-cell is measuring below 1,600 ohms of resistance in the light.

C. The cad-cell is measuring above 1,600 ohms of resistance in the light.

D. The cad-cell is measuring below 100,000 ohms of resistance in the dark.

16. When troubleshooting an oil-fired furnace that uses a cad-cell primary control, what two wires should be tested for line voltage that feeds the burner motor and ignition transformer?

A. the orange and white wires

B. the blue and orange wires

C. the black and blue wires

D. the white and black wires

17. A technician is servicing an oil-fired furnace that uses a stack mount relay. He notices that the burner motor is not operating. What two terminals must the technician place the voltmeter across to test for output voltage from the control to the burner motor?

A. terminals 1 and 2

B. terminals 1 and 3

C. terminals 2 and 3

D. terminals 3 and 4

18. A service technician performs a combustion efficiency test on an oil-fired furnace. She notices that there is too much of a draft. What can the technician do to decrease the amount of draft in the system?

A. She can close the barometric damper.

B. She can increase the size of the flue pipe.

C. She can add a chimney cap.

D. She can open the barometric damper.

19. A technician is servicing an oil-fired furnace and notices that there is a small amount of oil that drips from the nozzle when the system is de-energized. What can be added to the system to decrease the amount of dripping during shutdown?

A. a new nozzle

B. a delayed oil valve

C. a new in-line filter

D. a new nozzle assembly

20. How is a delayed oil valve wired in a heating circuit?

A. It is connected across the black and white wire on the cad-cell relay.

B. It is connected across terminals 2 and 3 on a stack mount relay.

C. It is connected across terminals 1 and 2 on a stack mount relay.

D. It is connected across the limit control.

21. Technician A says that a common CO_2 measurement for an oil-fired furnace is from 8.5 to 9.5%. Technician B says that a common CO_2 measurement for an oil-fired furnace is from 10 to 12%. Which technician is correct?

A. technician A

B. technician B

C. Both technicians are correct.

D. Both technicians are incorrect.

22. A technician arrives at a service call for an oil-fired furnace. He notices that the overfire draft reading is +.01″ WC. He then measures the breech draft and it is a −.04″ WC. What could be the problem with the furnace?

A. a plugged chimney

B. a plugged heat exchanger

C. insufficient primary air

D. All of the above

23. A common breech draft for an oil furnace is

A. −1″ to −2″ WC.

B. −.05″ to −.1″ WC.

C. −.03″ to −.06″ WC.

D. −.01″ to −.02″ WC.

24. When servicing an oil-fired furnace, a technician finds that the furnace is giving a smoke reading of 4. What is a possible problem that can cause this?

A. excessive combustion air

B. insufficient combustion air

C. there is no problem; the system should produce a smoke reading of 4.

D. the barometric damper opens too much.

25. What does a smoke reading of 0 indicate on an oil-fired furnace?

A. not enough of an over-the-fire draft

B. not enough primary air

C. too much primary air

D. no problem; good working condition

26. What is a possible problem if an oil-fired furnace has a stack temperature reading over 800°F?

A. nozzle is too large

B. excessive combustion air

C. too high of an oil pump pressure

D. All of the above

27. Using the "Cad-Cell Relay Diagram" (Figure 2–5), answer the following question. The system will not operate, the disconnect switch, the limit switch, and the thermostat are closed. There is a voltage reading of 120 volts across terminals "BK" and "WH." A slight leak is noticed in the combustion chamber. Why will the system not operate?

A. The slight leak in the combustion chamber causes the resistance of the cad-cell to decrease and not fire.

B. The slight leak in the combustion chamber causes the resistance of the cad-cell to increase and not fire.

C. The primary control must be condemned and replaced.

D. Both B and C

28. Using the "Cad-Cell Relay Diagram" (Figure 2–5), answer the following question. A technician notices that the system energizes for about 2 min, and then shuts down. A reading across terminals 1 and 2 on the LIM shows 120 volts. Another reading across terminals 1 and 2 on FS also shows 120 volts. What is the possible problem with the system?

A. The LIM is defective.

B. The cad-cell relay is defective.

C. The IBM is defective.

D. The FS is defective.

29. Using the "Cad-Cell Relay Diagram" (Figure 2–5), answer the following question. A technician sets the thermostat to call for heat and closes the disconnect switch. The system is sparking but does not fire. A further check shows that the burner motor is not in operation. What could be the problem with the system?

A. The cad-cell relay could be defective.

B. The cad-cell could be defective.

C. The burner motor could be defective.

D. Both A and B

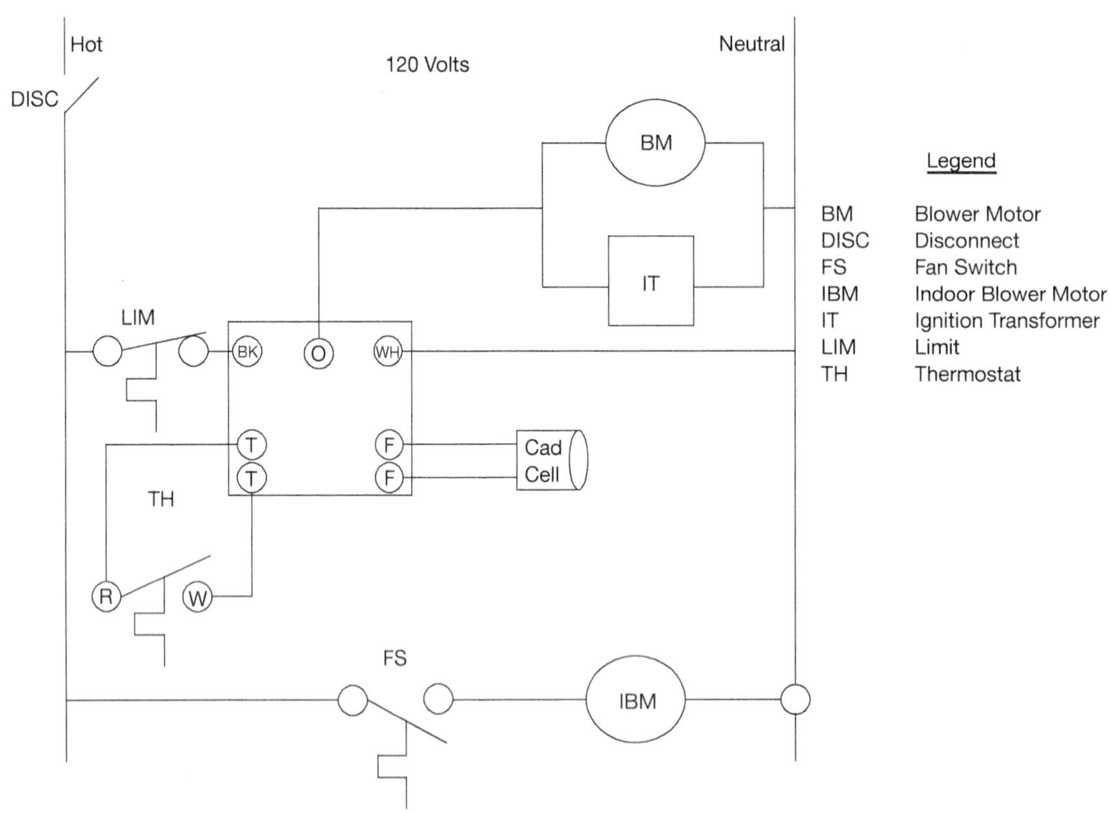

FIGURE 2–5
Cad-Cell Relay Diagram

30. The disconnect switch, the limit switch, and the thermostat are in the closed position. There is a spark at the electrodes, but no ignition. A further check shows that the burner motor is not operating. Where should a voltage reading be taken to further troubleshoot the system?

Use the "Stack Mount Relay Diagram" (Figure 2–6) to answer this question.

A. across terminals 1 and 2

B. across terminals 1 and 3

C. across terminals 1 and 4

D. across terminals 3 and 2

31. A technician notices that the system energizes for about 45 s and shuts down. Before it energizes, there is a satisfactory flame present in the combustion chamber. A reading across terminals 1 and 2 on the LIM shows 0 volts. Another reading across terminals 1 and 2 on the stack mount relay also shows 120 volts. What is the problem? Use the "Stack Mount Relay Diagram" (Figure 2–6) to answer this question.

A. The LIM is defective.

B. The stack mount relay is defective.

C. The IBM is defective.

D. The FS is defective.

32. A technician sets the thermostat to call for heat and closes the disconnect switch. The system is sparking but does not fire. A further check shows that the burner motor is not in operation. What could be the problem with the system? Use the "Stack Mount Relay Diagram" (Figure 2–6) to answer this question.

A. The stack mount relay could be defective.

B. The limit control could be defective.

C. The burner motor could be defective.

D. Both A and C

33. What two problems on a high-pressure gun burner can stop the flow of oil to the nozzle?

A. a defective ignition transformer and an inoperative oil pump

B. a broken coupling and a plugged in-line filter

C. an oil pump with too high of a pressure adjustment and a defective burner motor

D. a bad run capacitor on the oil pump and a plugged nozzle

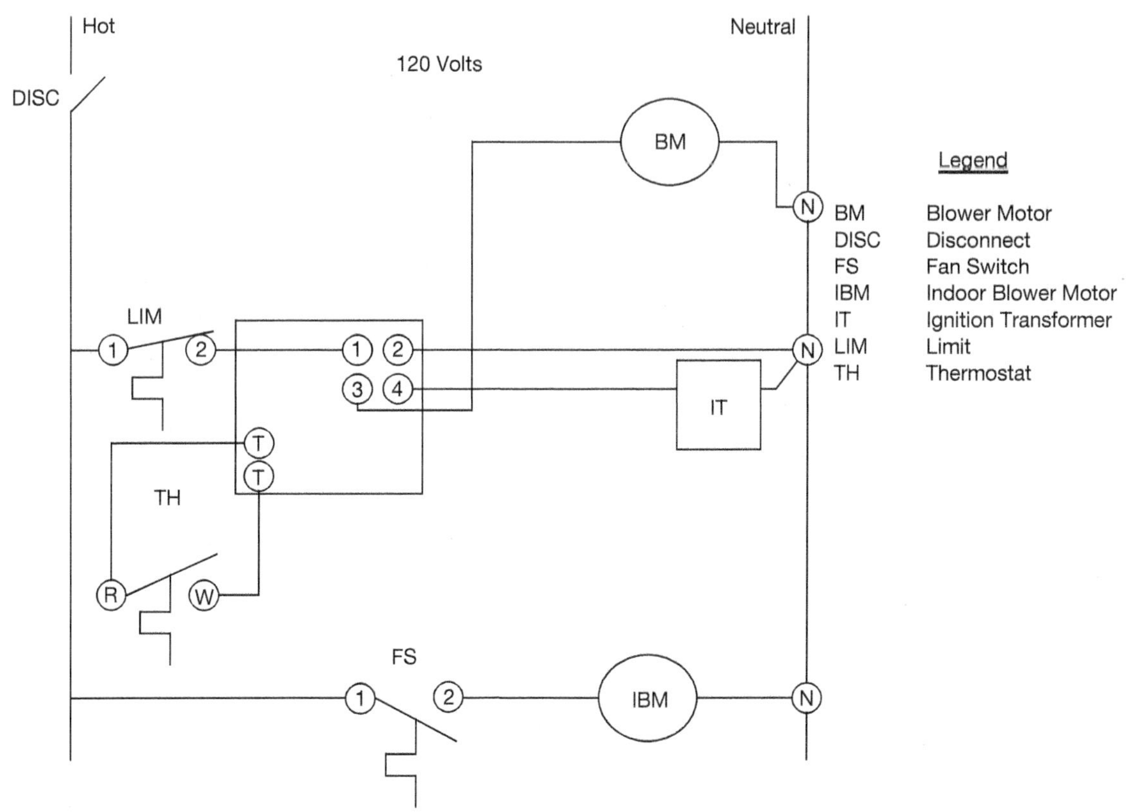

FIGURE 2–6
Stack Mount Relay Diagram

34. Technician A says that the distance the electrodes should be placed in front of a nozzle with an 80-degree spray angle is $\frac{1}{4}''$.

Technician B says that the distance the electrodes should be placed in front of a nozzle with an 80-degree spray angle is $\frac{1}{8}''$.

Which technician is correct?

A. technician A

B. technician B

C. Both technicians are correct.

D. Both technicians are incorrect.

35. On a stack mount relay, what position should the cold contacts be in on the startup of an oil-fired furnace?

A. the normally closed position

B. the normally open position

C. They can be in the open or closed position.

D. There are no cold contacts located on a stack mount relay.

36. When troubleshooting an oil-fired furnace that has an ignition problem, what are three tests that can be performed?

A. Test the ignition transformer, the burner motor, and the nozzle assembly.

B. Test the ignition transformer, the electrode assembly, and the blast tube assembly.

C. Test the ignition transformer, the nozzle assembly, and the blast tube assembly.

D. Test the ignition transformer, the electrode assembly, and the electrode adjustment.

37. When troubleshooting an oil-fired furnace that has an oil delivery problem, what are three tests that can be performed?

A. Test oil pump pressure, burner motor operation, and the nozzle assembly.

B. Test oil pump pressure, the electrode adjustment, and the blast tube assembly.

C. Test oil pump pressure, the electrode assembly, and the nozzle assembly.

D. Test oil pump pressure, the burner motor operation, and the blast tube assembly.

38. If a high-pressure gun-type oil burner has a sufficient spark and the oil pump pressure is adequate, what can cause the system not to fire properly?

A. a broken coupling

B. a plugged nozzle

C. a plugged in-line oil filter

D. a defective ignition transformer

39. What could cause a high-pressure gun-type oil burner to have a weak or insufficient spark?

A. cracked ceramic electrodes

B. improper electrode adjustment

C. faulty ignition transformer

D. All of the above

40. Which of the following could cause an oil burner to have insufficient oil?

A. plugged nozzle

B. bad oil pump

C. empty oil tank

D. All of the above

41. An oil furnace operating with a very high stack temperature and the CO level at 8% with no smoke is causing high fuel bills. The technician should

A. increase the primary air.

B. reduce the primary air.

C. replace the oil nozzle.

D. replace the fuel pump.

42. The first main safety concern when troubleshooting a no-heat call is to

A. check the oil nozzle.

B. replace the oil nozzle.

C. check the electrode gap.

D. check the combustion chamber for liquid oil present.

43. A technician diagnoses a furnace and finds that the burner operates for around 45 seconds then shuts down. The technician concludes that

A. there is too much oil pressure.

B. the oil nozzle is oversized.

C. the blower motor is on high speed.

D. the cad-cell is dirty or faulty.

44. To calculate temperature rise of an oil furnace, the formula is

A. indoor temperature minus outdoor temperature.

B. stack temperature minus return air.

C. stack temperature minus indoor temperature.

D. temperature of the supply air minus the return air temperature.

45. A technician finds that the cad-cell is reading a very high resistance while the burner is operating. The cause could be

A. low transformer voltage.

B. bad thermostat connections.

C. restricted oil nozzle.

D. dirtiness or sooting of the cell.

46. What would result if the burner oil flame is not detected by the cad-cell during a startup for heat?

A. The burner motor would not be affected.

B. The burner motor will shut down.

C. The burner will start and run shortly, then lock-out.

D. The primary control will allow operation.

THE FOLLOWING QUESTIONS MAY BE HELPFUL TO STUDY FOR THESE CERTIFICATION TESTS:
NATE—HG-SV, HG-IN, GS-SV; GS-IN, HO-SV, HO-IN.
ESCO INSTITUTE HVAC EXCELLENCE HYDRONIC HEAT SERVICE

HYDRONIC HEATING THEORY

1. Heat always flows from
   A. a warmer object to a cooler object.
   B. a cooler object to a warmer object.
   C. a lighter object to a darker object.
   D. a darker object to a lighter object.
2. _______ is the transfer of heat from one molecule to another when two objects are in contact with each other.
   A. Conduction
   B. Convection
   C. Radiation
   D. Evaporation
3. _______ is the transfer of heat through the movement of a fluid.
   A. Conduction
   B. Convection
   C. Radiation
   D. Evaporation
4. _______ is the transfer of heat by the use of waves through space.
   A. Conduction
   B. Convection
   C. Radiation
   D. Evaporation
5. _______ is when a liquid absorbs heat by changing its state from a liquid to a gas.
   A. Conduction
   B. Convection
   C. Radiation
   D. Evaporation
6. What is the definition of a Btu?
   A. It is the amount of heat required to raise one gallon of water 1°F.
   B. It is the amount of heat required to raise one pound of water 1°F.
   C. It is the amount of heat required to raise one cubic foot of water 1°F.
   D. It is the amount of heat required to raise one kilogram of water 1°C.
7. One gallon of water weighs
   A. 5.33 lb.
   B. 6.33 lb.
   C. 7.33 lb.
   D. 8.33 lb.

THE FOLLOWING QUESTIONS MAY BE HELPFUL TO STUDY FOR THESE CERTIFICATION TESTS:
NATE—HG-SV, HG-IN, GS-SV, GS-IN, OL-SV, OL-IN; HO-SV, HO-IN.
ESCO INSTITUTE HVAC EXCELLENCE HYDRONIC HEAT SERVICE

MECHANICAL CONTROLS

1. When is an automatic float-type air vent closed to the atmosphere?
   A. when the vent chamber is full of air
   B. when the vent chamber is full of water
   C. when the float is at the bottom of the chamber
   D. None of the above
2. A pressure-reducing valve drops city water pressure to
   A. 5 psi.
   B. 12 psig.
   C. 18 psig.
   D. 30 psig.
3. In what order should the main shutoff valve, the pressure-reducing valve, and the backflow preventer be placed in the main water feed line?
   A. backflow preventer, main shutoff, pressure-reducing valve
   B. main shutoff, pressure-reducing valve, backflow preventer
   C. pressure-reducing valve, main shutoff, backflow preventer
   D. main shutoff, backflow preventer, pressure-reducing valve
4. At what pressure does a pressure-relief valve on a hot water boiler open?
   A. 15 psi
   B. 20 psi
   C. 25 psi
   D. 30 psi

5. What is the purpose of the expansion tank located on a hot water boiler?
   A. It allows for the expansion of the water in the system.
   B. It is used as an air cushion.
   C. It is used to prevent excessive boiler pressures as water is heated.
   D. All of the above

6. What are common hot water boiler operating temperatures during winter months?
   A. 120–140°F
   B. 180–220°F
   C. 230–260°F
   D. 270–320°F

THE FOLLOWING QUESTIONS MAY BE HELPFUL TO STUDY FOR THESE CERTIFICATION TESTS:
NATE—HG-SV, HG-IN, GS-SV, GS-IN, OL-SV, OL-IN, HO-SV, HO-IN.
ESCO INSTITUTE HVAC EXCELLENCE HYDRONIC HEAT SERVICE

ELECTRICAL CONTROLS AND MECHANICAL COMPONENTS

1. Which of the following is the correct operation for a four-wire zone valve?
   A. power close–power open
   B. spring open–power close
   C. power open–spring close
   D. manual open–power close

2. What are the two red wires used for on a four-wire zone valve?
   A. the end switch
   B. the thermostat
   C. the zone valve motor
   D. the circulator pump

3. On a multiple-zone system that uses three zone valves, what controls current flow to the zone valve motor?
   A. the end switch
   B. the aquastat
   C. the thermostat
   D. the low limit

4. When a zone valve is fully open, which electrical control should always be closed?
   A. the aquastat
   B. the gas valve
   C. the end switch
   D. All of the above

5. Which boiler piping system has the first supplied radiation unit, the last to be returned?
   A. reverse return
   B. direct return
   C. series loop
   D. one-pipe system

6. Which boiler piping system has the first supplied radiation unit, the first to be returned?
   A. reverse return
   B. direct return
   C. series loop
   D. one-pipe system

7. Which of the following piping systems is easier to balance?
   A. series loop
   B. direct return
   C. reverse return
   D. multiple series loop

8. Which type of zone valve requires a single-pole, double-throw thermostat?
   A. a four-wire zone valve
   B. a five-wire zone valve
   C. a six-wire zone valve
   D. a seven-wire zone valve

9. Where must a circulator pump never be installed?
   A. between the pressure-reducing valve and the boiler
   B. between the zone valve and the boiler
   C. between the return header and the boiler
   D. between the supply header and the boiler

10. An aquastat reacts to boiler water
    A. pressure.
    B. temperature.
    C. altitude.
    D. All of the above

11. A reverse-acting low-limit control is used to
    A. maintain a maximum boiler water temperature.
    B. prevent circulator pump operation until a minimum boiler temperature is reached.
    C. energize the circulator pump when boiler water temperature drops to a preset temperature.
    D. All of the above

12. What is the purpose of a boiler's tridicator?

A. It is used to measure boiler temperature.

B. It is used to measure boiler altitude.

C. It is used to measure boiler pressure.

D. All of the above

13. The Btu/h rating on a pressure-relief valve

A. must never be more than the gross Btu output of the boiler.

B. must never be less than the gross Btu output of the boiler.

C. must never be the same as the gross Btu output of the boiler.

D. must always be less than the gross Btu output of the boiler.

14. When sizing a hot water boiler, the ________ rating should be equal to or slightly higher than the heat loss of the building.

A. net IBR

B. DOE

C. ASME

D. AGA

15. What is a standard rating per linear foot of $\frac{3}{4}''$ radiation for a hot water system operating at approximately 200°F?

A. 200 Btu

B. 400 Btu

C. 600 Btu

D. 800 Btu

16. Approximately how many pounds of water would be pumped in one hour if the flow rate through a system were 1 gal/min?

A. 330 lb

B. 500 lb

C. 650 lb

D. 800 lb

17. What type of valve must be located at the inlet of a terminal unit on a one-pipe supply-loop heating system?

A. a flow-diversion tee

B. a mixing tee

C. a zone valve

D. a pressure-reducing valve

18. The temperature difference between the water leaving the hot water boiler (supply) and the water entering the boiler (return) should be

A. 5°F.

B. 10°F.

C. 20°F.

D. 30°F.

19. Diaphragm-type expansion tanks are generally precharged to

A. 12 psi.

B. 15 psi.

C. 18 psi.

D. 20 psi.

20. Which of the following valves has the greatest resistance to flow?

A. a gate valve

B. a ball valve

C. a globe valve

D. a check valve

21. Which of the following is equivalent to 1 psi?

A. the height of 1 ft of a water column

B. the height of 1.85 ft of a water column

C. the height of 2 ft of a water column

D. the height of 2.31 ft of a water column

22. What type of valve must be located at the outlet of a terminal unit on a one-pipe supply-loop heating system?

A. a flow diversion tee

B. a mixing tee

C. a zone valve

D. a pressure-reducing valve

23. What component is commonly used in conjunction with an automatic air vent for the purpose of removing air from a boiler piping system?

A. an expansion tank

B. an air scoop

C. a monoflow tee

D. a check valve

24. Which of the following components would not be controlled by an aquastat controller?

A. a circulator pump

B. a gas valve

C. an oil burner motor

D. a zone valve

25. Which of the following agencies lists the pressure-relief valve on a hot water boiler?

A. net IBR

B. DOE

C. ASME

D. AGA

26. Which of the following components will prevent a hot water boiler from dry firing?

A. a low-water cut-out

B. a low-limit aquastat

C. a high-limit aquastat

D. All of the above

27. The outlet of a pressure-relief device must terminate within

A. 6″ from the floor.

B. 8″ from the floor.

C. 10″ from the floor.

D. 12″ from the floor.

28. Under no circumstances should the

A. pressure-reducing valve be placed on the outlet side of the circulator pump.

B. diameter of the discharge piping of the pressure-relief valve be less than the outlet diameter of the valve.

C. the aquastat be set above 180°F.

D. All of the above

29. What is a common "cold start" pressure of a hot water boiler?

A. 8 psi

B. 12 psi

C. 18 psi

D. 22 psi

30. What is the purpose of a flow control valve/flow check valve?

A. It prevents thermo-siphoning during the off cycle of the boiler.

B. It prevents the bypass line from overfeeding a boiler.

C. It prevents boiler water from backing into the city water.

D. It maintains boiler loop pressure.

31. When is a flow check valve recommended?

A. when boiler pressure exceeds 30 psi

B. when a system uses a power-open-spring-return-type zone valve

C. when a system uses separate circulators for each zone

D. when a system uses a power-open, power-close-type zone valve

32. A pressure switch that is used with an induced draft hot water boiler

A. proves the operation of the circulator pump.

B. proves the operation of the induced draft blower.

C. senses the gas pressure to the gas valve.

D. senses the gas pressure to the pilot valve.

33. What control is commonly used on hot water boilers that will protect against high stack temperatures or flue gas spillage?

A. limit control

B. flame roll-out control

C. vent safety control

D. an auxiliary limit control

34. What control is commonly used on hot water boilers to protect against a plugged heat exchanger?

A. a limit control

B. a flame roll-out control

C. a vent safety control

D. an auxiliary limit control

35. A bimetal flame sensor

A. proves the existence of a pilot flame.

B. proves the existence of the main gas flame.

C. uses a normally open set of contacts to control the pilot valve.

D. uses a normally closed set of contacts to control the main valve.

36. A flame rod sensor

A. proves the presence only of a pilot flame.

B. proves the presence only of a main burner flame.

C. proves the presence of a pilot or main burner flame.

D. proves the presence of the spark from an ignition system.

37. The circulator pump is most commonly located on the __________ of the boiler.

A. supply side

B. return side

C. header section

D. None of the above

38. Water quantity through a hot water heating system is rated in __________.

A. gpm

B. fpm

C. cfm

D. pressure difference

39. A circulator pump uses a(n) __________ to move the water through the piping system.

A. piston

B. impeller

C. vane

D. volute

40. The inlet to the circulator pump is known as the

A. impeller.

B. volute.

C. vane.

D. cylinder.

41. The pressure difference across the circulator pump is known as the

A. gpm pumped.

B. pump efficiency.

C. pump capacity.

D. pump head.

42. The pump head is measured in

A. psi.

B. gpm.

C. feet of water column.

D. cfm.

43. If the pump head is higher, the pump capacity will be

A. lower.

B. the same.

C. greater.

D. All of the above

44. Circulator pumps are rated in

A. voltage.

B. horsepower.

C. amperage.

D. wattage.

45. One horsepower is equal to

A. 33,000 watts.

B. 746 watts.

C. 200 watts.

D. 3.41 watts.

46. A direct drive pump often uses a ______ that connects the shaft of the pump motor to the shaft of the circulator pump.

A. flexible coupling

B. brass coupling

C. bronze coupling

D. None of the above

47. Name two types of bearings that are used in centrifugal pumps.

A. sleeve bearings and bronze bearings

B. sleeve bearings and bushings

C. sleeve bearings and ball bearings

D. ball bearings and bushings

48. Sleeve bearings are commonly made of

A. bronze.

B. brass.

C. stainless steel.

D. aluminum.

49. Which of the following bearing assemblies must be greased?

A. ball bearings

B. bronze bearings

C. sleeve bearings

D. brass bearings

50. Which of the following bearing assemblies can handle a larger load?

A. sleeve bearings

B. ball bearings

C. bronze bearings

D. None of the above

51. Which of the following are check valves that are commonly used on hot water heating systems?

A. swing check valve

B. ball check valve

C. magnetic check valve

D. All of the above

52. Which of the following check valves is a critically mounted check valve?

A. swing check valve

B. ball check valve

C. magnetic check valve

D. None of the above

53. Which of the following valves is used to decrease the amount of time it takes to fill a hot water boiler?

A. a PRV

B. a bypass valve

C. a reverse-acting purge valve

D. a flow control valve

54. Which of the following is an agency responsible for testing and certifying gas appliances?

A. ASME

B. AGA

C. DOE

D. AFUE

55. Which of the following is responsible for the design and construction features of hot water boilers?

A. ASME

B. AGA

C. DOE

D. AFUE

56. Which of the following is the efficiency rating of the appliance?

A. ASME

B. AGA

C. DOE

D. AFUE

57. Which of the following sets the standards and tests procedures to be followed by which boiler and appliance manufacturers certify the efficiency levels of their heating equipment?

A. ASME

B. AGA

C. DOE

D. AFUE

58. What two functions does the low-limit control perform?

A. It prevents the hot water boiler from overheating and starts the circulator pump.

B. It shuts the circulator pump down if the boiler temperature is too high, and starts the burners if the water temperature is too low.

C. It keeps the temperature of the boiler water to a minimum and controls the operation of the circulator pump.

D. All of the above

59. What terminals on the low-limit control operate the heating circuit?

A. "R" and "B"

B. "R" and "W"

C. "B" and "W"

D. None of the above

60. What terminals on the low-limit control operate the circulator pump circuit?

A. "R" and "B"

B. "R" and "W"

C. "B" and "W"

D. None of the above

61. Common operating pressures of a hot water boiler are

A. 5 to 10 psi.

B. 8 to 12 psi.

C. 18 to 25 psi.

D. 25 to 30 psi.

62. Common operating temperatures of a hot water boiler range from

A. 100 to 120°F.

B. 130 to 150°F.

C. 180 to 200°F.

D. 220 to 260°F.

63. What control directly operates the zone valve motor?

A. the thermostat

B. the high limit

C. the aquastat

D. the end switch

64. A zone valve circuit that uses a 40-VA transformer should use no more than

A. one zone valve.

B. two zone valves.

C. three zone valves.

D. four zone valves.

65. What is the purpose of the lever on the zone valve?

A. It is used to open the end switch.

B. It is used to open the zone valve.

C. It is used to energize the circulator pump.

D. It is used to bypass the thermostat.

66. Zone valves can be located

A. on the supply side of the piping only.

B. on the return side of the piping only.

C. on either the supply side or the return side of the piping system.

D. None of the above

67. A zone valve that uses a resistance heater instead of a synchronous motor

A. uses a wax substance to open the valve.

B. uses a bimetal to open the valve.

C. uses a monometal to open the valve.

D. uses a bimetal rod and tube to open the valve.

68. Nonelectric zone valves are located at the _________ of the baseboard section of each room when used.

A. inlet

B. outlet

C. bottom

D. top

69. A nonelectric zone valve is a _________ type of valve.

A. snap-action

B. reverse-acting

C. modulating

D. None of the above

70. Nonelectric zone valves

A. sense room temperature at the valve.

B. use an external thermostat.

C. do not sense temperature at the valve.

D. All of the above

71. Nonelectric zone valves

A. use a switching relay to close the end switch.

B. use a N.C. end switch.

C. do not use end switches.

D. operate by system pressure.

72. Nonelectric zone valves

A. must have the circulator pump in operation at all times.

B. control the circulator pump by its end switch.

C. control the circulator pump by the zone thermostat.

D. must shut down the circulator pump when it closes.

73. A heating thermostat

A. opens the circuit to the zone valve motor on a drop in temperature.

B. opens the zone valve motor circuit on a rise in temperature.

C. energizes the zone valve when its set point is below space temperature.

D. de-energizes the zone valve when its set point is above space temperature.

74. The heat anticipator

A. must be energized when the zone valve motor is de-energized.

B. controls system lag.

C. controls system overshoot.

D. is wired in parallel with the heating thermostat.

75. When adjusting a heat anticipator on a thermostat,

A. the heating and cooling control amperage must be measured.

B. the heating and cooling control voltage must be measured.

C. the cooling control amperage must be measured.

D. the heating control amperage must be measured.

THE FOLLOWING QUESTIONS MAY BE HELPFUL TO STUDY FOR THESE CERTIFICATION TESTS:
NATE—HG-SV, HG-IN, HO-SV, HO-IN, GS-SV, GS-IN.
ESCO INSTITUTE HVAC EXCELLENCE HYDRONIC HEAT SERVICE

TROUBLESHOOTING HYDRONIC HEATING SYSTEMS: ELECTRICAL DIAGRAMS

1. Which of the following will cause a hot water heating system to vent through the pressure-relief valve?

A. a faulty backflow preventor

B. a faulty low-water cut-out

C. a faulty pressure-reducing valve

D. a faulty zone valve

2. A customer complains that the pressure-relief valve keeps leaking water on his hot water heating system.

Technician A says there is a possibility that the expansion tank could be flooded with water.

Technician B says that the circulator pump may be at fault.

Which technician is more likely to be correct?

A. technician A

B. technician B

C. Both technicians are correct.

D. Both technicians are incorrect.

3. When troubleshooting a gas-fired hot water heating system that uses three zone valves, three zone thermostats, one circulator pump, and a boiler controller, it is found that two of the zones are keeping the rooms up to temperature but the third room is cold. Where would be the most likely place to start troubleshooting the system?

A. Check to make sure the boiler is firing.

B. Check the zone thermostat and zone valve of the third zone.

C. Check to make sure the circulator pump is operating.

D. Check for power to the system.

4. A customer complains that her hot water heating system is not properly heating the second floor of her two-story home. On arrival at the job, the technician finds that the system is a gas-fired boiler that uses two zone valves, two thermostats, and a boiler controller. The technician notices that the burners are operating and the boiler temperature reads 200°F What could be the possible problem with the system?

A. There could be air trapped in the piping system of the second story.

B. It could be a faulty zone thermostat.

C. It could be a faulty zone valve.

D. All of the above

5. A technician arrives on a job where a customer complains that his hot water heating system is not operating. The technician adjusts all three zone thermostats above room temperature before troubleshooting the system. At the boiler the technician measures voltage at the input to the boiler controller and finds there is 120 volts. The technician then measures voltage at the gas valve output and finds 0 volts, then notices that the circulator pump is not in operation. The technician bypasses the "TT" terminals at the boiler controller, the circulator pump starts, and the burners ignite. Which of the following components could cause this system fault?

A. a blown fuse in the 120-volt circuit

B. a faulty zone valve circuit transformer

C. a faulty zone valve

D. a faulty end switch on one of the zone valves

6. Customers complain that their hot water boiler is blowing off on the pressure-relief valve. Upon arriving on the job, the technician sees that the hot water heating system is a single-pipe series-loop system. A control relay, operated by one thermostat, energizes the burner and circulator pump and a high-limit controls the temperature in the boiler. The circulator pump seems to be in operation because the water pipes are extremely hot, but the burners do not seem to be shutting down. The temperature on the gauge reads 250°F. What could be the problem?

A. The thermostat is stuck in the closed position.

B. The control relay is faulty.

C. There is air in the piping system.

D. The high limit is stuck closed.

7. The purpose of a vent system on gas appliances is to

A. draw in air for combustion.

B. remove by-products of combustion.

C. mix air with combustible gases.

D. All of the above

8. A vent system consists of

A. a draft hood assembly.

B. a vertical riser.

C. horizontal connectors.

D. All of the above

9. What is the purpose of the draft hood?

A. It is used to dilute the flue gases with room air.

B. It creates a draft in the venting system.

C. It balances the air-to-gas ratio within the system.

D. None of the above

10. The draft hood is used to

A. increase flue gas temperatures.

B. lower flue gas temperatures.

C. increase the velocity of the flue gases.

D. increase the quantity of the flue gases.

11. The draft hood helps to

A. increase updrafts.

B. mix combustion by-products with fuel.

C. divert downdrafts.

D. remove carbon monoxide gases from the system.

12. Motorized vent dampers

A. close to the venting system during system operation.

B. open to the venting system during the off cycle.

C. open to the venting system during system operation.

D. are controlled by an end switch.

13. Motorized vent dampers

A. close to the venting system during system operation.

B. open to the venting systems during the off cycle.

C. close during the heating system's off cycle.

D. are controlled by an end switch.

14. During boiler operation

A. the vent damper motor is de-energized only.

B. the vent damper motor is energized only.

C. the end switch is open.

D. the vent damper motor can be either energized or de-energized.

15. During boiler off cycles,

A. the vent damper motor is de-energized only.

B. the vent damper motor is energized only.

C. the end switch is closed.

D. the vent damper motor can be either energized or de-energized.

16. An end switch in the vent damper

A. proves the motor is in the closed position and allows the heating system to energize.

B. closes the thermostat circuit.

C. proves the motor is in the open position and allows the heating system to energize.

D. opens the thermostat circuit during the off cycle.

17. Using the "Multiple Zone Valve System Diagram" (Figure 2–7), zone thermostat 1 is in the closed position while zone 2 and 3 are in the open position. The zone valve opens but there is no action at the circulator pump or the gas valve. A voltage of 24 volts is read across terminals "TT" on the aquastat controller. What could be the problem with the system?

A. a faulty aquastat boiler controller

B. a faulty zone valve motor

C. a faulty end switch

D. a faulty zone valve control transformer

18. Using the "Multiple Zone Valve System Diagram" (Figure 2–7), the power disconnect switch is in the closed position, and there is 120 volts at the primary side of the control transformer. The zone 2 and 3 thermostats are in the open position but the zone 1 thermostat is closed. There is no action in the control circuit or the aquastat controller. What are three possible faults that could cause this problem?

A. a faulty thermostat, aquastat controller, and thermostat

B. a faulty transformer, thermostat, and aquastat controller

C. a faulty zone valve motor, transformer, and aquastat controller

D. a faulty thermostat, transformer, or zone valve motor

19. Using the "Multiple Zone Valve System Diagram" (Figure 2–7), 120 volts is present at the aquastat boiler controller, and the zone 3 thermostat is in the closed position. The circulator pump is in operation but the gas valve is not opening. What are three possible problems with the system?

A. a faulty zone 3 thermostat, zone valve, and end switch

B. a faulty aquastat boiler controller, end switch, and gas valve

C. a high limit, aquastat controller, and gas valve

D. All of the above

20. Using the "Multiple Zone Valve System Diagram" (Figure 2–7), there is 120 volts at the aquastat boiler controller, and zone thermostat 2 is in the closed position. The circulator pump and the gas valve will not operate. What would be the next logical procedure when troubleshooting the system?

A. Measure voltage at the gas valve.

B. Measure voltage across the zone valve motor 2.

C. Make sure there is 24 volts across the end switch at zone valve 2.

D. Check for 120 volts at the secondary winding of the transformer.

21. Use the "Multiple Zone Valve System Diagram" (Figure 2–7) to answer the following question. A customer complains that the circulator pump is operating continually, and the zone 2 is not heating properly. The technician arrives on the job and sets the zone 1 and 3 thermostats to the open position. The technician sets the zone 2 thermostat to the closed position and places voltmeter leads across the zone valve motor and reads 24 volts. The technician also places leads across the end switch terminals and reads 24 volts. The gas valve has no voltage but the circulator pump reads 120 volts. What are two problems with this system?

A. The relay within the aquastat control that operates the circulator pump is stuck closed, and the high-limit control in the aquastat control is stuck open.

B. The zone valve 2 is faulty, and the high-limit control is stuck open.

C. The relay within the aquastat control that operates the circulator pump is stuck closed, and the zone valve 2 is stuck open.

D. All of the above

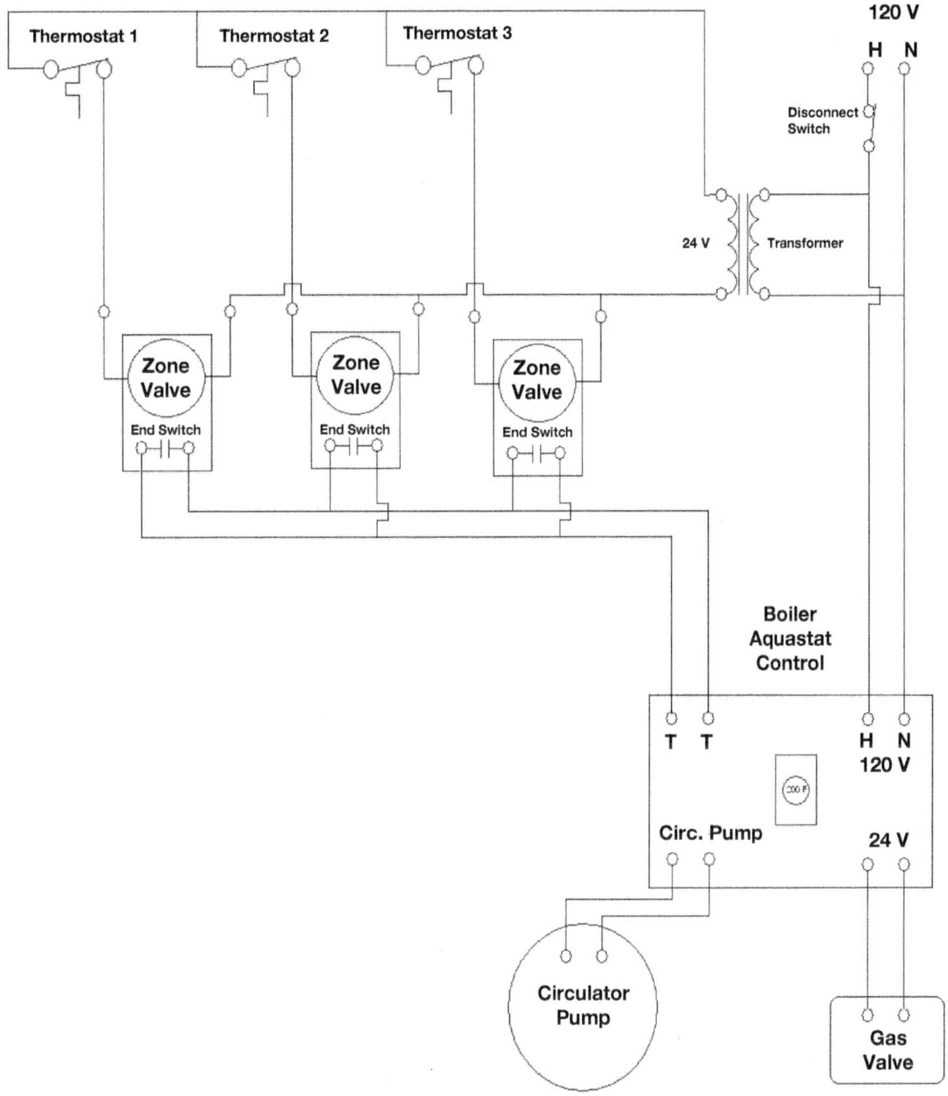

FIGURE 2–7

Multiple Zone Valve System Diagram

22. When using a 120-volt/24-volt 40-VA transformer, the 24-volt control amperage must never exceed

A. 1.67 amps.

B. 5.1 amps.

C. 33 amps.

D. 3.3 amps.

23. When using a 230-volt/24-volt 50-VA transformer, the 24-volt control amperage must never exceed

A. 9.58 amps.

B. 2.08 amps.

C. 104 amps.

D. 5 amps.

24. Using the "Oil-Fired Multiple Zone Valve System Diagram" (Figure 2–8), the burner motor and ignition transformer will not operate, and there is voltage at the 120-volt side of the aquastat boiler controller that feeds to the cad-cell relay. All thermostats are in the closed position and the circulator pump is operating. There is a voltage reading of 120 volts across terminals "BK" and "WH." A slight light leak is noticed in the combustion chamber. Why will the system not operate?

A. The light leak in the combustion chamber causes the resistance of the cad-cell to decrease and not fire.

B. The light leak in the combustion chamber causes the resistance of the cad-cell to increase and not fire.

C. The primary control must be condemned and replaced.

D. The light leak will not affect the operation of the system.

25. Use the "Oil-Fired Multiple Zone Valve System Diagram" (Figure 2–8) to answer the following question. A technician sets all thermostats to call for heat. The circulator pump is operating and the ignition system is sparking but does not fire. A further check shows that the burner motor is not operating. What could be the problem with the system?

A. The cad-cell relay could be defective.

B. The cad-cell could be defective.

C. The burner motor could be defective.

D. Both A and C

26. Using the "Oil-Fired Multiple Zone Valve System Diagram" (Figure 2–8), all zone thermostats are in the closed position, and there is 0 volts across terminals "TT" on the aquastat boiler controller; 120 volts is present to the control but the circulator pump and the gun burner are not operating. What could be the problem?

A. The cad-cell could be defective.

B. The boiler controller could be defective.

C. The zone valve transformer could be defective.

D. The cad-cell relay could be defective.

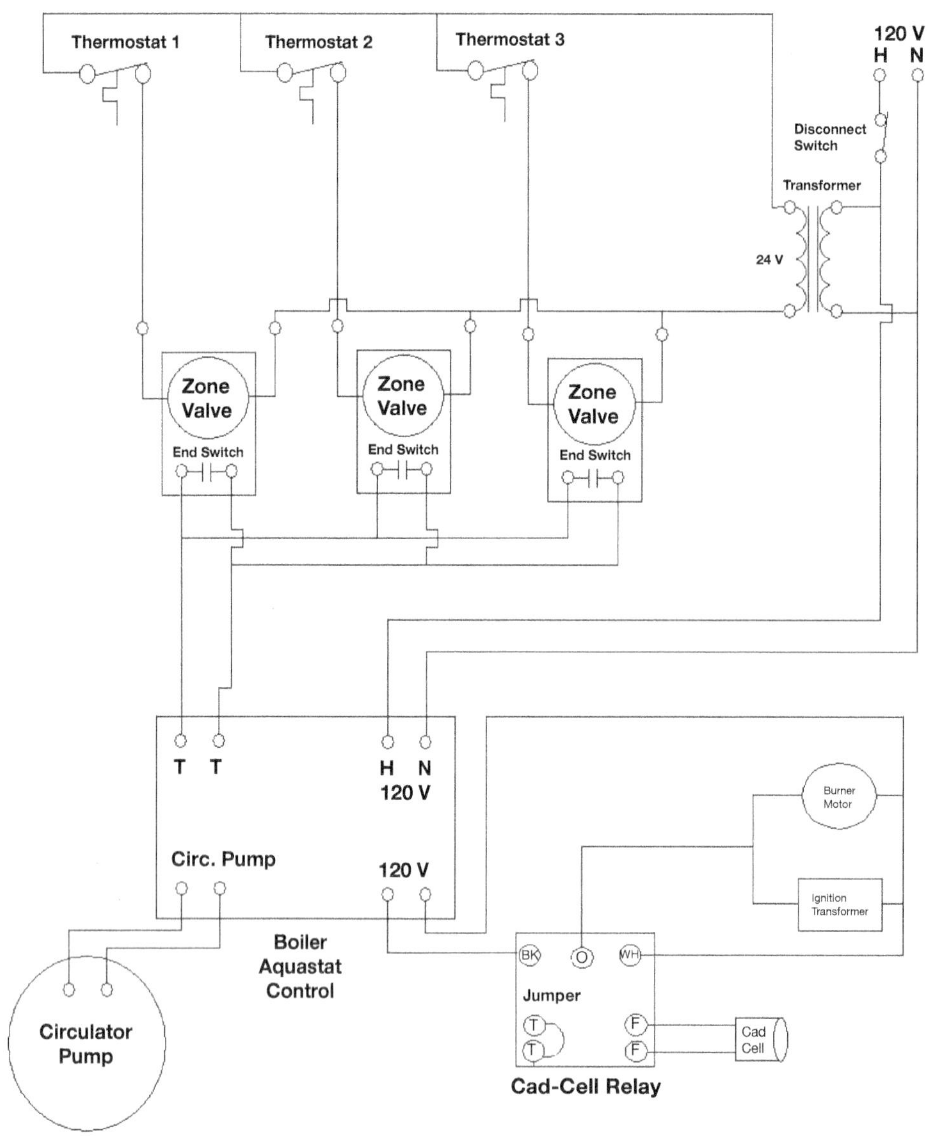

FIGURE 2–8

Oil-Fired Multiple Zone Valve System Diagram

27. Use the "Oil-Fired Multiple Zone Valve System Diagram" (Figure 2–8) to answer the following question.

Technician A says that when all zone thermostats are open there should be 0 volts present across terminals "TT" on the aquastat boiler controller.

Technician B says that when all zone thermostats are open there should be 24 volts present across terminals "TT" on the aquastat boiler controller.

Which technician is correct?

A. technician A

B. technician B

C. Both technicians are correct.

D. Both technicians are incorrect.

28. Using the "Oil-Fired Multiple Zone Valve System Diagram" (Figure 2–8), all thermostats are set to call for heat. The circulator pump is operating, and there is a flame in the combustion chamber. After 45 s the burner system shuts down on the safety control circuit. What could be the problem with the system?

A. The ignition transformer could be defective.

B. The cad-cell relay could be defective.

C. The cad-cell could be defective.

D. Both B and C

29. Using the "Oil-Fired Multiple Zone Valve System Diagram" (Figure 2–8), all thermostats are set to call for heat. The circulator pump and the burner motor are operating but there is no spark. After 45 s, the burner system shuts down on the safety control circuit. What could be the problem with the system?

A. The ignition transformer could be defective.

B. The system could be out of oil.

C. The cad-cell could be defective.

D. The boiler control could be faulty.

30. Using the "Oil-Fired Multiple Zone Valve System Diagram" (Figure 2–8), a technician sees that the "TT" terminals are jumped out on the cad-cell relay. He removes the jumper and starts the system. The zone 1 thermostat is closed and the circulator pump starts but there is no flame. The technician doesn't understand why there is a jumper across the "TT" terminals and believes there should be a thermostat connected. You, being the top technician at the firm, explain that

A. the technician is right, there should be a thermostat connected, so one should be connected.

B. the end switches should connect to the terminals that were jumped out, and the jumper should be across the "TT" terminals on the aquastat boiler controller.

C. normally a thermostat would be connected when the cad-cell relay is used on a forced-air heating system, but a jumper is required when zones are being used.

D. the "TT" terminals are always jumped out on a cad-cell relay, even when used on forced-air heating or series-loop systems.

31. Use the "Oil-Fired Multiple Zone Valve System Diagram" (Figure 2–8) to answer the following question. The homeowner is expanding his home and wants to add one more zone to the system. The system uses a 120-volt/24-volt 40-VA control transformer and each zone valve operates on 24 volts drawing .5 amps. The homeowner asks you to do the control wiring. What are you going to do?

A. Because the zone valves draw only .5 amps, simply add another zone valve to the control system.

B. Add a zone thermostat in series with the zone valve motor to the control system.

C. Add another 40-VA transformer before you add another zone valve circuit.

D. All of the above

32. Use the "Oil-Fired Multiple Zone Valve System Diagram" (Figure 2–8) to answer the following questions. A technician wants to troubleshoot zone valve 2 in the heat system. What would be the proper procedure?

A. Set the disconnect switch to the closed position, close zone valve 2, open zone valves 1 and 3, measure 24 volts across the zone valve motor and end switch, make sure the zone valve opens, and then measure 0 volts across the end switch when it closes.

B. Set the disconnect switch to the closed position, close zone valve 2, close zone valves 1 and 3, measure 24 volts across the zone valve motor and 0 volts across the end switch, make sure the zone valve opens, and then measure 24 volts across the end switch when it closes.

C. Set the disconnect switch to the closed position, close zone valve 2, open zone valves 1 and 3, measure 24 volts across the zone valve motor and 0 volts across the end switch, make sure the zone valve opens, and then measure 24 volts across the end switch when it closes.

D. Set the disconnect switch to the open position, close zone valve 2, close zone valves 1 and 3, measure 24 volts across the zone valve motor end switch, make sure the zone valve opens, and then measure 24 volts across the end switch when it closes.

33. Using the "Multiple Circulator Pump System Diagram" (Figure 2–9), zone 1 is calling for heat, the circulator pump is operating, but the boiler water temperature is 70°F. What three possibilities could cause this problem?

A. The high limit, transformer, or gas valve could be faulty.

B. The thermostat, gas valve, or switching relay could be faulty.

C. The switching relay, transformer, or high limit could be faulty.

D. The thermostat, transformer, or switching relay could be faulty.

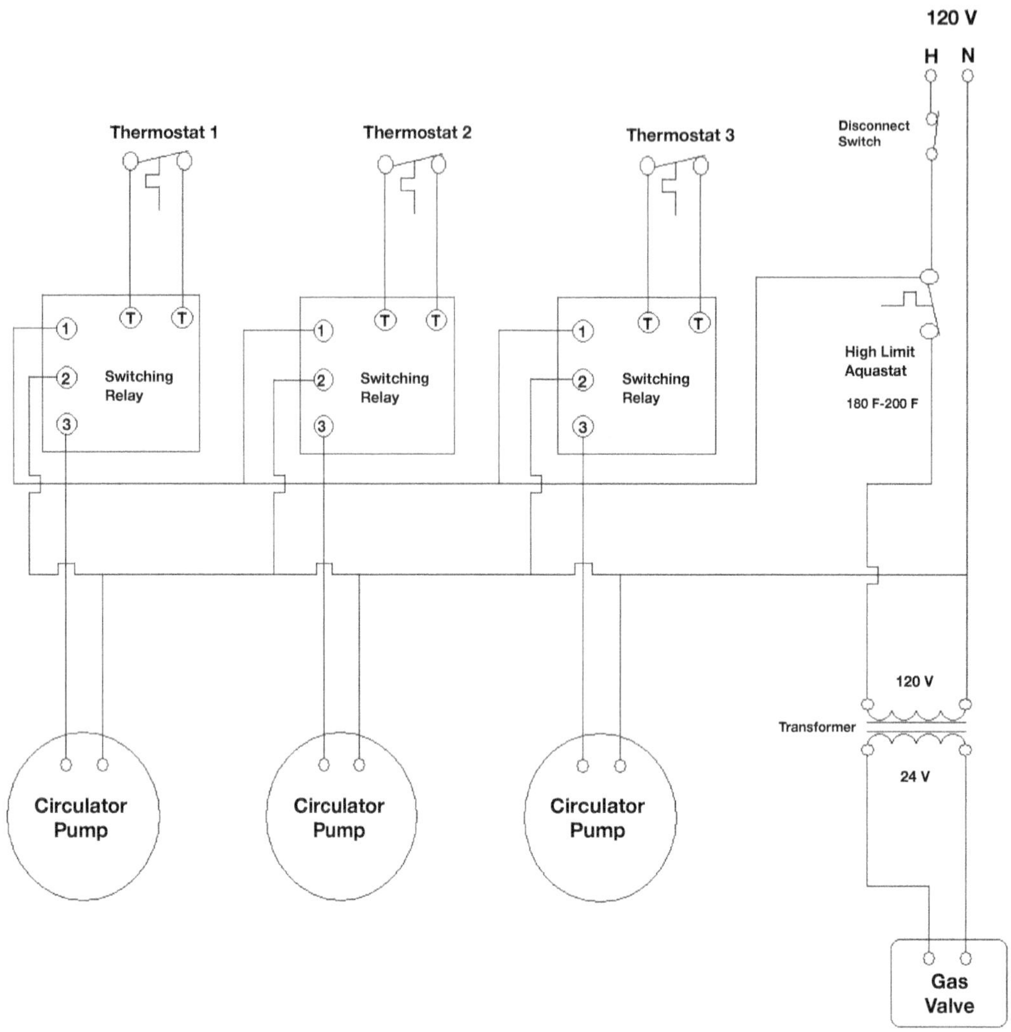

FIGURE 2–9

Multiple Circulator Pump System Diagram

34. Using the "Multiple Circulator Pump System Diagram," (Figure 2–9), on a call for heat from the zone 3 thermostat, the boiler water is warm but the circulator pump is not operating. A voltage reading from terminal "TT" on the switching relay reads 24 volts, and a voltage reading from the 1 and 2 terminals reads 120 volts, but when measuring terminals 2 and 3 there is 0 volts. What could be the problem?

A. a faulty thermostat

B. a faulty switching relay transformer

C. a faulty high-limit aquastat

D. a faulty circulator pump

35. Using the "Multiple Circulator Pump System Diagram" (Figure 2–9), the zone 2 thermostat is closed, and the zones 1 and 3 thermostats are open. The boiler water temperature is 190°F but the circulator pump is not operating. A voltage reading from 1 to 2 on the switching relay shows 120 volts, 1 to 3 shows 120 volts, and across "TT" is 0 volts. What could be the problem with the system?

A. a faulty circulator pump

B. a faulty zone 2 thermostat

C. a faulty switching relay

D. It could be all of the above.

36. Using the "Multiple Circulator Pump System Diagram" (Figure 2–9), all zone pumps are operating but the boiler water temperature is low, the high limit aquastat is in the closed position, and 120 volts is supplied across the transformer. Which of the following could be the problem with the system?

A. a shorted primary winding of the transformer

B. a blown fuse in the line voltage circuit

C. an open secondary winding of the transformer

D. All of the above

37. Using the "Multiple Circulator Pump System Diagram" (Figure 2–9), a properly operating switching relay, when energized, should have

A. 120 volts at terminals 1 and 2, 0 volts at "TT," and 120 volts at 2 and 3.

B. 120 volts at terminals 1 and 2, 24 volts at "TT," and 120 volts at 2 and 3.

C. 120 volts at terminals 1 and 2, 0 volts at "TT," and 0 volts at 2 and 3.

D. 0 volts at terminals 1 and 2, 24 volts at "TT," and 0 volts at 2 and 3.

38. Using the "Multiple Circulator Pump System Diagram" (Figure 2–9), what control will affect the operation of both the gas valve circuit and the circulator pump circuit?

A. the transformer

B. the high-limit aquastat

C. the switching relay

D. the disconnect switch

39. Using the "Multiple Circulator Pump System Diagram" (Figure 2–9), voltage at the transformer and no voltage out of the transformer can be caused by

A. a shorted primary winding only.

B. an open secondary winding only.

C. an open primary winding only.

D. an open primary winding or an open secondary winding.

40. Using the "Single-Pump Relay-Controlled System Diagram" (Figure 2–10), what two controls should have 24 volts present when the disconnect switch is in the closed position?

A. the relay coil and the thermostat

B. the secondary winding of the transformer and the limit switch

C. the vent damper and the secondary winding

D. the spill switch and the gas valve

41. Using the "Single-Pump Relay-Controlled System Diagram" (Figure 2–10), a customer complains that the hot water heating system is not operating properly. He notices that the circulator pump is operating but the burners are not igniting. A voltage check across terminals "G" and "C" reads 24 volts, and there is 0 volts across the "R1" contacts. Name two components that could be at fault.

A. a faulty relay coil and a faulty thermostat

B. faulty "R1" contacts and a bad high-limit switch

C. a bad thermostat and a faulty spill switch

D. a faulty vent damper and a bad high-limit switch

42. Use the "Single-Pump Relay-Controlled System Diagram" (Figure 2–10) to answer the following question. A customer complains that the hot water heating system is not operating properly. She notices that the circulator pump is operating but the burners are not igniting. A voltage check across terminals "G" and "C" reads 24 volts, and there is 0 volts across the "R1" terminal. Name two components that could be at fault.

A. a bad gas valve and an open flame roll-out sensor

B. "R1" contacts stuck open and a faulty spill switch

C. a bad gas valve and an open thermostat

D. a faulty relay coil and an open high limit

43. Using the "Single-Pump Relay-Controlled System Diagram" (Figure 2–10), how many load devices are in the 24-volt circuit?

A. 1

B. 2

C. 3

D. 4

44. Using the "Single-Pump Relay-Controlled System Diagram" (Figure 2–10), how many switching devices are in the 24-volt circuit?

A. 2

B. 3

C. 4

D. 5

45. Using the "Single-Pump Relay-Controlled System Diagram" (Figure 2–10), if the circulator pump is operating, what voltage reading should be across the "R2" contacts?

A. 0 volts

B. 120 volts

C. source voltage

D. 24 volts

46. Use the "Single-Pump Relay-Controlled System Diagram" (Figure 2–10) to answer the following question. While troubleshooting the system, the technician sets the thermostat to call for heat, and there is no action from the gas valve or circulator pump. A voltage check at the primary of the transformer reads 120 volts. A voltage check across the "R1" contacts of the relay shows 24 volts, and across the "R2" contact shows 120 volts. What could be the problem with the system?

A. a bad transformer or a bad relay coil

B. a bad relay coil or a bad thermostat

C. a faulty high limit or a bad transformer

D. an open flame roll-out or a faulty vent damper

47. Using the "Single-Pump Relay-Controlled System Diagram," (Figure 2–10), which control will protect the homeowner from a plugged heat exchanger?

A. the high limit

B. the flame roll-out

C. the spill switch

D. the vent damper

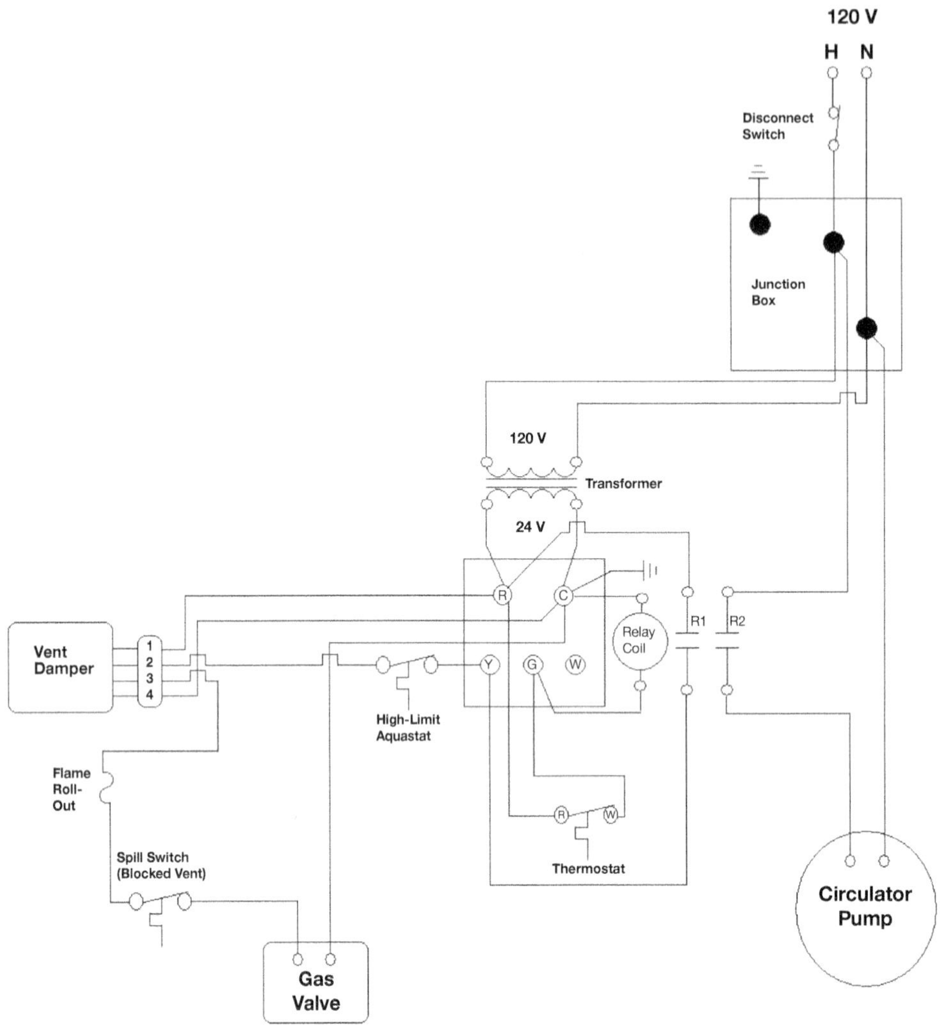

FIGURE 2–10

Single-Pump Relay-Controlled System Diagram

48. Using the "Single-Pump Relay-Controlled System Diagram" (Figure 2–10), which control will protect the homeowner from a plugged vent pipe?

A. the high limit

B. the flame roll-out

C. the spill switch

D. the vent damper

49. Using the "Single-Pump Relay-Controlled System Diagram" (Figure 2–10), which control will protect the boiler from overheating?

A. the high limit

B. the flame roll-out

C. the spill switch

D. the vent damper

50. Using the "Single-Pump Relay-Controlled System Diagram" (Figure 2–10), when the thermostat is calling for heat and the system is in operation, what voltage readings should be between terminals "R" and "C," "G" and "C," and "Y" and "C"?

A. "R" and "C" should be 24 volts, "G" and "C" should be 0 volts, and "Y" and "C" should be 24 volts.

B. "R" and "C" should be 24 volts, "G" and "C" should be 24 volts, and "Y" and "C" should be 0 volts.

C. "R" and "C" should be 24 volts, "G" and "C" should be 24 volts, and "Y" and "C" should be 24 volts.

D. "R" and "C" should be 0 volts, "G" and "C" should be 0 volts, and "Y" and "C" should be 24 volts.

51. Using the "Single-Pump Relay-Controlled System with Ignition Control Module Diagram" (Figure 2–11), the thermostat is set to call for heat, and 24 volts is present across terminals "R" and "C" and "G" and "C." The circulator pump is operating but there is 0 volts across terminals "Y" and "C." What could be the problem with the system?

A. The thermostat is faulty.

B. The relay coil is faulty.

C. The high limit is faulty.

D. The "R1" contact is faulty.

52. Using the "Single-Pump Relay-Controlled System with Ignition Control Module Diagram" (Figure 2–11), the thermostat is set for heat and the circulator pump is operating. The igniter on the ignition control system is sparking but there is no pilot. A voltage check across the 24-volt ground (GND) and 24-volt terminals reads 24 volts. A check across the main valve/pilot valve (MV/PV) and PV shows 24 volts. What could be the problem with the system?

A. The pilot valve is open.

B. The pilot line is plugged.

C. The pilot needs adjustment.

D. All of the above

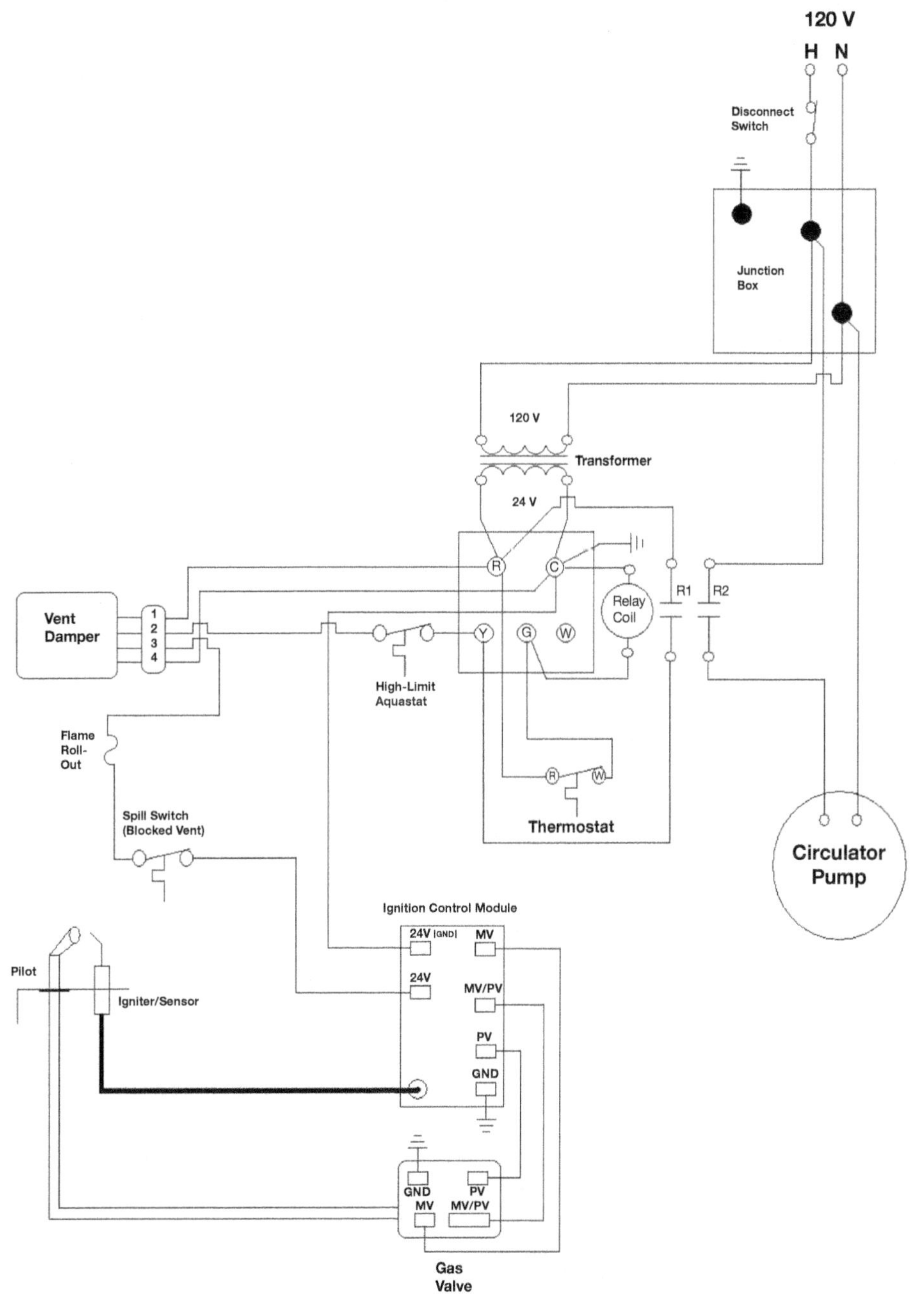

FIGURE 2–11

Single-Pump Relay-Controlled System with Ignition Control Module Diagram

53. Using the "Single-Pump Relay-Controlled System with Ignition Control Module Diagram" (Figure 2–11), the thermostat is set for heat and the circulator pump is operating but the burners are not lit. A voltage check from "Y" to "C" reads 24 volts, but 0 volts is present at terminals 24 volts GND and 24 volts. Which of the following could be the problem?

A. a closed limit

B. an open thermostat

C. a faulty vent damper

D. All of the above

54. Using the "Single-Pump Relay-Controlled System with Ignition Control Module Diagram" (Figure 2–11), the thermostat is set for heat and the circulator pump is operating but the burners are not lit. A voltage check from "Y" to "C" reads 24 volts but 0 volts is present at terminals 24 volts GND and 24 volts. Which of the following could be the problem?

A. an open spill switch

B. a faulty ignition module

C. a grounded ignition module

D. an open thermostat

55. Using the "Single-Pump Relay-Controlled System with Ignition Control Module Diagram" (Figure 2–11), the thermostat is set for heat and the circulator pump is operating. The pilot is lit but the burners are not. A voltage check at MV/PV and MV shows 0 volts. What could be the problem with the system?

A. a faulty gas valve

B. a faulty sensor

C. a faulty pilot valve

D. a faulty spill switch

56. Using the "Single-Pump Relay-Controlled System with Ignition Control Module Diagram" (Figure 2–11), the thermostat is set for heat and the circulator pump is operating. The pilot is lit but the burners are not. A voltage check at MV/PV and MV shows 0 volts. What could be the problem with the system?

A. a faulty ignition module

B. a faulty pilot valve

C. a shorted main gas valve

D. an open flame roll-out

57. Using the "Single-Pump Relay-Controlled System with Ignition Control Module Diagram" (Figure 2–11), the thermostat is set for heat and the circulator pump is operating. The pilot is lit but the burners are not. A voltage check at MV/PV and MV shows 0 volts. What could be the problem with the system?

A. a broken ground wire

B. a faulty main gas valve

C. a shorted pilot valve

D. a faulty vent damper

58. Using the "Multiple Zone Valve System Ignition Control Module and Vent Damper Diagram" (Figure 2–12), the zone 1 thermostat is calling for heat and the zone 2 and 3 thermostats are open. The problem with the system is that the circulator pump is operating but the burners are not lit and the water temperature is extremely low. Where would be the most logical place to start troubleshooting the system?

A. at zone valve 1

B. at the zone transformer

C. at the vent damper

D. at the circulator pump

59. Using the "Multiple Zone Valve System Ignition Control Module and Vent Damper Diagram" (Figure 2–12), the zone 2 thermostat is calling for heat and the zone 1 and 3 thermostats are open. The problem with the system is that the circulator pump is operating but the burners are not lit and the water temperature is extremely low. Where would be the most logical place to start troubleshooting the system?

A. at zone valve 2

B. at the zone transformer

C. at the circulator pump

D. at the 24-volt terminals on the boiler controller

60. Using the "Multiple Zone Valve System Ignition Control Module and Vent Damper Diagram" (Figure 2–12), the zone 2 thermostat is calling for heat and the zone 1 and 3 thermostats are open. 24 volts is present at the "TT" terminals. The circulator pump is de-energized and the burners will not light. What could be the problem with the system?

A. The zone transformer could be faulty.

B. The end switch could be faulty.

C. The zone valve motor could be faulty.

D. All of the above

61. Using the "Multiple Zone Valve System Ignition Control Module and Vent Damper Diagram" (Figure 2–12), if zone thermostats 1, 2, and 3 are all in the open position, what reading should be across the end switches of each zone valve?

A. 24 volts

B. 0 volts

C. 120 volts

D. 240 volts

62. Using the "Multiple Zone Valve System Ignition Control Module and Vent Damper Diagram" (Figure 2–12), if zone thermostats 1, 2, and 3 are all in the closed position, what reading should be across the end switches of each zone valve?

A. 24 volts

B. 0 volts

C. 120 volts

D. 240 volts

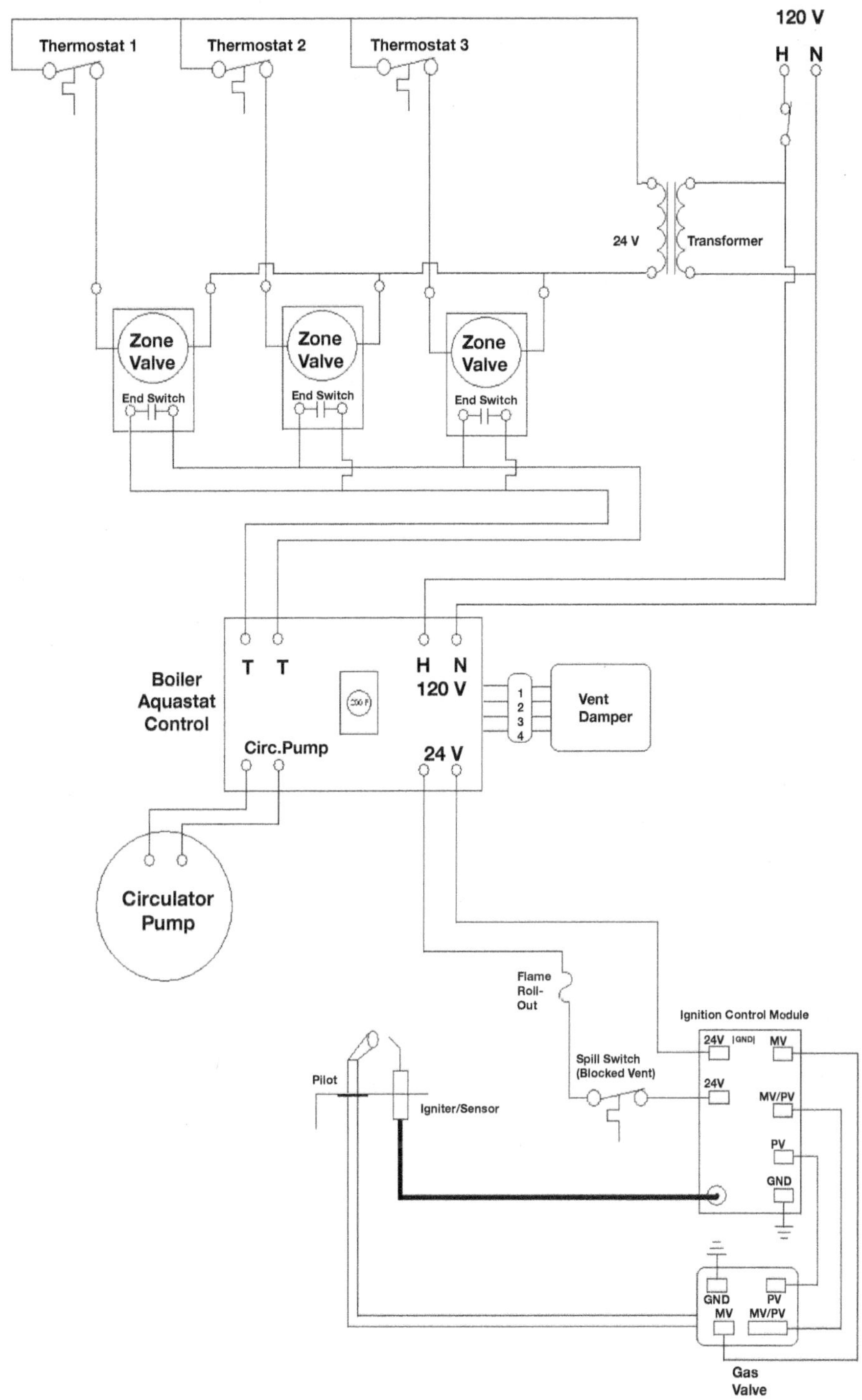

FIGURE 2–12

Multiple Zone Valve System Ignition Control Module and Vent Damper Diagram

63. Using the "Multiple Zone Valve System Ignition Control Module and Vent Damper Diagram" (Figure 2–12), zone thermostats 1, 2, and 3 are all in the closed position, and the circulator pump is operating but the burners are not. A voltage reading at the output of the aquastat boiler control reads 24 volts. What could be the problem with the system?

A. a faulty aquastat boiler control

B. a faulty gas valve

C. a faulty vent damper

D. a faulty zone valve end switch

64. Using the "Multiple Zone Valve System Ignition Control Module and Vent Damper Diagram" (Figure 2–12), zone thermostats 1, 2, and 3 are all in the closed position, and the circulator pump is operating but the burners are not. A voltage reading at the output of the aquastat boiler control reads 0 volts. A voltage check at the vent damper reads 24 volts across terminals 1 and 4. Terminals 2 and 4 also read 24 volts but at terminals 3 and 4 it reads 0 volts. The vent damper motor is opening the damper system but the burners will not operate. What could be the problem with the system?

A. a faulty damper motor

B. a faulty vent damper end switch

C. a faulty aquastat boiler

D. a faulty ignition control module

65. Using the "Multiple Zone Valve System Ignition Control Module and Vent Damper Diagram" (Figure 2–12), a call for heat from any zone thermostat starts the circulator pump but the burners are not lighting on the boiler. There is 24 volts present across the 24-volt terminals of the ignition control module but 0 volts present at the PV and MV/PV terminals. There is a spark but no pilot. What could be the problem with the system?

A. a faulty ignition control module

B. a faulty pilot valve

C. a faulty aquastat boiler control

D. an open flame roll-out

66. Using the "Multiple Zone Valve System Ignition Control Module and Vent Damper Diagram" (Figure 2–12), a call for heat from any zone thermostat starts the circulator pump but the burners are not lighting on the boiler. There is 24 volts present across the 24-volt terminals of the ignition control module and 24 volts is present at the PV and MV/PV terminals. There is a spark but no pilot. What could be the problem with the system?

A. a faulty ignition control module

B. a faulty pilot valve

C. a faulty aquastat boiler control

D. an open flame roll-out

67. Using the "Multiple Zone Valve System Ignition Control Module and Vent Damper Diagram" (Figure 2–12), a call for heat from any zone thermostat starts the circulator pump but the burners are not lighting on the boiler. There is 24 volts present across the 24-volt terminals of the ignition control module and 24 volts present at the PV and MV/PV terminals. The pilot has a sufficient flame but there is 0 volts at the MV/PV and MV terminals. What could be the problem with the system?

A. a faulty ignition control module

B. a faulty main gas valve

C. a faulty aquastat boiler control

D. a faulty pilot valve

68. Using the "Multiple Zone Valve System Ignition Control Module and Vent Damper Diagram" (Figure 2–12), a call for heat from any zone thermostat starts the circulator pump but the burners are not lighting on the boiler. There is 24 volts present across the 24-volt terminals of the ignition control module and 24 volts present at the PV and MV/PV terminals. The pilot has a sufficient flame and there is 24 volts at the MV/PV and MV terminals. What could be the problem with the system?

A. a faulty ignition control module

B. a faulty main gas valve

C. a faulty aquastat boiler control

D. a faulty pilot valve

69. Using the "Multiple Zone Valve System Ignition Control Module and Vent Damper Diagram" (Figure 2–12), a call for heat from any zone thermostat starts the circulator pump but the burners are not lighting on the boiler. There is 24 volts present across the 24-volt terminals of the ignition control module and 24 volts present at the PV and MV/PV terminals but the pilot is not lighting. There is no spark present but gas is flowing from the pilot line. What could be the problem with the system?

A. a faulty aquastat boiler control

B. a faulty pilot valve

C. a faulty ignition transformer

D. an open high limit

70. Use the "Multiple Zone Valve System Ignition Control Module and Vent Damper Diagram" (Figure 2–12) to answer the following question. Technician A says that there should be 24 volts present across the output of the aquastat boiler controller terminals and 24 volts present at the input to the ignition control module if the spill switch is in the open position.

Technician B says that there should be 24 volts present across the output of the aquastat boiler controller terminals and 0 volts present at the input to the ignition control module if the spill switch is in the open position.

Which technician is correct?

A. technician A

B. technician B

C. Both technicians are correct.

D. Both technicians are incorrect.

71. Use the "Multiple Zone Valve System Ignition Control Module and Vent Damper Diagram" (Figure 2–12) to answer the following question. Technician A says that there should be 24 volts present across the PV and MV/PV terminals at the ignition control module when the main burners are lit.

Technician B says that there should be 0 volts present across the PV and MV/PV terminals at the ignition control module when the main burners are lit.

Which technician is correct?

A. technician A is correct

B. technician B is correct

C. Both technicians are correct.

D. Both technicians are incorrect.

72. Use the "Multiple Zone Valve System Ignition Control Module and Vent Damper Diagram" (Figure 2–12) to answer the following question.

Technician A says that there should be 24 volts present across the MV and MV/PV terminals at the ignition control module when the main burners are lit.

Technician B says that there should be 0 volts present across the MV and MV/PV terminals at the ignition control module when the main burners are lit.

Which technician is correct?

A. technician A

B. technician B

C. Both technicians are correct.

D. Both technicians are incorrect.

73. Use the "Multiple Zone Valve System Ignition Control Module and Vent Damper Diagram" (Figure 2–12) to answer the following question.

Technician A says that when there is 24 volts present at any zone valve motor, there should be 24 volts present across the "TT" terminals at the aquastat boiler controller.

Technician B says that when there is 24 volts present at any zone valve motor, there should be 0 volts present across the "TT" terminals at the aquastat boiler controller.

Which technician is correct?

A. technician A

B. technician B

C. Both technicians are correct.

D. Both technicians are incorrect.

74. Use the "Multiple Zone Valve System Ignition Control Module and Vent Damper Diagram" (Figure 2–12) to answer the following question.

Technician A says that the method to properly test the ignition transformer is to ground a screwdriver and see if there is a $\frac{1}{4}''$ spark from the screwdriver to the output of the ignition transformer.

Technician B says a cracked ceramic on an electrode will prevent the pilot from lighting.

Which technician is correct?

A. technician A

B. technician B

C. Both technicians are correct.

D. Both technicians are incorrect.

75. Use the "Multiple Zone Valve System Ignition Control Module and Vent Damper Diagram" (Figure 2–12) to answer the following question.

Technician A says a corroded ground wire at the ignition control module could prevent the flame rectification system from proving the pilot.

Technician B says that a cracked ceramic on the flame sensor can prevent the flame rectification system from proving the pilot.

Which technician is correct?

A. technician A

B. technician B

C. Both technicians are correct.

D. Both technicians are incorrect.

76. Using the "Multiple Zone Valve System Ignition Control Module and Vent Damper Diagram" (Figure 2–12), which of the following will prevent the main burners from lighting once the pilot is lit?

A. a faulty flame sensor

B. a faulty main gas valve

C. a faulty ignition control module

D. all of the above

77. Using the "Multiple Zone Valve System Ignition Control Module and Vent Damper Diagram" (Figure 2–12), which of the following will cause the ignition control module to continue to spark once the pilot flame has been lit?

A. a faulty flame sensor wire

B. a faulty flame sensor

C. a faulty ignition control module

D. all of the above

78. Using the "Multiple-Pump System with Boiler Controller, Ignition Control Module, and Vent Damper Diagram" (Figure 2–13), the "XX" terminals of the switching relay are

A. normally opened.

B. normally closed.

C. opened when energized.

D. Both B and C

79. Using the "Multiple-Pump System with Boiler Controller, Ignition Control Module, and Vent Damper Diagram" (Figure 2–13), when one of the thermostats closes, what would be the next action of the system?

A. The burners would light.

B. The circulator pump would start.

C. The vent damper would close.

D. The igniter would spark.

80. Using the "Multiple-Pump System with Boiler Controller, Ignition Control Module, and Vent Damper Diagram" (Figure 2–13), what voltage should be across the "XX" terminals with all thermostats in the open position?

A. 24 volts

B. 0 volts

C. 120 volts

D. 240 volts

81. Using the "Multiple-Pump System with Boiler Controller, Ignition Control Module, and Vent Damper Diagram" (Figure 2–13), what voltage should be across the "TT" terminals with all thermostats in the open position?

A. 24 volts

B. 0 volts

C. 120 volts

D. 240 volts

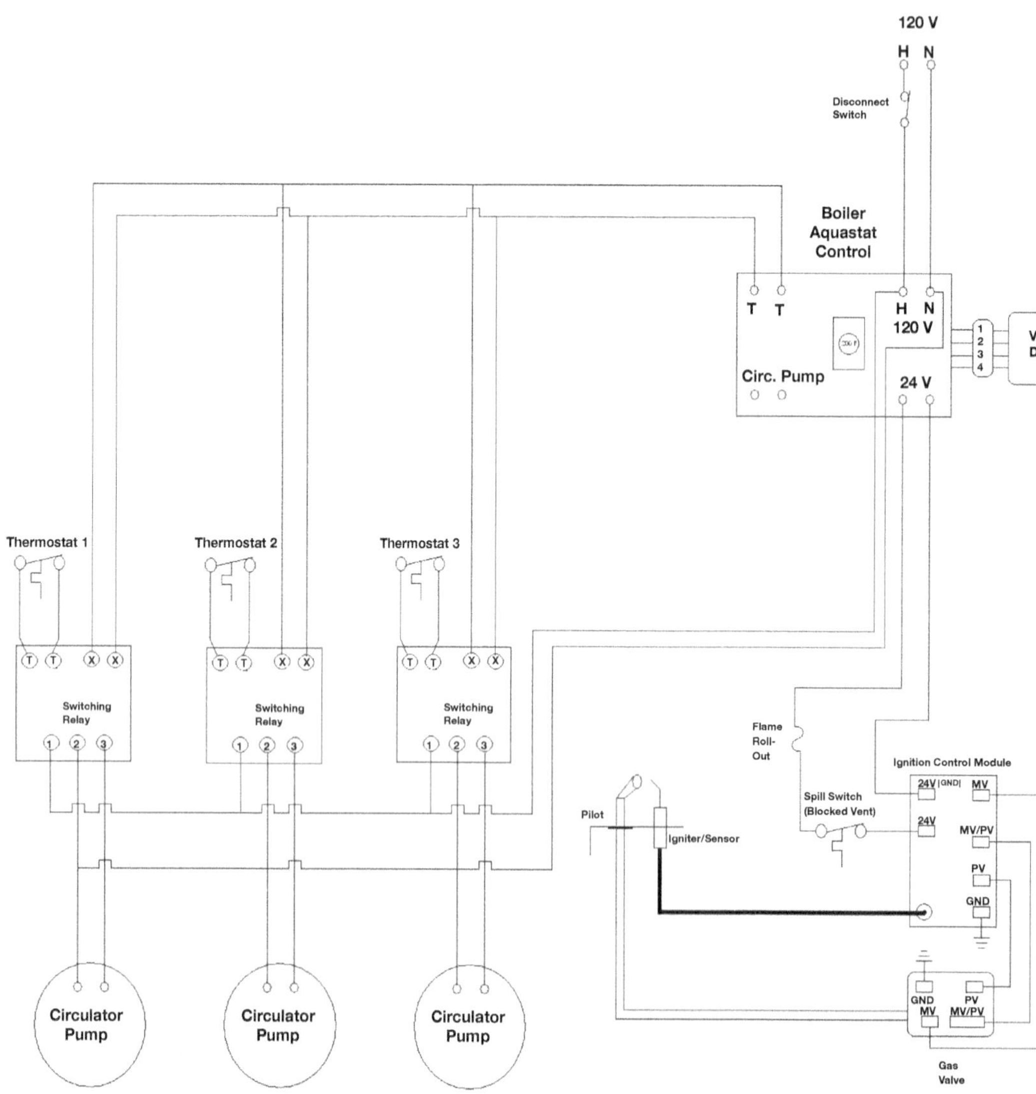

FIGURE 2–13

Multiple-Pump System with Boiler Controller, Ignition Control Module, and Vent Damper Diagram

82. Using the "Multiple-Pump System with Boiler Controller, Ignition Control Module, and Vent Damper Diagram" (Figure 2–13), if the vent damper was not opening and all of the circulator pumps were operating, which of the following would be the most logical test to perform when troubleshooting the system?

A. Check for 24 volts at the ignition control system.

B. Check for 120 volts at the input to the boiler aquastat control.

C. Check for voltage at terminals 1 and 4 at the vent damper.

D. Check for voltage across the high limit.

83. Using the "Multiple-Pump System with Boiler Controller, Ignition Control Module, and Vent Damper Diagram" (Figure 2–13), the zone 2 thermostat is calling for heat and the zone 1 and 3 thermostats are open. The burners fire but the circulator pump is not operating. What two terminals on the switching relay should feed 120 volts to the pump if the relay is operating correctly?

A. terminals "XX"

B. terminals 1 and 2

C. terminals 2 and 3

D. terminals 1 and 3

84. Using the "Multiple-Pump System with Boiler Controller, Ignition Control Module, and Vent Damper Diagram" (Figure 2–13), which of the following problems would allow the pilot to light but prevent the main gas valve from energizing?

A. an open limit

B. an open flame roll-out

C. a faulty end switch on the vent damper

D. a faulty flame sensor

85. Using the "Multiple-Pump System with Boiler Controller, Ignition Control Module, and Vent Damper Diagram" (Figure 2–13), which of the following problems would allow the pilot to light but prevent the main gas valve from energizing?

A. a faulty pilot valve

B. a faulty end switch on the vent damper

C. a faulty ignition control module

D. a faulty boiler aquastat control

86. Using the "Multiple-Pump System with Boiler Controller, Ignition Control Module, and Vent Damper Diagram" (Figure 2–13), which of the following problems would cause the igniter to spark but prevent the pilot from lighting?

A. a plugged pilot line

B. a faulty ignition control module

C. an open pilot valve coil

D. All of the above

87. Using the "Multiple-Pump System with Boiler Controller, Ignition Control Module, and Vent Damper Diagram" (Figure 2–13), there is 24 volts present at the ignition control module 24-volt terminals. The pilot valve is opening but the igniter does not seem to be able to light the pilot. Which of the following would cause this problem?

A. a weak spark

B. faulty pilot valve

C. a faulty flame roll-out

D. a faulty high limit

88. Using the "Multiple-Pump System with Boiler Controller, Ignition Control Module, and Vent Damper Diagram" (Figure 2–13), which of the following would cause the igniter to have a weak spark?

A. a faulty pilot gas valve

B. a faulty ground

C. a faulty control transformer

D. a faulty main gas valve

89. Using the "Multiple-Pump System with Boiler Controller, Ignition Control Module, and Vent Damper Diagram" (Figure 2–13), which of the following would cause the igniter to have a weak spark?

A. a faulty main gas valve

B. a high limit

C. a faulty ignition transformer

D. a faulty pilot gas valve

90. Using the "Multiple-Pump System with Boiler Controller, Ignition Control Module, and Vent Damper Diagram" (Figure 2–13), which of the following would cause the igniter to have a weak spark?

A. a cracked ceramic flame sensor

B. a faulty ignition transformer

C. a faulty igniter cable

D. All of the above

91. Using the "Multiple-Pump System with Boiler Controller, Ignition Control Module, and Vent Damper Diagram" (Figure 2–13), after the thermostats close and the circulator pumps start, what will be the next action of the system?

A. The "XX" terminals will open the circuit.

B. The N.O. contacts will close to the "TT" terminals.

C. The N.C. contacts will open to the "TT" terminals.

D. The pilot valve will energize.

92. Using the "Multiple-Pump System with Boiler Controller, Ignition Control Module, and Vent Damper Diagram" (Figure 2–13), after the thermostats close, the circulator pump starts, and the "XX" terminals close, what will be the next action of the system?

A. The ignition transformer will be energized.

B. The pilot valve will open.

C. The vent damper will open.

D. The high limit will close.

93. Using the "Multiple-Pump System with Boiler Controller, Ignition Control Module, and Vent Damper Diagram" (Figure 2–13), which of the following meters must be used to measure the current flow through the flame rectification system?

A. a VOM

B. a milliammeter

C. a microammeter

D. an ohmmeter

SECTION 3

Electrical Theory and Application

THE FOLLOWING QUESTIONS MAY BE HELPFUL IN STUDYING FOR THESE CERTIFICATION TESTS:

NATE—CR-SV, CR-IN, AC-SV, AC-IN, HG-SV, HG-IN, GS-SV, GS-IN, HP-SV, HP-IN, OL-SV, OL-IN, HO-SV, HO-IN, RC-SV, RC-IN, RL-SV, RL-IN.

ESCO INSTITUTE HVAC EXCELLENCE ELECTRICAL

ELECTRICAL SAFETY

1. After installing or servicing an electrical system,
   - **A.** always inspect the system for loose connections, improper grounding, and defective wiring.
   - **B.** properly color code all wires.
   - **C.** seal control and line voltage wires in flexible conduit.
   - **D.** seal control and line voltage wires in conduit.
2. What must always be observed when servicing electrical equipment?
   - **A.** the brand name of the breaker panel
   - **B.** labels and precautions
   - **C.** the type of fuse being used
   - **D.** the size of the equipment
3. All electrical installations and repairs must meet
   - **A.** NFPA guidelines.
   - **B.** building inspector approval.
   - **C.** BOCA approvals.
   - **D.** NEC guidelines.
4. The technician should always
   - **A.** listen to the customer's diagnosis of a system.
   - **B.** determine the last time the equipment has been serviced.
   - **C.** check the electrical panel for open breakers.
   - **D.** be aware of the voltage rating of the equipment being worked on.
5. When servicing or replacing electrical components, a technician
   - **A.** must de-energize the equipment.
   - **B.** must wear hand protection.
   - **C.** must use insulated tools only.
   - **D.** must disconnect the primary side of the transformer.
6. An electrical safety control
   - **A.** must not be permanently bypassed.
   - **B.** must always be installed on the discharge line.
   - **C.** must always be installed on the suction line.
   - **D.** is considered the primary control of the system.
7. Before repairing electrical equipment,
   - **A.** check with the local electrical inspector.
   - **B.** a voltage test should be done to ensure there is no power to the system.
   - **C.** use an ohmmeter to check all fuses.
   - **D.** an ammeter test is required.
8. Before starting equipment, the technician should first
   - **A.** check for proper voltage to the equipment.
   - **B.** check superheat.
   - **C.** check refrigerant charge.
   - **D.** check all safety devices.

9. When servicing large air conditioning equipment, the technician should

A. always ask for assistance.

B. always check superheat.

C. always place a power lock-out/tag-out on the electrical disconnect.

D. always inform the customer of the equipment's electrical and mechanical condition in great detail.

10. An electrocution caused by a grounded compressor on an air conditioning system is possible

A. if the power leads of the system are reversed.

B. if the fuse protecting the circuit is improperly sized.

C. if the high-side pressure increases, causing a current surge.

D. if grounding wires are improperly connected.

11. A GFCI is used to

A. prevent an electrocution caused by a short circuit.

B. prevent an electrocution caused by an open motor winding.

C. prevent an electrocution caused by a grounded load.

D. prevent an electrocution caused by high voltage.

12. What is commonly used with an extension cord to prevent electrocution when working with power tools?

A. a GFCI circuit breaker

B. a GFCI outlet

C. a GFCI toggle switch

D. None of the above

13. Before testing a capacitor, always

A. check for the correct mfd rating.

B. check for the correct voltage rating.

C. discharge the capacitor.

D. None of the above

14. An electrical shock is what happens when

A. the body resistance is below 100,000 ohms.

B. voltage flows through the body.

C. current flows through the body.

D. Both B and C

15. What type of an electric shock can cause heart damage?

A. one that is across both arms

B. one that is across the right arm and the right leg

C. one that is across both legs

D. one that is across the hand and the shoulder

16. When are people more susceptible to an electric shock?

A. when their skin is dry

B. when a GFCI is used

C. when their skin is moist

D. when wearing rubber-soled shoes

17. An electrocution as low as

A. 14 amps can kill or cause serious injury.

B. 10 amps can kill or cause serious injury.

C. 1 amp can kill or cause serious injury.

D. .025 amps can kill or cause serious injury.

18. When replacing a fuse on an HVAC system,

A. it is safe to use a fuse that is lower than the system's nameplate.

B. it is safe to use a fuse that is higher than the system's nameplate.

C. use a timed delay fuse that is equal to the system's nameplate.

D. use a fast-acting fuse that is equal to the system's nameplate.

19. When removing a fuse from a fused disconnect, a technician must use

A. an insulated fuse puller.

B. an insulated screwdriver.

C. insulated needle-nose pliers.

D. an insulated pry bar.

20. When performing an electrical repair,

A. always use magnetized hand tools.

B. always inspect hand tools for adequate insulation.

C. use an ohmmeter for testing the power supply.

D. use an ammeter for testing the power supply.

21. When working on a blower that has an exposed pulley and belt system, a technician should always

A. test for belt tightness while the system is in operation.

B. inspect the system for proper safety guards.

C. test belt tension when the system is in operation.

D. adjust the belt for a 2″ slack.

22. When working on electrical panels, a technician should

A. have rubber-soled shoes.

B. wear insulated gloves.

C. wear eye protection.

D. wear a face shield.

23. To prevent a personal injury,

A. always disconnect the power supply before inspecting fans and motors.

B. the inspection of fans and motors should always be done with power to the system.

C. fans and motors rarely need inspection.

D. None of the above

24. Never service machinery

A. when wearing loose clothing.

B. without the use of a safety helmet.

C. unless wearing gloves.

D. None of the above

25. If a technician has excessively long hair, he should

A. get it cut.

B. not work on machinery.

C. tie hair back or into a cap before working on machinery.

D. None of the above

26. The manufacturers of HVAC/R equipment provide wiring diagrams to

A. aid the owner of the equipment.

B. provide help in case of remodifying the wiring.

C. indicate the correct wiring and provide location of components.

D. follow federal requirements.

27. When working on electronic air cleaners,

A. all electrical power should be turned off.

B. the indoor blower motor must be de-energized.

C. the thermostat subbase should be placed on "fan auto."

D. the thermostat subbase should be placed on "fan manual."

28. All electrical equipment should be

A. tagged "high voltage."

B. turned off before servicing.

C. labeled.

D. color coded.

THE FOLLOWING QUESTIONS MAY BE HELPFUL IN STUDYING FOR THESE CERTIFICATION TESTS:
NATE—CR-SV, CR-IN, AC-SV, AC-IN, HP-SV, HP-IN, RC-SV, RC-IN, RL-SV, RL-IN.
ESCO INSTITUTE HVAC EXCELLENCE ELECTRICAL

ELECTRICAL THEORY

1. The negatively charged particle of an atom is the

A. proton.

B. electron.

C. neutron.

D. nucleus.

2. The positively charged particle of an atom is the

A. proton.

B. electron.

C. neutron.

D. nucleus.

3. Like-charged particles

A. attract each other.

B. repel each other.

C. have a negative charge.

D. have a positive charge.

4. Which material would be the best conductor of electrons?

A. glass

B. wood

C. air

D. copper

5. Which material would be the best insulator?

A. aluminum

B. copper

C. rubber

D. gold

6. When a conductor is moved through a magnetic field,

A. magnetism is increased.

B. reactance is decreased.

C. a power supply is induced.

D. frequency is reduced.

7. Electrical pressure is measured in

A. amps.

B. ohms.

C. volts.

D. watts.

8. The symbol used for electrical pressure is

A. "E."

B. "I."

C. "R."

D. "P."

9. Electron flow is measured in

A. amps.

B. ohms.

C. volts.

D. watts.

10. The symbol used for electron flow is

A. "E."

B. "I."

C. "R."

D. "P."

11. Electrical power is measured in

A. amps.

B. ohms.

C. volts.

D. watts.

12. The symbol used for electrical power is
   A. "E."
   B. "I."
   C. "R."
   D. "P."
13. Electrical resistance is measured in
   A. amps.
   B. ohms.
   C. volts.
   D. watts.
14. The symbol used for electrical resistance is
   A. "E."
   B. "I."
   C. "R."
   D. "P."
15. What would happen to the resistance of a wire if its length were doubled? It would
   A. stay the same.
   B. decrease by one-half.
   C. be twice as great.
   D. become four times as great.
16. A millivolt is best expressed as
   A. 1,000 volts.
   B. 100 volts.
   C. 1/1,000 of a volt.
   D. 1/100 of a volt.
17. A microamp is best expressed as
   A. 100,000 amps.
   B. 1,000,000 amps.
   C. 1/100,000 of an amp.
   D. 1/1,000,000 of an amp.
18. A megaohm is best expressed as
   A. 100,000 ohms.
   B. 1,000,000 ohms.
   C. 1/100,000 of an ohm.
   D. 1/1,000,000 of an ohm.
19. One amp is defined as
   A. one coulomb per second.
   B. one watt per second.
   C. one volt multiplied by one ohm.
   D. one watt divided by one ohm.
20. Power is best described as
   A. a large rate of inrush current.
   B. the rate at which electrons do work.
   C. amperage in an electrical circuit.
   D. resistance of the load in a circuit.
21. One horsepower is the same as
   A. 3.412 watts.
   B. 3,142 watts.
   C. 746 watts.
   D. 3.142 watts.
22. One watt is equivalent to
   A. 3.412 Btu.
   B. 3,412 Btu.
   C. 746 Btu.
   D. 3.142 Btu.
23. One kilowatt is equivalent to
   A. 3,412 Btu.
   B. 3,142 Btu.
   C. 1,000 Btu.
   D. 3.142 Btu.
24. When converting kW to Btu/h, the formula is
   A. volts × amps.
   B. watts × amps.
   C. watts × 3.413.
   D. amps × 3.413.

THE FOLLOWING QUESTIONS MAY BE HELPFUL IN STUDYING FOR THESE CERTIFICATION TESTS:
NATE—CR-SV, CR-IN.
ESCO INSTITUTE HVAC EXCELLENCE ELECTRICAL

MAGNETISM

1. All magnets have
   A. stator poles.
   B. north and south poles.
   C. stator, north, and south poles.
   D. reversing poles.
2. If a north pole of one magnet is placed near the north pole of another magnet, the poles will
   A. repel each other.
   B. attract each other.
   C. There will be no effect.
   D. A north pole does not exist on a magnet.

3. What flows between the two ends of a magnet?
   A. current
   B. voltage
   C. electrical charges
   D. lines of force
4. What is known as *magnetic lines of force*?
   A. flux
   B. permeability
   C. capacitance
   D. eddy current
5. A permanent magnet is
   A. a coil of wire that has current flowing through it.
   B. magnetism created by induction.
   C. a piece of material that has been magnetized.
   D. created by striking two pieces of soft iron material together.
6. An electromagnet is produced by
   A. running current through a coiled wire.
   B. running current through an iron core.
   C. moving an iron core through a coiled wire.
   D. moving a coiled wire over an iron core.
7. How is the magnetic field intensified in an electrical circuit?
   A. by increasing the size of the conductor
   B. by decreasing the size of the conductor
   C. by adding more turns to the conductor
   D. by decreasing the turns in the conductor
8. When current flows through the conductor of a solenoid coil, what happens to the iron core?
   A. It tries to move away from the magnetic field.
   B. It tries to center itself in the magnetic field.
   C. No action occurs in the iron core.
   D. The magnetic field tries to center itself around the iron core.
9. What component uses magnetism for operation?
   A. a pressure control
   B. an electric heating element
   C. a motor
   D. a thermal overload protector

THE FOLLOWING QUESTIONS MAY BE HELPFUL IN STUDYING FOR THESE CERTIFICATION TESTS:
NATE—CR-SV, CR-IN.
ESCO INSTITUTE HVAC EXCELLENCE ELECTRICAL

DIRECT CURRENT

1. In a DC circuit, current flows
   A. from negative to positive.
   B. from positive to negative.
   C. slower if the voltage is reduced in a circuit.
   D. faster if the resistance is increased in the circuit.
2. The three components needed in an electrical circuit are the
   A. path, volts, and power.
   B. current, path, and load.
   C. power source, path, and load.
   D. switch, power source, and load.
3. On a direct current power source,
   A. black represents negative and green represents positive.
   B. black represents positive and green represents negative.
   C. red represents positive and black represents negative.
   D. red represents negative and black represents positive.
4. Increasing the length of the path of an electrical circuit will
   A. decrease the total circuit resistance.
   B. increase the current.
   C. increase resistance.
   D. increase wattage.
5. The load in an electrical circuit is used to
   A. create electrical energy.
   B. convert electrical energy to heat, light, or magnetism.
   C. convert mechanical energy to electrical energy.
   D. create heat, light, or magnetism.
6. A switch in an electrical circuit is used to
   A. control current to a load.
   B. control the resistance of an electrical load.
   C. control the flow of voltage.
   D. control the wattage of a load.
7. When placed in an electrical circuit, an open switch will
   A. make the path to a load.
   B. break the path to a load.
   C. have no effect on the path to the load.
   D. temporarily interrupt voltage flow.

8. When placed in an electrical circuit, a closed switch will

A. make the path to a load.

B. break the path to a load.

C. have no effect on the path to the load.

D. temporarily interrupt voltage flow.

9. An example of a manually operated switch is a

A. temperature control.

B. pressure control.

C. defrost timer.

D. toggle switch.

10. An example of an automatically operated switch is a

A. lever type.

B. push button.

C. high-pressure control.

D. single-pole, double-throw toggle switch.

11. On what type of an electrical circuit can Ohm's law be used?

A. an inductive circuit

B. a capacitive circuit

C. a resistive circuit

D. an inductive, capacitive, and resistive circuit

12. Which is used to find amperage using Ohm's law?

A. $E = IR$

B. $I = E/R$

C. $R = E/I$

D. $I = RE$

13. Which is used to find resistance using Ohm's law?

A. $E = IR$

B. $R = EI$

C. $R = E/I$

D. $I = RE$

14. Which is used to find voltage using Ohm's law?

A. $E = IR$

B. $E = I/R$

C. $R = E/I$

D. $I = RE$

15. Which is used to find power using Watt's law?

A. $P = E/R$

B. $P = EI$

C. $P = E^2/I$

D. $P = E/R$

16. Which is used to find power using Watt's law?

A. $P = E/I$

B. $P = E^2/I$

C. $P = I^2R$

D. $P = E/R$

17. When three 15-ohm resistors are connected in series using a 120-volt power source, what is the total amperage of the circuit?

A. 8 amps

B. 3.2 amps

C. 2.7 amps

D. 375 milliamps

18. When two 20-ohm resistors are connected in parallel and the source voltage is 240 volts, what is the total amperage of the circuit?

A. 12 amps

B. 24 amps

C. 3 amps

D. 10 amps

19. When a 24-volt power supply is used in a circuit with 12 ohms of resistance, what is the total power rating of the circuit?

A. 2 watts

B. 288 watts

C. 24 watts

D. 48 watts

20. What is the total resistance of a circuit when a 5-ohm resistor, a 10-ohm resistor, and a 15-ohm resistor are wired in parallel with each other?

A. 30

B. 15

C. 4.55

D. 2.73

21. In a series circuit, which of the following is true?

A. Current is the same throughout the circuit.

B. Current increases as loads are added.

C. Resistance decreases as loads are added.

D. Voltage increases as loads are added.

22. In a series circuit, which of the following is *not* true?

A. Resistance decreases as loads are added.

B. Resistance increases as loads are added.

C. Voltage drops across loads wired in series.

D. Current is the same throughout the circuit.

23. In a series circuit that is using an electric heater as the load, what would cause an increase in circuit current?

A. a decrease in resistance

B. an increase in resistance

C. a decrease in voltage

D. a burnt contact

24. In a parallel circuit, which would be true?

A. Voltage is the same across each load.

B. Current is always the same across each load.

C. Total resistance increases as loads are added.

D. Voltage increases as loads are added.

25. In a parallel circuit, which of the following is *not* true?

A. Total resistance decreases as loads are added.

B. Voltage is the same across each load.

C. Current decreases as loads are added.

D. Current increases as loads are added.

THE FOLLOWING QUESTIONS MAY BE HELPFUL IN STUDYING FOR THESE CERTIFICATION TESTS:

NATE—CR-SV, CR-IN.

ESCO INSTITUTE HVAC EXCELLENCE ELECTRICAL

ALTERNATING CURRENT

1. How does alternating current get its name?

A. Alternating current uses different paths for electron flow.

B. Alternating current constantly changes direction in a circuit.

C. An alternating current circuit has a constant voltage fluctuation.

D. An alternating current circuit has a magnetic field when it flows through a circuit.

2. How is alternating current generated?

A. by coiling a wire around an iron core

B. by moving current through a conductor

C. by moving a conductor through a magnetic field

D. by moving a conductor near an iron core

3. What determines the frequency in an AC circuit?

A. the number of cycles that occur in a circuit in one second

B. the amount of current that flows from a given source

C. the force of the generator that is being used

D. the speed of a generator divided by four

4. Frequency is measured in

A. henrys.

B. reactance.

C. farads.

D. hertz.

5. What is the resistance of the passing of magnetic flux lines through a substance called?

A. inductance

B. impedance

C. reluctance

D. inductive reactance

6. What is the ability to store electrons in an electrical circuit called?

A. inductance

B. capacitance

C. reluctance

D. impedance

7. What is the resistance in an AC circuit due to the change of current flow called?

A. reactance

B. impedance

C. reluctance

D. inductance

8. What is the opposition to change in voltage in an AC capacitive circuit called?

A. capacitive reactance

B. capacitance

C. impedance

D. reactance

9. What is the generation of a voltage by the variance of the current flow rate in an AC circuit called?

A. reactance

B. capacitance

C. impedance

D. inductance

10. What is the opposition to current flow in an AC inductive circuit caused by magnetic lines of flux called?

A. inductance

B. impedance

C. inductive reactance

D. reactance

11. What is the total resistance of an AC circuit that is caused by inductive reactance, resistance, and capacitive reactance called?

A. inductance

B. impedance

C. reactance

D. reluctance

12. What is the relationship of current and voltage in an inductive circuit?

A. Voltage leads current.

B. Current leads voltage.

C. Voltage is in phase with current.

D. There is no relationship.

13. What is the relationship of current and voltage in a capacitive circuit?

A. Voltage leads current.

B. Current leads voltage.

C. Voltage is in phase with current.

D. There is no relationship.

14. What is the relationship of current and voltage in a resistive circuit?

A. Voltage leads current.

B. Current leads voltage.

C. Voltage is in phase with current.

D. There is no relationship.

15. One complete sine wave represents

A. the force of an AC circuit.

B. the action of the magnetic flux lines in an AC circuit.

C. a complete cycle of a generator.

D. the direction of the current through a conductor.

16. If the current of an AC circuit changes direction 100 times in 1 s, the frequency of the power source is

A. 100 Hz.

B. 200 Hz.

C. 60 Hz.

D. 50 Hz.

17. On a 120-volt, 60-Hz power supply, what leg must be fused?

A. the hot leg

B. the neutral leg

C. the hot and neutral leg

D. A fuse is not needed.

18. On a 240-volt, 60-Hz single-phase power supply, where is a fuse necessary?

A. in one hot leg

B. in both hot legs

C. in both hot legs and a neutral leg

D. in one hot leg and one neutral leg

19. When placing a voltmeter across two hot legs of a 240-volt three-phase power supply, what voltage should be present?

A. 120 volts

B. 480 volts

C. 240 volts

D. 277 volts

20. How many degrees are any two hot legs of a three-phase electrical system out of phase with each other?

A. 90 degrees

B. 100 degrees

C. 120 degrees

D. 360 degrees

21. On a three-phase power supply, what legs must be fused?

A. L1, L2, and neutral

B. L1, L2, and L3

C. L1, L3, and neutral

D. T1, T2, and T3

22. The condition indicating the direction of electron flow is known as

A. current direction.

B. voltage direction.

C. flux condition.

D. polarity.

23. On a 120-volt power supply, what color of wire is the hot leg?

A. white

B. black

C. red

D. green

24. On a 120-volt power supply, what color of wire is the neutral leg?

A. white

B. black

C. red

D. green

25. On a 120-volt power supply, what color of wire is the grounded leg?

A. white

B. black

C. red

D. green

26. The purpose of connecting a ground wire to a load is to

A. allow a secondary path for electrons during normal conditions.

B. protect the equipment against short circuits.

C. protect the equipment against an electrical short-to-ground.

D. allow a secondary path for electrons during a grounded condition.

THE FOLLOWING QUESTIONS MAY BE HELPFUL IN STUDYING FOR THESE CERTIFICATION TESTS:

NATE—CR-SV, CR-IN, AC-SV, AC-IN, HP-SV, HP-IN, HP-SV, HP-IN, RC-SV, RC-IN, RL-SV, RL-IN, GS-SV, AND GS-IN

ESCO INSTITUTE HVAC EXCELLENCE ELECTRICAL

POWER DISTRIBUTION

1. For residential areas, substations reduce the voltage from 120,000 volts down to

A. 8,400 volts.

B. 4,800 volts.

C. 5,200 volts.

D. 2,500 volts.

2. For commercial areas, substations reduce the voltage from 120,000 volts down to

A. 34,000 volts.

B. 43,000 volts.

C. 52,000 volts.

D. 25,000 volts.

3. A step-down transformer will reduce the voltage to a common low-voltage residential application of

A. 240 volts, single phase, 60 Hz.

B. 240 volts, three phase, 60 Hz.

C. 480 volts, single phase, 60 Hz.

D. 460 volts, three phase, 60 Hz.

4. Using a delta transformer, commercial applications will step the voltage down to

A. 240 volts, single phase, 60 Hz.

B. 240 volts, three phase, 60 Hz.

C. 480 volts, single phase, 60 Hz.

D. 208 volts, three phase, 60 Hz.

5. Using a wye transformer, commercial applications will step the voltage down to

A. 240 volts, single phase, 60 Hz.

B. 240 volts, three phase, 60 Hz.

C. 480 volts, single phase, 60 Hz.

D. 208 volts, three phase, 60 Hz.

6. When measuring voltage using a voltmeter across any two hot legs of a low-voltage wye transformer, it should read

A. 120 volts.

B. 240 volts.

C. 208 volts.

D. 460 volts.

7. When measuring the voltage on a low-voltage delta system, what reading would be obtained across any two hot legs?

A. 120 volts

B. 240 volts

C. 208 volts

D. 480 volts

8. Which three-phase system, when measuring from L1, L2, and L3 to ground, will read 120 volts?

A. a 120-volt, three-phase delta system

B. a 208-volt, three-phase delta system

C. a 480-volt, three-phase wye system

D. a 208-volt, three-phase wye system

9. Which three-phase electrical system will measure 120 volts when testing two of the hot leads to ground and 208 volts on the third hot leg to ground?

A. a 460-volt, three-phase delta system

B. a 240-volt, three-phase delta system

C. a 480-volt, three-phase wye system

D. a 208-volt, three-phase wye system

10. On the wye transformer diagram (Figure 3–1), what voltage reading should be from L1 to L2?

A. 240 volts

B. 120 volts

C. 208 volts

D. 277 volts

11. On the wye transformer diagram (Figure 3–1), what voltage reading should be from L1 to ground?

A. 240 volts

B. 120 volts

C. 208 volts

D. 480 volts

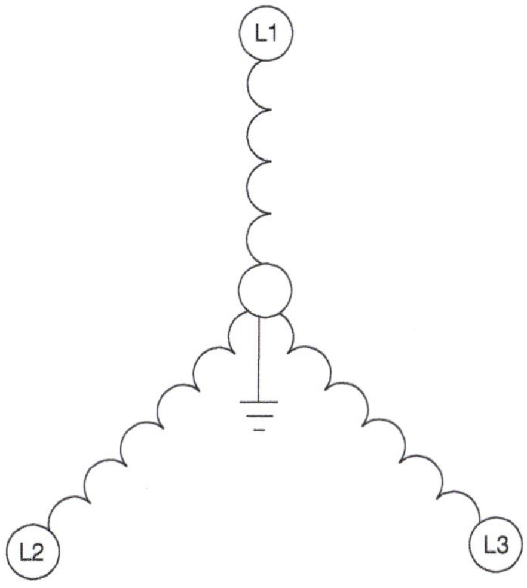

FIGURE 3–1
Low-Voltage Wye Transformer

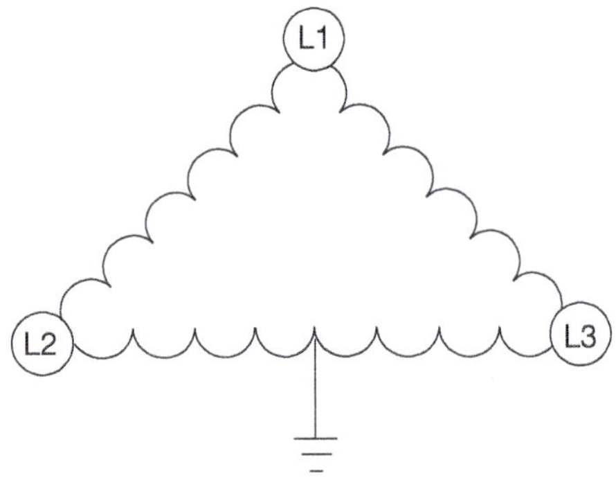

FIGURE 3–2
Low-Voltage Delta Transformer

12. On the delta transformer diagram (Figure 3–2), what voltage reading should be from L1 to L2?
   A. 240 volts
   B. 120 volts
   C. 208 volts
   D. 277 volts

13. On the delta transformer diagram (Figure 3–2), what voltage reading should be from L1 to ground?
   A. 240 volts
   B. 120 volts
   C. 208 volts
   D. 480 volts

14. When placing a voltmeter across any two legs on a 480-volt three-phase system, what voltage reading should be obtained?
   A. 240 volts
   B. 208 volts
   C. 480 volts
   D. 277 volts

15. What is the maximum voltage output of a 240-volt single-phase power supply?
   A. 250 volts
   B. 240 volts
   C. 277 volts
   D. 264 volts

THE FOLLOWING QUESTIONS MAY BE HELPFUL IN STUDYING FOR THESE CERTIFICATION TESTS:
NATE—CR-SV, CR-IN, AC-SV, AC-IN, GS-SV, GS-IN, HP-SV, HP-IN, RC-SV, RC-IN, RL-SV, RL-IN, HO-SV, AND HO-IN.
ESCO INSTITUTE HVAC EXCELLENCE ELECTRICAL

AC MOTORS

1. An AC motor is used to
   A. create electrical energy.
   B. convert electrical energy to mechanical energy.
   C. convert mechanical energy to electrical energy.
   D. create mechanical energy.

2. What is the maximum voltage that can be used on a 120-volt motor?
   A. 108 volts
   B. 120 volts
   C. 132 volts
   D. 135 volts

3. What is the minimum voltage that can be used on a 120-volt motor?
   A. 100 volts
   B. 108 volts
   C. 123 volts
   D. 135 volts

4. The shaft of a motor is connected to the
   A. stator.
   B. shell.
   C. drum.
   D. rotor.

5. The poles of a motor are also known as the
   A. stator.
   B. rotor.
   C. iron core.
   D. armature.

6. Increasing the poles of a motor
   A. increases the speed of the motor.
   B. decreases the speed of the motor.
   C. has no effect on the speed of the motor.
   D. changes the frequency of the motor.

7. A 240-volt, 60-Hz motor with four poles will operate at
   A. 3,600 rpm.
   B. 1,800 rpm.
   C. 1,200 rpm.
   D. 1,050 rpm.

8. A 240-volt, 60-Hz motor with six poles will operate at
   A. 3,600 rpm.
   B. 1,800 rpm.
   C. 1,200 rpm.
   D. 1,050 rpm.

9. The definition of horsepower is
   A. the amount of power to raise 550 lb in 1 s.
   B. the amount of power to raise 550 lb in 1 min.
   C. the amount of power to raise 746 lb in 1 s.
   D. the amount of power to raise 746 lb in 1 min.

10. One horsepower is equivalent to

A. 550 watts.

B. 550 kilowatts.

C. 746 watts.

D. 746 kilowatts.

11. On a single-phase motor, where would the highest resistance be found?

A. run to common

B. start to common

C. run to start

D. T1 to T2

12. On a single-phase motor, where would the lowest resistance be found?

A. run to common

B. start to common

C. run to start

D. T1 to T2

13. When is the start winding removed from the electrical circuit?

A. when the motor reaches 100% of its running speed

B. when the motor reaches 90% of its running speed

C. when the motor reaches 75% of its running speed

D. when the motor reaches 50% of its running speed

14. In a single-phase motor the start winding is

A. 45 degrees out of phase with the run winding.

B. 90 degrees out of phase with the run winding.

C. 120 degrees out of phase with the run winding.

D. in phase with the run winding.

15. In a single-phase motor the start and run winding are wired

A. in parallel with each other.

B. in series with each other.

C. in series parallel with each other.

D. in a delta configuration.

16. When testing the windings of a single-phase motor with an ohmmeter, the meter should always be set on

A. R × 1.

B. R × 100.

C. R × 1,000.

D. R × 10,000.

17. When checking the resistance of a single-phase compressor motor (Figure 3–3), a reading is obtained with an ohmmeter—from T1 to T2 is 3 ohms, from T2 to T3 is 5 ohms, and from T1 to T3 is 8 ohms. Which terminals are run, start, and common?

A. T1 is common. T2 is run. T3 is start.

B. T1 is start. T2 is common. T3 is run.

C. T1 is run. T2 is start. T3 is common.

D. T1 is run. T2 is common. T3 is start.

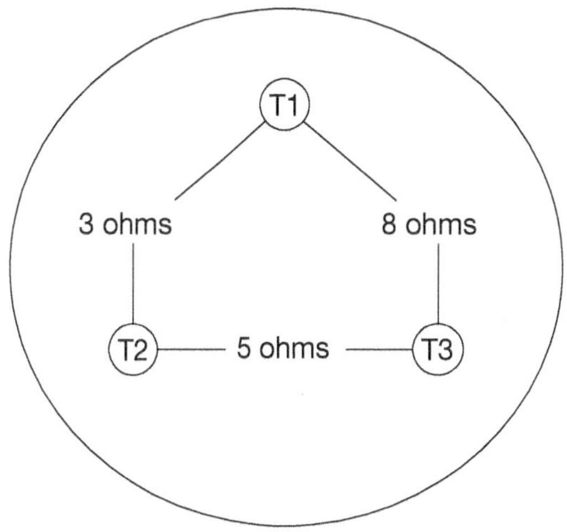

FIGURE 3–3

18. When measuring an open run winding, the ohmmeter will read

A. infinite ohms from start to common.

B. zero ohms from run to common.

C. infinite ohms from run to common.

D. zero ohms from start to common.

19. When measuring resistance on a compressor motor that has a grounded condition, the ohmmeter will read

A. infinite resistance from ground to T1 and T2 on the contactor.

B. a measurable resistance from ground to the bare copper wire at the power supply.

C. infinite resistance from ground to the compressor terminals.

D. a measurable resistance from ground to the compressor terminals.

20. When measuring a shorted start winding, the ohmmeter will read

A. infinite ohms from start to common.

B. lower resistance from run to common than from start to common.

C. infinite ohms from run to common.

D. lower resistance from start to common than from run to common.

21. When troubleshooting a single-phase motor that has a shorted run winding, it is most likely that

A. the motor will hum and not start.

B. there will be voltage to the motor and the motor will not run.

C. the motor will cycle on the external overload.

D. there will be no voltage to the motor.

22. The contacts of a start relay are

A. wired in series with the run winding.

B. wired in series with the start winding.

C. always normally opened.

D. never normally opened.

23. Why are start relays *always* external to a hermetically sealed compressor?

A. Start relays are not *always* external to a hermetically sealed compressor.

B. Arcing could cause an explosion in the compressor.

C. Arcing from the relay could cause acids to form in the compressor.

D. Current will not flow in a sealed system.

24. In what position are the contacts of a centrifugal switch when in use as a start relay?

A. closed until the motor starts, then they open

B. opened until the motor starts, then they close

C. closed and they remain closed until the motor stops

D. opened and they close after the motor has reached 75% of its running speed

25. A current relay operates off the

A. back voltage from the run winding.

B. back voltage from the start winding.

C. amperage from the run winding.

D. amperage from the start winding.

26. A current relay has a set of

A. normally closed contacts in series with the start winding.

B. normally open contacts in parallel with the start winding.

C. normally closed contacts in parallel with the start winding.

D. normally open contacts in series with the start winding.

27. A potential relay has a set of

A. normally closed contacts in series with the start winding.

B. normally open contacts in parallel with the start winding.

C. normally closed contacts in parallel with the start winding.

D. normally open contacts in series with the start winding.

28. A potential relay operates off the

A. back voltage from the run winding.

B. back voltage from the start winding.

C. amperage from the run winding.

D. amperage from the start winding.

29. What terminals connect to the contacts on a potential relay?

A. terminals 1 and 2

B. terminals 1 and 5

C. terminals 2 and 5

D. terminals 4 and 5

30. What terminals connect to the coil on a potential relay?

A. terminals 1 and 2

B. terminals 1 and 5

C. terminals 2 and 5

D. terminals 4 and 5

31. What is connected to terminal 1 on a potential relay?

A. the start winding

B. the run capacitor

C. the start capacitor

D. the common terminal

32. What is connected to terminal 2 on a potential relay?

A. the start winding

B. the run capacitor

C. the start capacitor

D. the common terminal

33. What is connected to terminal 5 on a potential relay?

A. the start winding

B. the run capacitor

C. the start capacitor

D. the common terminal

34. A solid-state relay uses

A. a PTC thermistor in parallel with the start winding.

B. a PTC thermistor in series with the start winding.

C. an NTC thermistor in parallel with the start winding.

D. an NTC thermistor in series with the start winding.

35. What does FLA mean?

A. frequency loaded amperage

B. frequency locked amperage

C. fully locked amperage

D. full-load amperage

36. What does LRA mean?

A. loaded rotor amperage

B. locked rotor amperage

C. loaded running amperage

D. lagged running amperage

37. On a single-phase compressor, external overload protectors are commonly located

A. on the run terminal.

B. on the start terminal.

C. on the common terminal.

D. across the run and common terminals.

38. Overload protectors are designed to sense

A. heat only.

B. current and voltage.

C. heat and current.

D. heat and voltage.

39. The type of overload protector that is located in the windings of a motor and will break the circuit to a contactor coil is a

A. line break external overload.

B. pilot duty external overload.

C. line break internal overload.

D. pilot duty internal overload.

40. The type of overload protector that is located in the windings of a motor and will break the circuit directly to the compressor is a

A. line break external overload.

B. pilot duty external overload.

C. line break internal overload.

D. pilot duty internal overload.

41. The type of overload protector that is located outside of a motor and will break the circuit directly to the contactor coil is a

A. line break external overload.

B. pilot duty external overload.

C. line break internal overload.

D. pilot duty internal overload.

42. When testing a compressor motor with an ohmmeter, what reading would be obtained if its internal overload protector is in the open position?

A. run to start—zero ohms

B. run to common—zero ohms

C. run to start—infinite ohms

D. run to common—infinite ohms

43. Capacitors are commonly rated in

A. microfarads.

B. microamps.

C. milliamps.

D. microhertz.

44. The VAC of a capacitor

A. can be over by 20% of its rating.

B. should never be over its rating.

C. is additive when wired in parallel.

D. is additive when wired in series.

45. What is the purpose of the 20,000-ohm, 2-watt resistor located on a start capacitor?

A. to discharge the capacitor during motor operation

B. to increase the microfarad rating by 20%

C. to level off the sine wave for a much smoother start

D. to discharge the capacitor after motor operation

46. The capacitor analyzer is used to measure

A. shorted or open capacitors and microfarad ratings.

B. microfarad ratings and the capacitor's power factor.

C. open capacitors, microfarad ratings, and the capacitors' power factor.

D. open and shorted capacitors, microfarad ratings, and the capacitor's power factor.

47. What type of relay would remove the start capacitor from the circuit?

A. a switching relay

B. a timed delay relay

C. a potential relay

D. a contactor

48. Which capacitor uses an oil to dissipate heat?

A. the run capacitor

B. the start capacitor

C. the paper-type capacitor

D. the flux capacitor

49. Which capacitor is taken out of the circuit once the motor is running?

A. the start capacitor

B. the run capacitor

C. the dual capacitor

D. a run and start capacitor

50. A run capacitor is in the circuit

A. until the motor reaches 75% of its operating speed.

B. for 2 s.

C. for the full cycle of the motor.

D. until the capacitor is discharged.

51. When wiring two 30-μ run capacitors in parallel, what is the microfarad rating on the complete circuit?

A. 60μ

B. 15μ

C. 90μ

D. 25μ

52. When wiring two 30-μ run capacitors in series, what is the microfarad rating on the complete circuit?

A. 60μ

B. 15μ

C. 90μ

D. 25μ

53. When testing a good capacitor with an analog ohmmeter, what reading should be obtained?

A. The needle should move toward zero and stay there.

B. The needle should stay on infinite resistance.

C. The needle should move toward zero and fall back toward infinite resistance.

D. The needle should move toward zero and fall back to a readable resistance and stay.

54. When testing a shorted capacitor with an analog ohmmeter, what reading should be obtained?

A. The needle should move toward zero and stay there.

B. The needle should stay on infinite resistance.

C. The needle should move toward zero and fall back toward infinite resistance.

D. The needle should move toward zero and fall back to a readable resistance and stay.

55. When testing an open capacitor with an analog ohmmeter, what reading should be obtained?

A. The needle should move toward zero and stay there.

B. The needle should stay on infinite resistance.

C. The needle should move toward zero and fall back toward infinite resistance.

D. The needle should move toward zero and fall back to a readable resistance and stay.

56. Which motor uses no start or run capacitor, but needs a switch to remove the start windings after startup?

A. split phase

B. CS motor

C. CSR motor

D. PSC motor

57. Which motor uses a run capacitor, but has no start capacitor for operation?

A. split phase

B. CS motor

C. CSR motor

D. PSC motor

58. Which motor uses a start capacitor, but has no run capacitor for operation?

A. split phase

B. CS motor

C. CSR motor

D. PSC motor

59. Which motor uses a run capacitor and a start capacitor for operation?

A. split phase

B. CS motor

C. CSR motor

D. PSC motor

60. If a single-phase electric motor is using 4.5 amps in a 220-volt circuit with a power factor of 0.91, it is using power at a rate of

A. 5 watts.

B. 94.5 watts.

C. 942 watts.

D. 1037 watts.

61. A three-phase motor is commonly used when the system requires

A. a motor to have a current draw higher than 15 amps.

B. a motor higher than 5 horsepower.

C. a motor with the voltage higher than 240 volts.

D. a motor with the frequency higher than 50 Hz.

62. How is the direction of a three-phase motor reversed?

A. by reversing any two hot legs

B. by removing and repositioning the rotor

C. by reversing the grounding tap and one hot leg

D. A three-phase motor cannot be reversed.

63. What is commonly used to start a three-phase motor?

A. a potential relay and a start capacitor

B. a current relay and a start capacitor

C. a potential relay

D. a contactor or a starter

64. How many degrees out of phase with each other are the windings of a three-phase motor?

A. 45 degrees

B. 90 degrees

C. 120 degrees

D. The windings are in phase with each other.

65. A PSC motor is a

A. shaded-type motor.

B. capacitor-start/capacitor-run type.

C. single-phase induction type motor.

D. three-phase type motor.

66. PSC motors have

A. run winding only.

B. start winding with a run capacitor.

C. start and run winding.

D. They must use both a run and start capacitors.

67. On a PSC motor, the run capacitor is

A. wired in series with the start winding.

B. wired in series with start capacitor.

C. wired in parallel with the start winding.

D. not used only when a start capacitor is included.

68. How many speeds will a PSC motor have?

A. two or three speeds

B. four speeds

C. can have two to five manually adjustable speeds

D. two speeds with a run and start capacitor

69. How efficient is a PSC motor?

A. 40%–60% efficient

B. 40%–50% efficient

C. 75%–80% efficient

D. 80%–85% efficient

70. What type of bearings do PSC motors have?

A. brass or nickle bearings

B. sleeve or ball bearing type

C. all use sleeve type bearing

D. all use ball bearing type

71. At what lowest range of rpm would sleevetype bearings start to overheat?

A. below 600 rpm

B. below 650 rpm

C. below 700 rpm

D. below 500 rpm

72. What is the typical rpm speed of PSC motors?

A. 400–600 rpm

B. 650–800 rpm

C. 700–825 rpm

D. 800–1,100 rpm

73. Which is the correct formula to calculate rpm?

A. 40 Hz × 120 cycles per second divided by number of poles

B. 60 Hz × 120 cycles per second divided by number of poles

C. 60 Hz × 60 cycles per second divided by the number of poles

D. 60 Hz divided by the number of poles in 120 cycles

74. What is the average constant fan power consumption for a PSC motor?

A. 350–400 watts

B. 515 watts

C. 625 watts

D. 725 watts

75. Variable speed and constant torque motors use a(n)

A. motor start winding, an electronic motor control, and a permanent magnet for the rotor.

B. run capacitor, centrifugal switch, and a permanent magnet for the rotor.

C. temperature fan control, permanent magnet, and a start capacitor.

D. electronic control motor control and run winding.

76. Variable speed and constant torque motors (ECM) are approximately

A. 60% efficient.

B. 70% efficient.

C. 80% efficient.

D. 95% efficient.

77. Variable speed and constant torque motor controllers convert single-phase AC power to

A. three-phase AC power.

B. three phase then back to single phase.

C. three-phase DC power.

D. Only variable motors can be converted to DC power.

78. All new HVAC equipment must be equipped with __________ during manufacturing.

A. ECM motors

B. electronic controls

C. four safety cut-out switches

D. CO-detecting devices

79. What affects the speeds of variable motors?

A. the physical furnace size

B. the horsepower of the motor

C. the type of motor controller

D. static pressure of the system

80. Variable speed motors operate at a low constant airflow that helps to

A. decrease humidity and increases the stagnation of the air.

B. increase humidity and allows for constant changes of airflow.

C. prevent dust collection and stagnation of the air and provides humidity control.

D. keep the motor speed changing from low to high speeds.

81. Variable speed motors are

A. brushless three-phase DC type motors.

B. brushless single-phase DC type motors.

C. brushless three-phase AC type motors.

D. brushless single-phase type motors changing from AC to DC.

82. Variable speed motors can display the actual

A. rpm.

B. cfm.

C. Hp.

D. Hertz.

83. VS motor power consumption is approximately

A. 42 watts.

B. 62 watts.

C. 83 watts.

D. 92 watts.

84. Variable speed motors start unloaded and operate slowly, which

A. will increase locked rotor amps.

B. will decrease locked rotor amps but place more stress on the blower wheel.

C. will keep the locked rotor amps constant with less stress on the blower wheel.

D. will reduce locked rotor amps and place less stress on the blower wheel.

85. Variable speed motors can be set to a specific CFM using a __________ and adjusts using system static pressure.

A. 6-pin harness

B. 8-pin harness

C. 10-pin harness

D. 16-pin harness

86. What type of a pin power supply connection does a variable speed and constant torque motor have?

A. 3-pin connection

B. 4-pin connection

C. 5-pin connection

D. 8-pin connection

87. What is the speed range of VS motors?

A. 100–180 rpm

B. 200–1,300 rpm

C. 800–1,200 rpm

D. 1200–1,600 rpm

88. Constant torque motors use

A. a 4-pin harness and operate on either 120 V or 230 V.

B. a 6-pin harness and operate on 120 V only.

C. an 8-pin harness and operate on 230 V only.

D. a 12-pin harness and operate on 120 V or 230 V.

89. What type of motor is a constant torque motor?

A. a brushless single-phase DC type

B. a brushless three-phase DC type

C. a brushless PSC 230 V type

D. a brushless CS/CR 230 V type

90. What are the output ranges for single-phase PSC, CT, and VS motors?

A. 1/3, $\frac{1}{2}$, and $\frac{3}{4}$ HP

B. 1/3 and $\frac{1}{2}$ HP

C. 1/3, $\frac{1}{2}$, $\frac{3}{4}$, and 1 HP

D. 1/3 and 1 HP

91. Input single-phase voltages for PSC, CT, and VS motors are

A. 120 V or 230 V.

B. 120 V only.

C. 230 V only.

D. 208 V only.

92. Speed ranges for CT motors are

A. 300–400 rpm.

B. 400–600 rpm.

C. 500–800 rpm.

D. 600–1,100 rpm.

93. What are the torques for CT motors that are preprogrammed?

A. 2–5 torques

B. 5–8 torques

C. 9–10 torques

D. 10–12 torques

94. VS motors are programmed with

A. temperature sensors.

B. bimetal sensing controls.

C. adjustable variable airflow or torque.

D. adjustable airflow that will increase humidity.

THE FOLLOWING QUESTIONS MAY BE HELPFUL IN STUDYING FOR THESE CERTIFICATION TESTS:
NATE—CR-SV, CR-IN, AC-SV, AC-IN, HP-SV, HP-IN, HG-SV, HG-IN, GS-SV, GS-IN, OL-SV, OL-IN, AD-SV, AD-IN.
ESCO INSTITUTE HVAC EXCELLENCE ELECTRICAL AND ELECTRIC HEAT

ELECTRICAL CONTROLS

1. A fuse is used in an electrical circuit to

A. protect the inductive load from an overcurrent condition.

B. protect against a ground fault condition.

C. protect the circuit wires from an overcurrent condition.

D. protect against open circuits.

2. How are fuses rated?

A. in amps only

B. in volts only

C. in amps with a minimum voltage rating

D. in amps with a maximum voltage rating

3. What will an ohmmeter measure when testing an open fuse?

A. zero resistance

B. infinite resistance

C. measurable resistance

D. continuity

4. What does the symbol in Figure 3–4 represent?

A. a cooling thermostat

B. a low-pressure switch

C. a heating thermostat

D. a high-pressure switch

5. What does the symbol in Figure 3–5 represent?

A. a reverse-acting pressure switch

B. a low-pressure switch

C. a heating thermostat

D. a high-pressure switch

6. What does the symbol in Figure 3–6 represent?

A. a cooling thermostat

B. a low-pressure switch

C. a heating thermostat

D. a high-pressure switch

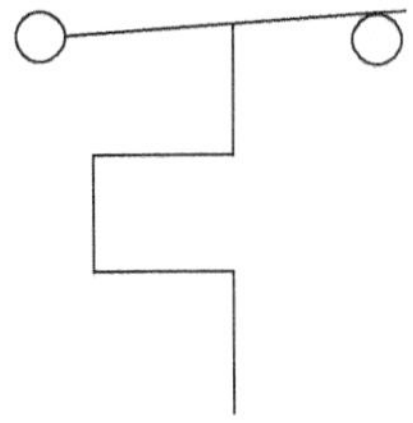

FIGURE 3–4

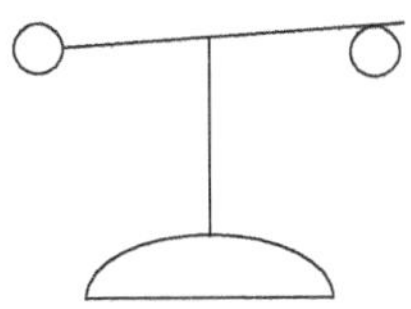

FIGURE 3–5

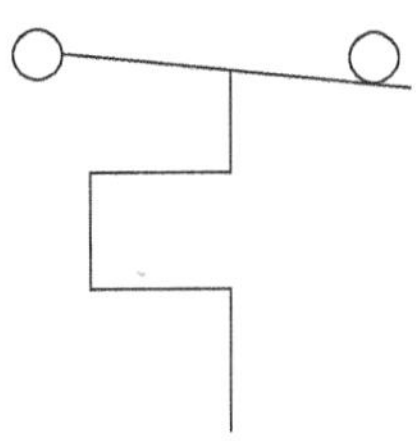

FIGURE 3–6

7. What does the symbol in Figure 3–7 represent?
   - **A.** a cooling thermostat
   - **B.** a low-pressure switch
   - **C.** a heating thermostat
   - **D.** a high-pressure switch
8. What does the symbol in Figure 3–8 represent?
   - **A.** a cooling thermostat
   - **B.** a thermal fuse
   - **C.** a limit switch
   - **D.** a fusible link
9. What does the symbol in Figure 3–9 represent?
   - **A.** a normally open delay on a make relay
   - **B.** a normally open delay on a break relay
   - **C.** a normally open thermal relay
   - **D.** a normally open timer contact
10. What does the symbol in Figure 3–10 represent?
    - **A.** a normally open delay on a make relay
    - **B.** a normally open delay on a break relay
    - **C.** a normally open thermal relay
    - **D.** a normally open timer contact

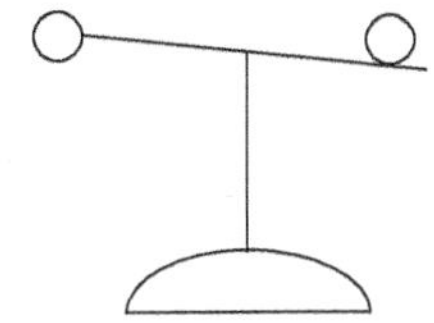

FIGURE 3–7

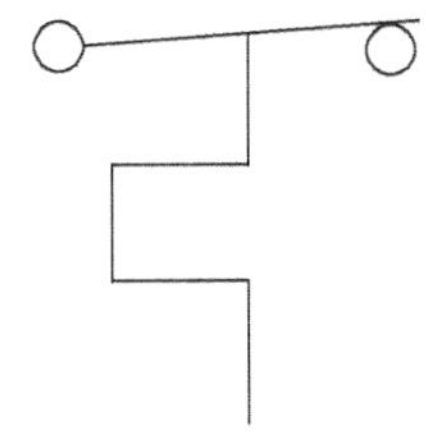

FIGURE 3–8

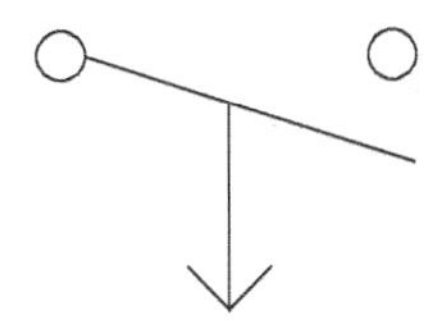

FIGURE 3–9

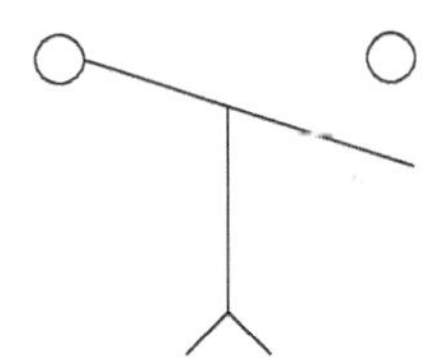

FIGURE 3–10

11. What type of sensing element is commonly found in fan and limit controls?
    - **A.** copper
    - **B.** helical invar
    - **C.** bimetal
    - **D.** mercury
12. What is used to sense space temperature in electronic temperature controls?
    - **A.** resistors
    - **B.** transistors
    - **C.** capacitors
    - **D.** thermistors

13. What is known about a liquid-charged remote-type temperature control?

A. Its bulb is both extremely sensitive and completely full of liquid.

B. Its bulb is extremely sensitive, has some gas present, and always has liquid present.

C. Its bulb is not extremely sensitive and is completely full of liquid.

D. Its bulb is not extremely sensitive, has some gas present, and always has liquid present.

14. When installing _______, its sensing bulb must always be the coldest part of the control.

A. the gas-charged remote temperature control

B. the cross-charged remote temperature control

C. the liquid-charged remote temperature control

D. the bimetal-type temperature control

15. A temperature control that uses a positive temperature coefficient thermistor (PTCT)

A. decreases resistance as temperature increases.

B. increases resistance as temperature increases.

C. must be level for proper operation.

D. must be installed near a supply air register for proper operation.

16. A heating thermostat

A. opens on a drop in temperature.

B. opens on a rise in temperature.

C. energizes a heating system when its set point is below space temperature.

D. de-energizes a heating system when its set point is above space temperature.

17. The heat anticipator

A. must be energized when the system is de-energized.

B. controls system lag.

C. controls system overshoot.

D. is wired in parallel with the heating thermostat.

18. When adjusting a heat anticipator on a thermostat,

A. the heating and cooling control amperage must be measured.

B. the heating and cooling control voltage must be measured.

C. the cooling control amperage must be measured.

D. the heating control amperage must be measured.

19. Choose the correct color code and its circuit for a heating and cooling thermostat.

A. Red—heating circuit, Yellow—cooling circuit, White—power source, Green—fan relay

B. Red—fan relay, Yellow—power source, White—heating circuit, Green—cooling circuit

C. Red—power source, Yellow—heating circuit, White—cooling circuit, Green—fan relay

D. Red—power source, Yellow—cooling circuit, White—heating circuit, Green—fan relay

20. How many wires does a two-stage heating and two-stage cooling thermostat commonly use?

A. 4

B. 6

C. 8

D. 10

21. What is the purpose of automatic changeover on a heating and cooling thermostat?

A. It automatically changes the thermostat's set point.

B. It automatically changes the thermostat's anticipation value.

C. It automatically changes from heating to cooling.

D. It automatically changes from first-stage heating to second-stage heating.

22. What is the definition of deadband?

A. Deadband is the point at which the thermostat is satisfied when in the heating cycle.

B. Deadband is the point at which the thermostat is satisfied when in the cooling cycle.

C. Deadband is the point at which the thermostat is satisfied when in the automatic cycle.

D. Deadband is the point at which the thermostat is satisfied between stages when using a two-stage thermostat.

23. An outdoor thermostat that is used on an electric heating system is commonly used to

A. energize electric heaters if the outdoor temperature drops below a predetermined temperature.

B. de-energize the system if the temperature rises above a predetermined temperature.

C. switch the system to emergency heat if the temperature drops below a predetermined temperature.

D. energize electric heaters if the outdoor temperature rises above a predetermined temperature.

24. A cooling thermostat

A. opens on a drop in temperature.

B. opens on a rise in temperature.

C. energizes a cooling system when its set point is above space temperature.

D. de-energizes a cooling system when its set point is below space temperature.

25. A cooling anticipator

A. must be de-energized when the system is de-energized.

B. controls system lag.

C. controls system overshoot.

D. is wired in series with the cooling thermostat.

26. Outdoor thermostats are used on air conditioning systems to

A. prevent short cycling of the compressor during the peak season.

B. prevent short cycling of the heating system during the air conditioning season.

C. de-energize the compressor during low ambient conditions.

D. energize the condenser fans during low ambient conditions.

27. What type of sensing element is the most accurate?

A. a helical bimetal

B. a refrigerant-filled sensing element

C. a snap-disc bimetal

D. a thermistor

28. What type of sensing element is used in a programmable thermostat?

A. a resistor

B. a thermistor

C. a transistor

D. a bimetal

29. What is known about the anticipators that are located on a programmable thermostat?

A. The heating has a fixed setting and the cooling can either be fixed or adjustable.

B. The heating either has a fixed setting or can be adjustable and the cooling has a fixed setting.

C. The heating has a fixed setting only and the cooling is also a fixed setting.

D. The heating has a fixed setting only and the cooling is adjustable.

30. When using a heat-pump-type programmable thermostat, what cycle commonly uses an LED light?

A. the heating cycle

B. the cooling cycle

C. the emergency heat cycle

D. the first- and second-stage heating and cooling cycles

31. When installing a programmable thermostat, what extra low-voltage wire is commonly added to the circuit?

A. a second power supply wire

B. the emergency heat circuit

C. the secondary blower circuit

D. the common from the transformer

32. How does a two-stage thermostat control a heat pump during the heating cycle?

A. The compressor is controlled by the first stage, and the auxiliary heaters are controlled by the second stage.

B. The auxiliary heaters are controlled by the first stage, and the compressor is controlled by the second stage.

C. The compressor and the auxiliary heaters are controlled by the first stage, and the gas-fired system is controlled by the second stage.

D. The gas-fired system is controlled by the first stage, and the compressor and auxiliary heaters are controlled by the second stage.

33. What is a battery used for on a programmable thermostat?

A. to energize the thermostat's LED lights

B. to control the automatic changeover cycle

C. to save the system's program if there is a loss of power

D. to energize the heat and cool anticipators

34. When using a programmable thermostat, what is "temperature setback"?

A. It is the temperature setting of the thermostat during an occupied time.

B. It is the temperature setting of the thermostat during an unoccupied time.

C. It is the point at which the temperature set point of the thermostat is between the heating and cooling cycle.

D. It is the point at which the thermostat's anticipation rate is set lower than the actual system current draw.

35. What is meant by the term *intelligent recovery*?

A. It is the feature that calculates the room temperature every 2 to 3 h.

B. It is the use of thermostat anticipation for the increase of energy conservation.

C. It is the gradual return of space temperature to the system's set point after a system setback.

D. None of the above

36. The override feature on a programmable thermostat

A. is used to temporarily bypass the heating control circuit.

B. is used to temporarily bypass the cooling control circuit.

C. is commonly accomplished by using the override or hold button.

D. All of the above

37. What is the purpose of the limit control on a heating system?

A. It controls the blower motor.

B. It limits the amount of gas that is brought to the heat exchanger.

C. It energizes the gas valve when the thermostat calls for heat and monitors space temperature.

D. It de-energizes the gas valve if the temperature of the heat exchanger becomes excessive.

38. The limit switch that is used on a forced-air gas-fired furnace is commonly

A. a closed bimetal switch.

B. an opened bimetal switch.

C. a closed pressure-sensitive switch.

D. an opened pressure-sensitive switch.

39. A limit switch that is used on a forced-air gas-fired furnace

A. can only be an automatic type of a reset.

B. can only be a manual type of a reset.

C. can be an automatic or a manual type of a reset.

D. must be replaced if the switch opens to the circuit.

40. What is the purpose of a low-pressure switch when it is used in an air conditioning system?

A. It makes the circuit to the compressor if the low-side pressure drops too low.

B. It breaks the circuit to the compressor if the high-side pressure drops too low.

C. It makes the circuit to the compressor if the high-side pressure drops too low.

D. It breaks the circuit to the compressor if the low-side pressure drops too low.

41. The low-pressure switch will

A. close on a rise in pressure and open on a drop in pressure.

B. open on a rise in pressure and close on a drop in pressure.

C. close if the system has lost its refrigerant charge.

D. open if the condenser has a loss of airflow.

42. What is the purpose of the high-pressure switch when it is used in an air conditioning system?

A. It makes the circuit to the compressor if the low-side pressure rises too high.

B. It breaks the circuit to the compressor if the high-side pressure rises too high.

C. It makes the circuit to the compressor if the high-side pressure rises too high.

D. It breaks the circuit to the compressor if the low-side pressure rises too high.

43. The high-pressure switch will

A. close on a rise in pressure and open on a drop in pressure.

B. open on a rise in pressure and close on a drop in pressure.

C. close if the system has an excessive refrigerant charge.

D. open if the air filter near the indoor coil becomes plugged.

44. A reverse-acting high-pressure switch is used to

A. energize the compressor when the pressure of the system is within operable limits.

B. de-energize the compressor when the system's pressure becomes excessive.

C. de-energize the condenser fans on a decrease of ambient temperature.

D. de-energize the condenser fans on an increase of ambient temperature.

45. What is the purpose of a relay?

A. It converts a power source from a high voltage to a low voltage.

B. It controls a higher-voltage circuit by a signal from a lower-voltage circuit.

C. It can only be used to control motor circuits.

D. It changes an AC power supply to a DC power supply by the signal from a low-voltage power supply.

46. The contacts of a relay must never exceed

A. 5 amps.

B. 10 amps.

C. 20 amps.

D. 40 amps.

47. When measuring the resistance of a relay coil and an infinite resistance is found, what is the problem with the relay?

A. The relay is in operable condition.

B. The relay coil is shorted.

C. The relay coil has continuity.

D. The relay coil is burnt open.

48. What must be taken into consideration when replacing a relay?

A. Replace the relay with the same coil voltage and switch contact configuration.

B. Replace the relay with the same coil amperage, contact voltage rating, and switch contact configuration.

C. Replace the relay with the same coil resistance, contact voltage rating, and mounting bracket.

D. Replace the relay with a higher coil resistance, the same amperage rating, and always add contacts for added load demands.

49. What is the major difference between a relay and a contactor?

A. A relay coil voltage is 24 volts and a contactor can be 24 or 120 volts.

B. A contactor has more resistance across its coil.

C. A contactor contact rating is higher than that of a relay.

D. A contactor can be used in motor circuits and a relay cannot.

50. When testing a contactor with an ohmmeter, a good operating contactor will read

A. zero resistance across the coil.

B. measurable resistance across the coil.

C. infinite resistance across the coil.

D. measurable resistance across the contacts.

51. What is the major difference between a contactor and a starter?

A. A starter incorporates built-in overload protectors and a contactor does not.

B. A starter is used on loads that are 460 volts, and a contactor is used on loads that are 220 volts.

C. The starter coil voltage ranges from 220 to 460 volts, and a contactor coil voltage is 24 volts.

D. The contacts of a contactor will change positions immediately after the coil is energized and the changing of the starter contacts is delayed.

52. When measuring the resistance of a starter coil, zero resistance is found. What is the problem with the starter?

A. The starter is in operable condition.

B. The starter coil is shorted.

C. The starter coil has no continuity.

D. The starter coil is burnt open.

53. Which type of overload protector is *not* incorporated in a starter?

A. a bimetal relay

B. a thermal relay

C. a molten-metal relay

D. a potential relay

54. When a timed delay relay is used during the heating cycle, what part of the circuit does it commonly control?

A. the gas valve circuit

B. the thermostat circuit

C. the blower circuit

D. the accessories circuit

55. Why would a timed delay relay circuit be used on an air conditioning system?

A. to prevent short cycling of the condenser fans

B. to prevent short cycling of the compressor

C. to prevent short cycling of the indoor blower

D. to delay the operation of the condenser fan until high-side pressure is established

56. What type of time delay is commonly used to de-energize a blower on a forced-air heating system?

A. a normally closed delay on make

B. a normally open delay on make

C. a normally closed delay on break

D. a normally open delay on break

57. What is the purpose of a sequencer?

A. It reduces inrush current by sequentially energizing and de energizing loads in an HVAC system.

B. It reduces holding current by sequentially energizing and de-energizing loads in an HVAC system.

C. It reduces inrush voltage by sequentially energizing and de-energizing loads in an HVAC system.

D. It reduces resistance by sequentially energizing and de-energizing loads in an HVAC system.

58. Sequencers are commonly found on

A. residential gas-fired furnaces.

B. commercial rooftop air conditioning systems.

C. residential air conditioning systems.

D. electric heating systems.

59. What component is often used in an electrical circuit as a sequencer?

A. a defrost timer

B. a magnetic switching relay

C. a microprocessor

D. a contactor

60. On what system is a defrost timer commonly found?

A. a heat pump system

B. a commercial air conditioning system

C. a residential air conditioning system

D. a cooling tower application

61. What are two methods that are used to control a defrost cycle?

A. defrost timers and switching relays

B. defrost timers and timed delay relays

C. defrost timers and sequencers

D. defrost timers and microprocessors

62. What is the purpose of a solenoid valve?

A. It controls the flow of current through a circuit.

B. It controls the flow of a fluid.

C. It allows a technician to read pressures on a manifold gauge set.

D. It controls the amount of refrigerant in an evaporator coil.

63. An automatic pump-down system uses

A. a high-pressure switch, a thermostat, and a solenoid valve in its circuit.

B. a high-pressure switch, a compound gauge, and a solenoid valve in its circuit.

C. a low-pressure switch, a solenoid valve, and a thermostat in its circuit.

D. a thermostat, a solenoid valve, and a reversing relay in its circuit.

64. What component is commonly used to control refrigerant flow in a heat pump system?

A. a reversing valve solenoid

B. a hot gas solenoid

C. a liquid-line solenoid

D. a bypass solenoid

65. What component is used to protect the compressor during low outdoor ambient conditions?

A. a subcooler

B. a crankcase heater

C. a thermal electric expansion valve

D. a de-superheat coil

66. What does the symbol in Figure 3–11 represent?

A. a fusible link

B. a thermal overload

C. an electric heater

D. a timed delay relay

FIGURE 3–11

67. What is the purpose of the crankcase heater?
    - A. to increase oil viscosity during low ambient conditions
    - B. to increase the efficiency of a system during low ambient conditions
    - C. to prevent refrigerant migration into the compressor during low ambient conditions
    - D. to decrease oil viscosity for proper lubrication of internal components
68. What is the Btu output of an electric heater that is rated at 1,500 watts?
    - A. 5,118 Btu
    - B. 4,710 Btu
    - C. 478 Btu
    - D. 440 Btu
69. What controls the electric heating elements in an air-to-air heat pump system?
    - A. the first stage of a two-stage thermostat
    - B. the second stage of a two-stage thermostat
    - C. the reversing valve relay
    - D. the anticipator relay
70. Where are auxiliary strip heaters commonly located on an air-to-air heat pump?
    - A. in the return air duct before the indoor blower
    - B. in the return air duct after the indoor blower
    - C. in the supply air plenum before the indoor coil
    - D. in the supply air plenum after the indoor coil
71. What does the symbol in Figure 3–12 represent?
    - A. a fuse
    - B. a fusible link
    - C. a thermal overload
    - D. a bimetal sensor
72. What component is commonly used as a flame roll-out sensor on a gas-fired furnace?
    - A. a diode
    - B. a transistor
    - C. an electric heating element
    - D. a fusible link

FIGURE 3–12

73. How is a fusible link wired when it is in an electric heating system?
    - A. in parallel with the heating element
    - B. in series with the heating element
    - C. in parallel with the heating element relay coil
    - D. in series with the heating element relay coil
74. A step-down transformer is designed to
    - A. take a lower amperage and increase it to a higher amperage.
    - B. take a higher amperage and decrease it to a lower amperage.
    - C. take a lower voltage and increase it to a higher voltage.
    - D. take a higher voltage and decrease it to a lower voltage.
75. On a step-up transformer, the turn ratio from the primary side to the secondary side will be
    - A. higher on the primary than on the secondary.
    - B. lower on the primary than on the secondary.
    - C. the same on the primary as on the secondary.
    - D. the same as a step-down transformer.
76. If a transformer uses a 240-volt primary winding that has 400 turns and a secondary that has 200 turns, what would be the voltage of the secondary winding?
    - A. 200 volts
    - B. 120 volts
    - C. 480 volts
    - D. None of the above
77. When replacing a 24-volt control transformer that has the VA rating of 40, what is the maximum amperage of the control circuit?
    - A. .06 amp
    - B. .6 amp
    - C. 1.67 amps
    - D. 2 amp
78. When there is voltage at the primary side of a transformer and no voltage at the secondary, what could be the problem with the transformer?
    - A. a bad secondary winding only
    - B. a bad primary winding only
    - C. a bad secondary or primary winding
    - D. a bad fuse at the power supply
79. When measuring an open secondary winding of a transformer, what measurement will be obtained?
    - A. an infinite resistance
    - B. a measurable resistance
    - C. a zero resistance
    - D. continuity

THE FOLLOWING QUESTIONS MAY BE HELPFUL IN STUDYING FOR THESE CERTIFICATION TESTS:
NATE—CR-SV, CR-IN, AC-SV, AC-IN, HP-SV, HP-IN, GS-SV, GS-IN, HG-SV, HG-IN, AD-SV, AD-IN, OL-SV, OL-IN.
ESCO INSTITUTE HVAC EXCELLENCE ELECTRICAL

SOLID-STATE ELECTRONICS

1. What meter must *never* be used when testing a circuit board?

A. an ammeter

B. a voltmeter

C. an ohmmeter

D. an oscilloscope

2. What is the purpose of a diode?

A. It converts direct current into alternating current.

B. It is used to increase the resistance of an electrical circuit.

C. It stores electrons in an electrical circuit.

D. It allows current to flow in one direction.

3. Which component is commonly used as a solid-state relay?

A. a transistor

B. a thermistor

C. a diode

D. a cathode

4. What solid-state component decreases its resistance as its temperature increases?

A. a cathode

B. an anode

C. a PTC thermistor

D. a NTC thermistor

5. Which component is used to store electrons in an electrical circuit?

A. a capacitor

B. a thermistor

C. a diode

D. a transistor

6. Which component is primarily used to drop voltage in an electrical circuit?

A. a diode

B. a resistor

C. a rectifier

D. a transistor

7. What component is used to convert alternating current to direct current?

A. a capacitor

B. a transistor

C. a rectifier

D. a thermistor

8. What component blocks DC and passes AC?

A. a cathode

B. an anode

C. a rectifier

D. a capacitor

THE FOLLOWING QUESTIONS MAY BE HELPFUL IN STUDYING FOR THESE CERTIFICATION TESTS:
NATE—CR-SV, CR-IN, AC-SV, AC-IN, HP-SV, HP-IN, GS-SV, GS-IN, HG-SV, HG-IN, OL-SV, OL-IN, HO-SV, AND HO-IN.
ESCO INSTITUTE HVAC EXCELLENCE ELECTRICAL

WIRING LAYOUTS AND ELECTRICAL DIAGRAMS

1. On a forced-air furnace, what loads are commonly energized with a 120-volt power supply?

A. the indoor blower, induced draft motor, and the transformer's secondary winding

B. the induced draft motor, the gas valve, and the indoor blower motor

C. the electronic air cleaner, the induced draft relay, and the humidifier motor

D. the induced draft motor, the indoor blower motor, and the transformer's primary winding

2. What is the proper method for connecting any electrical wires located in a junction box?

A. twisting together and connecting with a wirenut

B. twisting together and connecting with electrical tape

C. soldering together and covering with electrical tape

D. Wires must never be connected together in a junction box.

3. On a split-system condensing unit, what loads are commonly energized with a 240-volt power supply?

A. the compressor, the contactor, and the condenser fan

B. the condenser fan, the reverse-acting high-pressure control, and the compressor

C. the crankcase heater, the condenser fan cycle guard, and the compressor contactor

D. the compressor, the crankcase heater, and the condenser fan

4. On an HVAC package unit, what loads are commonly energized with a 240-volt power supply?
   A. the transformer's secondary winding, the contactor, the condenser fan, and the compressor
   B. the condenser fan, the induced draft motor, the compressor, and the indoor air blower
   C. the crankcase heater, the condenser fan relay coil, the compressor, and the thermostat
   D. the compressor, the crankcase heater, the condenser fan, and the fan relay coil
5. What is the control transformer output voltage for a residential split system?
   A. 24 volts
   B. 120 volts
   C. 208/240 volts
   D. 240 volts
6. What is the control transformer input voltage for a residential split system?
   A. 24 volts
   B. 120 volts
   C. 208/240 volts
   D. 240 volts
7. What is the control transformer input voltage for an HVAC package system?
   A. 24 volts
   B. 120 volts
   C. 208/240 volts
   D. 600 volts
8. What is the purpose of a schematic diagram?
   A. It shows the technician the physical positions of the system's components.
   B. It shows the technician the proper connections for the power supply.
   C. It shows the technician the color code of the wiring and the wire connections.
   D. It shows the technician the operational sequence of the system.
9. What do the vertical lines represent in an electrical schematic diagram?
   A. the power supply to the system
   B. the individual circuits in the electrical system
   C. all energized circuits
   D. all de-energized circuits
10. What do the horizontal lines represent in an electrical schematic diagram?
   A. the power supply to the system
   B. the individual circuits in the electrical system
   C. all energized circuits
   D. all de-energized circuits
11. A legend is used
   A. to explain the operational sequence in an electrical schematic.
   B. to identify the label of an electrical component.
   C. to locate electrical components in a wiring diagram.
   D. All of the above
12. The top portion of an electrical schematic is commonly
   A. the low-voltage circuit.
   B. the control voltage circuit.
   C. the line voltage circuit.
   D. the ignition voltage circuit.
13. The bottom portion of an electrical schematic is commonly
   A. the high-voltage circuit.
   B. the control voltage circuit.
   C. the line voltage circuit.
   D. the ignition voltage circuit.
14. What component is located between the high-voltage circuit and the low-voltage circuit?
   A. the control relay
   B. the contactor
   C. the transformer
   D. the potential relay
15. A pictorial diagram is used to
   A. show the system's operational sequence.
   B. show the actual wiring and location of system components.
   C. show the wire and fuse size.
   D. show the placement of the thermostat and the power supply wiring.
16. Which diagram is also known as a label or line diagram?
   A. an installation diagram
   B. a factual diagram
   C. a schematic diagram
   D. a pictorial diagram
17. Which diagram consists of an electrical schematic and a pictorial diagram?
   A. a factual diagram
   B. an installation diagram
   C. a label diagram
   D. a system diagram
18. Which diagram displays the system's fuse size and wire size, and shows the placement of the thermostat and power wiring?
   A. the schematic diagram
   B. the line diagram
   C. the installation diagram
   D. the ladder diagram

19. The heavy dashed lines on an installation diagram show

A. the line voltage field wiring.

B. the low-voltage field wiring.

C. the line voltage factory wiring.

D. the low-voltage factory wiring.

20. The heavy solid lines on an installation diagram show

A. the line voltage field wiring.

B. the low-voltage field wiring.

C. the line voltage factory wiring.

D. the low-voltage factory wiring.

21. The light dashed lines on an installation diagram show

A. the line voltage field wiring.

B. the low-voltage field wiring.

C. the line voltage factory wiring.

D. the low-voltage factory wiring.

22. The light solid lines on an installation diagram show

A. the line voltage field wiring.

B. the low-voltage field wiring.

C. the line voltage factory wiring.

D. the low-voltage factory wiring.

23. What is the purpose of the control labeled PR in Figure 3–13?

A. It directs current through the compressor's run winding.

B. It de-energizes the compressor if there is an increase of current through the compressor circuit.

C. It de-energizes the start capacitor after the compressor starts.

D. It energizes the run capacitor.

24. In Figure 3–13, what will cause the control ATH to close the circuit to the low speed?

A. an increase in ambient temperature

B. an increase in condenser pressure

C. a decrease in condenser pressure

D. a decrease in ambient temperature

25. In Figure 3–13, what will cause the compressor and condenser fan to become inoperative?

A. when CTH and LPS are in the closed position

B. when either CTH or LPS is in the opened position

C. when the pressure in the condenser reaches an unsafe level

D. when FR1 and FR2 have changed positions

26. In Figure 3–13, what will cause switch FS to close?

A. when terminals R and G close on the thermostat

B. when FR2 changes position

C. when the heat exchanger temperature increases

D. when the outdoor ambient temperature increases

27. In Figure 3–13, how many thermally activated switches are present?

A. 3

B. 9

C. 1

D. 2

28. In Figure 3–14, what switches must be in the closed position before the ignition system will energize?

A. FR1, IDR, PS

B. Fan Auto/On, PS

C. TH, PS, HL, SL

D. TH, Fan Auto/On, IDR

29. In Figure 3–14, what causes PS to close?

A. PV gas pressure

B. IDM pressure

C. MV gas pressure

D. IBM pressure

30. In Figure 3–14, with the Auto/On switch closed, what will happen to the fan circuit?

A. The FS switch changes position and operates the fan in the low speed.

B. The IDR switch changes position and operates the IDM.

C. The FR1 closes and operates the IBM on the high speed.

D. The FR2 closes and operates the IBM on the low speed.

31. In Figure 3–14, how many relay coils are being used?

A. 7

B. 5

C. 3

D. 1

32. How many switches are shown in Figure 3–14?

A. 8

B. 7

C. 6

D. 5

33. In Figure 3–15, what voltage should be present when testing across terminals TH and TR?

A. 240 volts

B. 120 volts

C. 24 volts

D. 0 volts

34. In Figure 3–15, when the thermostat is calling for heat, what two terminals should read 24 volts?

A. W and C

B. R and W

C. Gas 2 and C

D. GC and C

35. In Figure 3–15, what two terminals should be used to test for voltage to the fan motor during the heating season?

A. PR1 and PR2

B. COM and LO

C. COM and HI

D. Gas 1 and Gas 2

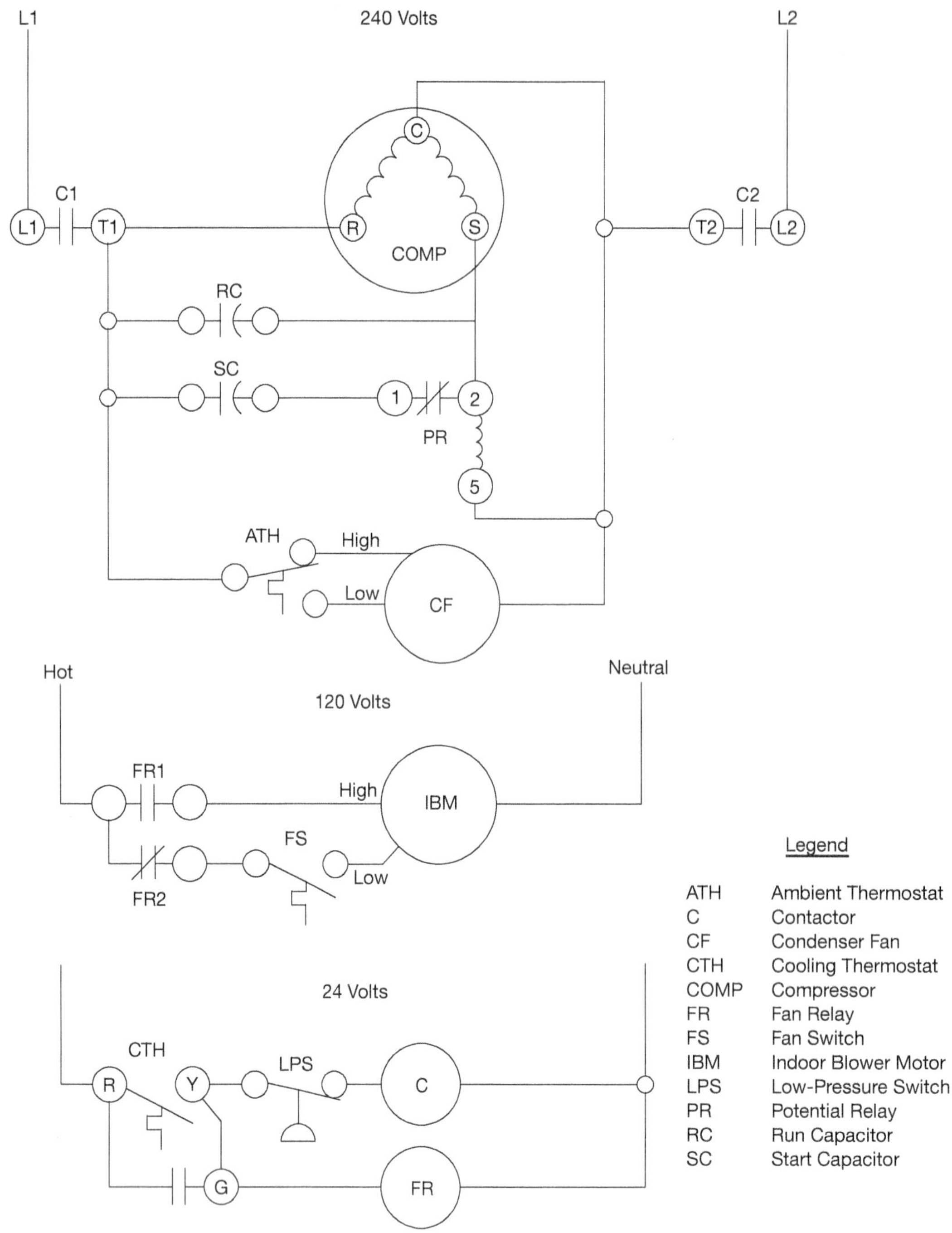

FIGURE 3–13

36. In Figure 3–15, the pilot is lit and the main gas valve is not energized. What action could be taken to test the operation of the main gas valve?

A. Bypass the yellow and the white wire on the sensor.

B. Bypass the yellow and the green wire on the sensor.

C. Bypass GAS 3 and GAS 2.

D. Bypass GAS 1 and GAS 2.

37. In Figure 3–15, what terminals are used to test the main gas valve for voltage?

A. GAS 2 and Ground

B. GAS 1 and GAS 2

C. GAS 1 and GAS 3

D. GAS 2 and GAS 3

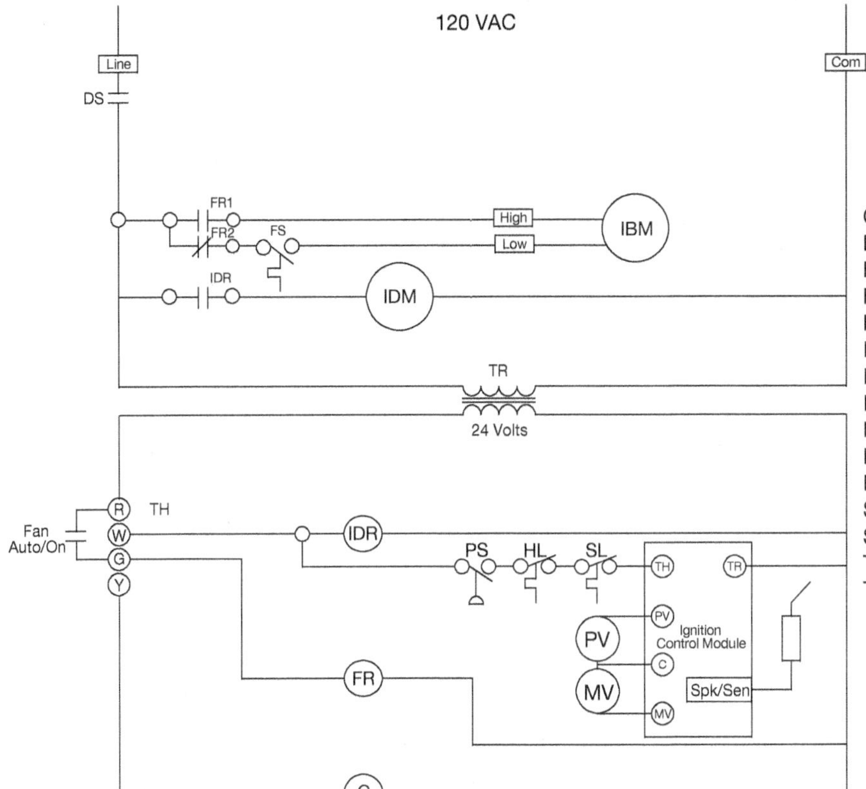

Legend

C	Contactor
DS	Door Switch
FR	Fan Relay
FS	Fan Switch
HL	High Limit
IBM	Indoor Blower Motor
IDM	Induced Draft Motor
IDR	Induced Draft Relay
MV	Main Valve
PS	Pressure Switch
PV	Pilot Valve
SL	Secondary Limit
Spk/Sen	Spark Ignitor/Flame Sensor
TH	Thermostat
TR	Transformer

FIGURE 3–14

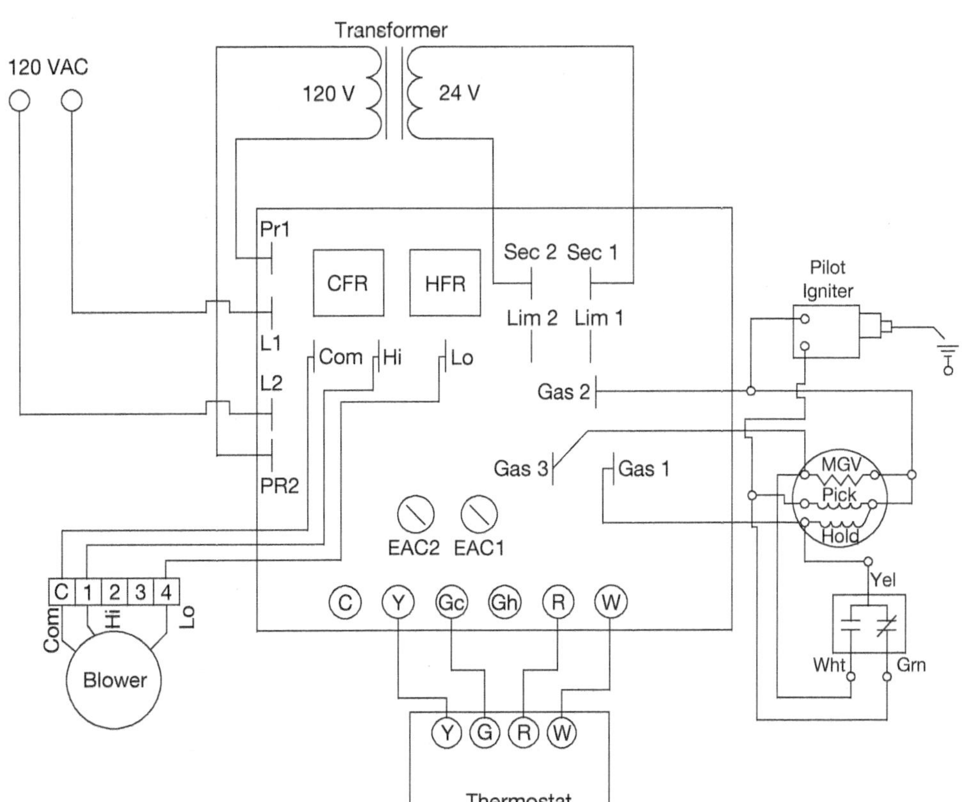

FIGURE 3–15

38. In Figure 3–16, what switches must be closed to energize the orange wire?

A. DISC, LIM, and FS

B. DISC, LIM, and IT

C. DISC, TH, and LIM

D. DISC and TH

39. In Figure 3–16, what can cause the system not to fire on startup?

A. The fan switch is in the closed position.

B. The cad-cell is positioned in the dark.

C. The cad-cell is positioned in the light.

D. Contacts R and W on TH are in the closed position.

40. In Figure 3–16, what is the first test that should be made when troubleshooting the system?

A. Test for resistance across the orange and white wires.

B. Test for voltage across the FF terminals.

C. Test for resistance across the TT terminals.

D. Test for voltage across the black and white wires.

41. In Figure 3–17, what is the purpose of the holding relay?

A. It holds in the emergency heat circuit.

B. It de-energizes the reversing valve circuit during a fault condition.

C. It de-energizes the contactor and energizes a light at the thermostat.

D. It de-energizes the control relay during a fault condition.

42. In Figure 3–17, what loads are controlled by the defrost relay, DR?

A. RV, ADR, APS, and TD1

B. ADR, TM, RV, and TD2

C. RV, ADR, APS, and TD1

D. ODT, TD2, TM, and RV

43. In Figure 3–17, what loads are energized during the emergency heating cycle if the outdoor thermostat ODT is closed from terminals 1 and 2?

A. TD1, H1, H2, FR, IFM, and R

B. TM, TD1, TD2, IFM, H1, H2, and H3

C. COMP, CF, RV, FR, IFM

D. ADR, FR, IFM, and H1

44. In Figure 3–17, when is the reversing valve, RV, energized?

A. during defrost only

B. during first-stage cooling only

C. during defrost and first-stage cooling

D. during defrost, first-stage cooling, and second-stage cooling

45. In Figure 3–17, what does the outdoor thermostat, ODT, control?

A. TD1 and TD2

B. FR and TD1

C. TD2

D. FR, TD1, and TD2

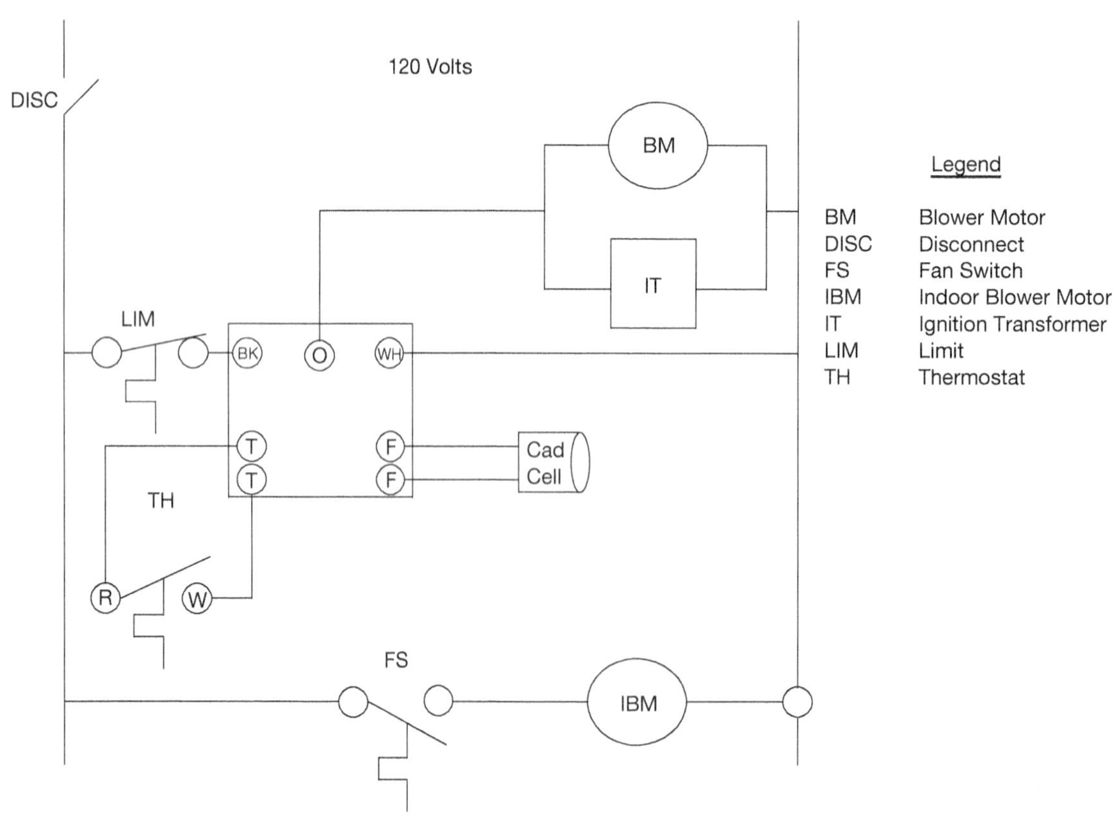

FIGURE 3–16

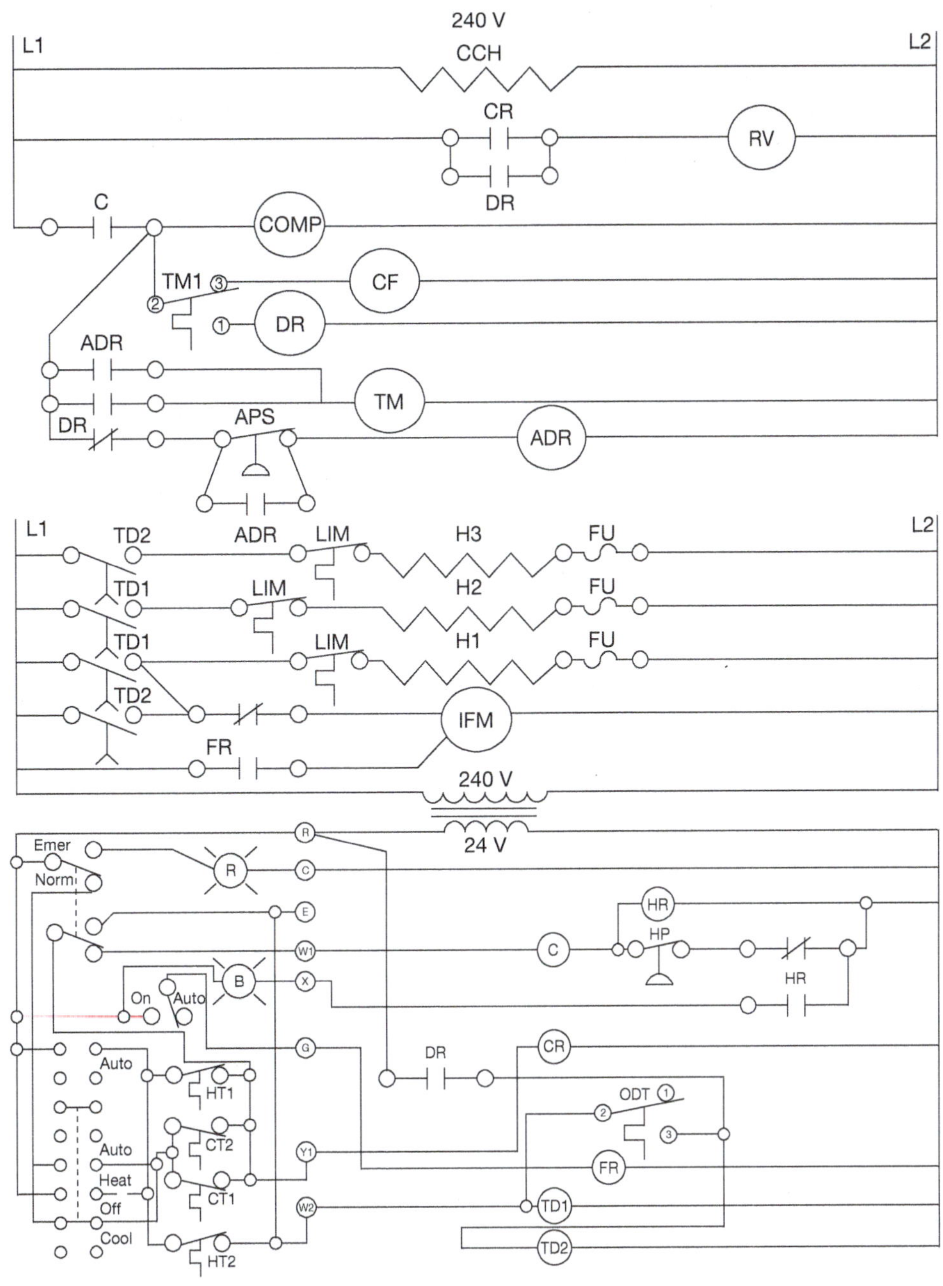

FIGURE 3–17

THE FOLLOWING QUESTIONS MAY BE HELPFUL IN STUDYING FOR THESE CERTIFICATION TESTS:
NATE—CR-SV, CR-IN, AC-SV, AC-IN, HP-SV, HP-IN, GS-SV, GS-IN, HG-SV, HG-IN, OL-SV, OL-IN.
ESCO INSTITUTE HVAC EXCELLENCE ELECTRICAL

ELECTRICAL METERS

1. The most accurate reading on an analog meter is

A. from the middle to the lower one-third of the scale.

B. in the middle of the scale.

C. below the middle of the scale.

D. from the middle to the upper two-thirds of the scale.

2. When using a digital meter that has the capability of displaying the proper scale to be used, this is called

A. autoramping.

B. autoranging.

C. prescaling.

D. self-scaling.

3. The meter used to measure the electrical pressure in a circuit is the
   A. ammeter.
   B. voltmeter.
   C. ohmmeter.
   D. wattmeter.
4. The meter used to measure the electrical current in a circuit is the
   A. ammeter.
   B. voltmeter.
   C. ohmmeter.
   D. wattmeter.
5. The meter used to measure the electrical resistance in a circuit is the
   A. ammeter.
   B. voltmeter.
   C. ohmmeter.
   D. wattmeter.
6. The meter used to measure the electrical power in a circuit is the
   A. ammeter.
   B. voltmeter.
   C. ohmmeter.
   D. wattmeter.
7. When measuring voltage to a load, the voltmeter must
   A. always be placed in series with the load.
   B. always be placed in parallel with the load.
   C. be in line with the switch controlling the load.
   D. be used with the load disconnected.
8. When testing for voltage across a switch located in a 120-volt circuit, what reading would be obtained if the switch were in the open position?
   A. 0 volts
   B. 60 volts
   C. 120 volts
   D. 240 volts
9. When testing for voltage across a switch located in a 120-volt circuit, what reading would be obtained if the switch were in the closed position?
   A. 0 volts
   B. 60 volts
   C. 120 volts
   D. 240 volts
10. A millivoltmeter is designed to measure
    A. 1,000 volts.
    B. 1 volt.
    C. 1/1,000 of a volt.
    D. 1/100,000 of a volt.
11. A voltmeter should
    A. always be set to the lowest scale when working with an unknown voltage.
    B. always be set to the highest scale when working with an unknown voltage.
    C. be set to a midscale when working with an unknown voltage.
    D. always be used with the power supply isolated from the circuit.
12. VOM stands for
    A. volt, ohm, meter.
    B. volt, ohm, megaohmmeter.
    C. volt, ohm, milliammeter.
    D. volt, ohm, microammeter.
13. An ammeter must
    A. always be placed in series with the load.
    B. always be placed in parallel with the load.
    C. be in parallel with the switch controlling the load.
    D. be used with the load disconnected.
14. When using a clamp-on ammeter, a wire is coiled 10 times and the clamp-on ammeter is placed around the coil of wire. What effect, if any, will this have on the amp reading?
    A. The amperage will not be affected.
    B. The amperage will multiply by 10.
    C. The amperage will divide by 10.
    D. The amperage will change depending on where the coil of wire is placed.
15. Which meter is more accurate?
    A. the clamp-on digital ammeter
    B. the clamp-on analog ammeter
    C. The accuracy of the analog is the same as the digital.
    D. the in-line analog ammeter
16. What is the purpose of the hold button located on a clamp-on ammeter?
    A. It automatically adjusts the meter scale.
    B. It retains the amperage reading.
    C. It keeps the meter on the amperage scale.
    D. It allows the meter to be suspended from the circuit being tested.
17. When using an ammeter,
    A. the circuit must always be disconnected.
    B. the circuit must always be energized.
    C. the circuit must always be isolated.
    D. the component must always be isolated.

18. An ammeter should

A. always be set to the lowest scale when working with an unknown amperage.

B. always be set to the highest scale when working with an unknown amperage.

C. never be used when working with unknown amperage.

D. always be used after testing the voltage.

19. A clamp-on ammeter must always measure the amperage by

A. clamping around one wire.

B. clamping around two wires.

C. clamping around one wire and the ground.

D. clamping around the ground wire.

20. When measuring resistance of a load, the ohmmeter must always

A. be placed in series with the load.

B. be placed in parallel with the load.

C. be used with the load disconnected.

D. Both B and C

21. When measuring the resistance across a switch, what reading would be obtained if the switch were closed?

A. 0 ohms

B. measurable resistance

C. infinite resistance

D. 1 megaohm

22. When measuring the resistance across a switch, what reading would be obtained if the switch were open?

A. 0 ohms

B. measurable resistance

C. infinite resistance

D. 1 megaohm

23. Which meter has its own built-in power supply?

A. the voltmeter

B. the ammeter

C. the ohmmeter

D. the wattmeter

24. Which meter must always be used with the power supply disconnected from the load?

A. the voltmeter

B. the ammeter

C. the ohmmeter

D. the wattmeter

25. A wattmeter is used to measure

A. the input power rating of the load.

B. the output power rating of the load.

C. the power factor input rating.

D. the power factor output rating.

26. If a motor has a power factor of 85% and a voltage of 120 volts with the amperage at 10 amps, the power rating would be

A. 1.2 kilowatts.

B. 1,000 watts.

C. 12 kilowatts.

D. 1,020 watts.

THE FOLLOWING QUESTIONS MAY BE HELPFUL IN STUDYING FOR THESE CERTIFICATION TESTS:
NATE—CR-SV, CR-IN, AC-SV, AC-IN, HP-SV, HP-IN, HG-SV, HG-IN, GS-SV, GS-IN, OL-SV, OL-IN, AD-SV, AD-IN
ESCO INSTITUTE HVAC EXCELLENCE ELECTRICAL

ELECTRICAL TROUBLESHOOTING

1. When testing an electrical circuit that is inoperable, what is commonly the first test that is performed?

A. an amperage test

B. a resistance test

C. a voltage test

D. a power test

2. When a relay or contactor coil is energized, what type of reading should be obtained across a normally closed set of contacts that is in an operating circuit?

A. 0 volts

B. source voltage

C. 0 ohms

D. circuit amperage

3. If source voltage is applied to an electrical load and the load is inoperable, what can be said about the load?

A. It has a short circuit.

B. It has a defective control circuit.

C. It has a grounded circuit.

D. It has an open circuit.

4. If a measurable voltage reading is obtained when testing across a normally open set of contacts on a relay that is energized, what is the problem with the relay?

A. There is no problem.

B. The contacts are burnt.

C. The contacts are closed.

D. The contacts are shorted.

5. When troubleshooting an electrical circuit using an ohmmeter,
   A. power must be supplied to the circuit.
   B. the meter must always be set to the R × 1 scale.
   C. the component being tested must be removed from the circuit.
   D. the circuit being tested must have current flow.
6. If a measurable resistance is obtained when testing across a normally closed set of contacts on a relay when using an ohmmeter, what is the problem with the relay?
   A. There is no problem.
   B. The contacts are burnt.
   C. The contacts are opened.
   D. The contacts are shorted.
7. When measuring the resistance of a load,
   A. start by setting the ohmmeter to the R × 100,000 scale.
   B. start by setting the ohmmeter to the R × 10,000 scale.
   C. start by setting the ohmmeter to the R × 1,000 scale.
   D. start by setting the ohmmeter to the R × 1 scale.
8. When measuring the resistance of a load, what type of resistance should be obtained if the load is operable?
   A. infinite resistance
   B. measurable resistance
   C. zero resistance
   D. no continuity
9. Which of the following resistance readings best displays the term *continuity*?
   A. an infinite and a zero resistance reading
   B. a measurable and an infinite resistance reading
   C. a measurable and a zero resistance reading
   D. an infinite resistance reading
10. When measuring the resistance of a shorted electrical load, what reading will be obtained?
    A. infinite resistance
    B. zero or very low measurable resistance
    C. 0 ohms only
    D. high measurable resistance
11. When measuring the resistance of an electrical load that has burnt open, what reading will be obtained?
    A. infinite resistance
    B. zero or very low measurable resistance
    C. 0 ohms only
    D. high measurable resistance
12. When measuring the resistance of a load that has a short to ground, it should be tested with the
    A. R × 1 scale.
    B. R × 10 scale.
    C. R × 1,000 scale.
    D. R × 1,000,000 scale.
13. When troubleshooting an electrical circuit using a clamp-on ammeter,
    A. the circuit must be de-energized and the ammeter must be clamped around two wires.
    B. the circuit must be energized and the ammeter must be clamped around two wires.
    C. the circuit must be energized and the ammeter must be clamped around one wire.
    D. the circuit must be de-energized and the ammeter must be clamped around one wire.
14. When testing a 24-volt circuit using an ammeter, what must be used to increase the amperage reading to a measurable level?
    A. a current multiplier
    B. a voltage rectifier
    C. a step-up transformer
    D. a potentiometer
15. What reading would be obtained when measuring the amperage of a short circuit?
    A. 0 amps
    B. steady measurable amperage
    C. excessive amperage
    D. low amperage
16. What reading would be obtained when measuring the amperage of an open circuit?
    A. 0 amps
    B. steady measurable amperage
    C. excessive amperage
    D. low amperage

SECTION 4

Airflow Components and Duct Fabrication

THE FOLLOWING QUESTIONS MAY BE HELPFUL IN STUDYING FOR THESE CERTIFICATION TESTS:
NATE—AD-SV, AD-IN, AC-SV, AC-IN, GS-SV, GS-IN, HP-SV, HP-IN, OL-SV, OL-IN.
ESCO INSTITUTE HVAC EXCELLENCE

AIRFLOW TOOLS AND MEASUREMENTS

1. A strobe tachometer is used to measure
   A. air speed.
   B. blower speed.
   C. air volume.
   D. static pressure.
2. A strobe tachometer's measurements are in
   A. fpm.
   B. rpm.
   C. cfm.
   D. inches WC.
3. A Pitot-tube has the capability of measuring
   A. static pressure.
   B. velocity pressure.
   C. velocity and static pressure.
   D. cfm.
4. The holes on the side of a Pitot-tube are used to measure
   A. static pressure.
   B. velocity pressure.
   C. velocity and static pressure.
   D. total pressure.
5. The hole on the end of a Pitot-tube is used to measure
   A. static pressure.
   B. velocity pressure.
   C. velocity and static pressure.
   D. total pressure.
6. The speed of air in a duct system is measured in
   A. cfm, cubic feet per minute.
   B. fpm, feet per minute.
   C. gph, gallons per hour.
   D. cfh, cubic feet per hour.
7. A Pitot-tube traverse is commonly used to
   A. take an average air velocity pressure reading in an air distribution system for the calculation of air volume.
   B. take an average static pressure reading in an air distribution system for the calculation of air velocity.
   C. take an average air volume reading in an air distribution system for the calculation of air velocity.
   D. None of the above
8. A Pitot-tube traverse
   A. must be taken in the main supply and return duct system only.
   B. must be taken in the individual branch runs only.
   C. must be taken in the main supply duct only.
   D. can be taken in the main supply, return, and individual branch runs.

9. A velometer measures

A. the quantity of air in a duct system.

B. the speed of the air in a duct system or at a diffuser.

C. the static pressure of the duct system.

D. None of the above

10. What instrument is commonly used with a velometer to measure the air velocity at a supply air register?

A. a Pitot-tube

B. a diffuser probe

C. an air scoop

D. a static ring

11. An anemometer measures

A. static pressure.

B. velocity in inches of water column.

C. velocity in feet per minute.

D. quantity of air in cubic feet per minute.

12. What are two common types of anemometers used for airflow testing?

A. a vane type and a Pitot type

B. a hot wire and a vane type

C. a hot wire type and a Pitot type

D. None of the above

13. What is considered static pressure in a duct system?

A. the speed of the air in the duct system

B. the pressure of the air flowing down the duct system

C. the pressure of the air pushing equally in all directions in the duct system

D. the total pressure in the duct system

14. What instrument must be used with a manometer to measure static, velocity, and total pressure?

A. a Pitot-tube

B. an incline gauge

C. an anemometer

D. a static disc

15. Which of the following formulas is correct?

A. Static Pressure/Velocity Pressure = Total Pressure

B. Velocity Pressure − Static Pressure = Total Pressure

C. Velocity Pressure × Static Pressure = Total Pressure

D. Static Pressure + Velocity Pressure = Total Pressure

16. A U-tube manometer is commonly used to measure pressures

A. below 1″ WC.

B. above 1″ WC.

C. up to 40″ WC.

D. None of the above

17. An incline manometer is commonly used to measure pressures

A. below 1″ WC.

B. above 1″ WC.

C. up to 40″ WC.

D. None of the above

18. What type of gauge uses a high- and low-pressure port to measure a pressure difference between two different pressure locations?

A. a high-pressure gauge

B. a low-pressure gauge

C. a differential pressure gauge

D. a high-/low-pressure gauge

19. Air volume is measured in

A. cfm.

B. fpm.

C. inches WC.

D. inches Hg.

20. Which of the following formulas is correct?

A. Area + Velocity = cfm

B. Velocity × Area = cfm

C. Velocity/Area = cfm

D. Area − Velocity = cfm

21. If an 8″ × 22″ supply air duct has air that is traveling at the speed of 1,000 fpm, what is the quantity of air that is moving through the system?

A. 176,000 cfm

B. 1,760 cfm

C. 1,222 cfm

D. 122 cfm

22. An 8″ × 12″ register has a velocity of 300 fpm with a free area of 75%. What is the quantity of air exiting the register?

A. 1,800 cfm

B. 2,880 cfm

C. 200 cfm

D. 150 cfm

23. An airhood is commonly used for measuring

A. cfm.

B. fpm.

C. static pressure.

D. velocity pressure.

24. Which of the following formulas is correct?

A. cfm = input(Btu/h)/1.08 × temperature difference

B. cfm = output(Btu/h) × 1.08/temperature difference

C. cfm = input(Btu/h)/1.08 + temperature difference

D. cfm = output(Btu/h)/1.08 × temperature difference

Airflow Calculations

V = velocity in fpm (feet per minute)
A = area of a duct in sq ft (square feet)
Q = quantity of air in cfm (cubic feet per minute)
Q = AV or (cfm = sq ft × fpm)
V = QA or (fpm = cfm/sq ft)
A = QV or (sq ft = cfm/fpm)

Calculation for Furnace Blower Output

cfm = output(Btu/h)/1.08 × temperature difference

Calculation for Weight of Air

Weight of air = W Volume = V Density = D

W = DV or W/D = V or W/V = D

25. Using the Duct Sizing Calculator (Figure 4–1), if the volume of air in a system is 1,600 cfm and the height of a rectangular duct is 8″ what should be the width of the duct?

A. 40″

B. 30″

C. 20″

D. 16″

26. Using the Duct Sizing Calculator (Figure 4–1), if the volume of air in a duct system is 1,600 cfm, what would the speed of the air be in the main trunk?

A. .10″ WC

B. 1.6″ WC

C. 1,100 fpm

D. 1,300 fpm

27. Using the Duct Sizing Calculator (Figure 4–1), if the quantity of air in a duct system is 1,600 cfm, what would be the static pressure of the duct per 100′?

A. .10″ WC

B. 1.6″ WC

C. 1,100 fpm

D. 1,300 fpm

28. Using the Duct Sizing Calculator (Figure 4–1), a 16″ round duct has the same face area as a

A. 14″ × 20″ rectangular duct.

B. 12″ × 19″ rectangular duct.

C. 10″ × 30″ rectangular duct.

D. 8″ × 40″ rectangular duct.

29. If the temperature difference across a heat exchanger is 70°F and the output of a furnace is 82,000Btu/h, what would be the cfm of the blower?

A. 950 cfm

B. 1,045 cfm

C. 1,085 cfm

D. 1,125 cfm

30. What do the vertical lines on the Friction Loss Chart (Figure 4–2) represent?

A. cfm

B. fpm

C. inches WC

D. inches duct diameter

31. What do the horizontal lines on the Friction Loss Chart (Figure 4–2) represent?

A. cfm

B. fpm

C. inches WC

D. inches duct diameter

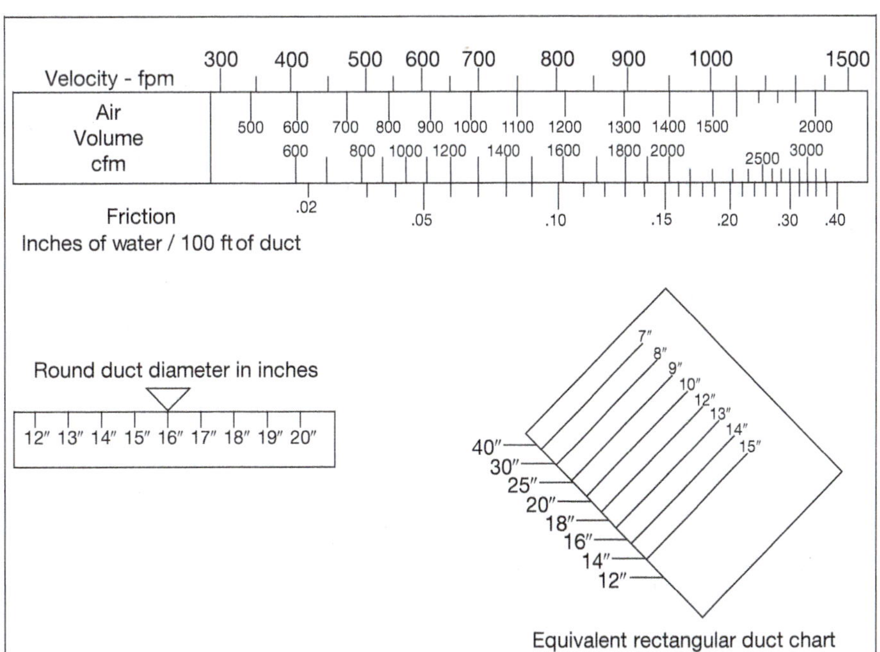

FIGURE 4–1

Duct Sizing Calculator

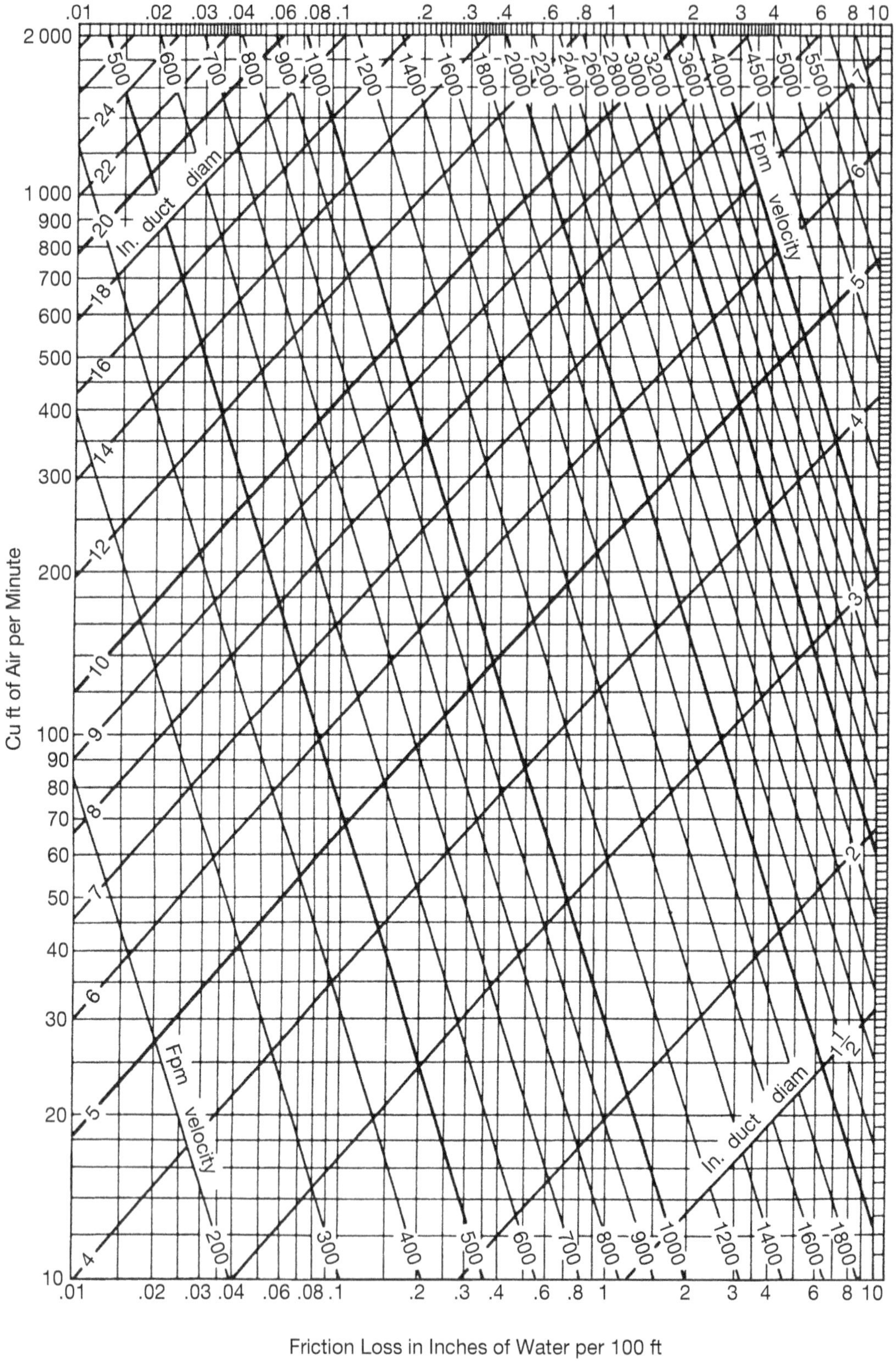

(Based on Standard Air of 0.075 lb per cu ft density flowing through average, clean, round, galvanized metal ducts having approximately 40 joints per 100 ft.)

Caution: Do not extrapolate below chart.

FIGURE 4–2

Friction Loss Chart

32. Using the Friction Loss Chart (Figure 4–2), what would the velocity of the air be at a 0.1 friction loss per 100 feet and 200 cfm of air?

A. 85 cfm

B. 550 cfm

C. 550 fpm

D. 650 fpm

33. Using the Friction Loss Chart (Figure 4–2), what would the diameter of the duct have to be to move 120 cfm at a velocity of 600 fpm?

A. 5″

B. 6″

C. 7″

D. 8″

34. Using the Friction Loss Chart (Figure 4–2), what would be the friction loss per 100′ if the speed of the air down a 5″ duct was 600 fpm?

A. .05

B. .08

C. 1.0

D. .15

THE FOLLOWING QUESTIONS MAY BE HELPFUL TO STUDY FOR THESE CERTIFICATION TESTS:
NATE—AD-SV, AD-IN, AC-SV, AC-IN, GS-SV, GS-IN, HP-SV, HP-IN, OL-SV, OL-IN.
ESCO INSTITUTE HVAC EXCELLENCE

BLOWERS AND FANS

1. Dry bearings can cause

A. excessive voltage to a motor.

B. an increase in efficiency.

C. an increase in amperage.

D. a decrease in amperage.

2. The sleeve in a sleeve bearing is commonly made of

A. copper.

B. bronze.

C. silver.

D. iron.

3. An oil-type motor typically requires oiling every

A. 1 to 3 months.

B. 3 to 6 months.

C. 12 months.

D. 18 months.

4. What type of bearing can handle a larger load?

A. sleeve bearing

B. bushing

C. ball bearing

D. axial bearings

5. Increasing the pitch of an axial fan blade will

A. increase the amperage of a fan motor.

B. decrease the amperage of a fan motor.

C. increase the torque of a fan motor.

D. increase the rpms of a fan motor.

6. Changing a condenser fan blade from a 4 to a 5 blade will

A. increase the amperage of a fan motor.

B. decrease the amperage of a fan motor.

C. increase the torque of a fan motor.

D. increase the rpms of a fan motor.

7. As the air volume is decreased over a fan, the motor amperage will

A. increase.

B. decrease.

C. stay the same.

D. Air volume cannot be changed.

8. Direct-drive condenser fans that are lower than 5 horsepower are commonly a

A. three-phase-type motor.

B. PSC-type motor.

C. CSIR-type motor.

D. CSCR-type motor.

9. How are condenser fan motors cycled during low ambient conditions?

A. by the use of a high-ambient-pressure switch

B. by the use of a low-pressure switch

C. by the use of a high-pressure switch

D. by the use of a reverse-acting high-pressure switch

10. Why is it necessary to cycle the condenser fans during low ambient conditions?

A. to prevent an increase in high-side pressure and the flooding of the evaporator coil

B. to prevent a decrease in high-side pressure and the flooding of the evaporator coil

C. to prevent an increase in high-side pressure and the starving of the evaporator coil

D. to prevent a decrease in high-side pressure and the starving of the evaporator coil

11. Belt drive blower motors lower than 5 horsepower commonly use a

A. three-phase-type motor.

B. split-phase-type motor.

C. CSIR-type motor.

D. PSC-type motor.

12. What is the color of the low speed on a 120-volt three-speed motor?

A. white

B. red

C. black

D. yellow

13. What is the color of the high speed on a 120-volt three-speed motor?

A. white

B. red

C. black

D. yellow

14. What is the color of the common lead on a 120-volt three-speed motor?

A. green

B. white

C. orange

D. black

15. A fan motor that operates at 1,050 rpm and has a pulley size of 6″ operates a blower that has a pulley size of 9″. What is the rpm of the blower?

A. 1,500

B. 1,575

C. 775

D. 700

16. How many cfms of air must a blower provide per ton of refrigeration effect?

A. 300

B. 400

C. 600

D. 800

17. How does an increased static pressure in a duct system affect the blower motor?

A. The amperage increases.

B. The amperage decreases.

C. The air quantity increases.

D. The air quantity decreases.

18. Name two common fans that are used on residential and light commercial heating and air conditioning systems.

A. a centrifugal and rotary type

B. an axial and reverse rotary

C. a squirrel cage and a centrifugal type

D. an axial and a squirrel-cage type

19. What is the FLA of a motor?

A. the starting amperage

B. the running amperage

C. the counterelectromotive force amperage

D. the fully locked amperage

20. What is the LRA of a motor?

A. the starting amperage

B. the running amperage

C. the counterelectromotive force amperage

D. the locked resistance amperage

21. Common external static pressure for a residential furnace blower is

A. $\frac{1}{2}''$ WC.

B. 1″ WC.

C. $1\frac{1}{2}''$ WC.

D. 2″ WC.

22. To reverse the direction of a three-phase motor,

A. the motor will have to be replaced.

B. the reverse speed taps will have to be changed.

C. reverse any two power leads at the motor terminals.

D. None of the above

THE FOLLOWING QUESTIONS MAY BE HELPFUL TO STUDY FOR THESE CERTIFICATIONS TESTS:
NATE—AD-SV, AD-IN, AC-SV, AC-IN, GS-SV, GS-IN, HP-SV, HP-IN, OL-SV, OL-IN.
ESCO INSTITUTE HVAC EXCELLENCE

AIR DUCT SYSTEMS AND FABRICATION

1. How often must a horizontal rectangular duct under 60″ wide be supported?

A. not less than every 5′

B. not less than every 8′

C. not exceeding every 10′

D. not exceeding every 12′

2. What is the minimum width of the galvanized steel hanger that supports a horizontal duct of 60″ or less?

A. $\frac{3}{4}''$

B. 1″

C. $1\frac{1}{4}''$

D. $1\frac{1}{2}''$

3. Duct joints must always

A. be sealed with a pressure-sensitive tape.

B. be sealed with mastics.

C. be sealed with gasket material.

D. be made airtight.

4. The crimp of a round duct joint must have a contact lap of at least

A. $1\frac{1}{2}''$.

B. 1″.

C. 2″.

D. $1\frac{3}{4}''$.

5. When connecting two sections of round duct,

A. the air flows from the female section of the duct to the male section.

B. a minimum of three sheet metal screws must be used per section.

C. screws or tape are sufficient to hold the sections together.

D. a minimum of two sheet metal screws must be used per section.

6. The reason for using a flex connector in the duct system is to

A. provide a leakproof section.

B. provide an area for temperature measurements.

C. eliminate noise transmissions from the air handler to the ductwork.

D. meet state code requirements.

7. Vibration isolators that are installed between metal ducts and mechanical equipment

A. must not exceed 10″ in length.

B. must not exceed 5″ in length.

C. must not exceed 3″ in length.

D. must be at least 2″ in length.

8. What is the minimum distance that a duct system is allowed to be installed from the ground?

A. flush

B. 2″

C. 1″

D. 4″

9. When a metal duct system is installed under a concrete slab,

A. the duct must be encased with at least 2″ of concrete.

B. the duct must be fabricated of stainless steel.

C. the duct must be aerated to prevent moisture buildup.

D. A duct is not allowed to be installed under a concrete slab.

10. When circular bands are used to connect sections of round duct,

A. the bands must be at least 4″ wide and use the same gauge metal as the duct system.

B. the bands must be at least 3″ wide and use the same gauge metal as the duct system.

C. the bands must be at least 2″ wide and use the same gauge metal as the duct system.

D. the bands must be at least 1″ wide and use the same gauge metal as the duct system.

11. The minimum total area of a supply air duct used on a forced-warm-air furnace

A. must be at least 1 sq in. per 1,000 Btu/h approved output rating.

B. must be at least 2 sq in. per 1,000 Btu/h approved output rating.

C. must be at least 4 sq in. per 1,000 Btu/h approved output rating.

D. must be at least 6 sq in. per 1,000 Btu/h approved output rating.

12. The minimum total area of a supply air duct used on a heat pump
    A. must be at least 1 sq in. per 1,000 Btu/h approved output rating.
    B. must be at least 2 sq in. per 1,000 Btu/h approved output rating.
    C. must be at least 4 sq in. per 1,000 Btu/h approved output rating.
    D. must be at least 6 sq in. per 1,000 Btu/h approved output rating.
13. When installing a metal air duct system on a roof,
    A. the duct system should have at least 1″ of insulation and be sealed with an approved weatherproof barrier.
    B. the duct system should have at least 3″ of insulation and be sealed with an approved weatherproof barrier.
    C. the maximum length of the duct system must not exceed 10′.
    D. the duct must be at least 28 gauge.
14. When installing a metal air duct in a crawl space or attic,
    A. the duct system should have at least 1″ of insulation and be sealed with an approved vapor barrier.
    B. the duct system should have at least 3″ of insulation and be sealed with an approved vapor barrier.
    C. the duct must have fire dampers installed.
    D. the duct must always be pitched toward the furnace.
15. The duct fitting that in most cases looks like a box and is connected to the air-discharge side of an air handler is the
    A. plenum.
    B. return.
    C. branch.
    D. diffuser.
16. In a single-dwelling unit, exposed rectangular ducts or plenums 14″ or less must be fabricated with a minimum of
    A. 24-gauge galvanized steel.
    B. 26-gauge galvanized steel.
    C. 28-gauge galvanized steel.
    D. 30-gauge galvanized steel.
17. In a single-dwelling unit, exposed rectangular ducts or plenums of more than 14″ must be fabricated with a minimum of
    A. 24-gauge galvanized steel.
    B. 26-gauge galvanized steel.
    C. 28-gauge galvanized steel.
    D. 30-gauge galvanized steel.
18. What is the purpose of a cross-break in a section of duct?
    A. to increase air volume
    B. to increase air velocity
    C. to decrease air currents caused by air velocity
    D. to increase its rigidity and decrease noise
19. Low-pressure duct systems are designed for a maximum static pressure of
    A. $\frac{1}{2}$″ WC.
    B. 1″ WC.
    C. 2″ WC.
    D. $2\frac{1}{2}$″ WC.
20. When air ducts are run through floors, what special provision must be made to meet code requirements?
    A. The air duct must be at least 5″ in diameter.
    B. There must be at least 2″ from any combustible surface.
    C. Shutoff dampers must be within 4″ of the floor.
    D. A firestop must be placed around the duct.
21. What type of duct system is one where the furnace is in a central location of the home and the trunk is branched from two sides of the furnace?
    A. an extended plenum
    B. a reducing extended plenum
    C. a perimeter loop
    D. a radial system
22. What type of duct system is commonly used in homes with concrete slab floors where a single pipe runs on the outside wall?
    A. an extended plenum
    B. a reducing extended plenum
    C. a perimeter loop
    D. a radial system
23. What type of duct system is commonly installed on the first floor where several ducts run from the furnace to the outside walls?
    A. an extended plenum
    B. a reducing extended plenum
    C. a perimeter loop
    D. a radial system
24. What type of duct system has the furnace located at one end of the house and the trunk size is reduced as it increases in distance from the furnace?
    A. an extended plenum
    B. a reducing extended plenum
    C. a perimeter loop
    D. a radial system
25. On an extended plenum, what is the maximum distance that a duct can extend without having a reduction in size?
    A. 8′
    B. 12′
    C. 16′
    D. 24′

26. Decreasing the size on a reducing extended plenum will

A. increase air velocity.

B. increase air volume.

C. decrease air volume.

D. reduce air turbulence.

27. What is the name of the section of duct that connects two different-sized ducts together?

A. a duct intersection

B. a duct transverse

C. a duct transition

D. a duct collar

28. Which tool is commonly used to decrease the diameter of the male fitting on a round duct?

A. the duct seamer

B. the crimping tool

C. the bar folder

D. the reducing tool

29. Where should a balancing damper be located on a branch run?

A. at the center point between the plenum and the register

B. as close to the register as possible

C. as close to the supply air plenum as possible

D. If the system is sized properly, a balancing damper is not required.

30. When adjusting a balancing damper located in a branch run, the damper is closed

A. when the adjusting handle is perpendicular to the duct.

B. when the adjusting handle is parallel to the duct.

C. when the adjusting handle is pointed 45 degrees above the duct.

D. when the adjusting handle is pointed 45 degrees below the duct.

31. A main supply air plenum must be located

A. at least 3″ from a combustible surface.

B. at least $2\frac{1}{2}$″ from a combustible surface.

C. at least 2″ from a combustible surface.

D. at least 1″ from a combustible surface.

32. What is the major difference between a register and a grill?

A. A register is adjustable and is used on the supply-air side of a system. A grill is not adjustable and is used on the return-air side of a system.

B. A register is not adjustable and is used on the supply-air side of a system. A grill is adjustable and is used on the return-air side of a system.

C. A grill is not adjustable and is used on the supply-air side of a system. A register is adjustable and is used on the return-air side of a system.

D. A grill is adjustable and is used on the supply-air side of a system. A register is not adjustable and is used on the return-air side of a system.

33. What type of supply-air register is required for bathrooms, kitchens, and utility rooms?

A. a floor register

B. a floor register with a return-air grill

C. a toe kick with a return-air grill

D. a toe kick or side wall register

34. In homes, what is most commonly used to return the conditioned space air back to the furnace?

A. prefabricated sheet-metal return air channels

B. wall studs and gypsum board

C. individual chases

D. 26-gauge return-air runs

35. What is the name for the metal duct connection shown in Figure 4–3?

A. hemmed "S" slip

B. drive slip

C. plain "S" slip

D. standing drive slip

36. What is the name for the metal duct connection shown in Figure 4–4?

A. hemmed "S" slip

B. drive slip

C. plain "S" slip

D. standing drive slip

37. What is the name for the metal duct connection shown in Figure 4–5?

A. hemmed "S" slip

B. inside slip joint

C. double "S" slip

D. standing "S" slip

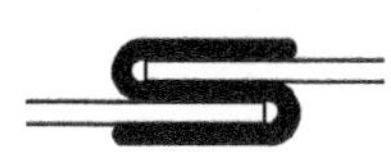

FIGURE 4–3

FIGURE 4–4

FIGURE 4–5

38. What is the name for the metal duct connection shown in Figure 4–6?

A. drive slip

B. inside slip joint

C. hemmed "S" slip

D. standing drive slip

39. What is the name for the metal duct connection shown in Figure 4–7?

A. hemmed "S" slip

B. inside slip joint

C. drive slip

D. standing drive slip

40. What is the name for the metal duct connection shown in Figure 4–8?

A. hemmed "S" slip

B. inside slip joint

C. double "S" slip

D. standing "S" slip

41. What is the name for the metal duct connection shown in Figure 4–9?

A. hemmed "S" slip

B. inside slip joint

C. double "S" slip

D. standing "S" slip

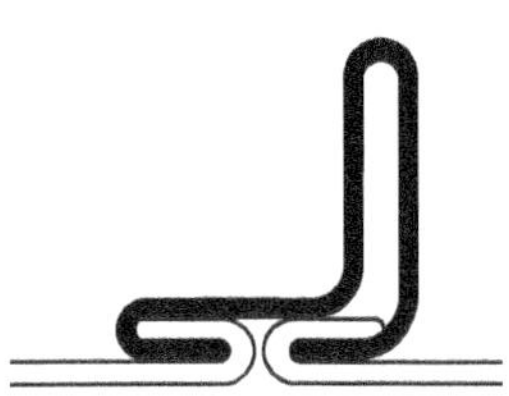

FIGURE 4–6

FIGURE 4–7

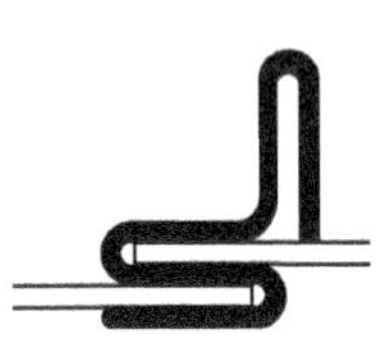

FIGURE 4–8

FIGURE 4–9

42. "S" slips are commonly connected

A. horizontally on the duct sections.

B. vertically on the duct sections.

C. by folding the end tabs.

D. using rivets every 2′.

43. Drive slips are commonly connected

A. horizontally on the duct sections.

B. vertically on the duct sections.

C. using sheet-metal screws.

D. using rivets every 2′.

44. The purpose for the tabs on drive cleats is to

A. prevent movement of the drive cleat.

B. provide a convenient section for installation.

C. prevent static loss.

D. prevent air loss.

45. What is the name for the metal duct connection shown in Figure 4–10?

A. hemmed "S" slip

B. inside slip joint

C. Pittsburgh lock seam

D. standing "S" slip

46. What is the purpose of a diffuser?

A. It is a covering over any opening through which air passes.

B. It is an outlet through which air discharges into a room.

C. It is a covering over any opening through which air passes that is equipped with a damper.

D. None of the above

47. What is the purpose of a register?

A. It is a covering over any opening through which air passes.

B. It is an outlet through which air discharges into a room.

C. It is a covering over any opening through which air passes that is equipped with a damper.

D. None of the above

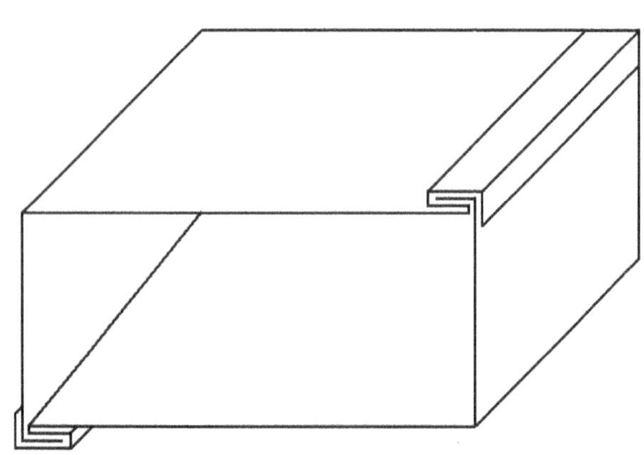

FIGURE 4–10

48. A floor register must resist without failure a concentrated load of

A. 150 pounds per 1 square inch of area.

B. 200 pounds per 1 square inch of area.

C. 200 pounds per 2 square inch of area.

D. None of the above

49. Duct construction must conform to the requirements of

A. ASHRAE Standards.

B. ACCA Manual D.

C. SMACNA Standards.

D. ACCA Manual J.

50. Which of the following is taken into consideration when determining the thickness of the duct material?

A. the velocity of the air through the duct system

B. the static pressure of the system

C. the temperature of the air in the duct system

D. air velocity

51. Which of the following tests for fire performance characteristics is used on duct systems?

A. NFPA 37

B. ANSI Z21

C. UL 181

D. ASHRAE Standards

52. Nonmetallic and metallic duct systems must be tested and classified in accordance with

A. ASHRAE Standards.

B. SMACNA Standards.

C. UL 181.

D. ANSI Z21.

53. What is the maximum discharge temperature permitted by industry standards for warm-air heating systems?

A. 200°F

B. 250°F

C. 300°F

D. 400°F

54. When gypsum board is used as an air shaft, the air temperature must not exceed

A. 90°F.

B. 125°F.

C. 150°F.

D. 200°F.

THE FOLLOWING QUESTIONS MAY BE HELPFUL TO STUDY FOR THESE CERTIFICATION TESTS:

NATE—AD-SV, AD-IN, GS-SV, GS-IN, AC-SV, AC-IN.

ESCO INSTITUTE HVAC EXCELLENCE

FLEXIBLE (FLEX) DUCT

1. When installing a flexible duct, it is important to always

A. support the duct every 10″.

B. fully extend the duct section to prevent the reduction of airflow.

C. use a supporting saddle material greater than 3″.

D. decrease the duct sag down to 2 inches per foot of section.

2. The radius of a flexible duct must not be less than

A. half of the diameter of duct being used.

B. one diameter of the duct being used.

C. two diameters of the duct being used.

D. three diameters of the duct being used.

3. Sheet-metal collars that are connected to a flexible duct must be a minimum of

A. $1\frac{1}{2}$″ in length.

B. 3″ in length.

C. 1″ in length.

D. 2″ in length.

4. Sheet-metal sleeves used for joining two sections of flexible duct must be a minimum of

A. 3″ in length.

B. 6″ in length.

C. 2″ in length.

D. 4″ in length.

5. A horizontal flexible duct must be supported

A. every 8′.

B. every 2′.

C. every 4′.

D. every 6′.

6. A vertical flexible duct must be supported

A. every 8′.

B. every 2′.

C. every 4′.

D. every 6′.

7. What is the minimum width of saddle material used to support a flexible duct?

   A. 1″

   B. $1\frac{1}{2}$″

   C. 2″

   D. 3″

8. What is the maximum sag that is allowed for flexible duct between supports?

   A. $\frac{1}{2}$ inch per foot

   B. 1 inch per foot

   C. $1\frac{1}{2}$ inches per foot

   D. 2 inches per foot

9. Code limits the length of a flexible air connector to

   A. 9 feet.

   B. 14 feet.

   C. 21 feet.

   D. 25 feet.

THE FOLLOWING QUESTIONS MAY BE HELPFUL TO STUDY FOR THESE CERTIFICATION TESTS:
NATE—AD-SV, AD-IN, GS-SV, GS-IN, AC-SV, AC-IN.

FIBERGLASS DUCTBOARD

1. Fiberglass ductboard is available in thicknesses of

   A. $\frac{1}{2}$″ and 1″.

   B. $\frac{1}{2}$″ and $1\frac{1}{2}$″.

   C. 1″ and $1\frac{1}{2}$″

   D. $1\frac{1}{2}$″ and $2\frac{1}{2}$″.

2. What is the maximum static pressure that can be obtained using fiberglass ductboard on the positive side of the system?

   A. $\frac{1}{2}$″ WC

   B. 1″ WC

   C. $1\frac{1}{2}$″ WC

   D. 2″ WC

3. What is the maximum static pressure that can be obtained using fiberglass ductboard on the negative side of the system?

   A. $\frac{1}{2}$″ WC

   B. 1″ WC

   C. $1\frac{1}{2}$″ WC

   D. 2″ WC

4. What is the maximum air velocity that can be used in a fiberglass ductboard system?

   A. 1,200 fpm

   B. 1,800 fpm

   C. 2,400 fpm

   D. 3,600 fpm

5. What is the maximum temperature that is permitted inside a fiberglass ductboard system?

   A. 200°F

   B. 250°F

   C. 300°F

   D. 350°F

6. What is the maximum temperature that is permitted on the outside surface of a fiberglass ductboard system?

   A. 150°F

   B. 200°F

   C. 100°F

   D. 250°F

7. When screws are used to fasten ductboard to a metal surface, which is the proper procedure?

   A. Use a #8 sheet-metal screw with a 2″ round washer.

   B. Use a #10 sheet-metal screw with a $2\frac{1}{2}$″ square washer.

   C. Use a #8 sheet-metal screw with a $2\frac{1}{2}$″ square washer.

   D. Use a #10 sheet-metal screw with a 2″ round washer.

8. Screws used to fasten ductboard material must be

   A. a minimum of $\frac{1}{2}$″ longer than the board thickness.

   B. a minimum of $\frac{1}{4}$″ longer than the board thickness.

   C. a minimum of $\frac{3}{4}$″ longer than the board thickness.

   D. a minimum of 1″ longer than the board thickness.

9. A shiplap joint flap must be stapled every

   A. 2″ on center.

   B. 3″ on center.

   C. 4″ on center.

   D. 5″ on center.

10. What is commonly used to cover a stapled shiplap joint flap?

    A. 2% silver tape

    B. duct tape

    C. aluminum foil tape

    D. mastic used with masking tape

11. Crosstabs are used to hold seams in position when staples are not used. A crosstab is made of
    A. a 5″ minimum length of aluminum foil tape that is placed across a taped joint every 15″.
    B. an 8″ minimum length of aluminum foil tape that is placed across a taped joint every 12″.
    C. a 10″ minimum length of aluminum foil tape that is placed across a taped joint every 15″.
    D. a 12″ minimum length of aluminum foil tape that is placed across a taped joint every 12″.

12. The minimum width of pressure-sensitive aluminum foil tape that is used on fiberglass ductboard is
    A. $1\frac{1}{2}''$.
    B. 2″.
    C. $2\frac{1}{2}''$.
    D. 3″.

13. Aluminum foil tape must overlap each side of a fiberglass ductboard joint by a minimum of
    A. 1″.
    B. $1\frac{1}{2}''$.
    C. 2″.
    D. $2\frac{1}{2}''$.

14. If aluminum foil tape is stored in an area less than 50°F,
    A. the tape is permanently damaged and should not be used.
    B. the tape can be used only with a mastic material.
    C. the tape must be preheated before it is used.
    D. additional glue must be added to the tape before use.

15. When using heat-activated tape, what temperature must the tape be heated to after being applied to the fiberglass ductboard?
    A. 300–400°F
    B. 400–500°F
    C. 450–550°F
    D. 550–600°F

16. What is the minimum width of heat-activated tape that is used on fiberglass ductboard?
    A. 2″
    B. $2\frac{1}{2}''$
    C. 3″
    D. $3\frac{1}{2}''$

17. What is the purpose of a tie rod in a fiberglass ductboard system?
    A. It is used to hang the duct from a ceiling.
    B. It is used to raise the duct off of the ceiling joists when placed in an attic.
    C. It is used to connect two sections of duct together.
    D. It is used to increase the rigidity of the ducts.

18. A tie rod used for fiberglass ductboard must be made of
    A. 18-gauge stainless steel wire.
    B. 16-gauge galvanized steel wire.
    C. 14-gauge stainless steel wire.
    D. 12-gauge galvanized steel wire.

19. A tie rod
    A. is commonly placed horizontally in a ductboard system.
    B. is commonly placed vertically in a ductboard system.
    C. can only be used with 1″-thick fiberglass ductboard systems.
    D. can only be used with $1\frac{1}{2}''$-thick fiberglass ductboard systems.

20. When a transverse joint is used in a fiberglass ductboard system, the tie rod must be placed
    A. 4″ from the end of the male joint.
    B. 4″ from the end of the female joint.
    C. 5″ from the end of the male joint.
    D. 5″ from the end of the female joint.

21. When a butt joint is used in a fiberglass ductboard system, the tie rod must be placed
    A. 3″ on either side of the joint.
    B. 4″ on either side of the joint.
    C. 5″ on either side of the joint.
    D. 6″ on either side of the joint.

22. When using a male and female shiplap joint in a fiberglass ductboard system,
    A. air must always flow from the female to the male end of the joint.
    B. air must always flow from the male to the female end of the joint.
    C. air can flow in either direction.
    D. Male shiplap joints are never used with female joints.

23. When are V-grooves most commonly used in fiberglass ductboard systems?
    A. when connecting two ductboard sections together
    B. on ductboard corners as a seam
    C. on single flat board sections to create a rectangular duct
    D. to connect supports to the ductboard system

24. What type of joint is *not* permitted on corners of a ductboard system?
    A. male-to-female shiplap joints
    B. square-cut-to-square-cut joints
    C. female-shiplap-to-square-cut joints
    D. V-grooves

25. In Figure 4–11, what type of corner is being used?
    A. male shiplap with a square cut
    B. male shiplap with a female shiplap
    C. female shiplap with a square cut
    D. square cut to square cut

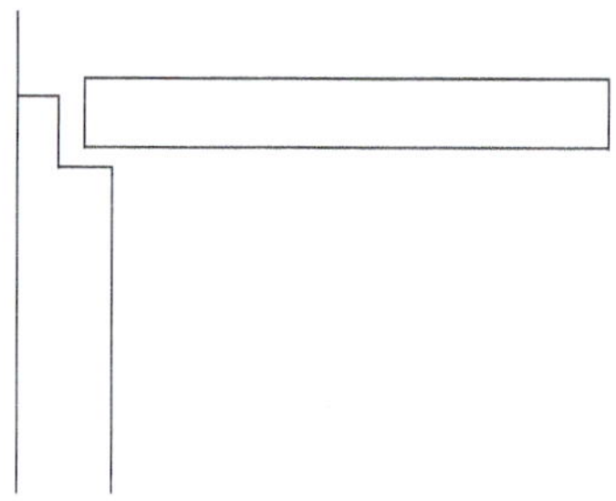

FIGURE 4–11

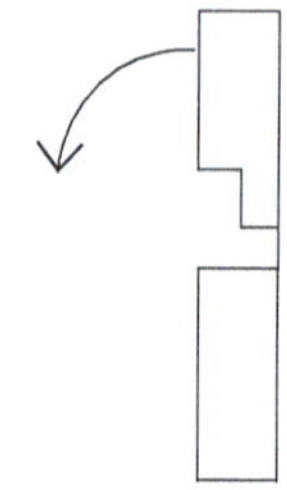

FIGURE 4–12

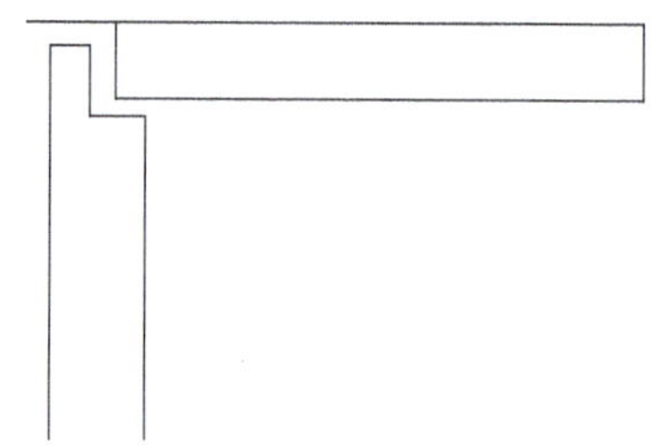

FIGURE 4–13

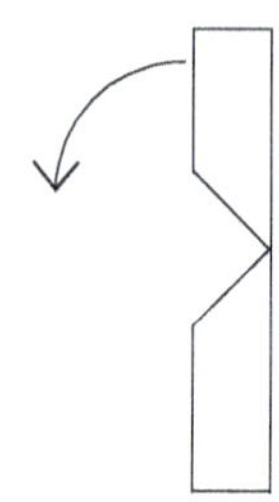

FIGURE 4–14

26. In Figure 4–12, what type of corner is being used?

A. male shiplap with a square cut

B. shiplap corner fold

C. female shiplap with a square cut

D. male shiplap with a female shiplap

27. In Figure 4–13, what type of corner is being used?

A. male shiplap and square cut with a flap

B. shiplap corner fold with a flap

C. female shiplap and square cut with a flap

D. male shiplap and a female shiplap with a flap

28. In Figure 4–14, what type of corner is being used?

A. male-to-female shiplap joint

B. square-cut-to-square-cut joint

C. female-shiplap-to-square-cut joint

D. V-groove corner fold

THE FOLLOWING QUESTIONS MAY BE HELPFUL TO STUDY FOR THESE CERTIFICATION TESTS:
NATE—AD-SV, AD-IN, GS-SV, GS-IN, AC-SV, AC-IN
ESCO INSTITUTE HVAC EXCELLENCE

AIR-SIDE COMPONENTS

1. Two general types of damper blades are

A. adjustable and nonadjustable.

B. parallel blade and opposing blade.

C. eccentric blade and opposing blade.

D. direct and indirect.

2. Which type of damper is known to give more uniformity of airflow through a system?

A. adjustable direct damper

B. parallel blade damper

C. eccentric damper

D. opposing blade damper

3. A mixed air sensor is commonly used in conjunction with

A. an economizer.

B. an ignition control system.

C. a compressor circuit.

D. an electric heating coil.

4. The main purpose of an economizer circuit is to

A. add a heat load to an air conditioning system during low load conditions.

B. introduce outside air for ventilation and free cooling.

C. ventilate unwanted air from a building.

D. bypass cooling coils during low load conditions.

5. What controls are commonly found in economizer circuits?
   A. enthalpy, air volume, and air velocity sensors
   B. humidistats, thermostats, and discharge air sensors
   C. pressure and temperature sensors
   D. enthalpy, discharge air, and mixed air sensors
6. What can be used to prevent the pressurization of a building?
   A. an economizer system
   B. relief dampers
   C. opposing-blade-type dampers
   D. face and bypass dampers
7. What is used to control the amount of air over a heating or cooling coil in a constant air volume system?
   A. an economizer system
   B. relief dampers
   C. opposing-blade-type dampers
   D. face and bypass dampers
8. How does a variable air volume (VAV) system operate?
   A. Air volume is reduced as the load is increased.
   B. Air volume is increased as the load increases.
   C. Air volume is increased as the room temperature decreases during cooling.
   D. Air volume is decreased as the room temperature increases during heating.
9. What is the primary purpose of a damper?
   A. to increase the velocity in a branch run
   B. to decrease velocity in a branch run
   C. to isolate a duct section
   D. to balance airflow in a duct system
10. What type of damper is required in a commercial building where a section of duct extends through two different rooms?
    A. a balancing damper
    B. a partitioning damper
    C. a remote damper
    D. a fire damper
11. What is used to open and close a fire damper?
    A. an electrical cam and switching mechanism
    B. a bimetal rod
    C. a fusible link
    D. an expansion-type joint
12. What is commonly used to open and close a smoke damper?
    A. a photoelectric sensor
    B. a cadmium sensor
    C. a fusible link
    D. an expansion joint
13. What is the purpose for proper ventilation?
    A. to bring the proper combustion air into a heating system
    B. to bring the proper primary and secondary air into a heating system
    C. so a system can extract unwanted dust particles
    D. to increase oxygen levels in the air
14. The temperature difference between conditioned space air and ventilation air must not exceed
    A. 5°F.
    B. 10°F.
    C. 20°F.
    D. 30°F.

THE FOLLOWING QUESTIONS MAY BE HELPFUL TO STUDY FOR THESE CERTIFICATION TESTS:
NATE—AD-SV, AD-IN, GS-SV, GS-IN, AC-SV, AC-IN.
ESCO INSTITUTE HVAC EXCELLENCE

AIR FILTERS

1. Where should a filter be placed in an HVAC system?
   A. in the return-air system upstream from a heat exchanger or a cooling coil
   B. in the return-air system downstream from a heat exchanger or a cooling coil
   C. in the supply-air system upstream from a heat exchanger or a cooling coil
   D. in the supply-air system downstream from a heat exchanger or a cooling coil
2. Air filters must be installed
   A. so that they are readily removable.
   B. so that they are always next to a return-air boot.
   C. within 6″ maximum from the furnace return-air opening.
   D. within 10″ maximum from the furnace return-air opening.
3. What aids an air distribution system in evenly disbursing air across a filter medium?
   A. an accordion-type air filter
   B. a nonadjustable damper
   C. turning vanes
   D. an adjustable damper

4. What is meant by the term *ionizing*?
   - **A.** a process of removing heat
   - **B.** charging a particle with a negative or positive charge
   - **C.** adding resistance to a circuit
   - **D.** coating a material with a solution
5. What are three common types of electronic air filters?
   - **A.** removable-plate type, ionizing-plate type, and charged-media-ionizing type
   - **B.** retractable-static type, ionizing type, and nonionizing type
   - **C.** ionizing-plate type, charged-media-nonionizing type, and charged-media-ionizing type
   - **D.** removable-plate type, retractable-static type, and ionizing-plate type
6. What type of electronic filter generates positive ions on wires and uses a high-voltage panel that holds the dust particles?
   - **A.** ionizing-plate type
   - **B.** retractable-static type
   - **C.** charged-media-nonionizing type
   - **D.** removable-plate type
7. What type of electronic filter uses a dielectric medium that is supported by an electrically charged grid?
   - **A.** ionizing-plate type
   - **B.** retractable-static type
   - **C.** charged-media-ionizing type
   - **D.** charged-media-nonionizing type
8. What type of electronic filter generates ions and uses a dielectric medium that is supported by an electrically charged grid?
   - **A.** ionizing-plate type
   - **B.** retractable-static type
   - **C.** charged-media-ionizing type
   - **D.** charged-media-nonionizing type
9. An electronic air filter removes particles as small as
   - **A.** 0.01 microns.
   - **B.** 0.001 microns.
   - **C.** 0.0001 microns.
   - **D.** 0.00001 microns.
10. Where is an electronic air filter commonly installed on a furnace?
    - **A.** ahead of the heat exchanger
    - **B.** downstream from the blower
    - **C.** in the supply side of the furnace
    - **D.** in the return side of the furnace
11. What is commonly used with electronic air filters to trap large dust particles?
    - **A.** a mesh screen after the electronic element
    - **B.** a prefilter before the electronic element
    - **C.** a prefilter after the electronic element
    - **D.** an electrostatic medium before the element
12. On average, the filters of an electronic air cleaner must be cleaned
    - **A.** once a week.
    - **B.** once every 2 months.
    - **C.** once every month.
    - **D.** once every year.
13. What is used to measure the pressure drop across a filter?
    - **A.** a water gauge manometer
    - **B.** a mercury gauge manometer
    - **C.** an anemometer
    - **D.** a velometer
14. Disposable filters should be changed at least
    - **A.** once a week.
    - **B.** once every 2 months.
    - **C.** once every month.
    - **D.** once every year.

THE FOLLOWING QUESTIONS MAY BE HELPFUL TO STUDY FOR THESE CERTIFICATION TESTS:
NATE—GS-SV, GS-IN, OL-SV, OL-IN.
ESCO INSTITUTE HVAC EXCELLENCE

HUMIDIFIERS

1. A plate-type humidifier, a rotating-drum-type humidifier, and a fan-powered humidifier are all considered
   - **A.** evaporative-type humidifiers.
   - **B.** atomizing-type humidifiers.
   - **C.** vaporizing-type humidifiers.
   - **D.** Both A and C
2. Which two of the following humidifiers use a bypass duct?
   - **A.** the plate-type and rotating-plate-type
   - **B.** the drum-type and plate-type
   - **C.** the flow-through type and the fan-powered type
   - **D.** the drum type and the flow-through type

3. What is the purpose of the bypass duct that is used on some humidifiers?

A. It allows the humidifier to be closed during the summer months.

B. It allows for the humidifier to be used only in the winter months.

C. It increases the efficiency of the humidifier.

D. All of the above

4. Which type of humidifier uses a solenoid valve to control water flow?

A. a flow-through type

B. a fan-powered type

C. a rotating-plate type

D. a rotating-drum type

5. Which type of humidifier uses a float valve to control water flow?

A. a plate type

B. a rotating-plate type

C. a rotating-drum type

D. All of the above

6. Which two of the following humidifiers are commonly mounted on the supply duct without the use of a bypass duct?

A. the plate type and fan-powered type

B. the drum type and the flow-through type

C. the flow-through type and the fan-powered type

D. the drum type and plate type

7. Which humidifier commonly uses a heating element?

A. the vaporizing type

B. the atomizing type

C. the evaporative type

D. the flow-through type

8. What is the average recommended relative humidity for a home during the winter months?

A. 10 to 20%

B. 35 to 45%

C. 65 to 75%

D. 85 to 95%

9. Why must the humidity level in a home be increased during the winter months?

A. This will cause the moisture on the skin to evaporate more rapidly, improving the comfort level.

B. This will cause the moisture on the skin to evaporate more slowly, improving the comfort level.

C. This will not affect the rate of the evaporation of moisture on the skin; it mainly has to do with a decrease of static electricity.

D. None of the above

10. The most common type of humidifier circuit for residential applications is

A. 120 volts/15 amps.

B. 24 volts/10 amps.

C. 24 volts/20 VA.

D. 120 volts/40 VA.

11. What is the purpose of the humidistat when used to control a humidifier?

A. It senses humidity and closes on a decrease.

B. It senses temperature and humidity and closes on a decrease.

C. It senses temperature and humidity and opens on a decrease.

D. It senses humidity and opens on a decrease.

12. What load must be energized before the humidifier circuit is energized?

A. the gas valve

B. the induced-draft blower

C. the indoor blower

D. All of the above

13. What is the purpose of a current-sensing relay?

A. It senses the high speed of the blower and energizes the humidifier transformer.

B. It always senses the low speed of the blower and energizes the 24-volt humidifier circuit.

C. It senses the heating speed of the blower and energizes the humidifier transformer or the 24-volt humidifier circuit.

D. A current-sensing relay is only used on the cooling circuit.

14. On modern furnace controls, the terminals marked H or HUM

A. energize the 120-volt primary side of the humidifier transformer only.

B. energize the 24-volt humidifier control circuit only.

C. can energize the 120-volt primary side of the transformer or the 24-volt humidifier control circuit.

D. None of the above

15. The humidifier is best operated with

A. a continuous fan.

B. an intermittent fan and short heating cycles.

C. an intermittent fan and long heating cycles.

D. an intermittent fan on high speed.

16. Which of the following will cause an increase of static electricity throughout the home?

A. a defective humidifier water valve

B. a bypass damper in the closed position during winter months

C. a faulty humidifier transformer

D. All of the above

THE FOLLOWING QUESTIONS MAY BE HELPFUL TO STUDY FOR THESE CERTIFICATION TESTS:
NATE—AD-SV, AD-IN, AC-SV, AC-IN, GS-SV, GS-IN, OL-SV, OL-IN.
ESCO INSTITUTE HVAC EXCELLENCE

INDOOR AIR QUALITY (IAQ)

1. What's the difference between a contaminant and a pollutant?
   A. A contaminant may or may not be a health risk and a pollutant is always a health risk.
   B. A pollutant may or may not be a health risk and a contaminant is always a health risk.
   C. There is no difference. Contaminants and pollutants are always health risks.
   D. There is no difference. Contaminants and pollutants are never health risks.
2. Respirable particles that are breathed deep into the lung are commonly
   A. 100 microns or less.
   B. 50 microns or less.
   C. 40 microns or less.
   D. 10 microns or less.
3. Asbestos is made from a mineral called chrysolite. If it is breathed into the lungs, it could cause
   A. scarring of the lung tissue.
   B. lung cancer.
   C. damage to the abdomen.
   D. All of the above
4. The first task a technician should do to improve air quality is to
   A. clean the ventilation system.
   B. close off return air, then clean ductwork.
   C. change the air filters.
   D. open all dampers to supplies.
5. Most inexpensive filters are
   A. approximately 7% efficient.
   B. approximately 50% efficient.
   C. approximately 75% efficient.
   D. approximately 90% efficient.
6. What is the efficiency of a common pleated filter?
   A. 5%
   B. 30%
   C. 80%
   D. 90%
7. What type of air filter would remove tobacco smoke particles efficiently?
   A. disposable filters
   B. electronic filters
   C. pleated filters
   D. carbon-type filters
8. Stratification refers to air that
   A. has been overheated.
   B. has been overcooled.
   C. settles in different temperature layers.
   D. has an extreme temperature rise.
9. Air must be kept moving
   A. to eliminate stratification.
   B. to prevent a pressure drop.
   C. to increase a positive building pressure.
   D. to decrease a positive building pressure.
10. Into which four groups are air contaminants classified?
    A. gases/vapors, inert particles, microorganisms, pollen
    B. gases/vapors, pollen, radon, organisms
    C. radon, gases/vapors, molds, pollen
    D. pollen, molds, inert particles, microorganisms
11. Man-made fibers, dust, and cigarette smoke are in which of the following groups?
    A. vapors
    B. gases
    C. inert particles
    D. organisms
12. Microorganisms include
    A. gases and bacteria.
    B. fungi, mold, and bacteria.
    C. vapors and mold.
    D. mold and fibers.
13. Which of the following contaminants greatly affect HVAC/R air distribution equipment?
    A. microorganisms
    B. inert particles
    C. gases and vapors
    D. pollens

14. Which of the following are considered to be pollen-type contaminants?

A. molds and bacteria

B. ragweed and molds

C. fungi and grass

D. grass and seasonal plants

15. Symptoms such as eye, nose, and throat irritation; dizziness; headaches; drowsiness; fatigue; and burning eyes are commonly caused by

A. influenza symptoms.

B. cold symptoms.

C. smoke inhalation symptoms.

D. sick building syndrome (SBS).

16. A satisfactory CO_2 measurement in an indoor environment would be

A. a maximum of 1,400 ppm.

B. a maximum of 1,200 ppm.

C. a maximum of 1,000 ppm.

D. a maximum of 800 ppm.

17. Indoor air contaminants are

A. commonly 2–5 times higher in quantity than outdoor air contaminants.

B. commonly 8–10 times higher in quantity than outdoor air contaminants.

C. commonly 12–14 times higher in quantity than outdoor air contaminants.

D. commonly the same quantity as outdoor air contaminants.

18. Which illness is commonly caused by untreated cooling-tower water?

A. influenza virus

B. Parkinson's disease

C. Legionnaires' disease

D. respiratory symptoms

19. When cleaning nonporous materials, what is commonly used to prevent the growth of mold?

A. a non-oil-based detergent

B. a solution of 90% water and 10% bleach

C. 50% bleach and 50% soap

D. a toxic solution of 50% bleach and 50% ammonia

20. Some air leakage into a building will help

A. decrease the temperature rise.

B. increase return-air temperatures.

C. reduce indoor air pollution.

D. increase indoor air pollution.

21. A high relative humidity within a building can

A. cause static electricity.

B. cause a heating system to overshoot.

C. decrease the supply-air temperature of a heating system.

D. produce mold.

22. A ventilation system will exhaust stale indoor air and

A. eliminate air pollution.

B. maintain a low relative humidity.

C. introduce fresh outside air.

D. maintain a high relative humidity.

23. Common types of ventilation systems are

A. passive and exhaust-only types.

B. nonheat recovery units and passive types.

C. balanced mechanical and heat recovery types.

D. All of the above

24. What must a public building have for proper ventilation?

A. an outside air supply and an exhaust system

B. closed dampers to the interior room

C. at least 50% outside air at all times

D. at least 40% outside air at all times

25. A ventilation system producing a slight positive pressure will

A. reduce heating demands.

B. minimize infiltration.

C. reduce cooling demands.

D. eliminate the need for makeup air.

26. What is a VOC (volatile organic compound)?

A. a highly evaporative chemical that gives off gases

B. a chemical that stays a liquid and has a high boiling point

C. a solid compound that is commonly found in the ground

D. None of the above

27. What symptoms can VOCs cause in a human being?

A. rashes and nausea

B. sinus congestion and drowsiness

C. headaches and respiratory problems

D. All of the above

28. The radioactive gas that comes from the earth's crust is known as

A. phosgene.

B. radon.

C. VOC compound.

D. formaldehyde.

29. Radon enters the home through

A. cracks, joints, and nontrapped drains.

B. sump pumps and water supplies.

C. underground service openings and crawl spaces.

D. All of the above

30. Why does radon commonly stay near ground level?

A. because it's lighter than air

B. because it's nine times heavier than air

C. because radioactive particles are commonly attracted to the earth's crust

D. None of the above

31. Radon is not considered harmful to humans until it
   A. decays.
   B. mixes with mold bacteria.
   C. mixes with oxygen.
   D. is breathed into the lungs.

32. Radon levels inside a building should be considered harmful when the average annual concentration in the normal living areas exceeds
   A. 300 becquerels (Bq) per cubic meter.
   B. 600 becquerels (Bq) per cubic meter.
   C. 800 becquerels (Bq) per cubic meter.
   D. 400 becquerels (Bq) per cubic meter.

33. What organic compounds are produced when materials containing carbon and hydrogen are burned?
   A. hydrogen oxide
   B. polynuclear aromatic hydrocarbons
   C. phosgene gases
   D. halogen gases

34. Outdoor sources of PAHs are commonly caused by
   A. wood-burning appliances.
   B. coal-burning equipment.
   C. internal combustion engines.
   D. All of the above

35. Indoor sources of PAHs are commonly caused by
   A. cigarette smoke.
   B. burning of food.
   C. fireplaces.
   D. All of the above

36. What is ozone?
   A. a stable oxygen molecule
   B. 1 oxygen atom and 1 hydrogen atom
   C. a makeup of oxygen and nitrogen molecules
   D. 3 oxygen atoms

37. How is ozone produced?
   A. when air is heated by the sun and the humidity is increased
   B. when chlorine is released to the atmosphere
   C. during an electrical storm
   D. Both A and B

38. The level of ozone inside the home
   A. is directly correlated with the levels outside the home.
   B. is commonly higher than the levels outside the home.
   C. is caused by heat-producing appliances.
   D. None of the above

39. Nitric oxide (NO) is a poisonous gas that is produced
   A. during low-temperature combustion.
   B. when there is a lack of primary air.
   C. when the combustion process has been delayed too long.
   D. during high-temperature combustion.

40. Nitric oxide is
   A. highly toxic.
   B. nontoxic.
   C. flammable.
   D. explosive.

41. Common indoor levels of nitric oxide are
   A. 0.25 ppm or less.
   B. 5 ppm or less.
   C. 20 ppm or less.
   D. 50 ppm or less.

42. Nitric oxide and nitrogen dioxide are produced when
   A. fuel-burning appliances are not vented.
   B. nitrogen and oxygen are mixed.
   C. carbon dioxide and oxygen are combined.
   D. not enough air is infiltrated into a home.

43. Once nitric oxide is released into the air, it quickly changes to
   A. carbon dioxide.
   B. carbon monoxide.
   C. nitrogen.
   D. nitrogen dioxide.

44. Carbon monoxide (CO) is a poisonous gas that is produced
   A. during complete combustion of gas products.
   B. during incomplete combustion of fossil fuels.
   C. during any burning of fuels, as it's a product of combustion.
   D. All of the above

45. Indoor sources of carbon monoxide are
   A. cracked heat exchangers and chimneys.
   B. unvented appliances.
   C. cigarette smoke.
   D. All of the above

46. Average indoor levels of carbon monoxide range from
   A. 0.5 to 5 ppm.
   B. 8 to 10 ppm.
   C. 10 to 15 ppm.
   D. 20 to 50 ppm.

47. If a home has excessive negative pressure, the combustion gases of an appliance could
   A. enter back into the home.
   B. produce carbon monoxide.
   C. produce ozone.
   D. Both A and B

48. What type of contaminant is formaldehyde?

A. a VOC

B. an inert particle

C. a microorganism

D. None of the above

49. Medium-density fiberboard (MDF) and particle board inside a home are common sources of which indoor contaminant?

A. formaldehyde

B. nitric acid

C. carbon monoxide

D. excessive CO_2

50. The rate at which formaldehyde gas is released in a home is directly proportional to

A. an increase in humidity and a decrease in temperature.

B. a decrease in humidity and an increase in temperature.

C. a decrease in humidity and a decrease in temperature.

D. an increase in humidity and an increase in temperature.

THE FOLLOWING QUESTIONS MAY BE HELPFUL TO STUDY FOR THESE CERTIFICATION TESTS:

NATE—AD-SV, AD-IN, AC-SV, AC-IN, GS-SV, GS-IN, OL-SV, OL-IN.

ESCO INSTITUTE HVAC EXCELLENCE

AIR DISTRIBUTION TROUBLESHOOTING

1. An air conditioning service call is made and the complaint is that the system does not seem to be cooling properly. It is noticed that the suction line and the evaporator coil are frosted up. The filter is checked and it is clean. It is suspected that the indoor blower is not operating at the proper speed.

Technician A says that the speed can be verified with a strobe tachometer to be sure that the rpms are as written on the nameplate.

Technician B says that the airflow can be measured by taking a static pressure reading.

Which technician is correct?

A. technician A

B. technician B

C. Both technicians are correct.

D. Both technicians are incorrect.

2. An airflow measurement taken in a duct system by drilling small holes in a vertical and horizontal section of a rectangular duct to find an average air velocity reading is known as

A. a Pitot-tube traverse.

B. an anemometer test.

C. an air quantity test.

D. a static pressure test.

3. To calculate velocity pressure when using a manometer, which of the following would be the proper formula?

A. static pressure + total pressure = velocity pressure

B. static pressure − total pressure = velocity pressure

C. total pressure − static pressure = velocity pressure

D. total pressure + static pressure = velocity pressure

4. What is the primary purpose of an isolation collar?

A. It is used to isolate the duct system from the furnace when it needs to be replaced.

B. It is used to eliminate furnace vibration through the duct system.

C. It is used to increase the air velocity in a duct system.

D. It is used to decrease the air velocity in a duct system.

5. During the cooling season, a complaint is made that a two-story home is too cool on the first floor and too warm on the second floor. What can be done to help the system cool properly?

A. The first-floor in-line dampers can be adjusted parallel to the duct.

B. The first-floor in-line dampers can be adjusted perpendicular to the duct.

C. The second-floor in-line dampers can be adjusted perpendicular to the duct.

D. None of the above

6. When installing sections of ducts such as plenums, trunk sections, and duct transitions, it is important to

A. attach the sections using drive cleats only.

B. attach the sections using S-slips only.

C. connect the duct sections using the proper S-slips and drive cleats loosely to prevent duct vibration.

D. connect the duct sections using the proper S-slips and drive cleats tightly to prevent duct vibration.

7. A service call is made stating that a family's furnace and duct system have recently been installed by a friend of the family. The problem is that there doesn't seem to be enough airflow out of the registers. After arriving at the job, the technician noticed that the furnace is 75,000 Btu/h and the air conditioning system is 2 tons of cooling. After further investigation, the technician notices that the main supply duct system is 60″ by 8″ by 25′ long. What is the problem with the system?

A. The system's static pressure is too low.

B. The system's main duct is too small.

C. The system's static pressure is too high.

D. The system's main duct is too large.

E. Both A and D

8. Technician A measures the airflow in a main supply duct by using a Pitot-tube traverse. His measurement is 1,200 fpm. The area of the duct at the point of measurement is 8″ × 30″. He says there is 2,000 cfm.

Technician B measures the airflow in a main supply duct by using a Pitot-tube traverse. His measurement is 1,200 fpm. The area of the duct at there is measurement is 10″ × 30″. He says there is 2,500 cfm.

Which technician is correct?

A. Technician A

B. Technician B

C. Both technicians are correct.

D. Both technicians are incorrect.

9. When should the dampers and registers of a new system be 100% open?

A. after the system has been fully balanced

B. on initial startup of the system

C. during the heating season only

D. during the cooling season only

10. When using the Duct Sizing Calculator (Figure 4–15), technician A uses 1,800 cfm as the air quantity with 900 fpm air velocity. He says that the static pressure is 0.10″ of water column.

When using the Duct Sizing Calculator, technician B uses 1,600 cfm as the air quantity with 1,100 fpm air velocity. He says that the static pressure is 0.10″ of water column.

Which technician is correct?

A. Technician A

B. Technician B

C. Both technicians are correct.

D. Both technicians are incorrect.

11. When using the Duct Sizing Calculator (Figure 4–15), the required volume of air is 1,600 cfm and air velocity is 1,100 fpm. What size of round and rectangular duct is needed?

A. 18″ and 8″ × 30″

B. 16″ and 10″ × 20″

C. 20″ and 12″ × 18″

D. 16″ and 14″ × 18″

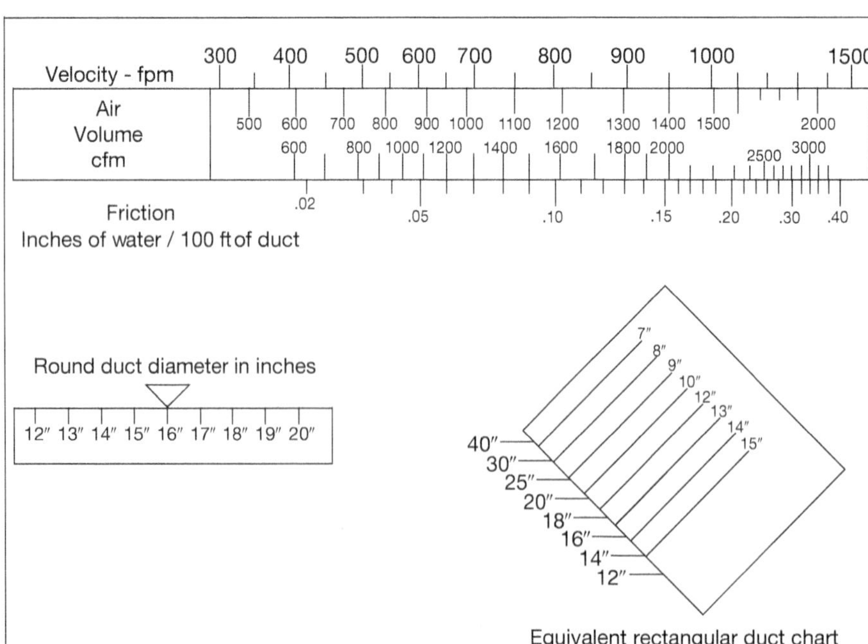

FIGURE 4–15
Duct Sizing Calculator

12. Using the Duct Sizing Calculator (Figure 4–15), what is the air velocity if the air quantity is 1,400 cfm?

A. 800 fpm

B. 950 fpm

C. 1,800 fpm

D. 2,000 fpm

13. Technician A says that an 8″ × 12″ register that has the velocity of 300 fpm with a free area of 75% has an air quantity of 150 cfm.

Technician B says that an 8″ × 12″ register that has the velocity of 300 fpm with a free area of 75% has an air quantity of 250 cfm.

Which technician is correct?

A. Technician A

B. Technician B

C. Both technicians are correct.

D. Both technicians are incorrect.

14. A service call is made on the complaint of not enough warm air out of the supply registers. The technician does an airflow calculation by measuring the temperature difference across the heat exchanger, which is 100°F. He then determines the furnace output, which is 90,000 Btu/h. With this information, he can determine the quantity of airflow out of the system. Which of the following formulas must be used?

A. cfm = [Output(Btu/h)]/(1.08 × temperature rise)

B. cfm = [Output(Btu/h) × 1.08]/(temperature rise)

C. cfm = [Output(Btu/h) × temperature rise]/1.08

D. cfm = [Output(Btu/h)]/(1.08 + temperature rise)

15. On the previous question, what is the quantity of air that is flowing in the system?

A. 833 cfm

B. 972 cfm

C. 8,333 cfm

D. 890 cfm

16. What would be the velocity of the air at a 0.2 friction loss per 100′ and 50 cfm of air? Use the Friction Loss Chart (Figure 4–16) to help answer this question.

A. 85 fpm

B. 200 fpm

C. 500 fpm

D. 600 fpm

17. What would the diameter of the duct have to be to move 300 cfm at the velocity of 600 fpm? Use the Friction Loss Chart (Figure 4–16) to help answer this question.

A. 6″

B. 8″

C. 10″

D. 12″

18. What would be the friction loss per 100′ if the speed of the air through an 8″ duct was 600 fpm? Use the Friction Loss Chart (Figure 4–16) to help answer this question.

A. .05

B. .08

C. 1.0

D. .15

19. Using the Friction Loss Chart (Figure 4–16), technician A says that to keep the static pressure of a duct at .1″ WC, with 250 cfm in the duct, the size of the duct must be 8″ round.

Using the Friction Loss Chart, technician B says that to keep the velocity of a 6″ duct to 500 fpm, the air quantity must be 100 cfm. Which technician is correct?

A. Technician A

B. Technician B

C. Both technicians are correct.

D. Both technicians are incorrect.

20. Which of the following formulas is used to calculate the area of a round duct?

A. diameter × circumference

B. radius × 3.14

C. 3.14 × radius^2

D. 3.14 × diameter^2

21. Technician A says that a blower over 5 horsepower is commonly a PSC-type motor. Technician B says that a blower over 5 horsepower is commonly a three-phase-type motor. Which technician is correct?

A. Technician A

B. Technician B

C. Both technicians are correct.

D. Both technicians are incorrect.

22. What is commonly used to cycle condenser fan motors on HVAC rooftop equipment during low ambient conditions?

A. a high-ambient-pressure switch

B. a low-pressure switch

C. a high-pressure switch

D. a reverse-acting high-pressure switch

23. Why is it necessary to cycle the condenser fans during low ambient conditions?

A. to prevent an increase in high-side pressure and the flooding of the evaporator coil

B. to prevent a decrease in high-side pressure and the flooding of the evaporator coil

C. to prevent an increase in high-side pressure and the starving of the evaporator coil

D. to prevent a decrease in high-side pressure and the starving of the evaporator coil

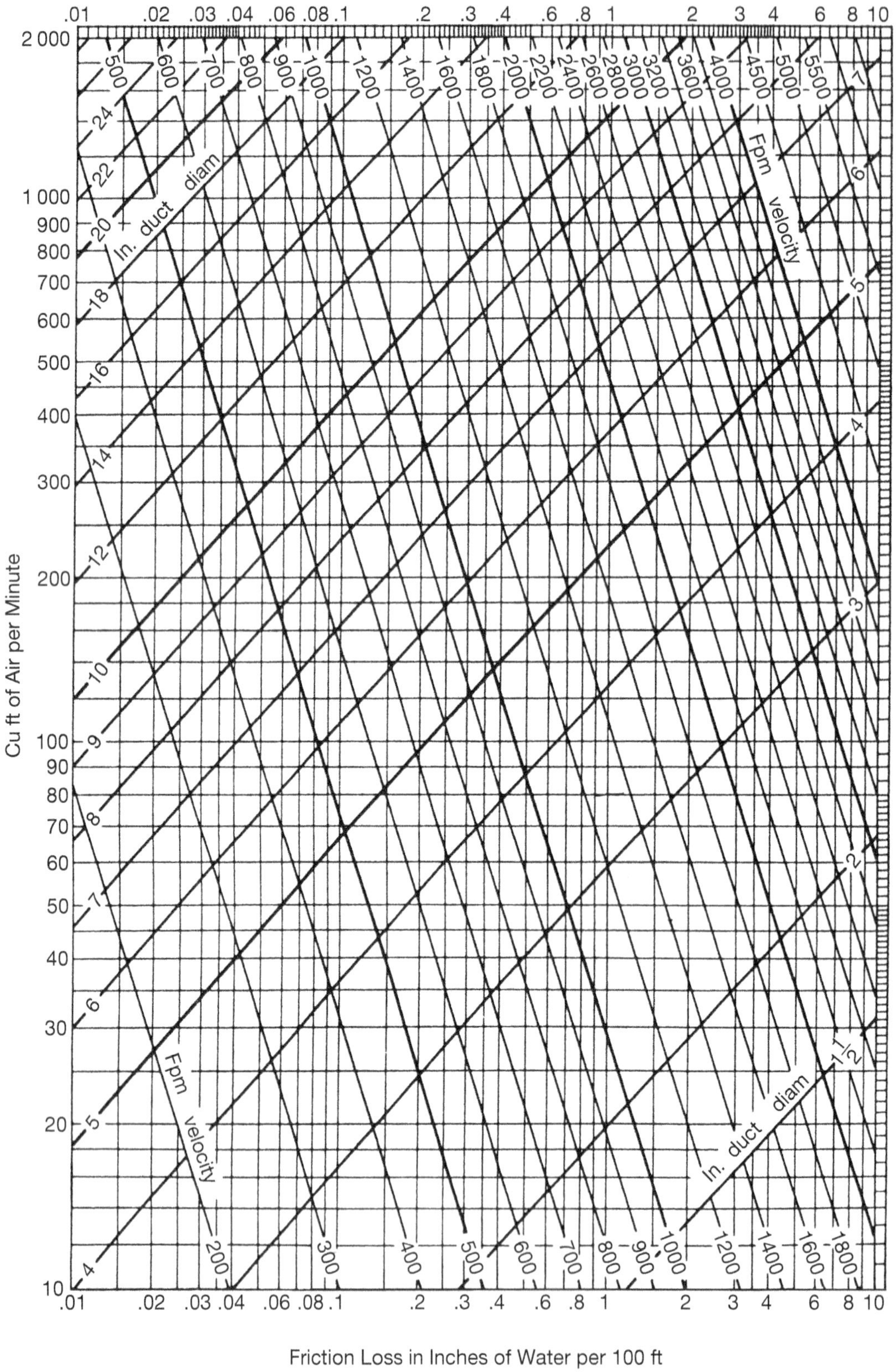

(Based on Standard Air of 0.075 lb per cu ft density flowing through average, clean, round, galvanized metal ducts having approximately 40 joints per 100 ft.)

Caution: Do not extrapolate below chart.

FIGURE 4–16

Friction Loss Chart

24. A belt-drive fan motor that runs at 1,050 rpm uses a 6″ pulley and operates a blower that has a 9″ pulley.

What is the rpm of the blower?

A. 1,500

B. 1,575

C. 775

D. 700

25. What quantity of air must the indoor blower provide per ton of refrigeration effect?

A. 300 cfm

B. 400 cfm

C. 600 cfm

D. 800 cfm

26. When the static pressure is increased in a duct system, how does this affect the blower motor?

A. The amperage increases.

B. The amperage decreases.

C. The air quantity increases.

D. The air quantity decreases.

27. Which of the following can cause an increase (more positive) of supply-air-duct static pressure and an increase (less negative) of return-air-duct static pressure?

A. a dirty air filter

B. too small of a return-air duct

C. too small of a supply-air duct

D. Both A and B

28. Which of the following can cause a decrease (less positive) of supply-air-duct static pressure and a decrease (more negative) of return-air-duct static pressure?

A. too large of a supply-air duct

B. too small of a return-air duct

C. too small of a supply-air duct

D. too large of a return-air duct

29. Which of the following can cause an increase (more positive) of supply-air-duct static pressure and a decrease (more negative) of return-air-duct static pressure?

A. too large of a supply-air duct and too small of a return duct

B. too small of a return-air duct and too small of a supply duct

C. too small of a supply-air duct and too large of a return duct

D. too large of a return-air duct and too large of a supply duct

30. Which of the following can cause a decrease (less positive) of supply-air-duct static pressure and an increase (less negative) of return-air-duct static pressure?

A. too large of a supply-air duct and too small of a return duct

B. too small of a return-air duct and too small of a supply duct

C. too small of a supply-air duct and too large of a return duct

D. too large of a return-air duct and too large of a supply duct

31. The minimum total area of a supply-air duct used on a forced-warm-air furnace

A. must be at least 1 sq in. per 1,000 Btu/h approved output rating.

B. must be at least 2 sq in. per 1,000 Btu/h approved output rating.

C. must be at least 4 sq in. per 1,000 Btu/h approved output rating.

D. must be at least 6 sq in. per 1,000 Btu/h approved output rating.

32. The minimum total area of a supply-air duct used on a heat pump

A. must be at least 1 sq in. per 1,000 Btu/h approved output rating.

B. must be at least 2 sq in. per 1,000 Btu/h approved output rating.

C. must be at least 4 sq in. per 1,000 Btu/h approved output rating.

D. must be at least 6 sq in. per 1,000 Btu/h approved output rating.

33. Technician A says that a smaller area in the supply-air trunk will increase the air velocity.

Technician B says that a smaller area in the supply-air trunk increases the air quantity.

Which technician is correct?

A. Technician A

B. Technician B

C. Both technicians are correct.

D. Both technicians are incorrect.

34. Which of the following will cause a decrease of supply-air static pressure?

A. a section of the return-air trunk being disconnected

B. a higher fan speed

C. a section of the supply-air trunk being disconnected

D. Both B and C

35. Technician A says that to increase the speed on a belt-drive blower, the adjusting pulley should be closed more.

Technician B says that to increase the speed on a belt-drive blower, the adjusting pulley should be opened more.

Which technician is correct?

A. Technician A

B. Technician B

C. Both technicians are correct.

D. Both technicians are incorrect.

36. What is the definition of *friction loss* in a duct system?

A. the resistance to airflow

B. the resistance to air pressure

C. the resistance to static pressure

D. None of the above

37. When a leaking duct system has been sealed,

A. static pressure will decrease.

B. static pressure will increase.

C. velocity increases.

D. cfm decreases.

38. An advantage of the use of long-radius fittings in duct systems is that it

A. reduces pressure drops.

B. decreases cfm.

C. increases static pressures.

D. decreases velocity.

39. The condition when temperature layers are throughout a room is called

A. wet-bulb temperature.

B. dry-bulb temperature.

C. stratification.

D. imbalance temperature.

40. When the return-air side of an air-handling system is decreased, how does this affect the blower motor?

A. The amperage increases.

B. The amperage decreases.

C. The air quantity increases.

D. The air quantity decreases.

41. A technician inspecting a motorized economizer should inspect

A. the operation of the damper in the free cooling mode.

B. the operation of the damper in the mechanical cooling mode.

C. the condition of the vanes to ensure there is no binding.

D. All of the above

42. A technician inspecting a duct system should inspect

A. the air filter.

B. the blower operation.

C. the duct systems for proper connections.

D. All of the above

THE FOLLOWING QUESTIONS MAY BE HELPFUL TO STUDY FOR THESE CERTIFICATION TESTS:
NATE—CR-SV, CR-IN, AD-SV, AD-IN, OL-SV, OL-IN, AC-SV, AC-IN, HP-SV, HP-IN, GS-SV, GS-IN.
ESCO INSTITUTE HVAC EXCELLENCE HEAT LOAD CALCULATION

TEMPERATURE, HUMIDITY, AND PSYCHROMETRICS

1. What is commonly used as a sensing element for an electronic thermometer?

A. a resistor

B. a transistor

C. a thermistor

D. a diode

2. A dry-bulb thermometer

A. measures sensible temperature of a substance.

B. senses the point at which moisture condenses in the air.

C. uses a saturated wick over the end of a thermometer.

D. senses the moisture content of the air.

3. Which instrument contains both the wet-bulb and the dry-bulb thermometer?

A. a sling psychrometer

B. a dial thermometer

C. a recording thermometer

D. an electronic thermometer

4. A sling psychrometer is used to determine

A. the absolute humidity of the air.

B. the relative humidity of the air.

C. the dewpoint temperature of the air.

D. the apparatus dewpoint temperature.

5. When using a wet-bulb thermometer, as the moisture content of the air decreases

A. the temperature of the bulb increases.

B. the temperature of the bulb decreases.

C. the temperature of the bulb stays the same.

D. moisture content does not affect the wet-bulb temperature.

6. A psychrometric chart is used to

A. measure air content.

B. measure air currents.

C. measure air properties.

D. measure air in SI units.

7. Which term refers to the temperature at which moisture in the air begins to condense?

A. dewpoint

B. enthalpy

C. saturation point

D. wet bulb

8. A property of substances that is a measurement of total heat content is called

A. dry bulb.

B. saturation temperature.

C. sensible heat.

D. enthalpy.

9. The term *humidity ratio* refers to

A. the weight of water vapor in each pound of vapor.

B. the weight of moisture measured in grains.

C. total relative humidity.

D. the difference of humidity and pounds of dry air.

10. The amount of cubic feet of the mixture per pound of dry air is best described as

A. total heat.

B. specific heat.

C. specific volume.

D. enthalpy.

11. On a psychrometric chart (Figure 4–17), which lines are vertical?

A. the dry-bulb temperature lines

B. the wet-bulb temperature lines

C. the dewpoint temperature lines

D. the relative humidity lines

12. On a psychrometric chart (Figure 4–18), which lines are horizontal?

A. the dry-bulb temperature lines

B. the wet-bulb temperature lines

C. the dewpoint temperature lines

D. the relative humidity lines

13. On a psychrometric chart (Figure 4–19), which lines are diagonal?

A. the dry-bulb temperature lines

B. the wet-bulb temperature lines

C. the dewpoint temperature lines

D. the relative humidity lines

14. On a psychrometric chart (Figure 4–20), which lines are curved?

A. the dry-bulb temperature lines

B. the wet-bulb temperature lines

C. the dewpoint temperature lines

D. the relative humidity lines

15. Complete air conditioning means to

A. cool air within a space.

B. dehumidify air within a space.

C. cool and dehumidify air within a space.

D. heat, humidify, cool, dehumidify, and filter air within a space.

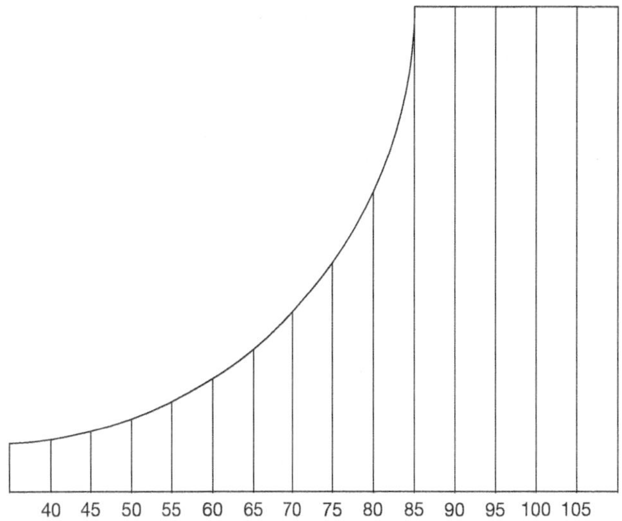

FIGURE 4–17

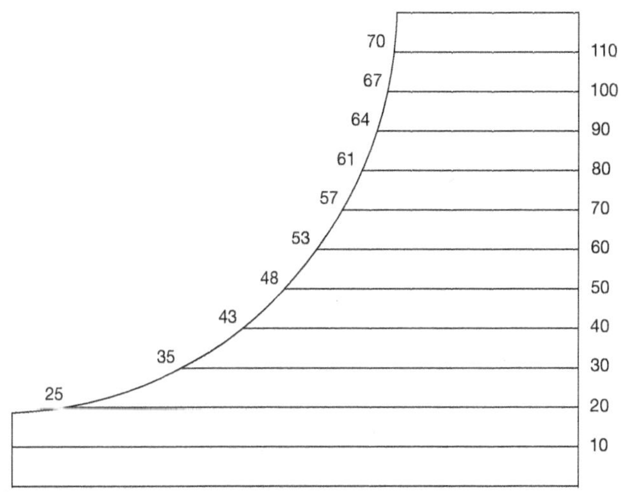

FIGURE 4–18

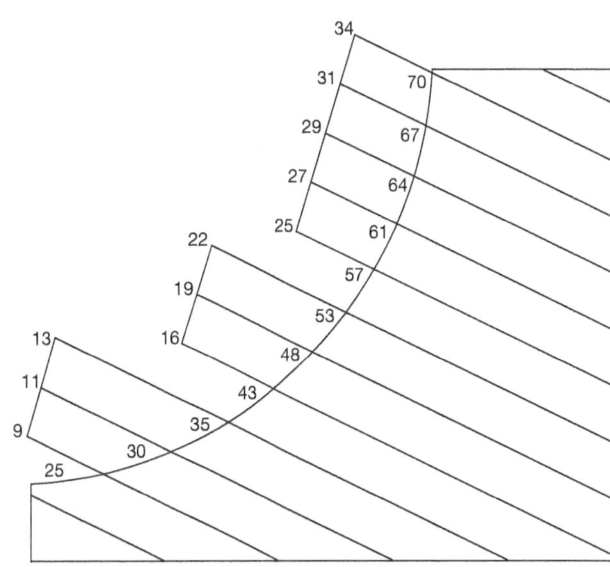

FIGURE 4–19

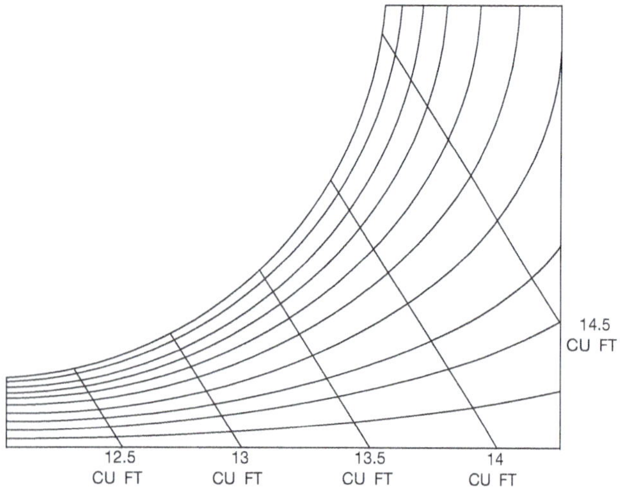

FIGURE 4–20

16. A typical comfort goal for the HVAC industry is maintaining air
   A. to an 80°F wet-bulb temperature at 70% relative humidity.
   B. to a 75°F dry-bulb temperature at 50% relative humidity.
   C. to a 65°F dry-bulb temperature at 40% relative humidity.
   D. to a 60°F wet-bulb temperature at 35% relative humidity.
17. What two factors decrease the rate of evaporation of moisture from a person to the surrounding air?
   A. a high dry-bulb temperature and a low relative humidity
   B. a low dry-bulb temperature and a high relative humidity
   C. a low dry-bulb temperature and a low relative humidity
   D. a high dry-bulb temperature and a high relative humidity

THE FOLLOWING QUESTIONS MAY BE HELPFUL TO STUDY FOR THESE CERTIFICATION TESTS:
NATE—CR-SV, CR-IN, AD-SV, AD-IN, GS-SV, GS-IN, OL-SV, OL-IN, AC-SV, AC-IN, HP-SV, HP-IN.
ESCO INSTITUTE HVAC EXCELLENCE HEAT LOAD CALCULATION

HEAT LOSS AND HEAT GAIN

1. What are the factors that can be used to determine the heat transfer through a wall?
   A. wall area, TD, and U-value
   B. wall TD length and properties
   C. wall height properties and C-value
   D. wall thickness, TD, and k-value
2. The R-value of a material is its ability to
   A. conduct heat flow through a wall.
   B. resist the flow of heat through a wall.
   C. conduct solar heat.
   D. conduct infiltration heat gain.
3. The higher the R-value of a material, the
   A. more resistive it is to heat transfer.
   B. less resistive it is to heat transfer.
   C. more conductive it is to heat transfer.
   D. more conductivity the material has.
4. To determine heat transfer through a wall using the R-value, use the following calculation:
   A. $Q = C + R \times A \times TD$.
   B. $Q = R \times A \times TD$.
   C. $Q = 1/R \times A \times TD$.
   D. None of the above
5. Wall material and thickness affect the heat transfer through the wall. There are two terms used when it comes to the construction of walls:
   A. R-value and conductance.
   B. U-value and conductance.
   C. conductance and conductivity.
   D. R-value and U-value.
6. Conductance is the ability of a material to transfer heat in Btu/hr/ft^2/F; it is known as the
   A. C-value.
   B. k-value.
   C. R-value.
   D. U-value.
7. *Conductance* is a term used to denote heat flow through
   A. nonhomogeneous materials such as glass blocks, hollow clay tiles, and concrete block.
   B. a 1″ thickness of homogeneous materials.
   C. all of the material in an entire wall.
   D. None of the above
8. Conductivity is the conductance of a material per unit of thickness; it is known as the
   A. C-value.
   B. k-value.
   C. R-value.
   D. U-value.

9. What is defined as the heat flow per hour (Btu/hr) through 1 ft^2 of 1″-thick homogeneous material when the temperature difference between the two faces is 1°F?
   A. thermal conductance
   B. thermal resistance
   C. thermal conductivity
   D. thermal negligence

10. Which of the following symbols shows the conductance of the air film of inside still air?
    A. Fi = C = 2.2
    B. Fo = k = .625
    C. Fi = C = 1.6
    D. Fi = k = 2.2

11. When calculating heat transfer through a wall, which of the following statements is true?
    A. C-values must be determined on all materials and added up to calculate the U-value.
    B. k-values must be determined on all materials and added up to calculate the U-value.
    C. F-values must be determined on all materials and added up to calculate the U-value.
    D. R-values must be determined on all materials and added up to calculate the U-value.

12. To determine overall heat transfer of a wall that uses four or five different materials, the
    A. U-factor is used to determine total heat transfer through the area of a wall.
    B. C-factor is used to determine total heat transfer through the area of a wall.
    C. k-factor is used to determine total heat transfer through the area of a wall.
    D. R-factor is used to determine total heat transfer through the area of a wall.

13. Which of the following is the formula for overall heat transfer through a wall?
    A. Q = A × R × TD
    B. Q = A × C × TD
    C. Q = A × U × TD
    D. Q = A × k × TD

14. What is the first step in determining the heat gain through an exterior wall or door?
    A. Determine the time of day to be considered.
    B. Determine the wall's U-value.
    C. Determine the CLTD.
    D. Determine the indoor design and the outdoor design conditions.

15. Which of the following is the formula that should be used to determine the heat gain through a sun-lit wall?
    A. C × A × TD = Btu/hr
    B. C × A × CLTD = Btu/hr
    C. U × A × CLTD = Btu/hr
    D. U × A × TD = Btu/hr

16. Which of the following is an important factor when determining the heat gain through a sun-lit wall?
    A. the time of day
    B. the direction the wall faces—N, S, E, W
    C. the U-value of the wall
    D. All of the above

17. Which two factors determine the heat gain through a window?
    A. the conduction of the window and the S.H.G.F.
    B. the resistance of the window and the CLTD
    C. the U- and k-value of the window
    D. All of the above

18. What is the first step in determining the heat gain through an exterior window?
    A. Determine the SC1.
    B. Determine the wall direction (N, S, E, W).
    C. Determine the time of day.
    D. None of the above

19. When a window has some type of interior or exterior shading, which of the following must be taken into consideration when calculating the heat gain through the window?
    A. the direction of the sun
    B. the CLTD
    C. the shade coefficient
    D. the U-value only

20. What is the first step in determining the heat gain through a roof?
    A. Determine the SC1.
    B. Determine the wall direction (N, S, E, W).
    C. Determine the roof construction and U-value.
    D. None of the above

21. Which of the following is the formula for determining the heat gain through a roof?
    A. Btu/hr = Factor × Area × U
    B. Btu/hr = Factor × Area × k
    C. Btu/hr = Factor × Area × R
    D. Btu/hr = Factor × Area × TD

22. Which of the following is considered the infiltration of a building?
    A. a quantity of air that enters a building
    B. a quantity of air that exits a building
    C. still air
    D. air in motion

23. Infiltration is always measured in

A. FPM.

B. GPM.

C. CFM.

D. CFH.

24. What is the minimum amount of ventilation air required for each person in a commercial building?

A. 15 CFM

B. 30 CFM

C. 40 CFM

D. 50 CFM

25. What is known as the "crack method"?

A. It is a method used to determine the CLTD of a window.

B. It is a method used to calculate sun rays through a window.

C. It is a method used to calculate the infiltration of a window.

D. It is a method used to calculate the Btu/min rating of a window.

26. Infiltration air always brings with it

A. sensible heat only.

B. latent heat only.

C. sensible and latent heat.

D. radiant heat.

27. To determine heat lost or gained into a building, take the required ventilation of the building and use the following formula to determine sensible heat loss or gain.

A. Sensible heat in Btu/hr = CFM × 1.08 or (1.1) × TD

B. Sensible heat Btu/hr = CFM × .68 × Grains of moisture difference

C. C × A × TD = Btu/hr

D. C × A × CLTD = Btu/hr

28. To determine heat lost or gained into a building, take the required ventilation of the building and use the following formula to determine latent heat loss or gain.

A. CFM × 1.08 or (1.1) × TD

B. CFM × .68 × Grains of moisture difference

C. C × A × TD = Btu/hr

D. C × A × CLTD = Btu/hr

29. Which of the following are considered sources of internal heat gain in a building?

A. lighting

B. appliances

C. motors and people

D. All of the above

30. Which of the following formulas is used to determine the heat gain due to incandescent light bulbs?

A. Incandescent bulb wattage × Number of bulbs × 4.13 Btu/watt = Btu/hr

B. Incandescent bulb wattage × Number of bulbs × 3.14 Btu/watt = Btu/hr

C. Incandescent bulb wattage × Number of bulbs × 3.41 Btu/watt = Btu/hr

D. Incandescent bulb wattage × Number of bulbs × 4.2 Btu/watt = Btu/hr

31. Which of the following formulas is used to determine the heat gain due to fluorescent light bulbs?

A. Fluorescent wattage × Number of bulbs per fixturee × Number of fixtures × 1.2 = total Watts × 4.31 = Btu/hr

B. Fluorescent wattage × Number of bulbs per fixture × Number of fixtures × 1.2 = total Watts × 3.41 = Btu/hr

C. Fluorescent wattage × Number of bulbs per fixture × Number of fixtures × 2.1 = total Watts × 3.41 = Btu/hr

D. Fluorescent wattage × Number of bulbs per fixture × Number of fixtures × 1.2 = total Watts × 3.14 = Btu/hr

32. Which of the following formulas is used to determine heat gain by motors?

A. H.P. × 2545 Btu/hr/hp = Btu/hr

B. H.P. × 2255 Btu/hr/hp = Btu/hr

C. H.P. × 4525 Btu/hr/hp = Btu/hr

D. H.P. × 5522 Btu/hr/hp = Btu/hr

33. When determining the amount of heat people give off, it is always based on activity levels, and people produce

A. sensible heat only.

B. latent heat only.

C. sensible and latent heat.

D. radiant heat.

34. What is the amount of heat that a $\frac{1}{2}$-hp motor produces?

A. 7215.5 Btu/hr sensible

B. 7215.5 Btu/hr latent and sensible

C. 1272.5 Btu/hr latent and sensible

D. 1272.5 Btu/hr sensible

35. What type of heat energy do lights produce?

A. sensible heat only

B. latent heat only

C. sensible and latent heat

D. radiant heat

PART III
Answers

SECTION 1

Refrigeration and Air Conditioning Systems and Tools

Personal Safety

1. D	2. A	3. C	4. C	5. A	6. B
7. D	8. C	9. D	10. A	11. C	12. D
13. B	14. D	15. A	16. C	17. A	18. B

Hand Tools

1. D	2. B	3. C	4. A	5. D	6. A
7. D	8. B	9. D	10. A	11. B	12. C
13. D	14. B	15. B	16. A	17. C	18. D
19. A	20. B	21. D	22. A	23. B	24. C
25. D	26. A	27. A	28. A	29. C	30. A
31. B	32. A	33. C	34. B		

Basic Construction

1. C	2. B	3. A	4. D	5. B	6. B
7. C	8. C	9. D	10. C	11. A	12. D
13. B	14. B	15. A			

Refrigerant Safety

1. C	2. D	3. B	4. D	5. D	6. B
7. C	8. A	9. C	10. D	11. C	12. D
13. B	14. D	15. C	16. A	17. C	18. A
19. B	20. A	21. C	22. D	23. C	24. D
25. C	26. A	27. B	28. A	29. C	30. C
31. C	32. A	33. D	34. C	35. C	36. A
37. B	38. B	39. C	40. A	41. C	42. B
43. D					

Refrigeration Theory

1. A	2. C	3. B	4. D	5. B	6. D
7. A	8. C	9. A	10. B	11. A	12. A
13. C	14. A	15. D	16. B	17. C	18. C
19. C	20. D	21. A	22. B	23. D	24. A
25. B	26. C	27. A	28. B	29. D	30. B
31. C	32. D	33. B	34. B	35. D	36. A
37. C	38. B	39. A			

Refrigerants

1. B	2. B	3. D	4. C	5. B	6. A
7. D	8. C	9. A	10. A	11. D	12. B
13. D	14. B	15. B	16. C	17. A	18. B
19. D	20. B	21. A	22. B	23. C	24. A
25. B	26. C	27. B	28. D	29. B	

Lubricants

1. A	2. D	3. B	4. A	5. D	6. C
7. A	8. C	9. A	10. B	11. D	12. A
13. C	14. A	15. C	16. A	17. B	18. C
19. B					

Compressors

1. C	2. B	3. D	4. D	5. B	6. B
7. C	8. B	9. C	10. C	11. A	12. C

Air-Cooled Condensers

1. B 2. D 3. D 4. C 5. C 6. B
7. D

Water-Cooled Condensers

1. C 2. B 3. A 4. D 5. B 6. C
7. D 8. C 9. C 10. D 11. C 12. D
13. A 14. D 15. C 16. B 17. A 18. C
19. A 20. B 21. C 22. D 23. C 24. B
25. C 26. A

Indoor Evaporator Coils

1. B 2. B 3. A 4. D 5. B 6. D
7. C 8. B 9. C 10. A 11. B 12. C
13. D 14. B 15. B 16. A 17. A 18. B
19. C 20. B 21. C 22. A 23. B 24. C

Metering Devices

1. C 2. B 3. A 4. D 5. C 6. D
7. C 8. D 9. C 10. B 11. A 12. C
13. C

Thermostatic Expansion Valves

1. D 2. D 3. B 4. D 5. C 6. C
7. C 8. A 9. B 10. A 11. A 12. B
13. C 14. D 15. B 16. C 17. A 18. D
19. A 20. C 21. B 22. D 23. C 24. B
25. A 26. D 27. C 28. B 29. C 30. D
31. A 32. D 33. B 34. D 35. B 36. C
37. A 38. C 39. B 40. C 41. A 42. B
43. C 44. A 45. B 46. D 47. B 48. A
49. C 50. D 51. B 52. D 53. B 54. C
55. A 56. C

Service Valves

1. B 2. A 3. B 4. C 5. B 6. A
7. B 8. D 9. C 10. C 11. D 12. A
13. B

Leak Detection

1. A 2. B 3. A 4. C 5. B 6. D
7. C 8. D 9. D 10. A 11. C 12. B
13. C 14. C 15. D

Recovery Equipment

1. A 2. A 3. C 4. D 5. A 6. B
7. D 8. B 9. D 10. A 11. D 12. C
13. B

Soldering and Brazing Safety

1. C 2. D 3. B 4. C 5. D 6. D
7. A 8. C 9. B 10. D 11. C 12. D
13. B

Soldering and Brazing

1. C 2. B 3. C 4. A 5. B 6. D
7. C 8. A 9. A 10. D 11. C 12. C
13. B 14. A 15. D 16. A 17. D 18. C
19. D 20. B 21. C 22. A

ESCO Institute-HVAC Excellence Brazing and Soldering Certification Employment Ready (ER)

1. D 2. B 3. C 4. B 5. A 6. C
7. A 8. D 9. D 10. A 11. A 12. D
13. C 14. B 15. C 16. D 17. B 18. A
19. C 20. A 21. A 22. C 23. A 24. A
25. D 26. D 27. C 28. B 29. A 30. D
31. A 32. C 33. B 34. C 35. C 36. C
37. B 38. D 39. B 40. D 41. A 42. C
43. A 44. C 45. A 46. B 47. B 48. D
49. A 50. C

Manifold Gauges

1. C 2. D 3. B 4. B 5. C 6. C
7. A 8. D 9. A 10. B 11. C

System Evacuation

1. A 2. C 3. B 4. C 5. C 6. D
7. A 8. C 9. C 10. B 11. B 12. B

ESCO Institute-HVAC Excellence System Recovery and Evacuation Certification Employment Ready (ER)

1. A	2. D	3. A	4. B	5. C	6. D
7. B	8. A	9. A	10. A	11. D	12. C
13. B	14. B	15. A	16. C	17. C	18. B
19. B	20. D	21. D	22. B	23. C	24. A
25. B	26. B	27. B	28. C	29. B	30. D
31. C	32. C	33. D	34. B	35. D	36. D
37. A	38. B	39. A	40. B	41. C	42. C
43. A	44. B	45. B	46. C	47. B	48. B
49. C	50. B				

Refrigerant Charging

1. A	2. B	3. B	4. C	5. D	6. A
7. C	8. A	9. C	10. D	11. A	

ESCO Institute-HVAC Excellence Basic Refrigeration and Charging Procedures Certification Employment Ready (ER)

1. C	2. B	3. C	4. D	5. B	6. D
7. A	8. C	9. B	10. A	11. A	12. A
13. C	14. A	15. C	16. B	17. C	18. A
19. A	20. A	21. C	22. D	23. A	24. B
25. D	26. B	27. B	28. A	29. B	30. D
31. A	32. D	33. D	34. B	35. A	36. A
37. B	38. A	39. B	40. C	41. D	42. B
43. B	44. D	45. C	46. C	47. A	48. A
49. C	50. C				

EPA Section 608 Certification Core Section

1. A	2. C	3. B	4. C	5. A	6. B
7. A	8. C	9. B	10. A	11. B	12. A
13. C	14. C	15. B	16. A	17. C	18. C
19. A	20. B	21. B	22. B	23. C	24. A
25. C	26. D	27. C	28. B	29. A	30. B
31. C	32. A	33. C	34. A	35. B	36. C
37. B	38. B	39. C	40. B	41. C	42. C
43. A	44. A	45. C	46. C	47. C	48. A
49. B	50. C	51. B	52. C	53. A	54. A
55. B	56. C	57. A	58. B	59. A	60. B
61. A	62. A	63. C	64. A	65. C	

Type I Certification

1. B	2. A	3. C	4. B	5. A	6. C
7. B	8. B	9. C	10. C	11. C	12. B
13. A	14. C	15. A	16. C	17. A	18. B
19. B	20. B	21. A	22. A	23. A	

Type II Certification

1. C	2. C	3. C	4. C	5. B	6. A
7. A	8. B	9. B	10. B	11. C	12. A
13. A	14. C	15. B	16. A	17. A	18. C
19. A	20. A	21. C	22. A	23. C	24. A
25. C	26. B	27. A	28. A	29. C	30. C
31. C	32. C	33. C	34. C	35. A	36. C
37. B	38. C	39. B	40. C		

Type III Certification

1. C	2. B	3. B	4. B	5. C	6. A
7. A	8. A	9. B	10. A	11. A	12. B
13. A	14. C	15. B	16. B	17. A	18. C
19. C	20. C	21. C	22. C	23. C	24. A
25. C	26. B	27. C	28. B	29. C	

Refrigerant Piping

1. D	2. B	3. D	4. C	5. D	6. A
7. B	8. B	9. D	10. D	11. B	12. C
13. D	14. D	15. B	16. D	17. D	18. D
19. C	20. A	21. C	22. C	23. B	24. B
25. D	26. A	27. C	28. B	29. A	30. D
31. A	32. D	33. C	34. D	35. D	36. B
37. B	38. C	39. C			

Refrigeration Accessories

1. D	2. D	3. A	4. D	5. C	6. C
7. C	8. A	9. D	10. C	11. B	12. C
13. C	14. C	15. D	16. D	17. D	18. A
19. D	20. D	21. D	22. B	23. B	24. A
25. C	26. C	27. C	28. C	29. B	30. A
31. B	32. C	33. B	34. C	35. A	

Medium- and Low-Temperature Refrigeration

1. C	2. D	3. B	4. A	5. C	6. B
7. D	8. D	9. A	10. A	11. C	12. C
13. D	14. B	15. C	16. C	17. C	18. B
19. D	20. C	21. A	22. A	23. B	24. B
25. B	26. C	27. C	28. A	29. B	30. C
31. D	32. A	33. A	34. D	35. D	36. A
37. C	38. D				

Defrost Systems

1. C	2. B	3. A	4. A	5. D	6. D
7. D	8. B	9. C	10. A	11. D	12. C
13. A	14. B	15. A	16. A	17. D	18. C
19. D	20. D	21. D	22. A	23. B	24. C
25. D	26. C				

Commercial Refrigeration Electrical Diagrams

1. C	2. A	3. C	4. A	5. C	6. B
7. D	8. B	9. B	10. D	11. A	12. D
13. C	14. D	15. A	16. D	17. B	18. D
19. B	20. B	21. C	22. B	23. B	24. C
25. D	26. D	27. C	28. A	29. B	30. C
31. C	32. D	33. D	34. D	35. B	36. A
37. B	38. C	39. D	40. C	41. D	42. C
43. B	44. D	45. D	46. B	47. B	48. A
49. A	50. D	51. C	52. D	53. D	54. D
55. B	56. A	57. C	58. D	59. D	60. A
61. C	62. D				

Commercial Refrigeration Troubleshooting

1. D	2. C	3. B	4. B	5. D	6. A
7. B	8. C	9. C	10. C	11. B	12. D
13. D	14. C	15. C	16. C	17. A	18. A
19. D	20. D	21. C	22. B	23. B	24. A
25. B	26. A	27. D	28. A	29. D	30. B
31. C	32. B	33. A	34. C	35. D	36. C
37. D	38. C	39. D	40. B	41. C	42. B
43. D	44. B	45. C	46. C	47. B	48. D
49. D	50. B	51. A	52. D	53. C	54. A
55. B	56. C	57. B	58. A	59. C	60. B
61. B	62. B	63. A	64. C	65. B	66. D
67. D	68. D	69. C	70. A	71. D	72. D
73. C	74. A	75. B	76. B	77. B	78. C
79. A	80. C	81. C	82. B	83. C	84. C
85. D	86. C	87. C	88. C	89. D	90. B
91. B	92. C	93. B	94. C	95. C	96. B
97. B	98. D	99. C	100. A	101. A	102. D
103. D	104. D	105. A	106. D	107. D	108. D
109. A	110. B	111. B	112. D	113. D	114. C
115. B	116. A	117. B	118. D	119. A	120. D
121. C	122. B	123. C	124. C	125. A	126. B
127. D	128. B	129. A	130. A	131. D	132. A
133. D	134. C	135. B	136. B	137. A	138. B
139. C	140. B	141. B	142. C	143. A	144. D
145. B	146. C	147. A	148. A	149. A	150. B
151. D					

Air Conditioning Components

1. A	2. B	3. D	4. C	5. C	6. C
7. D					

8.

1. B	2. H	3. C	4. I	5. E	6. J
7. F	8. G	9. D	10. A		

9. A

10.

1. A	2. D	3. B	4. C	5. H	6. F
7. I	8. E	9. J	10. G		

11. B	12. C	13. A	14. C	15. C	16. A
17. C	18. D	19. A	20. A	21. D	22. C
23. B	24. D	25. A	26. C	27. B	28. C
29. A	30. A	31. B	32. C	33. C	34. B
35. C	36. D	37. B			

Air Conditioning Troubleshooting

1. C	2. A	3. C	4. B	5. A	6. B
7. D	8. B	9. C	10. D	11. B	12. B
13. D	14. A	15. D	16. B	17. A	18. C
19. D	20. B	21. D	22. B	23. D	24. B
25. D	26. B	27. D	28. A	29. B	30. D
31. A	32. A	33. B	34. A	35. D	36. C
37. D	38. D	39. C	40. D	41. D	42. C
43. D	44. D	45. B	46. B	47. D	48. B
49. C	50. D	51. C	52. B	53. D	54. B
55. D	56. A	57. B	58. C	59. D	60. B
61. D	62. B	63. A	64. A	65. B	66. C
67. B	68. D	69. C	70. D	71. D	72. A
73. D	74. B	75. B	76. B	77. D	78. B
79. C	80. A	81. D	82. A	83. B	84. D
85. D	86. A	87. A	88. B	89. A	90. A
91. B	92. D	93. C	94. D	95. A	96. C
97. B	98. A	99. D	100. D		

Heat Pumps

1.

1. C	2. H	3. E	4. B	5. D	6. F
7. G	8. A				

2. C	3. A	4. B	5. C	6. C	7. D
8. B	9. A	10. A	11. D	12. B	13. C
14. A	15. C	16. D	17. C	18. B	19. D
20. D	21. B	22. B	23. B	24. B	25. C
26. B	27. D	28. D	29. D	30. B	31. D
32. C	33. A	34. D	35. B	36. C	37. D
38. C	39. D	40.B	41. A	42. D	

Heat Pump Troubleshooting

1. C	2. D	3. A	4. D	5. A	6. B
7. A	8. A	9. C	10. C	11. B	12. C
13. B	14. D	15. B	16. C	17. D	18. D
19. D	20. B	21. B	22. B	23. D	24. D
25. D	26. D	27. D	28. C	29. C	30. B
31. D	32. B	33. C	34. A	35. B	36. C
37. B	38. A	39. D	40. A	41. C	42. D
43. D	44. A	45. A	46. D	47. C	48. C
49. D					

SECTION 2

Gas and Oil Forced-Air and Hot Water Heating Systems

Fundamentals of Gas Combustion

1. B	2. B	3. D	4. D	5. A	6. B
7. A	8. D	9. B	10. A	11. C	12. C
13. B	14. A	15. D	16. B		

Gas Furnace Controls and Components

1. A	2. D	3. C	4. B	5. C	6. A
7. B	8. B	9. D	10. B	11. C	12. D
13. D	14. A	15. B	16. B	17. D	18. D
19. D	20. C	21. B	22. A	23. B	24. C
25. A	26. A	27. C	28. A	29. B	30. B
31. C	32. D	33. C	34. B	35. B	36. D
37. A	38. C	39. C	40. C	41. B	42. D
43. D	44. C	45. C	46. B		

Combustion Air

1. C	2. D	3. C	4. C	5. D	6. B
7. C	8. A	9. B	10. B	11. C	12. B
13. B	14. B	15. D	16. A	17. B	18. C
19. C	20. A	21. C	22. A	23. C	24. A
25. D					

Vents and Chimneys

1. A	2. C	3. D	4. A	5. B	6. B
7. B	8. D	9. A	10. D	11. A	12. D
13. C	14. B	15. A	16. D	17. A	18. B
19. B	20. D	21. C	22. D	23. C	24. B
25. C	26. C	27. B	28. D	29. C	30. D
31. A	32. C	33. B	34. C	35. A	36. A
37. B	38. A	39. B	40. A	41. D	42. C
43. D	44. A	45. B	46. A	47. B	48. B
49. A	50. C				

Gas Piping

1. A	2. B	3. C	4. A	5. D	6. A
7. B	8. B	9. A	10. C	11. B	12. C
13. D	14. D	15. D	16. D	17. B	18. B
19. A	20. D	21. D	22. A	23. D	24. D
25. B	26. C	27. D	28. A	29. A	30. B
31. C	32. B	33. D	34. C	35. D	36. C

ESCO Institute-HVAC Excellence Carbon Monoxide Safety Certification Employment Ready (ER)

1. B	2. A	3. C	4. A	5. C	6. D
7. D	8. A	9. A	10. D	11. B	12. C
13. D	14. D	15. D	16. D	17. D	18. C
19. B	20. C	21. D	22. A	23. C	24. A
25. B	26. A	27. A	28. D	29. B	30. C
31. D	32. A	33. A	34. B	35. D	36. A
37. C					

Gas Furnace Troubleshooting

1. C	2. D	3. C	4. C	5. B	6. A
7. D	8. D	9. A	10. B	11. C	12. C
13. C	14. C	15. A	16. B	17. B	18. C
19. A	20. D	21. A	22. A	23. A	24. B
25. B	26. A	27. D	28. A	29. D	30. B
31. B	32. C	33. B	34. D	35. C	36. A
37. C	38. A	39. D	40. C	41. B	42. C
43. B	44. A	45. D	46. D	47. D	48. A
49. B	50. A	51. A	52. A	53. C	54. D
55. B	56. C	57. B	58. B	59. A	60. D
61. B	62. D	63. B	64. C	65. D	66. C
67. A	68. C	69. D	70. A	71. A	72. C
73. C	74. B	75. C	76. D	77. A	78. C
79. C	80. A	81. B	82. B	83. D	84. D
85. D	86. A	87. C	88. C	89. B	90. D
91. B	92. A	93. D	94. A	95. C	96. C
97. D	98. D	99. C	100. B	101. C	102. A
103. B	104. C	105. D	106. A	107. D	108. D
109. B	110. C	111. C	112. D	113. B	114. A
115. D	116. C	117. A	118. B	119. C	120. D

Oil Furnaces

1. C	2. D	3. B	4. C	5. A	6. A
7. B	8. D	9. C	10. C	11. B	12. B
13. C	14. A	15. C	16. B	17. C	18. A
19. D	20. B	21. A	22. D	23. D	24. B
25. B	26. C	27. B	28. B	29. C	30. A
31. B	32. A	33. D	34. C	35. B	36. B
37. D	38. C	39. C	40. D	41. A	

Oil Furnace Troubleshooting

1. C	2. D	3. B	4. D	5. C	6. A
7. D	8. D	9. B	10. C	11. A	12. B
13. C	14. A	15. C	16. A	17. C	18. D
19. B	20. B	21. A	22. B	23. C	24. B
25. C	26. D	27. A	28. D	29. C	30. D
31. B	32. D	33. B	34. B	35. A	36. D
37. A	38. B	39. D	40. D	41. B	42. D
43. D	44. D	45. D	46. C		

Hydronic Heating Theory

1. A	2. A	3. B	4. C	5. D	6. B
7. D					

Mechanical Controls

1. B	2. B	3. D	4. D	5. D	6. B

Electrical Controls and Mechanical Components

1. C	2. A	3. C	4. C	5. A	6. B
7. C	8. B	9. A	10. B	11. B	12. D
13. A	14. A	15. C	16. B	17. A	18. C
19. A	20. C	21. D	22. B	23. B	24. D
25. C	26. A	27. A	28. B	29. B	30. A
31. C	32. B	33. C	34. B	35. A	36. C
37. B	38. A	39. B	40. B	41. D	42. C
43. A	44. B	45. B	46. A	47. C	48. A
49. A	50. B	51. D	52. A	53. B	54. B
55. A	56. D	57. C	58. C	59. A	60. B
61. C	62. C	63. A	64. C	65. B	66. C
67. A	68. A	69. C	70. A	71. C	72. A
73. B	74. C	75. D			

Troubleshooting Hydronic Heating Systems: Electrical Diagrams

1. C	2. A	3. B	4. D	5. B	6. D
7. B	8. D	9. A	10. B	11. C	12. C
13. C	14. D	15. D	16. C	17. C	18. D
19. C	20. B	21. C	22. A	23. B	24. A
25. C	26. B	27. B	28. D	29. A	30. C
31. C	32. A	33. A	34. A	35. C	36. C
37. A	38. D	39. D	40. C	41. D	42. A
43. C	44. D	45. A	46. B	47. B	48. C
49. A	50. C	51. D	52. D	53. C	54. A
55. B	56. A	57. A	58. C	59. D	60. D
61. A	62. B	63. B	64. B	65. A	66. B
67. A	68. B	69. C	70. B	71. A	72. A
73. B	74. C	75. C	76. D	77. D	78. A
79. B	80. A	81. A	82. C	83. C	84. D
85. C	86. D	87. A	88. B	89. C	90. D
91. B	92. C	93. C			

SECTION 3

Electrical Theory and Application

Electrical Safety

1. A	2. B	3. D	4. D	5. A	6. A
7. B	8. A	9. C	10. D	11. C	12. B
13. C	14. C	15. A	16. C	17. D	18. C
19. A	20. B	21. B	22. A	23. A	24. A
25. C	26. C	27. A	28. B		

Electrical Theory

1. B	2. A	3. B	4. D	5. C	6. C
7. C	8. A	9. A	10. B	11. D	12. D
13. B	14. C	15. C	16. C	17. D	18. B
19. A	20. B	21. C	22. A	23. A	24. C

Magnetism

1. B	2. A	3. D	4. A	5. C	6. A
7. C	8. B	9. C			

Direct Current

1. A	2. C	3. C	4. C	5. B	6. A
7. B	8. A	9. D	10. C	11. C	12. B
13. C	14. A	15. B	16. C	17. C	18. B
19. D	20. D	21. A	22. A	23. A	24. A
25. D					

Alternating Current

1. B	2. C	3. A	4. D	5. C	6. B
7. D	8. A	9. D	10. C	11. B	12. A
13. B	14. C	15. C	16. D	17. A	18. B
19. C	20. C	21. B	22. D	23. B	24. A
25. D					
26. D					

Power Distribution

1. B	2. A	3. A	4. B	5. D	6. C
7. B	8. D	9. B	10. C	11. B	12. A
13. C	14. C	15. D			

AC Motors

1. B	2. C	3. B	4. D	5. A	6. B
7. B	8. C	9. A	10. C	11. C	12. A
13. C	14. B	15. A	16. A	17. D	18. C
19. D	20. D	21. D	22. B	23. C	24. A
25. C	26. D	27. A	28. B	29. A	30. C
31. C	32. A	33. D	34. B	35. D	36. B
37. C	38. C	39. D	40. C	41. B	42. D
43. A	44. B	45. D	46. D	47. C	48. A
49. A	50. C	51. A	52. B	53. C	54. A
55. B	56. A	57. D	58. B	59. C	60. C
61. B	62. A	63. D	64. C	65. C	66. C
67. A	68. C	69. A	70. B	71. D	72. D
73. B	74. B	75. A	76. C	77. C	78. A
79. D	80. C	81. A	82. B	83. C	84. D
85. D	86. C	87. B	88. A	89. B	90. C
91. A	92. D	93. A	94. C		

Electrical Controls

1. C	2. D	3. B	4. C	5. D	6. A
7. B	8. C	9. B	10. A	11. C	12. D
13. A	14. A	15. B	16. B	17. C	18. D
19. D	20. B	21. C	22. C	23. A	24. A
25. B	26. C	27. D	28. B	29. B	30. C
31. D	32. A	33. C	34. B	35. C	36. D
37. D	38. A	39. C	40. D	41. A	42. B
43. B	44. C	45. B	46. C	47. D	48. A
49. C	50. B	51. A	52. B	53. D	54. C
55. B	56. D	57. A	58. D	59. C	60. A
61. D	62. B	63. C	64. A	65. B	66. C
67. C	68. A	69. B	70. D	71. B	72. D
73. B	74. D	75. B	76. B	77. C	78. C
79. A					

Solid-State Electronics

1. C	2. D	3. A	4. D	5. A	6. B
7. C	8. D				

Wiring Layouts and Electrical Diagrams

1. D	2. A	3. D	4. B	5. A	6. B
7. C	8. D	9. A	10. B	11. B	12. C
13. B	14. C	15. B	16. D	17. A	18. C
19. A	20. C	21. B	22. D	23. C	24. D
25. B	26. C	27. A	28. C	29. B	30. C
31. C	32. A	33. C	34. A	35. B	36. A
37. D	38. C	39. C	40. D	41. C	42. B
43. A	44. D	45. C			

Electrical Meters

1. D	2. B	3. B	4. A	5. C	6. D
7. B	8. C	9. A	10. C	11. B	12. C
13. A	14. B	15. A	16. B	17. B	18. B
19. A	20. D	21. A	22. C	23. C	24. C
25. B	26. D				

Electrical Troubleshooting

1. C	2. B	3. D	4. B	5. C	6. B
7. D	8. B	9. C	10. B	11. A	12. D
13. C	14. A	15. C	16. A		

SECTION 4

Airflow Components and Duct Fabrication

Airflow Tools and Measurements

1. B	2. B	3. C	4. A	5. D	6. B
7. A	8. D	9. B	10. B	11. C	12. B
13. C	14. A	15. D	16. B	17. A	18. C
19. A	20. B	21. C	22. C	23. A	24. D
25. B	26. C	27. B	28. B	29. C	30. C
31. A	32. D	33. B	34. D		

Blowers and Fans

1. C	2. B	3. B	4. C	5. A	6. A
7. B	8. B	9. D	10. D	11. C	12. B
13. C	14. B	15. D	16. B	17. A	18. D
19. B	20. A	21. A	22. C		

Air Duct Systems and Fabrication

1. C	2. B	3. D	4. A	5. B	6. C
7. A	8. D	9. A	10. D	11. B	12. D
13. D	14. A	15. A	16. C	17. B	18. D
19. C	20. D	21. A	22. C	23. D	24. B
25. D	26. A	27. C	28. B	29. C	30. A
31. D	32. A	33. D	34. B	35. C	36. A
37. B	38. D	39. C	40. D	41. C	42. A
43. B	44. A	45. C	46. B	47. C	48. C
49. C	50. B	51. C	52. C	53. B	54. B

Flexible (Flex) Duct

1. B	2. B	3. D	4. D	5. C	6. D
7. B	8. A	9. B			

Fiberglass Ductboard

1. C	2. D	3. D	4. C	5. B	6. A
7. B	8. A	9. A	10. C	11. B	12. C
13. A	14. C	15. D	16. C	17. D	18. D
19. B	20. B	21. A	22. B	23. C	24. B
25. C	26. B	27. C	28. D		

Air Side Components

1. B	2. D	3. A	4. B	5. D	6. B
7. D	8. D	9. D	10. D	11. C	12. A
13. D	14. B				

Air Filters

1. A	2. A	3. C	4. B	5. C	6. A
7. C	8. C	9. C	10. D	11. B	12. B
13. A	14. C				

Humidifiers

1. A	2. D	3. D	4. A	5. D	6. A
7. A	8. B	9. B	10. C	11. A	12. D
13. C	14. C	15. C	16. D		

Indoor Air Quality (IAQ)

1. A	2. D	3. D	4. C	5. A	6. B
7. D	8. C	9. A	10. A	11. C	12. B
13. B	14. D	15. D	16. C	17. A	18. C
19. B	20. C	21. D	22. C	23. D	24. A
25. B	26. A	27. D	28. B	29. D	30. B
31. A	32. C	33. B	34. D	35. D	36. D
37. C	38. A	39. D	40. A	41. A	42. A
43. D	44. B	45. D	46. A	47. D	48. A
49. A	50. D				

Air Distribution Troubleshooting

1. A	2. A	3. C	4. B	5. B	6. D
7. E	8. C	9. B	10. B	11. B	12. B
13. A	14. A	15. A	16. D	17. C	18. B
19. C	20. C	21. B	22. D	23. D	24. D
25. B	26. A	27. C	28. B	29. B	30. D
31. B	32. D	33. A	34. C	35. A	36. A
37. B	38. A	39. C	40. B	41. D	42. D

Temperature, Humidity, and Psychrometrics

1. C	2. A	3. A	4. B	5. B	6. C
7. A	8. D	9. B	10. C	11. A	12. C
13. B	14. D	15. D	16. B	17. D	

Heat Loss and Heat Gain

1. A	2. B	3. A	4. C	5. C	6. A
7. A	8. B	9. C	10. C	11. D	12. A
13. C	14. D	15. C	16. D	17. A	18. B
19. C	20. C	21. A	22. A	23. C	24. A
25. C	26. C	27. A	28. B	29. D	30. C
31. B	32. A	33. C	34. D	35. A	